APPLICATIONS OF COMPUTATIONAL MECHANICS IN
GEOTECHNICAL ENGINEERING

BALKEMA – Proceedings and Monographs
in Engineering, Water and Earth Sciences

PROCEEDINGS OF THE 5TH INTERNATIONAL WORKSHOP ON APPLICATIONS OF COMPUTATIONAL MECHANICS IN GEOTECHNICAL ENGINEERING
GUIMARÃES / PORTUGAL / 1–4 APRIL 2007

Applications of Computational Mechanics in Geotechnical Engineering

Edited by

L.R. Sousa & M.M. Fernandes
Faculty of Engineering, University of Porto, Portugal

E.A. Vargas Jr.
Department of Civil Engineering, Catholic University of Rio de Janeiro, Brazil

R.F. Azevedo
Department of Civil Engineering, Federal University of Viçosa, Brazil

Taylor & Francis
Taylor & Francis Group
LONDON / LEIDEN / NEW YORK / PHILADELPHIA / SINGAPORE

Taylor & Francis is an imprint of the Taylor & Francis Group, an informa business

© 2007 Taylor & Francis Group, London, UK

Typeset by Charon Tec Ltd (A Macmillan Company), Chennai, India
Printed and bound in Great Britain by CPI Bath Press (CPI Group), Bath

Published by: Taylor & Francis/Balkema
                 P.O. Box 447, 2300 AK Leiden, The Netherlands
                 e-mail: Pub.NL@tandf.co.uk
                 www.balkema.nl, www.taylorandfrancis.co.uk, www.crcpress.com

ISBN 13: 978-0-415-43789-9

Applications of Computational Mechanics in Geotechnical Engineering – Sousa,
Fernandes, Vargas Jr & Azevedo (eds)
© 2007 Taylor & Francis Group, London, ISBN 978-0-415-43789-9

Table of contents

Applications of Computational Mechanics in Geotechnical Engineering – Sousa,
Fernandes, Vargas Jr & Azevedo (eds)
© 2007 Taylor & Francis Group, London, ISBN 978-0-415-43789-9

Preface

Computational mechanics techniques have gained acceptance in the solution of complex geotechnical problems. The School of Engineering of University of Minho, the Faculty of Engineering of University of Porto and the Civil Engineering Departments of the Catholic University of Rio de Janeiro and the Federal University of Viçosa organized at Guimarães, April 1 to 4, 2007, the 5th International Workshop on Applications of Computational Mechanics in Geotechnical Engineering.

The previous editions of the Workshop took place in Rio de Janeiro, 1991 and 1994, in Porto, 1998, and in Ouro Preto, 2003. These meetings started as a result of a joint research project involving organizations from Brazil and Portugal. It was purpose of these events to join renowned researchers in an informal meeting where people could exchange views and point out perspectives on the relevant subjects.

The Workshop provided an excellent opportunity to hold high level discussions and to define novel approaches for the solution of relevant engineering problems. The meeting attracted researchers, academics, students, software developers, and professionals of all areas of Geotechnics. During the event the following topics were addressed:

Constitutive models
Computational models
Artificial intelligence
Underground structures
Soil and rock excavations
Foundations
Ground reinforcement
Environmental geotechnics
Oil geomechanics
Embankments and rail track for high speed trains

Brazilian and Portuguese researchers participated actively in the event, as well as well known invited international experts. This book contains 42 papers from 18 countries. The content of the book will be of use to engineers and researchers.

L.R. Sousa
M.M. Fernandes
E.A. Vargas Jr.
R.F. Azevedo

Applications of Computational Mechanics in Geotechnical Engineering – Sousa,
Fernandes, Vargas Jr & Azevedo (eds)
© 2007 Taylor & Francis Group, London, ISBN 978-0-415-43789-9

Acknowledgements

The Organizing Committee of the 5th International Workshop on Applications of Computational
Mechanics in Geotechnical Engineering is especially grateful to:

Escola de Engenharia da Universidade do Minho
Faculdade de Engenharia da Universidade do Porto
Pontifícia Universidade Católica do Rio de Janeiro
Universidade Federal de Viçosa
Fundação para a Ciência e Tecnologia

The publication of the proceedings has been partially funded by
Fundação Calouste Gulbenkian

Constitutive models

Applications of Computational Mechanics in Geotechnical Engineering – Sousa,
Fernandes, Vargas Jr & Azevedo (eds)
© 2007 Taylor & Francis Group, London, ISBN 978-0-415-43789-9

Estimation of strength and deformation parameters of jointed rock masses

S.A. Yufin, E.V. Lamonina & O.K. Postolskaya
Moscow State University of Civil Engineering, Moscow, Russian Federation

ABSTRACT: Filling the existing gap between experimental and numerical techniques in rock engineering is a challenging endeavor. This goal can be reached by combining accumulated experience and factual data with the capabilities of recent computer methods. As one of the first steps in the creation of a comprehensive numerical model for jointed and layered rock, a series of FEM numerical experiments was conducted in 2D and 3D for jointed rock blocks using Mohr-Coulomb, multilaminate and Hoek-Brown models. Compression strength and deformation parameters of blocks were evaluated for different joint spacing, joint thickness and joint inclination angles. The results fit well into the multitude of independent experimental, analytical and numerical data. Practical implementation for a tunnel in jointed rock is shown.

1 INTRODUCTION

Rock masses are complicated and controversial research objects. They are formed in various geological and climatic conditions under the influence of variety of different factors and represent discrete multi-component composite natural formations. They are of heterogeneous and, frequently, anisotropic nature due to the fact that almost any rock is layered and jointed which directly affects its strength and deformability.

Modern numerical techniques provide capabilities for simulation of real rock masses and estimating parameters of the media weakened with intersecting joints without influence of the scale factor. However, available recommendations primarily consider extreme types of discontinuities such as faults and large joints on one side and networks of thin unfilled fissures on the other side. Filled joints require further improvement of relevant modeling techniques.

2 COMPARISON OF DIFFERENT APPROACHES TO MODELING JOINTED ROCK

There are three basic ways to represent jointed media in numerical simulation within the Finite Element Analyses:

1. Joints are modeled as contacts using contact elements (Goodman et al 1968);
2. Jointed mass is replaced with equivalent solid medium and is simulated with a multilaminate model (Commend et al 1996, Zienkiewicz & Pande 1977);
3. Individual joints are represented as a layer of standard 4-node finite elements (in two-dimensional simulation) or 8-node finite elements (in three-dimensional simulation).

The last-numbered approach to model jointed media implies that the layer containing a joint includes also a pinnate weakness zone around this joint, i.e. the parameters of the elements layer should take into account both the properties of the joint filler itself and those of the solid rock

Table 1.   Rock sample parameters.

Parameter	Value
Parameters of solid rock	
Modulus of elasticity E_i, GPa	25
Poisson's ratio v_i	0.2
Uniaxial compressive strength σ_{ci}, MPa	100
Uniaxial tensile strength σ_{ti}, MPa	10
Specific cohesion c_i, MPa	25
Angle of internal friction ϕ_i, °	41
Size of solid rock block a, m	1
Parameters of joints	
Joint width Δa, m	0.001
Relative joint width α	0.001
Relative rock contact area ξ	$4 \cdot 10^{-5}$
Specific cohesion c_j, kPa	50
Angle of internal friction ϕ_j, °	35

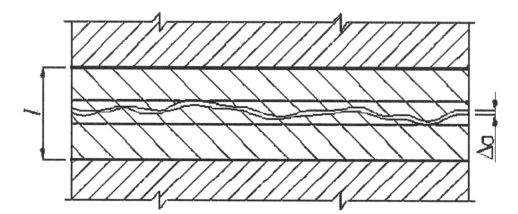

Figure 1.   Determination of the width of the layer representing the joint.

blocks. Both parameters can be taken into account through averaging the properties within the elements layer. To do so the asymptotic method of averaging proposed in (Vlasov, 1990) can be used.

Let us consider a jointed rock sample with parameters given in Table 1.

Let us determine the unknown modulus of joint deformation Ej, taking into account the relative rock contact area (Rats 1973):

$$E_j = \xi E_i = 4 \cdot 10^{-5} \cdot 25000 \text{ MPa} = 1 \text{ MPa} \tag{1}$$

Let us take the relative width of elements layer simulating the joint, with the weakness zone taken into account, equal to $l = 10$ cm (Figure 1).

We determine the modulus of deformation for the layer containing the joint from the relationship proposed in (Vlasov 1990):

$$Em = \frac{E_j \cdot E_i}{E_j + \alpha E_i} = 962 \text{MPa} \tag{2}$$

Performing a series of numerical experiments with the joint being simulated by a contact element requires determination of the shear and normal stiffness of the joints.

From the relationship

$$Ej = \Delta l k_n,\tag{3}$$

where $\Delta l = l/a$ is a relative width of layer simulating the joint, we can determine the normal stiffness of the joint:

$$k_n = E_j /\Delta l = 962 \text{ MPa} /0.1 = 9620 \text{ MPa}\tag{4}$$

Then, the shear stiffness of the joint (Ruppeneit 1975):

$$k_S = 0.4 k_n = 0.4 \cdot 9620 \text{ MPa} = 3848 \text{ MPa}\tag{5}$$

The model of elastoplastic medium with the Mohr-Coulomb criterion of failure (6) is most often used in simulation of joints with layers of elements, while the Hoek-Brown failure criterion (7) was used for solid rock blocks.

$$|\tau| = \sigma_n tg\varphi + c\tag{6}$$

where τ is limit shear stress, σ_n is normal stress, c is cohesion, and φ is the angle of internal friction.

$$\sigma_1 = \sigma_3 + \sigma_{ci}\sqrt{m_i \frac{\sigma_3}{\sigma_{ci}} + s}\tag{7}$$

where σ_1, σ_2, σ_3 are principal stresses, σ_{ci} is the sample's uniaxial compressive strength, and m_i, s are parameters determined for each individual rock.

In order to expand the applicability of the multilaminate model and to compare methods of representation of jointed rock masses numerically, the third simulation approach proposes using multilaminate model of the material for the layer replacing the joint. Thus, we will perform four series of rock sample examination in the numerical experiment. To obtain a more comprehensive representation, let us consider samples with various dip angles of joints for each series.

The representativity is accounted for through the definition of the minimum acceptable number of blocks in the sample which meets the criteria of quasi-continuity and quasi-homogeneity (Ukhov 1975):

$$n_k = \frac{2[(100 + k)(A+1)M_T B - 100(M_a A + 2M_T B)]}{k(A+1)(M_a A + 2M_T B)},\tag{8}$$

where $k = 5\%$ is the adopted accuracy, $a = 1$ m is the distance between joints, $l = 0.1$ m is the width of the layer simulating the joint, E_a, μ_a, E_m, μ_m are moduli of deformation and Poisson ratios of the materials forming blocks and joints, E_L is the average modulus of deformation, $A = a/l = 1/0.1 = 10$, $B = Ea/ET = 25.9$, $Ma = 1 - 2\mu a = 0.6$, $MT = 1 - 2\mu T = 1$.

Number of blocks equals to 5 in each direction. The sample under study is schematically shown in Figure 2. β, the rotation angle of joint system No. 2 from the horizontal line, is taken equal to: $\beta = 0°$ (0°; 90°), $\beta = 5°$ (5°; 85°), $\beta = 20°$ (20°; 70°), $\beta = 40°$ (40°; 50°), while the angle between the joint systems is constant and equals 90°.

The following series of numerical simulation for jointed rock samples were performed under the conditions of uniaxial compression:

Series I – solid rock blocks were simulated with material with the Hoek-Brown criterion, the joints were represented by contact elements with normal and shear stiffness.

5

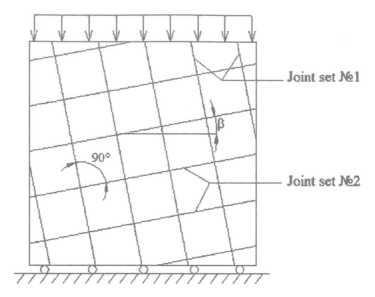

Figure 2.   Layout of the jointed rock sample.

Series II – joints in the sample were not specially simulated, the behavior of the sample was studied on the standard orthogonal grid and described using the multilaminate model. Strata were oriented under an angle equal to the dip angle of the joints β.

Series III – joints were presented as finite element layers with the width corresponding to the width of joint opening Δa, the weakness zone being taken into account. Solid rock material is modeled in accordance with the Hoek-Brown criterion, while joint material is assigned the Mohr-Coulomb criterion.

Series IV – joints are presented as a finite element layer, the same as in Series III. Solid blocks were simulated using the Hoek-Brown criterion, joint material – using multilaminate model, the dip angles of the weakened layers were taken equal to the dip angles of joint systems No. 1 and No. 2.

Z_SOIL PC® FEM computer code (Z_SOIL 2006) was used throughout this work. Based on the results of calculations, a diagram was constructed showing the change of compressive strength of jointed samples for each series in the experiment (Figure 3).

Considerable variation of the Series II curve from others gives evidence that it is unpractical to use this model to describe material having the proposed geometric characteristics (in particular, when the size of solid rock blocks considerably exceed the width of joint opening). Besides, it is obvious that the applicability of this or that model is dependent on the dip angle of the fissures.

The proposed combined materials model consisting of the multilaminate model for joints and material with the Hoek-Brown criterion for solid rock implemented in Series IV of the numerical experiment allows obtaining results which are similar in nature to those received using other methods to simulate a jointed sample.

Let us compare the values of the deformation modulus of the jointed sample cut by orthogonal joint systems having the dip angle of $\beta = 0°$ with the analytically obtained values of the deformation modulus. The sample represented by the multilaminate model in pure form was not considered due to considerable variation of results of this series of the numerical experiment in comparison with all the other series.

The deformation moduli of the jointed sample for different series were equal to:
Series I – $\bar{E} = 5768.6$ MPa,
Series III – $\bar{E} = 5330.0$ MPa,
Series IV – $\bar{E} = 5733.3$ MPa.

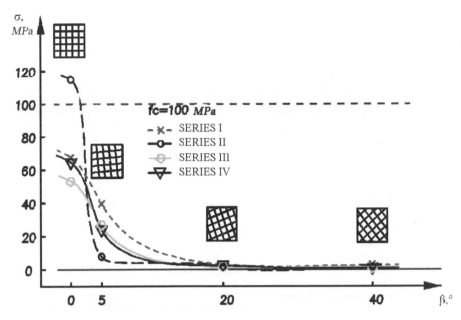

Figure 3. Dependencies of compressive strength of the rock sample on β in each series of experiment.

Effective modulus of deformation was defined on the basis of relationship (1):

$$E_\perp = \frac{E_j E_i}{E_j + \alpha E_i} = 6946.8\,M\varPi a \qquad (9)$$

The comparison demonstrated that the minimal variation of analytical and numerical data was observed in the results of Series I (simulating joints via a contact element) and Series IV (simulating joints via the layer represented by a multilaminate model) – 20.4% and 21.2%, respectively. The deformation modulus obtained in Series III when using the Mohr-Coulomb failure criterion in the layer simulating a joint is significantly different from the others. The variation was 30.3%. The results of the comparison suggest the conclusion that it is possible to use the multilaminate model of the material alongside with the contact element to simulate joints in the rock sample. When the solid rock blocks are simulated using Hoek-Brown criterion, the accuracy of the numerical simulation results are comparable to those received analytically.

3 NUMERICAL ANALYSIS OF THE EFFECT OF THE JOINT STEPPING IN THE SAMPLE ON ITS STRENGTH PARAMETERS

Evaluation of the efficiency of the proposed model of jointed mass based on the combination of existing models of continua, i.e. the elastoplastic model based on the Hoek-Brown failure criterion for solid rock blocks and the multilaminate model for the layer replacing a joint having a pinnate weakness zone, was performed through the comparison of the results obtained in numerical and laboratory experiments. This study investigated the effect of the joint stepping in a jointed sample on its effective strength parameters. The parameters of solid rock blocks and joint materials are given in Table 2.

A three-dimensional numerical analysis was performed. The sample was cut by three orthogonal joint systems parallel to sample sides. The size of the joint stepping was: s = 0, 1/8, 2/8, 3/8, 4/8

Table 2a. Parameters of solid rock blocks.

Parameter	Value
Modulus of elasticity E_i, GPa	5.34
Poisson's ratio v_i	0.19
Uniaxial compressive strength σ_{ci}, MPa	17.13
Uniaxial tensile strength σ_{ti}, MPa	2.49
Specific cohesion c_i, MPa	4.67
Angle of internal friction ϕ_i, °	33
Modulus of elasticity E_i, GPa	25.4

Table 2b. Parameters of joints.

Parameter	Value
Modulus of elasticity E_i, GPa	25.4
Poisson's ratio v_i	0
Relative width of the layer simulating the joint Δl	0.1
Specific cohesion c_j, kPa	0
Angle of internal friction ϕ_j, °	37

of the block size. Due to symmetry further increase of the stepping will lead to repetitive results. Finite element meshes for considered stepping sizes are given in Figure 4.

The loading to the samples was applied as linearly increasing upper edge displacements. The moment of failure of the sample was considered reached when the maximal value of vertical stress in any element approached the value of the uniaxial compressive strength of solid rock $\sigma_{ci} = 17.13$ MPa. At this moment of calculation the values of effective stress were determined in the sample.

The numerical experiment resulted in evaluation of uniaxial compressive strength values for block samples. Dependency of vertical effective stresses on the deformation of samples with stepping 's' = 0, 1/8, 2/8, 3/8, 4/8 is presented in diagrams on Figure 5.

For convenience of comparing the results of the numerical experiment against the laboratory data (Singh et al 2002) the limit strength of the samples was determined as a percentage of the ultimate strength of the solid block material $\sigma_{cr} = \sigma_{cj}/\sigma_{ci} \cdot 100\%$. Graphics are presented in Figure 6.

The average strength of jointed samples is 65% of the strength of solid rock. Certain strength variation is observed for the curve depicting numerical simulation results depending on the size of the blocks stepping: from 10.18 MPa (59.43%) for stepping 's' = 1/8 to 13.29 MPa (77.58%) for stepping 's' = 0. The jointed sample has the maximum strength value (13.29 MPa) if there is no joint displacement. As the stepping increases, the strength decreases to the value of 10.18 MPa ('s' = 1/8) and then increases to 11.47 MPa ('s' = 4/8). The nature of the curve obtained as a result of laboratory experiments shows also that as the stepping increases there is certain strength decrease from 68% (for 's' = 0) to 56% (for 's' = 1/8) with its subsequent increase to 68% (for 's' = 4/8). The strength increase at the half-width solid block's stepping was forecast and it is explained by the staggered distribution of the stress concentrators (joint intersection locations) one under another without any pronounced direction of weakened surfaces exposed to failure. As a result, the strength of a sample with stepping 's' = 4/8 is almost equal to the strength of a sample without blocks stepping along the joint 's' = 0. In other cases stress concentrators are located with certain displacement in each next row and form the surface of failure.

Variation between the curves obtained in laboratory and numerical experiments has values from 0.86% ('s' = 3/8) to 9.58% ('s' = 0).

Based on the close correspondence of data obtained as a result of numerical simulation and laboratory research the conclusion may be drawn that the proposed joint model using multilaminate

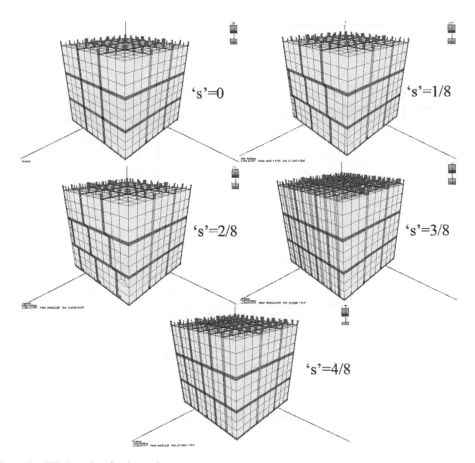

Figure 4.   FEM meshes for the analyses.

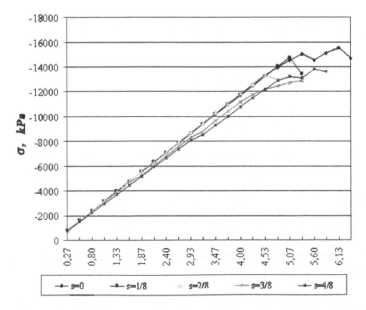

Figure 5.   Dependency of vertical stresses on jointed sample deformation for different joint stepping.

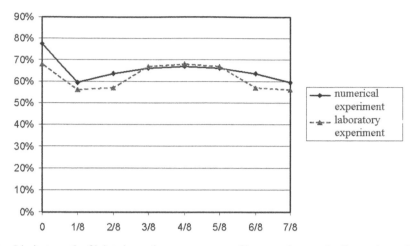

Figure 6. Limit strength of jointed samples as percentage of intact rock strength. Comparison of numerical and laboratory data.

Table 3. Material properties

Property	Value	
	Granite	Joints
Poisson's ratio, ν	0.2	0
Cohesion c, κPa	55	0.9
Friction angle φ, °	21	21
Uniaxial compressive strength σ_c, κPa	3000	–
Uniaxial tensile strength σ_t, κPa	330	–
Young modulus E, κPa	$308 \cdot 10^3$	$308 \cdot 10^3$

material model can be applied to determine the stress-strain state of the jointed sample. Such simulation results can be extrapolated to the entire jointed rock mass provided the representativity requirements are met.

4 SHALLOW TUNNEL IN JOINTED ROCK

Practical implementation of the proposed approach to model jointed rock masses is illustrated in comparison with results of the independent research. The structure in question is a four lanes motor tunnel. The tunnel span is d = 12 m, the height is h = 12 m too, the minimal depth is approximately 15 m. On the slope at the distance approximately equal to 38 m there is a motorway.

The containing rock is weak granite with several systems of tectonic discontinuities. The inclination angles between these joint systems approximately equal 90°, thus, to make a numerical model we assume these systems orthogonal. The joint spacing is 3 m and the relative thickness is $\alpha = 0.01$. The area along these discontinuities is being considered as a weak zone due to the presence of joints of the smaller order. Owing to it, for modeling, we believe the thickness of the weak layers equal to 0.1 m.

The other parameters of the joints and of the intact rock are presented in Table 3.

The geometry of the minimal-depth cross-section and some features of the problem setup incidentally resemble rather closely the problem described by E. Hoek. This provides the basis for a qualitative comparison. However, E. Hoek considered a continuous rock mass with averaged properties while we model jointed rock mass.

10

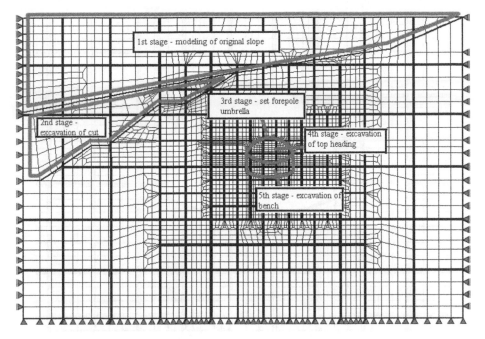

Figure 7.   Modeling stages and the FEM mesh.

The intact rock is assigned material with the Hoek-Brown criterion, the joints are represented with the multilaminate model. Directions of strata correspond to directions of joint systems.

Due to complexity of representation of a very fine net of joints and limitations on the number of nodes and elements in the FEM mesh the real frequency of joints was introduced only in the central part of the domain at the distances of some 15 m around the tunnel. Further on in the directions towards the domain boundaries sizes of intact rock blocks were gradually increased.

The modeling consisted of 5 stages:

1. Creation of the natural slope by removing part of the rock mass to evaluate the natural stress state.
2. Creation of the cut for the motorway.
3. Modeling of the forepolling reinforcement consisted of steel tubes 12 m long, with inner diameter 100 mm, outer diameter 114 mm placed at the distance of 0.5 m from each other in the tunnel vault.
4. Excavation of the tunnel top heading with shotcrete reinforcement 30 cm thick.
5. Excavation of the bench.

The boundary conditions were applied in a usual way (Figure 7). The forepolling umbrella was represented with the element layer of composite material with $E = 1288$ MPa, which is analogous to that accepted in the paper of E. Hoek.

Figure 7 presents the FEM mesh with modeling stages and Figure 8 – central part of this mesh with real frequency of joints.

Modeling of the cut for the motorway revealed stress and displacement distribution in the slope (Figure 9). The maximal displacement of the slope is equal to 7 cm. All results of the further stages were referenced to this initial condition.

After the excavation of the top heading (Figure 10) a relatively small surface settlement approximately 8 mm occurs. The character of deformation considerably depends on the direction of the joint systems, in particular it is visible below the excavated portion of the tunnel. Maximal absolute displacements at this stage reached 24 mm.

11

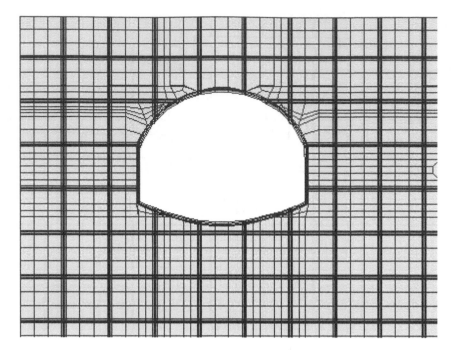

Figure 8.   Tunnel in jointed rock.

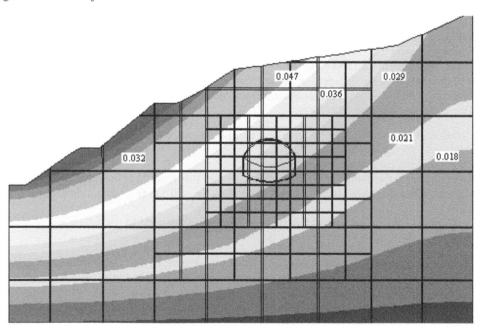

Figure 9.   Displacements distribution after excavation of the cut.

Figure 11 shows displacements at the final excavation stage. The maximal value is 25 mm.

E. Hoek in his work used the following parameters for continuous rock mass:

GSI = 25; Hoek-Brown constant $m_i = 8$; intact rock compressive strength $\sigma_{ci} = 3$ MPa; damage factor $D = 0.3$; material constant $m_b = 0.342$; material constant $s = 0.0001$; material constant $\alpha = 0.531$; deformation modulus $E = 308$ MPa.

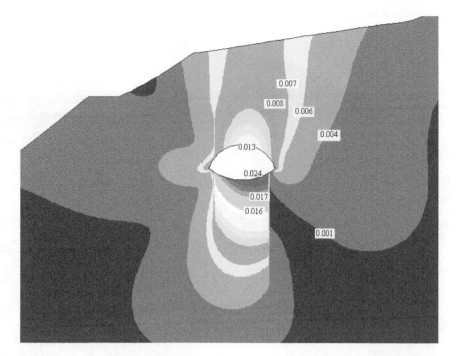

Figure 10.   Displacements after top heading excavation.

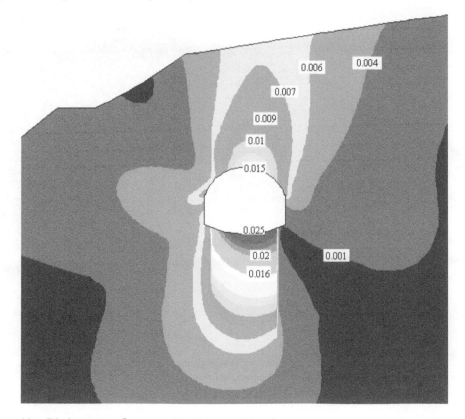

Figure 11.   Displacements after excavation of the tunnel bench.

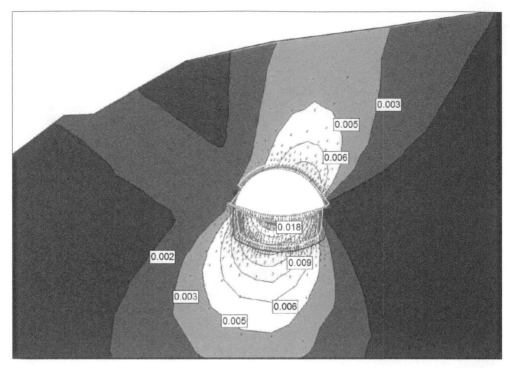

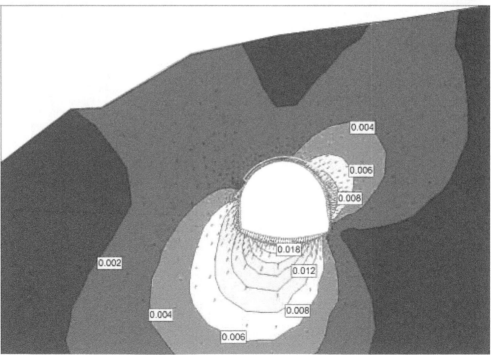

Figure 12.   Displacements distribution in continuous rock mass (E. Hoek).

The comparison of results obtained for the model of jointed rock with independent modeling of continuous media with averaged properties shows good qualitative correspondence. Existing differences are due to existence of joints, decreasing strength of the rock mass and governing directions of intact rock blocks displacements. The obvious similarity in the behavior of different models and software provides the basis for future implementation of the proposed approach in practice.

5 CONCLUSION

Complexity of the behavior of jointed rock masses demands special approaches to the design for each specific case. Herein one of possible approaches to numerical modeling of rock masses with filled joints of certain width is proposed. Representation of joints and surrounding fractured zone with the layer of finite elements and the multilaminate model of joint material showed good correspondence with results of independent laboratory experiments. This proves practical acceptability of the proposed model of jointed rock.

Comparison of numerical analyses of a tunnel driven in jointed rock mass and in continuous medium shows considerable dependency of the rock mass deformation on its structure, where joints govern the overall behavior of the rock mass.

REFERENCES

Commend S., Truty A., Zimmermann T. 1996. Numerical simulation of failure in elastoplastic layered media: theory and applications. *LSC-DGC-EPFL*, report 1996-S1 (unpublished).

Goodman R.E., Taylor R.L., Brekke T.L. 1968. A model for the mechanics of jointed rock. *Proc. ASCE*. Vol. 94. No. EM3.

Hoek E. Numerical modeling for shallow tunnels in weak rock. *http:www.rockscience.com/hoek/pdf/numerical modeling of shallow tunnels.pdf*

Hoek E. & Brown E.T. 1980. *Underground excavation in rock*. Institution of Mining and Metallurgy, U.K.

Rats. M.V. 1973. *Structural Models in Engineering Geology*. – Nedra, Moscow, pp. 216 (in Russian).

Ruppeneit K.V. 1975. *Deformability of Jointed Rock Masses*. – Nedra, Moscow (in Russian).

Singh M., Rao K.S. & Rammamurthy T. 2002. Strength and deformation behaviour of a jointed rock mass. *Rock Mech. and Rock Engng* 35 (1), 45–64

Ukhov S.B. 1975. Rock Foundations of Hydraulic Structures. – Energy, Moscow, pp. 264 (in Russian).

Vlasov A.N. 1990. Determination of effective deformation characteristics for laminated and jointed rock. – Abstract of a thesis …Ph.D. in Engineering –Moscow Civil Engineering Institute, Moscow (in Russian).

Zienkiewicz O.C., Pande G.N. 1977. Time dependent multi-laminate model of rocks—a numerical study of deformation and failure of rock masses. *Int J Num Anal Meth Geomech*; 1:219–47

Z_Soil 2006. *User Manual*. Zace Services Ltd Report 1985–2006. Lausanne: Elmepress International.

Applications of Computational Mechanics in Geotechnical Engineering – Sousa,
Fernandes, Vargas Jr & Azevedo (eds)
© 2007 Taylor & Francis Group, London, ISBN 978-0-415-43789-9

Simulation of fracture flow on single rock joint

K. Kishida, P. Mgaya & T. Hosoda
Department of Urban Management, Kyoto University, Kyoto, Japan

ABSTRACT: The groundwater flow in rock fractures is typically three-dimensional flow. Therefore, solving the 3D-Navier Stokes equations will estimate such flow accurately. The solution of
Navier-Stokes equations under the complicated fractures geometry is not computationally straight
forward and impractical. The Parallel Plate model (the cubic law) and the Local Cubic Law (LCL)
sometimes referred to as the Reynolds equations. However, these models are too simplified and
their applicability is limited. In this study, a fundamental flow model (2D0model) is developed. The
developed model is applied to simulate flow in both idealized fracture geometry with sinusoidal
aperture variation and measured aperture fields of the rock joint. Then, it is discussed to validity
and applicability of developed model.

1 INTRODUCTION

Flow in fractures was initially estimated using the conceptually parallel plate model (Snow, 1965).
In parallel plate model an individual fracture is represented by two infinite smooth parallel plates
separated by constant distance (aperture) between them. The flow is assumed to be laminar with a
parabolic velocity profile across the aperture. This led to the well-know cubic law (Witherspoon
et al., 1980) relating fluid flux to aperture as follows:

$$Q = W \frac{D^3}{12\mu} \Delta p \qquad (1)$$

where Q is the volumetric flow rate, W is the width of the fracture perpendicular to flow, D is the
aperture size, μ is the fluid viscosity and p is the fluid driving pressure.

The important implication of the cubic law is that the fluid is characterized by the separation
distance (aperture) although the velocity varies across the distance. Real fractures, however, have
rough (irregular) surface walls. Therefore, variable apertures in addition, locations exist where the
two surfaces of the fracture may come into contact, thereby creating zero aperture size.

This paper is concerned with developing the depth averaged flow model for estimating flows
in single fractured joint of the laboratory experiments (Kishida et al., 2005). The authors were
motivated to perform this research basing on the fact that the herein derived model includes inertia
term, viscous term and fracture surface variation components which could not be incorporated
in the previous models (i.e. cubic law and Reynolds equation). We believe this approach avoids
the restrictions involved in the using the Reynolds equation and is computationally tractable. We
also expect that, by including the friction variation component, the resistance to flow is correctly
estimated and therefore better flow characteristics can be determined. Moreover, the difference
between the governing equations for the 2D-model and the LCL is discussed. Then, simulations
of fluid flow in measured aperture fields of a standard permeability test specimens are conducted
using both, the LCL and the 2D-model. Through these simulations a direct comparison between
the LCL and 2D-model is presented. Note that the comparison between the experimental and the
2D-model simulation can be obtained in the study by Kishida et al. (2005)

2 PREVIOUS STUDY ON FLOW IN A SINGLE FRACTUER

Here, the previous study on flow in a single fracture is briefly introduced.

As mentioned above, the basic concept of flow in a fracture was parallel plate mode (Snow, 1965). In this model, an individual fracture is represented by two infinite smooth parallel plates separated by constant distance (an aperture). The flow is assumed to be laminar with a parabolic velocity profile across the aperture and be constant pressure distribution across the aperture. This led to simplification of the NS-equations to the well-known "cubic law" relating fluid flux to the aperture as shown in Eq.1. Regardless of the parallel plate model being simple, it does not include any effects, namely, aperture variation, contacted area on fractures, the inertia term, characteristic length of aperture variation, the velocity distribution across the aperture, pressure distribution across the aperture.

In considering the effect of aperture variation, a simplified form of NS-equations the Reynolds equation has always been used as follows (Brown, 1987; Zimmerman, et al., 1991; Renshaw, 1995; Zimmerman & Bodvarsson, 1996);

$$\frac{\partial}{\partial x}\left(D_{(x,y)}^3 \frac{\partial p}{\partial x} \right) + \frac{\partial}{\partial y}\left(D_{(x,y)}^3 \frac{\partial p}{\partial y} \right) = 0 \tag{2}$$

where p is the pressure and $D_{(x,y)}$ is the aperture size at coordinates x, y.

Theoretically speaking, solving the NS-equations under complicated fracture surfaces will provide details on pressure and flow velocity distributions in fractures and avoid restrictions involved in using the cubic law and Reynolds equation and thus estimate the flow in fractures more correctly. The solution of NS-equations, however, is by no means computationally straight forward.

Recently, flow in fractures has always been estimated by using the Reynolds equation (Local Cubic Law-LCL) which replaced the traditional Cubic Law (Nicholl et al., 1999; Brush & Thomson, 2003). Regardless of the fact that the solution of the Reynolds equation can simply be obtained in comparison to the full NS-equations, its applicability is restricted because of the assumptions which govern its derivation. In fact, with other assumptions, the most restrictive ones are: the viscous terms must dominate the inertia term and the velocity distribution along the axis perpendicular to the fracture's plane is parabolic.

In previous studies (Zimmerman & Bodvarsson, 1996; Brush & Thomson, 2003), it was proposed that in order for the viscous terms to dominate the inertia terms, the Reynolds number, Re, must be kept less than 1. On the other hand for maintaining the parabolic velocity distribution across the aperture, Zimmerman and Brodvarsson (1996) suggested that the wavelength of aperture variation $\lambda \leq 0.4$ while Brown (1987) suggested $\lambda \leq 0.03$.

The use of the LCL also requires that one should decide on which definition for the link transmissivity to be used. In previous studies, it was revealed that different definition for the link transmissivity can result into different estimate of the discharge through the fracture (Nicholl et al., 1999; Brush & Thomson, 2003). Therefore, it is not clear still on whether a unique definition of the link transmissivity can best represent the link transmissivity for all kinds of the aperture fields. On the other hand, since flows in fractures are dominated by channelization (i.e., flow tends to follow paths with large apertures), the effect of the inertia terms may differs from one fracture to the other. It is therefore, unclear how much small the Reynolds number should be less that 1 for the inertia terms in x-y plane (the velocity derivatives $\partial/\partial x$ and $\partial/\partial y$) to be negligible.

3 NUMERICAL MODEL

3.1 *Governing equations*

The basic equations governing the flow model herein developed, consists of continuity and momentum equations of plane 2-D flows, obtained by integrating the 3-D NS-equations (Batchelor, 1967).

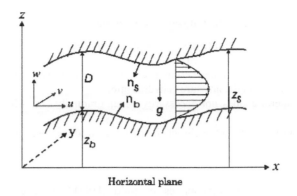

Horizontal plane

Figure 1.   Conceptual flow model in a single fracture.

Eqs 3 and 4 based on Cartesian coordinate system. The conceptual flow model is shown in Figure 1.

$$\nabla \cdot u = 0 \tag{3}$$

$$\frac{\partial u}{\partial t} + (u \cdot \nabla)u = F - \frac{1}{\rho}\nabla P + \frac{\mu}{\rho}\nabla^2 u \tag{4}$$

Continuity equation

$$\frac{\partial UD}{\partial x} + \frac{\partial VD}{\partial y} = 0 \tag{5}$$

Momentum equations, x-direction

$$\frac{\partial(UD)}{\partial t} + \beta\frac{\partial(U^2 D)}{\partial x} + \beta\frac{\partial(UVD)}{\partial y} = -D\frac{\partial}{\partial x}\left(\frac{P_D}{\rho}\right) - gD\frac{\partial(z_b + D)}{\partial x} + \frac{\partial}{\partial x}\left(vD\frac{\partial U}{\partial x}\right)$$
$$+ \frac{\partial}{\partial y}\left(vD\frac{\partial U}{\partial y}\right) - \frac{\tau_{bx}}{\rho}\sqrt{1+\left(\frac{\partial z_b}{\partial x}\right)^2 + \left(\frac{\partial z_b}{\partial y}\right)^2} - \frac{\tau_{sx}}{\rho}\sqrt{1+\left(\frac{\partial z_b}{\partial x}\right)^2 + \left(\frac{\partial z_b}{\partial y}\right)^2} \tag{6}$$

and y-direction

$$\frac{\partial(VD)}{\partial t} + \beta\frac{\partial(UVD)}{\partial x} + \beta\frac{\partial(V^2 D)}{\partial y} = -D\frac{\partial}{\partial y}\left(\frac{P_D}{\rho}\right) - gD\frac{\partial(z_b + D)}{\partial y} + \frac{\partial}{\partial x}\left(vD\frac{\partial V}{\partial x}\right)$$
$$+ \frac{\partial}{\partial y}\left(vD\frac{\partial V}{\partial y}\right) - \frac{\tau_{by}}{\rho}\sqrt{1+\left(\frac{\partial z_b}{\partial x}\right)^2 + \left(\frac{\partial z_b}{\partial y}\right)^2} - \frac{\tau_{sy}}{\rho}\sqrt{1+\left(\frac{\partial z_b}{\partial x}\right)^2 + \left(\frac{\partial z_b}{\partial y}\right)^2} \tag{7}$$

where τ_{bx}, τ_{by}, τ_{sx} and τ_{sy} are shear vectors on the bottom and the top walls, (U, V) are averaged velocities in x and y directions, P_D is the pressure at the top wall, β is a momentum correction factor (in this study, $\beta = 1.2$), v is coefficient of kinematics viscosity, D is aperture depth and g is gravitational acceleration. The wall shear stresses are calculated from the resistance law of laminar flow as follows:

$$\frac{\tau_{by}}{\rho} = \frac{\tau_{sy}}{\rho} = \frac{6vV}{D} \quad ; \quad \frac{\tau_{bx}}{\rho} = \frac{\tau_{sx}}{\rho} = \frac{6vU}{D}. \tag{8}$$

19

The developed model is based on the following assumptions;

- The fracture is free from external forces (no deformation occurs),
- Neither slip nor flow occur at the fracture walls,
- The velocity distribution across the aperture (i.e. in $z-$ direction) is parabolic,
- Pressure distribution across the aperture is hydrostatic,
- The variation of aperture does not necessarily occur gradually; the effect of wavelength is considered, and
- The inertia terms in $x - y$ are not negligible in comparison to the viscous terms.

3.2 Procedure of numerical

The standard numerical method for incompressible fluids (HSMAC) is used (Hosoda et al, 1993). It is assumed that at the initial condition the hydraulic variables M $(= UD)$, N $(=VD)$ and pressure P_D at time $t = n.\Delta t$ are known. Then, the hydraulic variables M^* and N^* at time step $t = (n + 1)\Delta t$ are calculated, explicitly. This procedure utilizes the concept of pressure (P_D) correction as follows;

$$P^*_{D\,i,j} = P^n_{Di,j} + \delta P^*_D \qquad (9)$$

where

$$\delta^* P_{D\,i,j} = -\frac{\omega \varepsilon^*_{i,j}}{2gD_{i,j}\Delta t\left(\dfrac{1}{\Delta x^2} + \dfrac{1}{\Delta y^2}\right)}, \omega = 0.5 \quad \text{and}$$

$$\varepsilon^* = \frac{M^*_{i+1,j} - M^*_{i,j}}{\Delta x} + \frac{N^*_{i,j+1} - N^*_{i,j}}{\Delta y}.$$

M^* and N^* are then recalculated using the new value of P^*_D. This process is repeated until the criteria for the error $|\varepsilon^*|$ is satisfied, then, M^* and N^* are considered to be the hydraulic variables at time step $t = (n + 1)\Delta t$ (i.e. M^{n+1}, N^{n+1}).

4 ONE DIMENSIONAL SIMULATION

4.1 Sinusoidal aperture variation

One of the simplest aperture profile function that captures some of the geometrical properties of the "roughed surface walled" fracture is a duct with a sinusoidal aperture perturbation in Figure 2. In this study, we consider two sinusoidal surfaces put together separated by the constant average

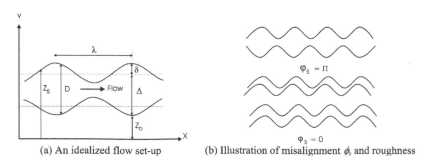

(a) An idealized flow set-up   (b) Illustration of misalignment ϕ_s and roughness

Figure 2. Geometry of sinusoidal rough-wall channel.

aperture (Δ). Different fracture configuration is obtained by imposing phase angle (ϕ_s) to the equation defining the top sinusoidal surface (Z_s). The phase angles, $\phi_s = 0$ and π, are for in phase and out phase configuration, respectively (Figure 2(b)).

The equations defining the top wall (Z_s), the bottom wall (Z_b) and the aperture (D) are expressed as follows:

$$Z_s = \delta \cos\left(\frac{2\pi x}{\lambda} + \phi_s\right) + \Delta \tag{10}$$

$$Z_b = \delta \cos\left(\frac{2\pi x}{\lambda}\right) \tag{11}$$

$$D = Z_s - Z_b \tag{12}$$

where δ is the amplitude (surface roughness) of the top and the bottom surfaces and λ is the wave length of the surfaces' roughness variations.

4.2 Governing equations

The governing equations consist of 1-D continuity and momentum equations reduces from 2-D equations (Eqs.5, 6 and 7) assuming that, the x axis is chosen so as to coincide with the microscopic pressure gradient.

Continuity equation

$$\frac{\partial(UD)}{\partial x} = 0 \tag{13}$$

Momentum equation

$$\beta \frac{\partial(U^2 D)}{\partial x} = -D \frac{\partial}{\partial x} \frac{P_D}{\rho} - gD \frac{\partial(Z_s + D)}{\partial x} + \frac{\partial}{\partial x}\left(\nu D \frac{\partial U}{\partial x}\right)$$
$$- \frac{\tau_{bx}}{\rho}\sqrt{1 + \left(\frac{\partial Z_b}{\partial x}\right)^2} - \frac{\tau_{sx}}{\rho}\sqrt{1 + \left(\frac{\partial Z_s}{\partial x}\right)^2} \tag{14}$$

4.3 Method of solution

The solution of 1-D model is obtained from analytical and numerical integration approaches under simple hydraulic conditions and some assumptions. It is assumed that the flow is under the fully developed steady state condition. We also consider relatively small amplitude to wave length ratio of the order less than 0.0125. Based on these assumptions, we utilize the concept of linear response for pressure variation (Patel, 1991), which is defined by the following expression.

$$P_D = P_0 + P' \tag{15}$$

$$\frac{P'}{\rho g} = \delta_p \cos\left(\frac{2\pi x}{\lambda} + \phi_p\right) + C \tag{16}$$

where P_0 is the pressure for undisturbed flow (parallel plate flow), P' is the wave induced pressure, δ_p is the amplitude of the pressure variation and ϕ_p is the pressure phase angle by which the maxima precede the wave crest.

(i) Analytical solution

The expressions of the idealized sinusoidal profiles (Z_s, Z_b) and the wavy induced pressure are substituted in Eq.14. After rearranging the like terms containing $\cos(nx)$, $\sin(nx)$, $\cos(2nx)$, $\sin(2nx)$, $\cos(3nx)$, $\sin(3nx)$, $\cos(4nx)$ and $\sin(4nx)$, Eq.14 is written in the following general form.

$$f_1 \cos(nx) + f_2 \sin(nx) + f_3 \cos(2nx) + f_4 \sin(2nx)$$
$$+ f_5 \cos(3nx) + f_6 \sin(3nx) + f_7 \cos(4nx) + f_8 \sin(4nx) + f_0 = 0 \tag{17}$$

where $n = 2\pi$ and $f_k = f(q, \phi_s, \phi_p, \delta_p, \lambda, \Delta$ and $S)$.

The solution of Eq.17 is obtained by equating the coefficients of the constant terms, sins and cosine of Eq.14 to zero, yielding a set of nine equations is obtained. Since three unknowns have to be solved in this case, equations f_0, f_1 and f_2 are considered.

$$\left. \begin{array}{l} f_0 : a_1 q + a_2 \delta_p \cos\phi_p + a_3 \delta_p \sin\phi_p + a_4 = 0 \\ f_1 : b_1 q^2 + b_2 q + b_3 \delta_p \cos\phi_p + b_4 \delta_p \sin\phi_p + a_5 = 0 \\ f_2 : c_1 q^2 + c_2 q + c_3 \delta_p \cos\phi_p + c_4 \delta_p \sin\phi_p + c_5 = 0 \end{array} \right\} \tag{18}$$

Keeping the pressure gradient of undisturbed flow (S) constant, a set of equations 18 is solved for q, δ_p and pressure phase angle ϕ_p. The constant C in the expression of pressure variation is obtained by applying the pressure boundary condition: at $x = 0$ and $P' = 0$ to Eq.16 from which the constant C is $-\delta_p \cos\phi_p$.

(ii) 1-Dintegral model

Eq.14 is first non-dimension using the following non-dimensional parameters;

$$D' = \frac{D}{\Delta}; \quad x' = \frac{x}{\lambda}; \quad q' = \frac{q}{q_0}; \quad Z_s' = \frac{Z_s}{\Delta}; \quad Z_b' = \frac{Z_b}{\Delta} \tag{19}$$

where $q = UD$ and $q_0 = (\Delta^3 gS)/(12\nu)$ is the discharge of the equivalent parallel plate flow with aperture size equal to the average aperture size (Δ).

Then, the 1-D integral model is obtained through integrating the non-dimension form of Eq.14 over one period of sinusoidal aperture variation as follows;

$$\beta \left(\frac{q_0}{\Delta} \right)^2 q'^2 \int_0^1 \frac{1}{D'^3} \frac{\partial D'}{\partial x'} dx' - \frac{6\nu q_0 \lambda}{\Delta^3} q' \int_0^1 \frac{1}{D'^3} \left[\sqrt{1 + \left(\frac{\partial Z_b'}{\partial x'} \right)^2 \left(\frac{\Delta}{\lambda} \right)^2} + \sqrt{1 + \left(\frac{\partial Z_s'}{\partial x'} \right)^2 \left(\frac{\Delta}{\lambda} \right)^2} \right] dx'$$
$$+ gS\lambda + \frac{\nu q_0}{\Delta \lambda} q' \left[\int_0^1 \frac{1}{D'^3} \left(\frac{\partial D'}{\partial x'} \right)^2 dx' - \int_0^1 \frac{1}{D'^2} \frac{\partial^2 D'}{\partial x'^2} dx' \right] = 0 \tag{20}$$

where S is constant pressure gradient $\left(-\dfrac{\partial}{\partial x'} \left(\dfrac{P_0}{\rho} \right) \right)$ of undisturbed parallel plate flow.

We further rewrite Eq.20 into a more general form, for the sake of the following discussion on the effects of few fracture parameters on flow.

$$\beta \left(\frac{q_0}{\Delta} \right)^2 q'^2 I_1 + gS\lambda - \frac{6\nu q_0 \lambda}{\Delta^3} q' I_2 + \frac{\nu q_0}{\Delta \lambda} q' I_3 = 0 \tag{21}$$

where,
$$I_1 = \int_0^1 \frac{1}{D'^3} \frac{\partial D'}{\partial x'} dx' \quad , \quad I_2 = \int_0^1 \frac{1}{D'^3} \left[\sqrt{1 + \left(\frac{\partial Z_b'}{\partial x'} \right)^2 \left(\frac{\Delta}{\lambda} \right)^2} + \sqrt{1 + \left(\frac{\partial Z_s'}{\partial x'} \right)^2 \left(\frac{\Delta}{\lambda} \right)^2} \right] dx' \quad \text{and}$$

$$I_3 = \int_0^1 \frac{1}{D'^3} \left(\frac{\partial D'}{\partial x'} \right)^2 dx' - \int_0^1 \frac{1}{D'^2} \frac{\partial^2 D'}{\partial x'^2} dx'.$$

22

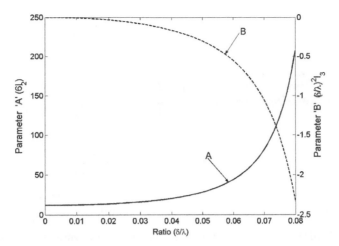

Figure 3. Contribution of parameters I_2 and I_3 on resistance to flow ($\Delta = 0.002$ mm and phase angle $\phi_s = \pi$).

Since Eq.20 is not practical, the results are obtained through numerical integration. It should be noted that, when the aperture variation is defined by a periodic function, the integration 'term I_1' is zero. For parallel plate flow, $I_2 = 2$ and parameters I_1 and I_3 vanish. Under this condition, Eq.22 reduces to cubic law of parallel plate flow. We, then, investigate the effect of parameters I_2 and I_3 on hydraulic conductivity of a single fracture by first rewriting Eq.22 in more tractable form as follows;

$$q' = \frac{12}{\left[6I_2 - \left(\frac{\Delta}{\lambda} \right)^2 I_3 \right]} \tag{22}$$

Based on Figure 3 and Eq.22, it is evident that wall friction plays a great role on flow resistance in this case of the parameter, I_2. The parameter, I_2, depends on the amplitude, δ, and the characteristic length, λ (wave-length) of the fracture's aperture variation. The larger the magnitude of parameter, I_2, is, the higher the resistance to flow occurs.

4.4 1-D results and comparison

Since we considered small amplitude to wave-length ratio of the order smaller than 0.0125, it is expected that the flow behavior is close to unidirectional laminar flow (Patel et al., 1991). In this section, therefore, we compare the results of the 2-D numerical model, 1-D analytical solution and 1-D integral model. Also the comparison is made with Zimmerman et al. (1991) results based on Reynolds equation of the sinusoidal aperture variation of 1-D model, which assumes that, the resistance due to each aperture element are in series Eq.23.

$$q' = \frac{\left(1 - \left(\frac{2\delta}{\Delta} \right)^2 \right)^{5/2}}{1 + \frac{1}{2} \left(\frac{2\delta}{\Delta} \right)^2} . \tag{23}$$

where $q' = (d_h / \Delta)^3$.

The variation of hydraulic conductivity for different phase angles (different fracture geometry), while keeping the amplitude-to-wavelength ratio, $(2\delta/\lambda) = 0.002$, is shown in Figure 4(a). It is observed that the geometry with phase angle, $\phi_s = 0.0$rad, has less resistance to hydraulic

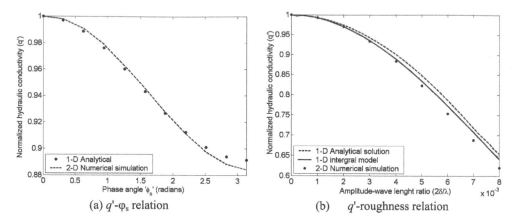

(a) q'-φ_s relation

(b) q'-roughness relation

Figure 4. Normalized hydraulic conductivity plotted against fractures parameters.

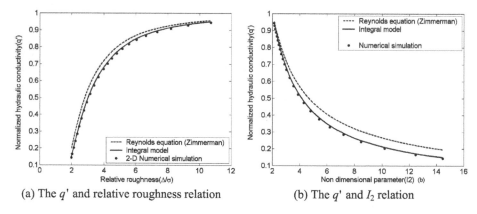

(a) The q' and relative roughness relation

(b) The q' and I_2 relation

Figure 5. Non-dimensional hydraulic conductivity plotted against fracture parameters.

conductivity $(d_h/\Delta)^3 \approx 1.0$. This implies that the resistance to flow is approximately equal to that of the parallel plate flow with aperture size, $\Delta = 2$ mm. Ge (1997) reported as similar results which the effect of tortuosity was not taken into account. Thereafter, the hydraulic conductivity decreases with increase of phase angle 'ϕ_s' and become minimum when the phase angle $\phi_s = \pi$ (out of phase configuration). There exists good agreement between 2-D numerical model 1-D analytical solution and 1-D integral model results. This trend is in agreement with the results obtained by Brown and Stockman (1995), who considered the numerical solution of the Reynolds equation.

Of a particular interest is the observation that increase of wall roughness expressed by amplitude-to-wavelength ratio, $2\delta/\lambda$, from 0.0 to 0.08, while keeping the phase angle $\phi_s = \pi$ caused a proportionate decrease of hydraulic conductivity as depicted in Figure 4(b). This is in accordance with the practical effects of fracture's surface roughness, that is, increases of amplitude-to-wavelength ratio of the sinusoidal variation while keeping the average aperture, Δ, constant, has an implication on an increase of fracture surface roughness which results into an increase of resistance to flow (Ge, 1997).

In order to make comparison with results of Zimmerman et al. (1991), the hydraulic conductivity was plotted against a non-dimensional roughness parameter (Δ/σ) and the skin friction parameter, I_2, for $\phi_s = \pi$. Figure 5 shows the hydraulic conductivity and relative roughness relation used by the Reynolds equation (Zimmerman, et al., 1991), the 1-D integral and the 2-D numerical simulation. It is observed that the Reynolds equation overestimates the magnitude of hydraulic conductivity as was expected in both cases.

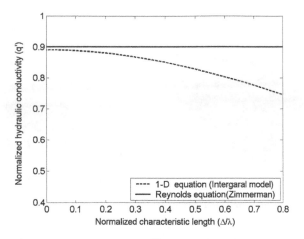

Figure 6. Dependence of hydraulic conductivity on characteristic length (λ) of aperture variation ($\Delta = 2$ mm, $\delta = 0.2$ mm).

A clear difference is observed when normalized hydraulic conductivity is plotted against the skin friction parameter, I_2, in Figure 5. The difference of the results keeps on increasing for higher values of the parameter, I_2, (i.e. when the roughness increases). This implies that when the effect of wall roughness becomes significant, Reynolds equation tend to overestimate the magnitude of hydraulic conductivity.

In Figure 6, the effect of characteristic length to hydraulic conductivity is expressed as a ratio Δ/λ for keeping consistence with the previous studies (Brown, 1987; Zimmerman & Bodvarsson, 1996). It is observed that the results of Reynolds equation based on the equation derived by Zimmerman and Bodvarsson (1996) is not influenced by the variation of the characteristic length, while the results of 1-D model herein considered shows sensitivity.

4.5 *Discussion*

On basis of the flow results of an idealized rock joint with sinusoidal aperture variation, it is evident that the developed 2-D model is not restricted by the characteristic length of aperture variation (λ). The effect of the characteristic length is included in the terms expressing the surfaces' variation of the rock joint (∇Z_s and ∇Z_b). This is in contrary to the Reynolds equation which has been reported to give reasonably accurate results up to a certain value of the ratio (Δ/λ). Zimmerman et al. (1991) proposed that Reynolds equation can be used to estimate flows in fracture as long as the ratio (Δ/λ) < 0.4, while Brown (1987) proposed (Δ/λ) < 0.03. However, in real fractures, the criteria proposed by Zimmerman et al. (1991) were not always met (Ge, 1997; Gentier et al., 1989). Therefore, the applicability of Reynolds equation is significantly restricted.

It should be noted that the dependence of hydraulic conductivity on characteristic length shown in Figure 6, is not aimed to give the quantitative value below which the Reynolds equation is applicable, but rather to show how the 2-D model herein developed takes into account the effect of characteristic length.

5 NUMERICAL SIMULATIONS OF FLOW IN MEASURED APERTURE OF ROCK JOINT

The 2-D model was used in the simulation of the two dimensional computations of fluid flow within the measured aperture fields of the experimental rock joints. The simulated domain consisted

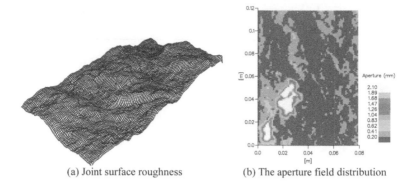

(a) Joint surface roughness　　　　　(b) The aperture field distribution

Figure 7.　Examples of the measured joint surface roughness and aperture distribution.

of specimens with dimension, 80×120 mm. The conditions for simulation were based on the experimental set up explained below.

5.1　The measured aperture fields

The key to successful simulation of flow on rock joints is the application of the correct aperture distribution and geometry of joint surface roughness. In this research work, the geometry of joint surface roughness and the aperture distribution on the rock joint during the shearing process are obtained through the mechanical shear model (Kishida et al., 2001). This model utilizes the discrete profiling data of the joint surface roughness. The topography of the upper and lower surfaces of fracture specimens was measured using the 3-D lasers-type profiling procedure at the interval of 0.25 mm. The 3-D profiling system used consisted of X-Y positioning table (with positioning accuracy of ± 15 μm, and repositioning accuracy of ± 3 μm) and a laser scanner (spot dimensions of 45×20 μm^2 with measurement range of ± 8 mm) with maximum resolution of 0.5 μm for elevation measurements. The roughness geometry of the fracture surfaces and the aperture distribution of the rock fracture during the shearing process were obtained through the use of shear mechanical model (Kishida et al., 2001). Figure 7 shows examples of the surface roughness and aperture field distribution considered in this study.

5.2　Description of laboratory experiments

The simultaneous direct shear and permeability experiments were carried out in consideration of joint surface roughness and material properties under constant normal confining conditions (Kishida et al., 2004). The experiments were conducted according to the standard method used in several studies for permeability test, where shear displacement, dilation, shear stress, normal stress, transmissivity and pore pressure can be determined. In this study, the hydraulic head and the normal confining stress were kept on constant at 1.0 m and 1.0 MPa, respectively. The permeability tests were performed at each predetermined shear displacement up to 3 mm.

5.3　Comparison of 2-D model to experimental results

In this section, comparison is made between the 2-D numerical model and experimental results using the effective transmissivity (T) of each joint, as its use does not require knowledge of the mean aperture. Flow rate (Q) through each fractured rock joint of the specimens was computed using the numerical procedure explained previously. The effective transmissivity is calculated based on the following equation;

$$T = \frac{QL}{W\Delta H} \tag{24}$$

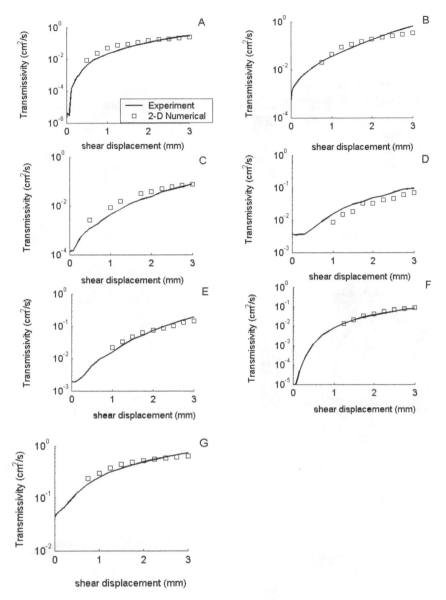

Figure 8. Comparison between 2-D simulation and experimental results on transmissivity of fractured rock joint.

where L and W are the length and the width of specimen, respectively, and ΔH is Hydraulic head.

Figure 8 shows the experimental and simulation results of shear displacement and transmissibility relation. It is observed that for specimens A, B, C, E, F and G the results of 2-D flow model exhibit fairly good agreement with those obtained through experiments. For specimen D, however, there exists a considerable disagreement for entire range of shear displacement, where the 2-D model underestimates the transmissivity.

In fractures, fluids tend to follow paths with bigger aperture sizes. Therefore, one of the important measures of the models applicability is how best it simulates the channeling effect with respect to the aperture size distribution. Figure 9 shows the flow vector distribution simulated by the 2-D model at 2.5 mm shear displacement.

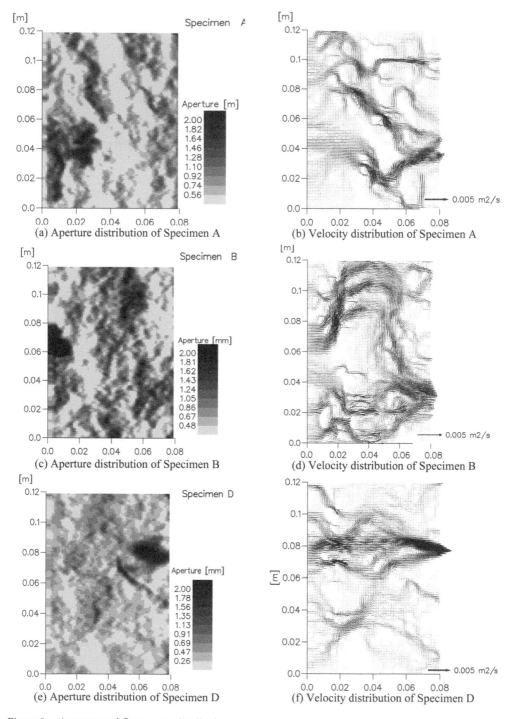

Figure 9. Aperture and flow vector distribution.

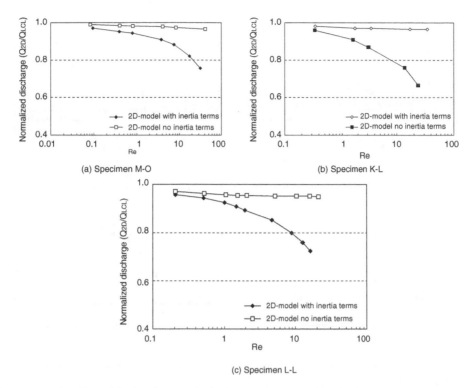

(a) Specimen M-O

(b) Specimen K-L

(c) Specimen L-L

Figure 10. The effect of the inertia terms in the 2D-model expressed by the dependence of the flow ratio Q_{2D}/Q_{LCL} on the Reynolds number (Re).

5.4 *Discussion*

Normally, during permeability tests in saturated rock fractures, using the shear mechanical model, transmissivity tend to increase with shear displacement. This is based on the fact that, the average mechanical aperture increases as result of dilation (Yeo et al., 1998). This phenomenon is also observed in the results presented in Figure 8. One possible explanation of the model to underestimate transmissivity for specimen D is that, it is strongly believed that, the aperture distribution was underestimated which resulted into underestimating of the transmissivity.

As shown in Figure 9, the brightest color refers to the largest aperture size and vice versa. It is observed that the model has been able to simulate the channeling effect across the rock joint of specimens with comparison to aperture distribution.

5.5 *Significance of the inertia terms*

The effect of the inertia terms is investigated using three different aperture distributions from different specimens in this case at shear displacement of 0.5 mm, namely specimens M-O, K-L and L-L. The effect of the inertia terms is expressed here as a function of the Reynolds number estimated using Eq. 8. This is based on the fact that, if the geometric conditions of the fracture are kept constant, the strength of the inertia terms depends on the hydraulic conditions which can be expressed by the Reynolds number.

Figure 10 shows the relationship between the hydraulic conductivity simulated by the 2D-model with and without the inertia terms and the Reynolds number. The hydraulic conductivity is expressed as the ratio of the discharge simulated by the 2D-model to that of the LCL (Q_{2D}/Q_{LCL}), which makes possible, not only investigate the effect of inertia terms, but also indirectly make comparison with LCL simulations. It is observed that as the Reynolds number increases, the hydraulic conductivity

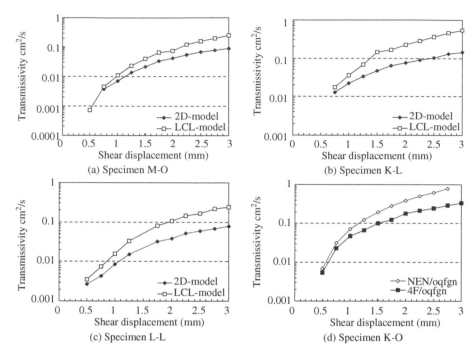

Figure 11. The transmissivity-shear displacement relation estimated by the 2D-model and LCL for four different specimens.

for 2D-model with inertia terms decreases. However, for the 2D-model without inertia terms the hydraulic conductivity almost remain unchanged with increase of Reynolds number. This implies that the inertia terms become important as the Reynolds number increases. Similar results were reported in previous studies (Nicholl et al., 1999; Zimmerman & Bodvarsson, 1996). In fact, it interesting to note that even when the inertia terms is neglected in the 2D-model, the ratio Q_{2D}/Q_{LCL} is not unity (i.e. the results from the 2D-model is not the same as that of LCL). We believe this deviation is caused by the terms other than the inertia terms which are in the 2D-model (such as $\partial Z_s/\partial x$, $\partial Z_s/\partial y$ $\partial Z_b/\partial x$ and $\partial Z_b/\partial x$). However, the effect of these terms seems to be small.

Note that for the type of aperture fields considered here, the maximum deviation of 2D-model without inertia terms from the LCL simulations is about 4%, while for the 2D-model with inertia terms is 34%.

In previous studies (Zimmerman & Bodvarsson, 1996), it was suggested that the inertia terms can be negligible as long as Re < 1. However, based on results in the Figure 10, it is evident that the importance of the inertia terms differs from one fracture to the other. Example for aperture field of specimen K-L at Re = 1, the deviation of results for 2D-model with and without inertia terms is about 7% while for specimens M-O and L-L is about 3%. Therefore from these results it can be concluded that even at Re < 1 depending on the type of fracture, in real fracture, the inertia terms may be important. It is also important to note that at around Re = 1, the maximum deviation of the results between the LCL and the 2D-model is about 10%.

5.6 Hydraulic conductivity

In this section, a comparison between LCL and 2D-model simulation results is presented. Firstly, the flow rate, Q, obtained for each specimen at predetermined shear displacement is used to calculate the effective transmissivity, T, using Eq. 24.

Figure 11 shows the variation of the hydraulic conductivity of different fractures with advancement of shearing process, expressed by transmissivity-shear displacement relations. It is observed that in all specimens the transmissivity estimated by both the 2D-model and LCL increase with the advancement of shear displacement. This is based on the fact that, during shearing the aperture size increases as result of dilation and therefore discharge increases (Yeo, 1998). In all specimens, also it is observed that the degree of disagreement between the 2D-model and the LCL results increases with advancement of the shear displacement, with the LCL overestimating the transmissivity.

Based on the result presented in Section 5.5, it is evident that this disagreement is likely to be caused by the inertia terms, since the effect of other terms seems to be small, refer to Figure 10. In Figure 11, it can also be observed that the difference in results of the 2D-model and the LCL diminishes with decrease of transmissivity which implies decrease of the Reynolds number. Example for specimen L-L, at shear displacement 3 mm, the difference in results is at a factor of 3.4 with $Re \approx 95$ while at shear displacement $= 0.5$ mm the difference in results is at a factor of 1.5 with $Re \approx 15$. The same trend can be observed in all four specimens. Therefore, it can be concluded that the difference between LCL and 2D-model is mainly due to the effect of inertia forces.

6 CONCLUSION

In fact, the use of Reynolds equation in simulating flow in fractures has several advantages that: it is computationally easier and analytical solution for simple geometry can be obtained. However, when applied to the measured fracture apertures, where zero apertures (contact regions) are obvious, the Reynolds equation has severe limitations. Thus the Reynolds equation can only be used successfully in unobstructed areas, whereas the obstructed areas have to be treated differently (Zimmerman et al., 1991). This problem is avoided by using 2-D flow model developed in this work, and therefore, the channeling phenomena obtained is sufficiently accurate.

In this study, we have applied the 2-D flow model to simulate flow in a single saturated rock fractures. The model has shown its applicability in simulating the flows herein considered especially for big range of shear displacement. However, there exists noticeable disagreement between the experiment and simulation results for higher shear displacement of few fractures. The authors believe that the disagreement is caused by the effect of aperture deformation that could not be properly accommodated for in aperture and roughness distribution. On the other hand since the model takes into consideration the effect of inertia, viscosity, wall shear and characteristic length of aperture variation, the model introduced in this study can serve a purpose in testing other models.

Moeover, flow simulations in a measured aperture fields are conducted. Through simulations a comparison between the depth averaged flow model (2D-model) and the Local Cubic Law (LCL) is presented. It is observed that, the main difference between the LCL and the 2D-model is due to the inertia terms included in the later model.

Based on the aperture fields considered in this study, the following conclusions can be made: (i) using the same hydraulic conditions, the effect of inertia terms differs from one fracture to the other, (ii) depending on the aperture distribution within the fracture, at $Re \leq 1$, the inertia terms can be important (iii) for the flow simulation of laboratory experiments, where the Reynolds number can not be kept much smaller that 1, and where the effect of inertia terms is not guaranteed to be negligible, the depth averaged flow model can improve the simulated results in comparison to the LCL model.

REFERENCES

Batchelor, G. K. 1967. An Introduction to fluid dynamics, pp.147–150. Cambridge University Press.
Brown, S. R. 1987. Fluid flow through rock joint: The effect of surface roughness. *J. Geophys. Res.* 92, B2: 1337–1347.
Brown, S. R. and H. W. Stockman. 1995. Applicability of Reynolds equation for modeling fluid flow between rough surfaces. *Geophysical Research Letters*, 22, No. 22: 2537–2540.
Brush, D. J. and N. R. Thomson. 2003. Fluid flow in synthetic rough-walled fractures: Navier-Stokes, Stokes, and local cubic law simulations, *Water Resour. Res.,* 39(4), 1085, doi: 10.1029/2002WR001346.

Ge, S.. 1997. A governing equation for fluid flow in rough fractures. *Water Resources Research*, 33, No.1: 53–61.

Gentier, S., D. Billaux and L. van Vliet. 1989. *Laboratory testing of the voids of a fracture*. Rock Mech. Rock Eng., 22: 149–157.

Hosoda, T., K. Inoue, and A. Tada. 1993. Hydraulic transient with propagation of interface between open-channel free surface flow and pressurized flow. *Proc. Int. Symp. On Comp. Fluid Dynamics*, Sendai, Vol. 1: 291–296.

Kishida, K., T. Adachi, and K. Tsuno. 2001. Modeling of the shear behavior of rock joints under constant normal confining conditions. *Rock Mechanics in the National Interest, Proceedings of the 38th U.S. Rock Mechanics Symposium*, Elsworth, Tinucci & Heasly, (eds). 791–798. Rotterdam: Balkema.

Kishida, K., T. Hosoda and A. Yamamoto. 2004. Estimation of the Hydro-Mechanical Behavior on Rock Joints through the Shear Model. *Gulf Rock 2004, Proceedings of the 6th North American Rock Mechanics Symposium*.

Kishida, K., P. Mgaya and T. Hosoda 2005. Application of depth averaged flow model in estimating the flow in a single rock fracture. In *proceedings of the 40th U. S. Rock Mechanics Symposium*, June 25–29, 2005– Anchorage, Alaska.

Nicholl, M. J., H. Rajaram, R. J. Glass, and R. Detwiler. 1999. Saturated flow in a single fracture: Evaluation of Reynolds equation in measured aperture field, *Water Resour. Res.*, 35(11), 3361–3373.

Patel, V.C., J.T. Chon and J.Y.Yoon. 1991. Laminar flow over wavy walls. *J. of Fluid Eng.*, 113: 574–578.

Renshaw, C. E. 1995. On the relationship between mechanical and hydraulic apertures in rough-walled fractures. *J. Geophys. Res.* 100, No.B12: 24629–24636.

Snow, D. T. 1965. A Parallel plate model of fractures Permeable media, Ph.D. Dissertation, University of Califonia, Berkeley, Califonia.

Witherspoon, P.A., Wang, J.S.Y., Iwai. K. and Gale, J. E. 1980. Validity of cubic law for fluid flow in a deformable rock fracture. *Water Resour. Res., 16, 1016–1024.*

Yeo, I. W., M.H. De Freitas and R. W. Zimmerman. 1998. Effect of shear displacement on the aperture and permeability of a rock fracture. *Int. J. Rock Mech. Min. Sci. Geomech. Abstr.* 35, No.8: 1051–1070.

Zimmerman, R. W., S. Kumar, and G.S. Bodvarsson. 1991. Lubrication theory analysis of the permeability of rough-walled fractures. *Int. J. Rock Mech. Min. Sci. Geomech. Abstr.* 28, No.4: 325–331.

Zimmerman R. W. and G. S. Bodvarsson. 1996. Hydraulic conductivity of rock fractures. *Transport in Porous Media* 23: 1–30.

Applications of Computational Mechanics in Geotechnical Engineering – Sousa,
Fernandes, Vargas Jr & Azevedo (eds)
© 2007 Taylor & Francis Group, London, ISBN 978-0-415-43789-9

Numerical simulations of materials with micro-structure: limit analysis and homogenization techniques

P.B. Lourenço
University of Minho, Department of Civil Engineering, Guimarães, Portugal

ABSTRACT: Continuum-based numerical methods have played a leading role in the numerical solution of problems in soil and rock mechanics. However, for stratified soils and fractured rocks, a continuum assumption often leads to difficult parameters to define and over-simplified geometry to be realistic. In such cases, approaches that consider the micro-structure of the material can be adopted. In this paper, two of such approaches are detailed, namely limit analysis incorporating fractures and individual blocks, and elastoplastic homogenization of layered soils.

1 INTRODUCTION

The present paper addresses the inelastic behavior of heterogeneous media, such as soils and rocks. This problem can be, basically, approached from two directions. A possible direction is to gather, collate and interpret extensive experimental data and, ultimately, manipulate it in the form of master curves in terms of non-dimension variables. One step further can be to seek empirical analytical expressions that fit the experimental data. However, this approach will not be followed here. Even tough such an approach can be useful, the results are limited to the conditions under which the data were obtained. New materials and/or application of a well known material in different loading conditions lead to a new set of costly experimental programs. Here, a more fundamental approach, with a predictive nature, is sought using two different tools, namely limit analysis and homogenization techniques.

2 LIMIT ANALYSIS

A continuum assumption can lead to difficult parameters to define and over-simplified geometry to be realistic, particularly in the case of fractured rocks. At least in such case, discrete representations of fractures and individual blocks must be adopted. Different approaches are possible to represent these heterogeneous media, namely, the discrete element method (DEM), the discontinuous finite element method (FEM) and limit analysis (LAn).

The explicit formulation of a discrete (or distinct) element method is detailed in the introductory paper by Cundall and Strack (1985). The discontinuous deformation analysis (DDA), an implicit DEM formulation, was originated from a back-analysis algorithm to determine a best fit to a deformed configuration of a block system from measured displacements and deformations, Shi and Goodman (1985). The relative advantages and shortcomings of DDA are compared with the explicit discrete element method and the finite element method (Hart, 1991), even if significant developments occurred in the last decade, particularly with respect to three-dimensional extension, solution techniques, contact representation and detection algorithms. The typical characteristics of discrete element methods are: (a) the consideration of rigid or deformable blocks (in combination with FEM); (b) connection between vertices and sides/faces; (c) interpenetration is usually possible; (d) integration of the equations of motion for the blocks (explicit solution) using the real damping coefficient (dynamic solution) or artificially large (static solution). The main advantages are an adequate formulation for large displacements, including contact update, and an independent mesh for each block, in case of deformable blocks. The main disadvantages are the need of a large number of contact points required for accurate representation of interface stresses and a rather time consuming analysis, especially for 3D problems.

The finite element method remains the most used tool for numerical analysis in solid mechanics and an extension from standard continuum finite elements to represent jointed rock was developed in the early days of non-linear mechanics, Goodman *et al.* (1967). On the contrary, limit analysis received far less attention from the technical and scientific community for jointed rock, while being very popular for soil problems, Fredlund and Krahn (1977). Still, limit analysis (kinematic approach) has the advantage of being a simple tool, while having the disadvantages that only collapse load and collapse mechanism can be obtained and loading history can hardly be included. Here, a novel implementation of limit analysis is detailed and applied to an illustrative example.

2.1 General formulation

The limit analysis formulation for a rigid block assemblage detailed next adopts the following basic hypotheses: the joint between two blocks can withstand limited compressive stresses and the Coulomb's law controls shear failure, which features non-associated flow given by a zero dilatancy angle.

The static variables, or generalized stresses, at a joint k are selected to be the shear force, V_k, the normal force, N_k, and the moment, M_k, all at the center of the joint. Correspondingly, the kinematic variables, or generalized strains, are the relative tangential, normal and angular displacement rates, δn_k, δs_k and $\delta \theta_k$ at the joint center, respectively. The degrees of freedom are the displacement rates in the x and y directions, and the angular change rate of the centroid of each block: δu_i, δv_i and $\delta \omega_i$ for the block i. In the same way, the external loads are described by the forces in x and y directions, as well as the moment at the centroid of the block. The loads are split in a constant part (with a subscript c) and a variable part (with a subscript v): f_{cxi}, f_{vxi}, for the forces in the x direction, f_{cyi}, f_{vyi}, for the forces in the y direction, and m_{ci}, m_{vi}, for the moments. These variables are collected in the vectors of generalized stresses $\mathbf{Q}$, generalized strains δq, displacements rates δu, constant (dead) loads $\mathbf{F}_c$, and variable (live) loads $\mathbf{F}_v$. Finally, the load factor α is defined, measuring the amount of the variable load vector applied. The load factor is the limit (minimum) value that the analyst wants to determine and is associated with collapse.

With the above notation, the total load vector $\mathbf{F}$ is given by

$$\mathbf{F} = \mathbf{F}_c + \alpha \mathbf{F}_v \tag{1}$$

2.2 Yield function

The yield function at each joint is composed by the crushing-hinging criterion and the Coulomb criterion. For the crushing-hinging criterion, it is assumed that the normal force is equilibrated by a constant stress distribution near the edge of the joint, see Figure 1a. Here, a is half of the length of a joint and w is the width of the joint normal to the plane of the block. The stress value is f_{cef}, given by Eq. (2), borrowed from concrete limit analysis (Nielsen, 1998). Here, f_c is the compressive strength of the material, and ν is an effectiveness factor, which takes into account reductions in the compressive strength due to the fact that limit analysis assumes a rigid-plastic behavior, while, in fact, softening occurs. Eq. (3) is an expression for the effectiveness factor commonly used for concrete (Nielsen, 1998), where f_c is expressed in N/mm^2.

$$f_{cef} = \nu f_c \tag{2}$$

$$\nu = 0.7 - \frac{f_c}{200} \tag{3}$$

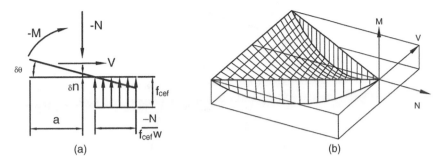

(a)   (b)

Figure 1. Joint failure: (a) generalized stresses and strains for the crushing-hinging failure mode; (b) geometric representation of a half of the yield surface.

The constant stress distribution hypothesis leads to the yield function φ given by Eq. (4), related to the equilibrium of moments; note that N_k represents a non-positive value. The Coulomb criterion is expressed by Eq. (5), related to the equilibrium of tangential forces. Here, μ is the friction coefficient or the tangent of friction angle at the joint. The equilibrium of normal forces is automatically ensured by the rectangular distribution of normal stresses. It is noted that the complete yield function is composed by four surfaces, two surfaces given by Eq. (4) and two surfaces given by Eq. (5), in view of the use of the absolute value operator. Figure 1b represents half of the yield surface $(M < 0)$, while the other half $(M > 0)$ is symmetric to the part shown.

$$\varphi_{1,2} \equiv N_k \left(a_k + \frac{N_k}{2 f_{cef} w_k} \right) + |M_k| \leq 0 \tag{4}$$

$$\varphi_{3,4} \equiv \mu N_k + |V_k| \leq 0 \tag{5}$$

2.3 *Flow rule*

Figure 1a illustrates also the flow mode corresponding to crushing-hinging, in agreement with the normality rule. It is noted that, for the Coulomb criterion, the flow consists of a tangential displacement only. The flow rule at a joint can be written, in matrix form, as given by Eq. (6), and, in a component-wise form, as given by Eq. (7), in which the joint subscripts have been dropped for clarity. Here, $\mathbf{N}_{0k}$ is the flow rule matrix at joint k and $\delta\boldsymbol{\lambda}_k$ is the vector of the flow multipliers, with each flow multiplier corresponding to a yield surface, and satisfying Eqs. (8-9). These equations indicate that plastic flow must involve dissipation of energy, Eq. (8), and that plastic flow cannot occur unless the stresses have reached the yield surface, Eq. (9). For the entire model, the flow rule results in Eq. (10), where the flow matrix $\mathbf{N}_0$ can be obtained by assembling all joints matrices.

$$\delta\mathbf{q}_k = \mathbf{N}_{0k}\delta\boldsymbol{\lambda}_k \tag{6}$$

$$\begin{bmatrix} \delta s \\ \delta n \\ \delta \theta \end{bmatrix} = \begin{bmatrix} 0 & 0 & -1 & 1 \\ a\left(1 - \dfrac{N}{f_{cef} w}\right) & a\left(1 - \dfrac{N}{f_{cef} w}\right) & 0 & 0 \\ -1 & 1 & 0 & 0 \end{bmatrix} \begin{bmatrix} \delta\lambda_1 \\ \delta\lambda_2 \\ \delta\lambda_3 \\ \delta\lambda_4 \end{bmatrix} \tag{7}$$

$$\delta\boldsymbol{\lambda}_k \geq \mathbf{0} \tag{8}$$

$$\boldsymbol{\varphi}_k^{\mathrm{T}} \delta\boldsymbol{\lambda}_k = 0 \tag{9}$$

$$\delta\mathbf{q} = \mathbf{N}_0 \delta\boldsymbol{\lambda} \tag{10}$$

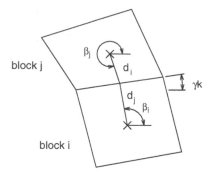

Figure 2.   Representation of main geometric parameters.

2.4 *Compatibility*

Compatibility between joint k generalized strains, and the displacement rates of the adjacent blocks i and j, is given in Eq. (11), being the vector δu_i defined in Eq. (12) and the compatibility matrix $C_{k,i}$, given in Eq. (13). Similarly, the vector δu_j and the matrix $C_{k,j}$ can be obtained. In this last equation γ_k, β_i, β_j, are the angles between the x axis and, the direction of joint k, the line defined from the centroid of block i to the center of joint k, and the line defined from the centroid of block j to the center of joint k, respectively. Variables d_i, d_j, represent the distances from the center of joint k to the centroid of the blocks i and j, respectively, see Figure 2.

$$\delta \mathbf{q}_k = \mathbf{C}_{k,j}\delta \mathbf{u}_j - \mathbf{C}_{k,i}\delta \mathbf{u}_i \tag{11}$$

$$\delta \mathbf{u}_i^T \equiv [\delta u_i \quad \delta v_i \quad \delta \omega_i] \tag{12}$$

$$\mathbf{C}_{k,i} = \begin{bmatrix} \cos(\gamma_k) & \sin(\gamma_k) & -d_i\sin(\beta_i - \gamma_k) \\ -\sin(\gamma_k) & \cos(\gamma_k) & d_i\cos(\beta_i - \gamma_k) \\ 0 & 0 & 1 \end{bmatrix} \tag{13}$$

Compatibility for all the joints in the model is given by Eq. (14), in which the compatibility matrix $\mathbf{C}$ is obtained by assembling the corresponding matrices for the joints of the model.

$$\delta \mathbf{q} = \mathbf{C}\delta \mathbf{u} \tag{14}$$

2.5 *Equilibrium*

Applying the contragredience principle, the equilibrium requirement is expressed by Eq. (15).

$$\mathbf{F}_c + \alpha \mathbf{F}_v = \mathbf{C}^T \mathbf{Q} \tag{15}$$

2.6 *The mathematical programming problem*

The solution to a limit analysis problem must fulfill the previously discussed principles. In the presence of non-associated flow, there is no unique solution satisfying these principles and the actual failure load corresponds to the mechanism with a minimum load factor (Baggio and Trovalusci, 1998). The proposed mathematical description results in the non-linear programming (NLP) problem expressed in Eqs. (16–22). Here, Eq. (16) is the objective function and Eq. (17) guarantees both compatibility and flow rule. Eq. (18) is a scaling condition of the displacement rates that ensures the existence of non-zero values. This expression can be freely replaced by similar equations, as, at collapse, the displacement rates are undefined and it is only possible to determine their relative values. Equilibrium is given by Eq. (19), and Eq. (20) is the expression of the yield condition, which together with the flow rule, Eq. (21), must fulfill Eq. (22).

Minimize: α (16)

Subject to:

$$\mathbf{N}_o \delta \lambda - \mathbf{C} \delta \mathbf{u} = \mathbf{0} \tag{17}$$

$$\mathbf{F}_v^T \delta \mathbf{u} - 1 = 0 \tag{18}$$

$$\mathbf{F}_c + \alpha \mathbf{F}_v = \mathbf{C}^T \mathbf{Q} \tag{19}$$

$$\varphi \leq \mathbf{0} \tag{20}$$

$$\delta \lambda \geq \mathbf{0} \tag{21}$$

$$\varphi^T \delta \lambda = 0 \tag{22}$$

This set of equations represents a case known in the mathematical programming literature as a Mathematical Problem with Equilibrium Constraints (MPEQ) (Ferris and Tin-Loi, 2001). This type of problems is hard to solve because of the complementarity constraint, Eq. (22). The solution strategy adopted here is that proposed by Ferris and Tin-Loi (2001). It consists of two phases, in the first, a Mixed Complementarity Problem (MCP), constituted by Eqs. (17–22) is solved. This gives a feasible initial solution. In the second phase, the objective function, Eq. (16), is reintroduced and Eq. (22) is substituted by Eq. (23). This equation provides a relaxation in the complementarity constraint, turns the NLP easier to solve, and allows to search for smaller values of the load factor. The relaxed NLP problem is solved for successively smaller values of ρ to force the complementarity term to approach zero.

$$-\varphi^T \delta \lambda \leq \rho \tag{23}$$

It must be said that trying to solve a MPEQ as a NLP problem does not guarantee that the solution is a local minimum (Ferris and Tin-Loi, 2001). Nevertheless, this procedure seems computationally efficient and provides solutions of better quality than other strategies tried previously, see Orduña and Lourenço (2001).

2.7 *Computer implementation*

The main objective of this work is to develop an analysis tool suitable to be used in practical engineering projects. For this reason, the computer implementation is done resorting to AutoCAD, in which the geometry of the blocky medium is drawn. The application, developed with Visual Basic for Applications within AutoCAD (ActiveX and VBA 1999), extracts the geometric data and pre-processes it. Complementary data is added such as volumetric weight, block thickness (typically 1 m for 2D geotechnical problems), friction coefficient and compressive strength. The limit analysis is done within the modeling environment GAMS (GAMS, 1998), which allows the user to develop large and complex mathematical programming models, and solve them with state of the art routines. The previously mentioned CAD package is used as post-processor to draw failure mechanisms.

2.8 *Application*

Validation of the formulation and the implementation of the 2D model is given in Orduña and Lourenço (2003), while the extension to 3D is fully detailed in Orduña and Lourenço (2005a,b). Here, to demonstrate the applicability of limit analysis to geotechnical problems, an example of

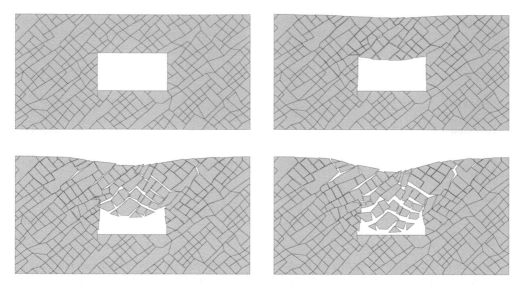

Figure 3. Evolution of a possible collapse mechanism in a tunnel excavated in a jointed rock mass.

shallow depth tunnel in fractured rock under increasing surface loads is simulated in Figure 3, see Jing (1998) for additional details. The problem geometry is formed by a rectangular domain containing two sets of random fractures. A rectangular tunnel is assumed to be excavated at the centre of the area. A vertical downward load due to self-weight appears and the other three boundaries are fixed in their respective normal directions. This problem has no practical implications and just serves as a demonstration.

3 HOMOGENIZATION TECHNIQUES

The chief disadvantage of the methods addressed in the previous section is the requirement for knowing the exact geometry of fracture systems in the problem domain. This condition can rarely be met in practice. This difficulty, however, exists for all mathematical models including those based on a continuum approach. The difference is how different approaches deal with this intrinsic difficulty. The continuum approach chooses to trade the geometrical simplicity over the material complexity by introducing complex constitutive laws and properties for an equivalent continuum through a proper homogenization process, which is only valid on the foundation of a "representative elementary volume"(REV). If this REV cannot be found or becomes too large (larger than the size of the excavation concerned, for example) then the continuum approach cannot be justified. But, in several cases, the continuum approach can be of relevance. Here, the problem of layered natural or reinforced soils is addressed by means of homogenization techniques.

3.1 *General considerations*

Composite materials consist of two or more different materials that form regions large enough to be regarded as continua and which are usually bonded together by an interface. Two kinds of information are needed to determine the properties of a composite material: the internal constituents' geometry and the physical properties of the phases. Many natural and artificial materials are immediately recognized to be of this nature, such as: laminated composites (as used in the aerospace and tire industry), alloys, porous media, cracked media (as jointed rocked masses), masonry, laminated wood, etc. However a lot of composite materials are normally assumed homogeneous. For example, this is the case of concrete, even if at a meso-level aggregates and matrix are already recognizable, and metals, in fact polycrystalline aggregates.

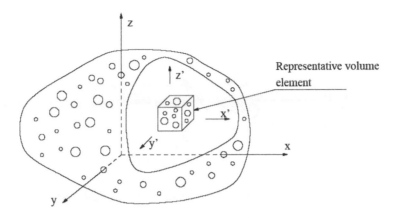

Figure 4.   Representative volume element in a composite body.

The existence of randomly distributed constituents was one of the first problems to be discussed by researchers, see Paul (1960), Hashin and Shtrikman (1962) and Hill (1963). The Hashin-Shtrikman bounds were derived from a variational principle and knew some popularity but they are only valid for arbitrary statistically isotropic phase geometry, see Figure 4. An additional condition for the practical use of these bounds is that the constituents stiffness ratios are small, see Hashin and Shtrikman (1962) and Watt and O'Connell (1980). They cannot obviously provide good estimates for extreme constituent stiffness ratios such as one rigid phase or an empty phase (porous or cracked media). This precludes the use of the above results for inelastic behavior where extremely large stiffness ratios will be found.

The basis underlying the above approach is however important. It is only natural that techniques to regard any composite as an homogeneous material are investigated, provided that the model size is substantially larger than the size of the inhomogenities. In fact all phenomena in a composite material are described by differential equations with rapidly oscillating coefficients. This renders numerical solutions practically impossible since too small meshes must be considered.

Consequently, effective constitutive relations of composite structures are of great importance in the global analysis of this type of media by numerical techniques (e.g. the Finite Elements Method). For this reason the rather complicated structure of a particular composite type is replaced by an effective medium by using a technique known as homogenization method or effective medium theory. Several mathematical techniques of different complexity are available to solve the problem, Sanchez-Palencia (1980, 1987), Suquet (1982) but for periodic structures the method of "Asymptotic Analysis" has renown some popularity, see also Bakhvalov and Panasenko (1989) and Bensoussan et al. (1978). The extension of these methods to inelastic behavior is however not simple. Knowledge of the actual macro- or homogenized constitutive law requires knowledge of an infinite number of internal variables, namely the whole set of inelastic micro-strains. However some simplified models with piecewise constants inelastic strains can be proposed (as carried out in this paper).

The objective of the homogenization process is to describe the composite material behavior from the behavior of the different components. This problem has known a large importance in the last decade due to the increasing use and manufacture of composites. To solve the present problem some simplifications are necessary and attention is given only to the special case of laminated or layered composites. This means that the geometry of the phases are known, which simplifies enormously the task of finding equivalent elastic properties for the medium. The composite material examined is made of laminae stacked at various orientations, the only restriction being imposed on the periodicity of the lamination along the structures (see Figure 5), which is usually fulfilled in numerous engineering applications. Examples include aerospace industries, laminated wood, stratified or sedimentary soils, stratified rock masses, etc.

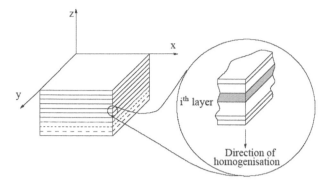

Figure 5.   The periodic laminated composite material and the structure of the unit cell.

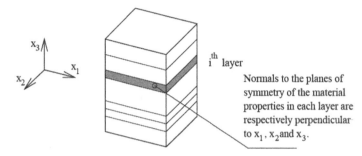

Figure 6.   Unit cell. Representative prism for system of parallel orthotropic layers.

3.2   *Theory*

In the previous section a reference is made to the mathematical techniques of homogenization of periodic structures. The asymptotic expansion of the displacement and stress fields is used for example in Paipetis *et al.* (1993) to obtain the elastic stiffness of a homogenized continuum. The solution is however presented in an explicit form for each coefficient of the stiffness matrix. This solution is not practical for computational implementation and almost precludes the use of homogenization techniques for inelastic behavior.

Here, the approach originated in the field of rock mechanics will be discussed, see Salamon (1968) and Gerrard (1982), and recast in matrix form. For that purpose the layered material shown in Figure 5 is considered again, built from a repeating system of parallel layers, each of which consists of orthotropic elastic material. This means that the material properties of each layer have three mutually perpendicular planes of symmetry. The layers are aligned so that each set of three planes are respectively perpendicular to a single set of Cartesian coordinates (x_1, x_2 and x_3). The notation selected for the axes is intentionally different than the usual (x, y, z axes) to draw the attention to the fact that the x_1, x_2 and x_3 axes are placed on the material symmetry axes. The unit cell, in the form of a representative prism, is shown in Figure 6. The homogenization direction (x_3) is normal to the layering planes and, hence, for each layer, one of the planes of elastic symmetry is parallel to the layering planes. In this direction the dimension of the prism is defined by the periodicity of the structure and in the other two directions a unit length can be assumed.

It is assumed that the system of layers remains continuous after deformation and that no relative displacement takes place in the interfaces between layers. The latter assumption simplifies the problem to great extent but, at least mathematically, discontinuities in the displacements field can be incorporated, see Bakhalov and Panasenko (1989). The prism representative of the composite material is further assumed to be subjected to homogeneous distributions of stress and strain, meaning that the volume of the prism must be sufficiently small to make negligible, in the equivalent medium, the variation of stresses and strains across it. The objective is to obtain a macro- or homogenized constitutive relation between homogenized stresses σ and homogenized strains ε,

$$\sigma = D^h \varepsilon , \tag{24}$$

where D^h is the homogenized stiffness matrix.

3.3 Elastic formulation

The macro-constitutive relation is obtained from the micro-constitutive relations. For the *ith* layer the relation between the (micro-)stress σ_i and the (micro-)strains ε_i is given by the (micro)stiffness matrix D_i and reads

$$\sigma_i = D_i \varepsilon_i . \tag{25}$$

In order to establish the relationships between stresses and strains in a homogenized continuum, the following averaged quantities in an equivalent prism are defined:

$$\sigma_j = \frac{1}{V} \int_V (\sigma_j)_i \, dV = \frac{1}{V} \sum_i \int_{V_i} (\sigma_j)_i \, dV \quad ; \quad \varepsilon_j = \frac{1}{V} \int_V (\varepsilon_j)_i \, dV = \frac{1}{V} \sum_i \int_{V_i} (\varepsilon_j)_i \, dV \tag{26}$$

where V is the volume of the representative prism and j indicates all the components of the stress and strain tensors.

Auxiliary stresses (t_1, t_2 and t_{12}) and auxiliary strains (e_3, e_{13} and e_{23}) are introduced as a deviation measure of the *ith* layer stress/strain state from the averaged values. Under the assumptions given above, the stress components for the *ith* layer read

$$\sigma_{1i} = \sigma_1 + t_{1i} \quad ; \quad \sigma_{2i} = \sigma_2 + t_{2i} \quad ; \quad \sigma_{3i} = \sigma_3$$
$$\tau_{12i} = \tau_{12} + t_{12i} \quad ; \quad \tau_{13i} = \tau_{13} \quad ; \quad \tau_{23i} = \tau_{23} \tag{27}$$

and the strain components for the *ith* layer read

$$\varepsilon_{1i} = \varepsilon_1 \quad ; \quad \varepsilon_{2i} = \varepsilon_2 \quad ; \quad \varepsilon_{3i} = \varepsilon_3 + e_{3i}$$
$$\gamma_{12i} = \gamma_{12} \quad ; \quad \gamma_{13i} = \gamma_{13} + e_{13i} \quad ; \quad \gamma_{23i} = \gamma_{23} + e_{23i} \tag{28}$$

Now let the thickness of the *ith* layer be h_i and the normalized thickness p_i be defined as

$$p_i = \frac{h_i}{L} \tag{29}$$

where L is the length of the representative prism in the x_3 direction. Substitution of Eqs. (27) in Eqs. (26.1) yields the following conditions for the auxiliary stress components:

$$\sum p_i t_{1i} = 0 \quad ; \quad \sum p_i t_{2i} = 0 \quad ; \quad \sum p_i t_{12i} = 0 \tag{30}$$

Similarly, from Eqs. (28) and Eqs. (26.2), the auxiliary strain components are constraint by

$$\sum p_i e_{3i} = 0 \quad ; \quad \sum p_i e_{13i} = 0 \quad ; \quad \sum p_i e_{23i} = 0 \tag{31}$$

Note that, if the conditions given in the above equations are fulfilled, then the strain energy stored in the representative prism and the equivalent prism are the same.

A vector of auxiliary stresses is now defined as t_i, given by

$$t_i = (t_{1i}, t_{2i}, t_{3i}, t_{12i}, t_{23i}, t_{13i})^T , \tag{32}$$

41

and the vector of auxiliary strains $\mathbf{e}_i$ is defined as

$$\mathbf{e}_i = (e_{1i}, e_{2i}, e_{3i}, e_{12i}, e_{23i}, e_{13i})^T . \tag{33}$$

It is noted that half of the components of the auxiliary stress and strain vectors are zero, see Eqs. (27,28). If the non-zero components are assembled in a vector of unknowns, $\mathbf{x}_i$, defined as

$$\mathbf{x}_i = \mathbf{P}_t \mathbf{t}_i + \mathbf{P}_e \mathbf{e}_i , \tag{34}$$

where the projection matrix into the stress space, $\mathbf{P}_t$, and the projection matrix into the strain space, $\mathbf{P}_e$, read

$$\mathbf{P}_t = diag\{0,1,1,0,1,0\} \quad \text{and} \quad \mathbf{P}_e = diag\{1,0,0,1,0,1\} \tag{35}$$

for homogenization along the axis x_1,

$$\mathbf{P}_t = diag\{1,0,1,0,0,1\} \quad \text{and} \quad \mathbf{P}_e = diag\{0,1,0,1,1,0\} \tag{36}$$

for homogenization along the axis x_2 and

$$\mathbf{P}_t = diag\{1,1,0,1,0,0\} \quad \text{and} \quad \mathbf{P}_e = diag\{0,0,1,0,1,1\} \tag{37}$$

for homogenization along the axis x_3.
Now, the auxiliary stresses can be redefined as

$$\mathbf{t}_i = \mathbf{P}_t \mathbf{x}_i \tag{38}$$

and the auxiliary strains as

$$\mathbf{e}_i = \mathbf{P}_e \mathbf{x}_i \tag{39}$$

The stresses and strains in the *ith* layer read

$$\sigma_i = \sigma + \mathbf{t}_i \quad \text{and} \quad \varepsilon_i = \varepsilon + \mathbf{e}_i . \tag{40}$$

Incorporating these equations in Eq. (25) leads to

$$\sigma + \mathbf{t}_i = \mathbf{D}_i (\varepsilon + \mathbf{e}_i) . \tag{41}$$

By using Eqs. (38, 39), further manipulation yields

$$\mathbf{x}_i = (\mathbf{P}_t - \mathbf{D}_i \mathbf{P}_e)^{-1} (\mathbf{D}_i \varepsilon - \sigma) , \tag{42}$$

which can be recast as

$$0 = \sum_i p_i \mathbf{x}_i = \sum_i p_i (\mathbf{P}_t - \mathbf{D}_i \mathbf{P}_e)^{-1} (\mathbf{D}_i \varepsilon - \sigma) , \tag{43}$$

This equation yields, finally, the relation between averaged stresses and strains, as

$$\sigma = \mathbf{D}^h \varepsilon , \tag{44}$$

42

Here, the homogenized linear elastic stiffness matrix, D^h, reads

$$D^h = \left[\sum_i p_i (P_t - D_i P_e)^{-1} \right]^{-1} \sum_i p_i (P_t - D_i P_e)^{-1} D_i . \tag{45}$$

Once the averaged stresses and strains are known, also the stresses and strains in the *ith* layer can be calculated. Using Eq. (42), the vector of unknowns x_i can be calculated. Algebraic manipulation yields

$$\sigma_i = T_{ti} \sigma \tag{46}$$

and

$$\varepsilon_i = T_{ei} \varepsilon \tag{47}$$

Here, the transformation matrices T_{ti} and T_{ei} read

$$\begin{aligned} T_{ti} &= I + P_t (P_t - D_i P_e)^{-1} (D_i D^{h^{-1}} - I) \\ T_{ei} &= I + P_e (P_t - D_i P_e)^{-1} (D_i - D^h) \end{aligned} \tag{48}$$

where I is the identity matrix.

3.4 *Elastoplastic formulation*

For plasticity the form of the elastic domain is defined by a yield function $f < 0$. Loading/ unloading can be conveniently established in standard Kuhn-Tucker form by means of the conditions

$$\lambda \geq 0, f \leq 0 \text{ and } \lambda f = 0 \tag{49}$$

where $\dot{\lambda}$ is the plastic multiplier rate. Here it will be assumed that the yield function is of the form

$$f(\sigma, \kappa) = \Phi(\sigma) + \overline{\sigma}(\kappa) \tag{50}$$

where κ is introduced as a measure for the amount of hardening or softening and Φ, $\overline{\sigma}$ represent generic functions. The usual elastoplastic relation hold: the total strain increment $\Delta\varepsilon$ is decomposed into an elastic, reversible part $\Delta\varepsilon^e$ and a plastic, irreversible part $\Delta\varepsilon^p$

$$\Delta\varepsilon = \Delta\varepsilon^e + \Delta\varepsilon^p \tag{51}$$

the elastic strain increment is related to the stress increment by the elastic constitutive matrix D as

$$\Delta\sigma = D\Delta\varepsilon^e \tag{52}$$

and the assumption of associated plasticity yields

$$\Delta\varepsilon^p = \Delta\lambda \frac{\partial f}{\partial \sigma} . \tag{53}$$

The scalar κ introduced before reads, in case of strain hardening,

$$\Delta\kappa = \sqrt{2/3 (\Delta\varepsilon^p)^T Q \Delta\varepsilon^p} , \tag{54}$$

where the diagonal matrix $Q = diag\{ 1, 1, 1, \frac{1}{2}, \frac{1}{2}, \frac{1}{2}, \frac{1}{2} \}$, and, in case of work hardening,

$$\Delta\kappa = \frac{1}{\sigma} \sigma^T \Delta\varepsilon^p , \tag{55}$$

43

The integration of the constitutive equations above is a problem of evolution that can be regarded as follows. At a stage n the total strain and plastic strain fields as well as the hardening parameter(s) (or equivalent plastic strain) are known:

$$\{\varepsilon_n, \varepsilon_n^p, \kappa_n\} \quad \text{given data.} \tag{56}$$

Note that the elastic strain and stress fields are regarded as dependent variables that can be always be obtained from the basic variables through the relations

$$\varepsilon_n^e = \varepsilon_n - \varepsilon_n^p \quad \text{and} \quad \sigma_n = D\varepsilon_n^e. \tag{57}$$

Therefore, the stress field at a stage $n+1$ is computed once the strain field is known. The problem is strain driven in the sense that the total strain ε is trivially updated according to the exact formula

$$\varepsilon_{n+1} = \varepsilon_n + \Delta\varepsilon_{n+1}. \tag{58}$$

It remains to update the plastic strains and the hardening parameter(s). These quantities are determined by integration of the flow rule and hardening law over the step $n \to n+1$. In the frame of a fully implicit Euler backward integration algorithm this problem is transformed into a constrained optimization problem governed by discrete Kuhn-Tucker conditions as shown by Simo $et\ al$ (1988).

It has been shown in different studies, e.g. Ortiz and Popov (1985) and Simo and Taylor (1986), that the implicit Euler backward algorithm is unconditionally stable and accurate for J2-plasticity.

This algorithm results in the following discrete set of equations:

$$
\begin{aligned}
\varepsilon_{n+1} &= \varepsilon_n + \Delta\varepsilon_{n+1} \\
\sigma_{n+1} &= \sigma^{trial} - \Delta\lambda_{n+1} D \frac{\partial f}{\partial \sigma}\bigg|_{n+1}, \\
\varepsilon_{n+1}^p &= \varepsilon_n^p + \Delta\lambda_{n+1} D \frac{\partial f}{\partial \sigma}\bigg|_{n+1} \\
\kappa_{n+1} &= \kappa_n + \Delta\kappa_{n+1}
\end{aligned}
\tag{59}
$$

in which $\Delta\kappa_{n+1}$ results from Eq. (54) or Eq. (55) and the elastic predictor step returns the value of the elastic trial stress σ trial

$$\sigma^{trial} = \sigma_n + D\Delta\varepsilon_{n+1}, \tag{60}$$

Application of the above algorithm to the homogenized material equivalent to the layered material shown in Figure 6 results in increasing difficulties. The homogenized material consists of several layers with individual elastic and inelastic properties. As given in Eqs. (46–47), different stresses and strains are obtained for each layer, which are again different from the average stresses and strains.

Due to this feature and the fact that inelastic behavior is considered for the layers, for a given strain increment in the equivalent material it is not possible to calculate immediately the correspondent strain increments for each layer. This precludes the use of a standard plasticity algorithm due to the fact that the algorithm is strain driven. An equivalent return mapping must be carried out, in which all the layers are considered simultaneously. The equivalent return mapping is, therefore, a return mapping in all the layers (or a return mapping in terms of average strains and average stresses). This has some similitude with the fraction model, Besseling (1958).

Here only the simple case of J2-plasticity is addressed, being the reader addressed to Lourenço (1995) for other cases. The Von Mises yield criterion can be written in a matrix form and reads

$$f = (^3/_2 \sigma^T P \sigma)^{1/2} - \overline{\sigma}(\kappa), \tag{61}$$

where the projection matrix P reads

$$P = \begin{bmatrix} {}^2/_3 & -{}^1/_3 & -{}^1/_3 & 0 & 0 & 0 \\ -{}^1/_3 & {}^2/_3 & -{}^1/_3 & 0 & 0 & 0 \\ -{}^1/_3 & -{}^1/_3 & {}^2/_3 & 0 & 0 & 0 \\ 0 & 0 & 0 & 2 & 0 & 0 \\ 0 & 0 & 0 & 0 & 2 & 0 \\ 0 & 0 & 0 & 0 & 0 & 2 \end{bmatrix} . \tag{62}$$

For this yield surface, holds the well-known relation

$$\Delta\kappa = \Delta\lambda \ , \tag{63}$$

both for a strain or work hardening rule. From the standard algorithm shown above, cf. eq. (59), it is straightforward to obtain the update of the stress vector σ_{n+1} as

$$\sigma_{n+1} = A \sigma^{trial}, \tag{64}$$

where the mapping matrix A reads

$$A = \left[I + \frac{3\Delta\lambda_{n+1}}{2\overline{\sigma}_{n+1}} D P \right]^{-1}, \tag{65}$$

Note that the mapping matrix is only a function of $\Delta\lambda_{n+1}$, i.e. $A = A(\Delta\lambda_{n+1})$. If the individual layers are considered, Eq. (64) for the ith layer must be rewritten as

$$\sigma_{n+1,i} = A_i \sigma_i^{trial}, \tag{66}$$

where the mapping matrix, A_i, is a function of $\Delta\lambda_{n+1,i}$ and the elastic trial stress σ_i^{trial}, cf. Eq. (60), reads

$$\sigma_i^{trial} = \sigma_{n,i} + D_i \Delta\varepsilon_{n+1,i}, \tag{67}$$

If all the layers are now considered simultaneously, the question arises whether one particular layer is elastic or plastic, but this will be considered trivial in the present note. In fact, in case of an elastic layer, Eq. (67) remains valid if the mapping matrix is redefined as the identity matrix (i.e. $\Delta\lambda_{n+1} = 0$),

$$A_i = I \quad \text{(if layer is elastic).} \tag{68}$$

Inserting Eq. (40) in Eq. (66), the latter can be rewritten in terms of average stresses and strains (σ and ε) and auxiliary stresses and strains of the ith layer (t_i and e_i) at stage $n + 1$ as

$$\sigma_{n+1} + t_{n+1,i} = A_i(\sigma_n + D_i\Delta\varepsilon_{n+1} + D_i\Delta e_{n+1,i}) \ . \tag{69}$$

If a modified trial stress for the ith layer is defined as

$$\overline{\sigma}_i^{trial} = \sigma_{n,i} + D_i \Delta\varepsilon_{n+1}, \tag{70}$$

(note that the average strain is included in the expression instead of the strain of the ith layer), Eq. (69) reads

$$\sigma_{n+1} + t_{n+1,i} = A_i(\overline{\sigma}_i^{trial} + D_i\Delta e_{n+1,i}) \ . \tag{71}$$

Inserting Eqs. (38,39) in this equation and further manipulation, see Lourenço (1996), leads to the update of the average stress as

$$\sigma_{n+1} = \left[\sum_i p_i (\boldsymbol{P}_t - \boldsymbol{A}_i \boldsymbol{D}_i \boldsymbol{P}_e)^{-1} \right]^{-1} \sum_i p_i (\boldsymbol{P}_t - \boldsymbol{A}_i \boldsymbol{D}_i \boldsymbol{P}_e)^{-1} \boldsymbol{A}_i \overline{\sigma}_i^{-trial} .$$ (72)

Note that the summation above is extended to all the layers (elastic and plastic). The return mapping procedure is now established in a standard way. From the average stresses it is possible to calculate the stresses in the *ith* layer. At this point it is clear that a system of non-linear equations in $\Delta\lambda_{n+1}$'s can be built: $f_i(\Delta\lambda_{n+1,1}, \ldots, \Delta\lambda_{n+1,j}) = 0, i = 1, \ldots, j$. This system of non-linear equations can be solved by any mathematical standard technique and has so many unknowns (and equations) as the number of plastic layers. Finally, when convergence is reached the state variables (ε, ε_p and κ) can be updated. For a discussion of the algorithm implementation and the calculation of the consistent tangent operator the reader is referred to Lourenço (1996).

3.5 Extension to a Cosserat continuum

Physical arguments in favor of a Cosserat approach in granular materials have been put forward in Mühlhaus and Vardoulakis (1987). However, the very important feature of a Cosserat continuum is the introduction of a characteristic length in the constitutive description, thus rendering the governing set of equations elliptic while allowing for localization of deformation in a narrow, but finite band of material. The excessive dependence of standard continua upon spatial discretization can be then obviated or, at least, considerably reduced.

The implementation carried out is based on the original work of de Borst (1991, 1993) on the two-dimensional elastoplastic Cosserat continuum and the extension of Groen *et al.* (1994) to the three-dimensional case, see Lourenço (1996) for details.

3.6 Validation

In the following examples the described algorithm is tested by considering a material built from two layers, viz. a "weak" and a "strong" layer. Different elastic and inelastic behavior is assumed for both layers. Large scale calculations are carried out in (average) plane strain and three-dimensional applications. The objective is to assess the performance of the theory in a different class of problems ranging from highly localized to distributed failure modes. For this purpose, different calculations of the same model with increasing number of layers are carried out (note that layer in this section will be generally assumed as the repeating pattern, i.e. one layer contains two different materials, a weak and a strong layer). Convergence to the homogenized solution must be found upon increasing the number of layers.

The same fictitious material, described in Lourenço (1996), is used throughout this section. In the large calculations, 2D quadratic elements (8-noded quads) with 2×2 Gauss integration are used and 3D quadratic elements (20-noded bricks) with 3×3 Gauss integration are used. For the elastoplastic calculations a (local) full Newton-Raphson method is used in the return mapping algorithm with a 10^{-7} tolerance for all equations and a (global) full Newton-Raphson method is used to obtain convergence at global level. The tolerance is set to an energy norm of 10^{-4}.

The first example is a rigid foundation indented into a stratified soil (see Figure 7). Only half of the mesh is considered due to symmetry conditions. This example is meant to assess the performance of the homogenization technique under plane strain conditions in situations not only constrained by the material behavior, but also by geometric constraints. Clearly a localized failure is obtained around the indentation area.

A comparison is given between the 2D (plane strain) homogenized continuum and three different 2D (plane strain) layered discretizations: 8, 16 and 32 equivalent material layers (see Figure 8). Note that the zero normal strain condition introduces a three dimensional state of stress, meaning that all the calculations can be carried out under plane strain conditions. The results obtained in a

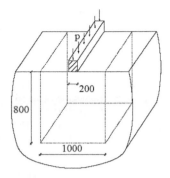

Figure 7.   Geometry of rigid foundation indentation test.

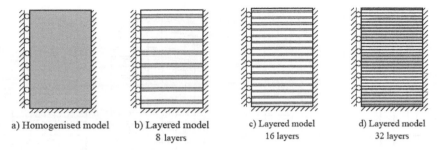

a) Homogenised model   b) Layered model   c) Layered model   d) Layered model
                 8 layers   16 layers   32 layers

Figure 8.   Different discretizations, homogenized and layered.

Table 1.   Linear analysis results for $d = 1.0$ mm. Homogenized vs. layered model.

	Homogenized	Layered		
		8 layers	16 layers	32 layers
p [kN/mm]	4.056	4.170	4.074	4.020
Error	reference	2.8%	0.4%	−0.9%

linear analysis are given in Table 1 for a vertical displacement d of the foundation with a value of 1.0 mm and the different discretizations. Very good agreement is found.

The results obtained in a non-linear analysis for the different discretizations are illustrated in Figure 9. Note that all the nodes under the foundation are tied to have equal displacements. Direct displacement control of the foundation is applied in 30 equally spaced steps of 0.1 mm. The results show indeed that a convergence exists to the homogenised 2D solution with an increasing number of layers. A plateau of the force-displacement diagram was not found, even if in one the examples the calculation was carried out until a later stage (d = 6.0 mm). This can be due to a bad performance of the elements but, in principle, see Groen (1994), 8-noded elements with reduced integration do not show volumetric locking. The author believes that the behavior found can be explained by the geometric constraints. It is reasonable to assume that under the above circumstances the "weak layers" show (quasi) infinite hardening due to the three dimensional state of stress induced (towards the volumetric axis). From the results obtained it is clear that such behavior occurs with minor plastification of the "strong layers".

47

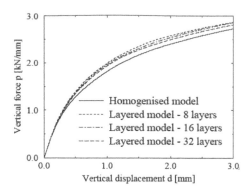

Figure 9.   Force-displacement diagrams for different discretisations.

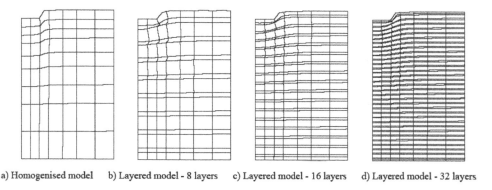

a) Homogenised model   b) Layered model - 8 layers   c) Layered model - 16 layers   d) Layered model - 32 layers

Figure 10.   Deformed mesh for different discretizations ($\times 15$).

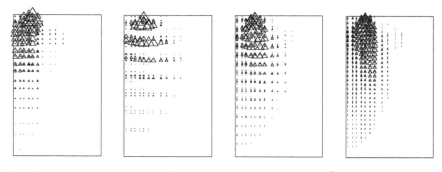

Figure 11.   Equivalent plastic strain for different discretizations ($>0.5 \times 10^{-4}$).

Fig. 10 shows the deformed meshes and Fig. 11 illustrates the plastic points at failure for the different discretizations. The deformation patterns of the layered models show some bending of the strong layers. This cannot be reproduced in the homogenized model.

It is important to point out that the foundation was assumed to be located in the middle plane of the strongest material. For examples with a small number of layers, due to the extremely large stress gradients around the geometrical discontinuities, some differences can be expected.

The collapse of a 45° slope subjected to its own weight (g) is now considered (see Figure 12). Again different discretizations are considered: the plane strain homogenized model and the plane strain layered model (4, 8 and 16 layers). The results of the maximum vertical displacement in

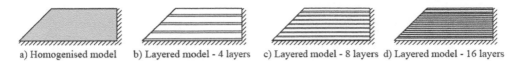

a) Homogenised model   b) Layered model - 4 layers   c) Layered model - 8 layers   d) Layered model - 16 layers

Figure 12.   Different discretizations, homogenized and layered.

Table 2.   Linear analysis results for $g = 1.0\,\text{N/mm}^3$. Homogenized vs. layered model

	Homogenized	Layered		
		4 layers	8 layers	16 layers
Disp. y [mm]	−25.47	−27.88	−26.70	−26.09
Error	reference	9.5%	4.8%	2.4%

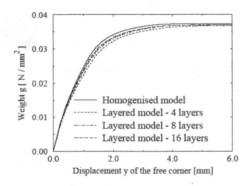

Figure 13.   Force-displacement diagrams for different discretizations.

the slope obtained in a linear analysis are given in Table 2 for a weight value g with a value of $1.0\,\text{N/mm}^3$ and the different discretisations. Good agreement is found.

The results obtained in a non-linear analysis for the different discretizations are illustrated in Figure 13. Indirect displacement control of the top left corner is applied in 30 equally steps of 0.2 mm. In this case even for a small number of layers the difference is minimal (1.8%). Figure 14 shows the deformed meshes and it is remarkable that such a close agreement is found in respect to the shear band formation. Figure 15 illustrates the plastic points at failure for the homogenized model in the two different materials (these results agree well with the layered model). The plastic points of the different materials are not shown in a single picture because different scales are used. The inelastic behavior of the weak material close to the bottom support is substantially larger than the diagonal shear failure.

It is important to point out that the bottom layer was assumed to be from the weak material. This has some influence on the results, especially if only a few layers are considered. This assumption is also the reason for the force-displacement diagram of the layered models being less stiff than the homogenized result (remember that the slope is subjected to self weight loading). If the bottom layer was assumed to be from the strong material, the force-displacement diagram of the layered models would be stiffer than the homogenized result.

Finally, a last example is considered in order to assess the formulation of the homogenization continuum for truly three-dimensional finite elements. For this purpose, the collapse of an embankment subjected to its own weight (g) is analyzed. Again different discretizations are considered: the 3D homogenized model and the 3D layered model with 4 and 8 equivalent material layers

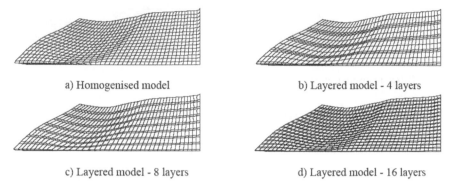

a) Homogenised model

b) Layered model - 4 layers

c) Layered model - 8 layers

d) Layered model - 16 layers

Figure 14.   Deformed mesh for different discretizations ($\times$ 15).

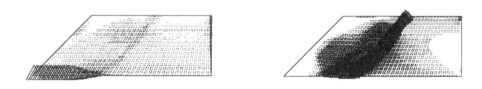

a) Homogenised model - Weak layer ($\times$ 260)   b) Homogenised model - Strong layer ($\times$ 6700)

Figure 15.   Equivalent plastic strain for different discretizations ($>0.5 \times 10^{-4}$).

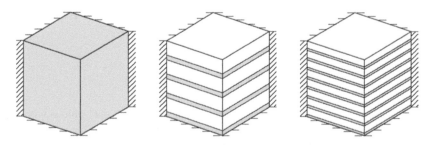

Figure 16.   Different discretizations, homogenized and layered.

Table 3.   Linear analysis results for $g = 1.0\,\text{N/mm}^3$ (displacement of the free corner). Homogenized vs. layered model

	Homogenized	Layered	
		4 layers	8 layers
Disp. z [mm]	−29.06	−30.29	−29.63
Error	reference	4.2%	2.0%

(see Figure 16). The results obtained in a linear analysis are given in Table 3 for a weight g with a value of $1.0\,\text{N/mm}^3$ and the different discretizations. Good agreement is found.

The results obtained in a non-linear analysis for the different discretizations are illustrated in Figure 17. Indirect displacement control of the free corner is applied in 20 equally spaced steps of 0.3 mm. The results show indeed that a convergence exists to the 3D homogenised solution with

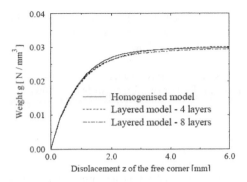

Figure 17. Force-displacement diagrams for different discretizations.

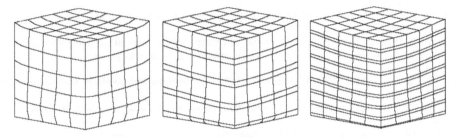

Figure 18. Total deformed mesh for different discretizations (×10).

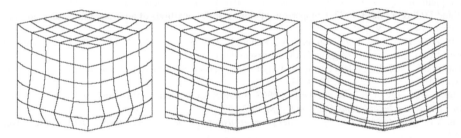

Figure 19. Incremental deformed mesh for different discretizations (×300).

an increasing number of layers. In this case even for a small number of layers the difference is minimal (1.5%). Figures 18 and 19 show the deformed meshes at failure (total and incremental) for the different discretizations. Note that, even for the incremental displacements, a good agreement is found between the 3D homogenised model and the 3D layered model (8 layers). As in the previous example, it is important to point out that the bottom layer was assumed to be from the weak material.

4 CONCLUSIONS

The difficulties involved the numerical modeling of non-homogeneous continuum media are well known. Different approaches are possible to represent heterogeneous media, namely, the discrete element method, the finite element method and limit analysis. Here, two approaches have been detailed.

The first approach considered is limit analysis of fracture or slip lines, together with individual rigid blocks. The adopted formulation assumes, besides the standard Coulomb friction law, a compression limiter and non-associated flow. This results in a non-linear optimization problem that can be successfully solved with state of the art techniques. An example of application to the excavation of a tunnel in fractured rock is shown.

The second approach considered is the elastoplastic homogenization of layered media. A matrix formulation is proposed with straightforward extension to inelastic problems. Several examples of validation and application to stratified or reinforced soils are shown.

REFERENCES

ActiveX and VBA developer's guide. 1999. Autodesk.

Baggio, C. & Trovalusci, P. 1998. Limit analysis for no-tension and frictional three-dimensional discrete systems. *Mechanics of Structures and Machines* 26(3): 287–304.

Bakhvalov, N. & Panasenko, G. 1989. *Homogenisation: averaging processes in periodic media*. Dordrecht : Kluwer Academic Publishers.

Bensoussan, A., Lions, J.-L. & Papanicolaou, G. 1978. *Asymptotic analysis of periodic structures*. Dordrecht : Kluwer Academic Publishers.

Besseling, J.F. 1958. A theory of elastic, plastic and creep deformations of an initially isotropic material showing anisotropic strain-hardening, creep recovery and secondary creep. *J. Appl. Mech.* 22: 529–536.

Cundall, P.A. & Strack O.D.L. 1979. Discrete numerical-model for granular assemblies. *Geotechnique* 29 (1): 47–65.

de Borst, R. 1991. Simulation of strain localisation: A reappraisal of the Cosserat continuum, *Engng. Comput.* 8: 317–332.

de Borst, R. 1993. A generalisation of J2-flow theory for polar continua, *Comp. Meth. Appl. Mech. Engng.* 103: 347–362.

Ferris, M.C. & Tin-Loi, F. 2001. Limit analysis of frictional block assemblies as a mathematical program with complementarity constraints. *International Journal of Mechanical Sciences* 43: 209–224.

Fredlund D.G. & Krahn J. 1977. Comparison of slope stability methods of analysis. *Canadian Geotechnical Journal* 14 (3): 429–439.

GAMS A user's guide. 1998. Brooke, A., Kendrick, D., Meeraus, A., Raman, R. & Rosenthal, R.E. http://www.gams.com/docs/gams/GAMSUsersGuide.pdf.

Gerrard, C.M. 1982. Equivalent elastic moduli of a rock mass consisting of orthorhombic layers. *Int. J. Rock. Mech. Min. Sci. & Geomech.* 19: 9–14.

Goodman, R.E., Taylor, R.L. & Brekke T.L. 1968. A model for mechanics of jointed rock. *Journal of the Soil Mechanics and Foundations Division* 94 (3): 637–659.

Groen, A.E. 1994. *Improvement of low order elements using assumed strain concepts*. Delft : Delft University of Technology.

Groen, A.E., Schellekens, J.C.S. & de Borst, R. 1994. Three-dimensional finite element studies of failure in soil bodies using a Cosserat continuum. In: H.J. Siriwardane & M.M. Zaman (eds), *Computer Methods and Advances in Geomechanics*: 581–586. Rotterdam: Balkema.

Hart, R.D. 1991. General report: an introduction to distinct element modelling for rock engineering. In: W. Wittke (eds), *Proceedings of the of 7th Congress on ISRM*: 1881–1892. Rotterdam: Balkema.

Hashin, Z. & Shtrikman, S. 1962. On some variational principles in anisotropic and nonhomogeneous elasticity. *J. Mech. Phys. Solids* 10: 335–342.

Hill, R. 1963. Elastic properties of reinforced solids: Some theoretical principles. *J. Mech. Phys. Solids* 11: 357–372.

Jing, L. 1998. Formulation of discontinuous deformation analysis (DDA) – an implicit discrete element model for block systems. *Engineering Geology* 49: 371–381.

Lourenço, P.B. 1995. *The elastoplastic implementation of homogenization techniques: With an extension to masonry structures*. Delft : Delft University of Technology.

Lourenço, P.B. 1996. A matrix formulation for the elastoplastic homogenisation of layered materials, *Mechanics of Cohesive-Frictional Materials* 1: 273–294.

Mühlhaus, H.-B. & Vardoulakis, I. 1987. The thickness of shear bands in granular materials, *Geotechnique* 37: 271–283.

Nielsen, M.P. 1999. *Limit Analysis and Concrete Plasticity*. Boca Raton : CRC Press.

Orduña, A. & Lourenço, P.B. 2001. Limit analysis as a tool for the simplified assessment of ancient masonry structures. In P.B. Lourenço & P. Roca (eds) *Proc. 3rd Int. Seminar on Historical Constructions*: 511–520. Guimarães : Universidade do Minho.

Orduña, A. & Lourenço, P.B. 2003. Cap model for limit analysis and strengthening of masonry structures. *J. Struct. Engrg.* 129(10), 1367–1375.

Orduña, A. & Lourenço, P.B. 2005a. Three-dimensional limit analysis of rigid blocks assemblages. Part I: Torsion failure on frictional joints and limit analysis formulation. *Int. J. Solids and Structures* 42(18–19): 5140–5160.

Orduña, A. & Lourenço, P.B. 2005b. Three-dimensional limit analysis of rigid blocks assemblages. Part II: Load-path following solution procedure and validation. *Int. J. Solids and Structures* 42(18-19): 5161–5180.

Ortiz, M. & Popov, E.P. 1985. Accuracy and stability of integration algorithms for elastoplastic constitutive relations. *Int. J. Numer. Methods Engrg.* 21: 1561–1576.

Paipetis, S.A., Polyzos, D. & Valavanidis, M. 1993. Constitutive relations of periodic laminated composites with anisotropic dissipation. *Arch. Appl. Mech.* 64: 32–42.

Paul, B. 1960. Prediction of elastic constants of multiphase materials. *Trans. of AIME* 218: 36–41.

Salamon, M.D.G. 1968. *Elastic moduli of stratified rock mass. Int. J. Rock. Mech. Min. Sci.* 5: 519–527.

Sanchez-Palencia, E. 1980. *Non-homogeneous media and vibration theory*. Berlin: Springer.

Sanchez-Palencia, E. 1987. *Homogenization techniques for composite media*. Berlin: Springer.

Shi, G. & Goodman, R.E. 1985. Two dimensional discontinuous deformation analysis. *Int. J. Numerical Analyt. Meth. Geomech.* 9, 541–556.

Simo, J.C. & Taylor, R.L. 1986. A return mapping for plane stress elastoplasticity. *Int. J. Numer. Methods Engrg.* 22: 649–670.

Simo, J.C., Kennedy, J.G. & Govindjee, S. 1988. Non-smooth multisurface plasticity and viscoplasticity. Loading/unloading conditions and numerical algorithms. *Int. J. Numer. Methods Engrg.* 26: 2161–2185.

Suquet, P. 1982. *Plasticité et homogénéisation*. PhD Dissertation. Paris : Université de Paris.

Watt, J.P. & O'Connell, R.J. 1980. An experimental investigation of the Hashin-Shtrikman bounds on twophase aggregate elastic properties. *Physics of Earth and Planetary Interiors* 21: 359–370.

Applications of Computational Mechanics in Geotechnical Engineering – Sousa,
Fernandes, Vargas Jr & Azevedo (eds)
© 2007 Taylor & Francis Group, London, ISBN 978-0-415-43789-9

Modeling water flow in an unsaturated compacted soil

K. V. Bicalho

Federal University of Espirito Santo, Brazil

ABSTRACT: Accurate simulations of unsaturated water flow are essential for many practical applications, including clay liners in waste containment and compacted clay cores in earth dams. The modeling of water flow through compacted soils requires that the unsaturated hydraulic constitutive relationships, namely the soil water retention and the hydraulic conductivity functions, of the soil be defined. Over the years a number of unsaturated hydraulic constitutive functions have been suggested. Most of these functions were developed to describe the movement of water in a porous medium containing continuous air channels. This paper shows that the functions proposed for modeling water flow with continuous air channels should not be applied to the case of systems involving discontinuous air phase. Hydraulic constitutive functions for soils with discontinuous air phase are proposed. Numerical simulations show the ability of the proposed constitutive relationships to model infiltration through compacted soils.

1 INTRODUCTION

Knowledge about water flow in the unsaturated zone is important for many practical applications, including clay liners in waste containment and compacted clay cores in earth dams. Unsaturated flow processes are in general complicated and difficult to describe quantitatively, since they require that the complex unsaturated hydraulic constitutive, UHC, relationships, namely the soil water retention and the hydraulic conductivity functions, of the soil be defined. For this reason, the development of theoretical and experimental methods for defining these relationships has become one of the most important and active topics of research.

Due to the relative difficulty associated with the measurement of unsaturated hydraulic conductivity functions, the soil-water retention curve, SWRC, is often used as the basis for the prediction of the unsaturated hydraulic conductivity function (Burdine 1953, Kunze et al. 1968, Mualem 1976, Fredlund et al. 1994). The SWRC for a soil is the relationship between the amount of water in the soil and the capillary pressure for the soil. In geotechnical engineering practice and the petroleum literature, it is customary to plot capillary pressure as a function of the degree of saturation. A number of equations have been proposed to empirically produce the best fit to the SWRC data (Brooks & Corey 1964, van Genuchten 1980). Modeling efforts have also been made to estimate the SWRC from particle size distribution (Arya & Paris 1981) or pore size distribution data of the soil (Fredlund & Xing 1994). Hysteresis of the capillary pressure-saturation relationship may complicate the measurement and use of the SWRC data. Although most of the models were developed to represent the boundary drying SWRC, they also are applied routinely to describe wetting SWRC data required for simulating an infiltration process (Chiu & Shackelford 1998). However, at higher degrees of saturation on the wetting cycle the air phase becomes discontinuous with isolated air pockets within the pore space which have no connection over macroscopic distances (occluded air bubbles or entrapped air). Considering that there is no means of measuring the pore air pressures in an occluded sample, which will in any case vary depending on the size of the bubble, and which, as saturation is approched, will probably reach very high values greatly in excess of the atmospheric, the SWRC for discontinuous air phase is not well defined.

Current understanding of the process of air entrapment during infiltration suggests that, in nature, significant number of occluded air bubbles will be formed in the process (often occur). Therefore, the simulation of infiltration events with the main drying SWRC can be misleading. To the author's knowledge, none of the existing predictive UHC functions has been developed or evaluated for a porous medium containing a discontinuous air phase. The following is an attempt to show that the UHC functions derived for soils with continuous air channels, in which the air is assumed to exist at constant atmospheric pressure, should not be used for simulating infiltration process. The paper presents UHC functions for soils with discontinuous air phase. A numerical solution of Richards' equation for one-dimensional transient water flow was used in this paper to evaluated UHC functions proposed for continuous and discontinuous air phase. For simplicity, the word "water" corresponds to liquid water, unless stated otherwise.

2 THEORY

The governing equation for one-dimensional vertical water flow in an unsaturated soil is given by Richards (1931) as:

$$\frac{\partial(\theta_w)}{\partial t} = \frac{\partial}{\partial z}[k_w(\theta_w)\frac{\partial(h_{pw})}{\partial z}] + \frac{\partial k_w(\theta_w)}{\partial z}$$
(1)

where θ_w is the volumetric water content, $k_w(\theta_w)$ is hydraulic conductivity of the water phase, h_{pw} is the pressure head, t is time, and z is elevation, assumed positive upward. The volumetric water content is expressed as $\theta_w = n_p S$, where n_p is the soil porosity and S is the saturation corresponding to θ_w. For an essentially incompressible soil, the change in θ_w is proportional to the change in S since n_p remains constant. The Richards' equation is based on Darcy's law and the continuity equation. The solution of the equation requires the knowledge of the initial and boundary conditions and the UHC relations (i.e., the SWRC relating the θ_w with the h_{pw} and the $k_w(\theta_w)$ function). Equation 1 is a partial differential equation which is highly non-linear due to the non-linear constitutive relationships, and exact analytical solutions for specific boundary conditions are extremely difficult to obtain. Therefore, Richards' equation can be solved analytically only for a very limited number of cases. A numerical solution of the Richards's equation is described in this paper and its algorithm was coded in a computer program that was used to analyze the experimental results.

It is usual to put Equation 1 into a form where the independent variable is either θ_w or h_{pw}. The constitutive relationship between θ_w and h_{pw} allows for conversion of one form of the equation to another. Equation 1 is often called a mixed form of Richards' equation. Celia et al. (1990) demonstrated that numerical solutions for the unsaturated flow equation could be substantially improved by using the so-called mixed form of Richards' equation. Both the θ_w-based and h_{pw}-based forms of the equation present some problems. These difficulties include poor mass balance and associated poor accuracy in h_{pw}-based equation (due to the inequality of the terms $\partial\theta_w/\partial h_{pw}$ ($\partial h_{pw}/\partial t$) and $\partial\theta_w/\partial t$ in discrete forms and the highly nonlinear nature of the $\partial\theta_w/\partial h_{pw}$ term), and restricted applicability of θ_w-based models. The use of the so-called mixed form of Richards' equation leads to significant improvement in numerical solution performance, while requiring no additional computational effort. Therefore, the mixed form of Richards' equation is used in this paper.

2.1 *Capillary pressure-saturation relationship*

The capillary pressure-saturation relationship is highly hysteretic with respect to the wetting and drying processes. A curve is obtained starting with a completely saturated condition. This is called a primary boundary SWRC on drying condition. A different curve is obtained by starting with a sample containing only the nonwetting phase (i.e., completely dry) and allowing it to imbibe the

56

wetting phase. The latter curve is called a primary boundary SWRC on wetting condition. These two boundary curves belong to an infinite family of curves that might be obtained by starting at any particular saturation and either increasing or decreasing saturation (Bicalho et al. 2005b).

Based on analysis of a large number of experimental data, Brooks & Corey (1964) presented the following empirical expression for describing the relationship between capillary pressure, P_c, and saturation, S, for continuous air channels ($S < S_m$):

$$P_{c1}(S) = \frac{P_b}{(S_e)^{\frac{1}{\lambda}}} \qquad (2)$$

where $P_{c1}(S)$ is the relationship between P_c and S considering continuous air channels, P_b and λ are characteristic constants of the medium and S_m is the correspondent degree of saturation where the air start to be entrapped. In this case, the capillary pressure (i.e., $u_a - u_w$) is equal to the negative pressure of water relative to atmospheric air pressure (i.e., $-(u_w - u_a)$) and S_e is the effective saturation, defined as:

$$S_e = \frac{(S - S_r)}{(1 - S_r)} \qquad (3a)$$

where S_r is the residual saturation.

Corey (1994) recommends that the value of S_e where the soil typically is not completely saturated due to the presence of entrapped air (infiltration processes) should be redefined as:

$$S_e = \frac{(S - S_r)}{(S_m - S_r)} \qquad (3b)$$

Corey (1994) indicates that $0.80 \leq S_m \leq 0.92$ for infiltration (wetting processes), which is consistent with the range of S_m values reported by Chiu & Shackelford (1998). The value of S_m is considered the maximum field saturation. In the considered approach the value of hydraulic conductivity correspondent to S_m is considered the maximum hydraulic conductivity value for the infiltration (wetting) processes. However, Bloomsburg & Corey (1964) show that the porous medium will eventually become fully saturated by imbibition even though there is no liquid flowing through the porous material. Moreover, compacted soils at great depths below the phreatic surface (i.e., subjected to higher pressures) may have the saturation in the occluded zone increased to full saturation. Therefore, it is necessary to define the $P_c - S$ relationship in the occluded zone (i.e., $S \geq S_m$).

The author (Bicalho 1999; Bicalho et al. 2000), assuming that the volume of the air-water mixture is compensated by adding water (i. e., there is no variation in the volume of void space), has proposed the following theoretical relationship between u_w and S considering occluded air bubbles. The relationship, derived using the theory proposed by Schuurman (1966), has been verified by comparison with experimental data and it is given by:

$$u_w = u_a - (u_{a0} - u_{w0})[\frac{(1 - S_0)}{(1 - S)}]^{1/3}, \ S \geq S_m \qquad (4a)$$

$$u_a = \frac{[1 + S_0(H - 1)]u_{a0}}{S(H - 1) + 1} \qquad (4b)$$

u_w and S are the variables in Equation 4 while u_{w0}, u_{a0} and S_0 are the initial values of water pressure, air pressure and degree of saturation, respectively, and H is the dimensionless coefficient of solubility or Henry's constant (equals 0.02 at 20 °C). The equations parameters are physically meaningful and can be easily determined (Bicalho 1999). The initial air pressure is assumed to be the atmospheric pressure. A compacted cohesive soil contains air that may initially be under higher pressure due to the compaction process. However, if the soil is not immediately sealed, the air pressure will soon become atmospheric (Hilf 1956).

Equation 4 is in terms of absolute pressures, and it is determined using Boyle's law and Henry's law of solubility, and the Kelvin equation for pressure difference across the interface between the air and water phases. Schuurman (1966) also assumes that all spherical air bubbles are of the same size and the number of air bubbles within the pore space is constant. According to Henry's law the amount of dissolved air only depends on the air pressure in the bubbles. Hilf (1956) concluded that a uniform distribution of dissolved air requires an identical air pressure in all the bubbles. Since the interfacial tension is assumed constant, the equilibrium condition demands that free air bubbles in a uniform fluid have the same radius.

For the fully saturated state (S = 1) the function has a singularity with water pressure tending towards −∞. This indicates that prior to their disappearance, the bubbles become unstable, collapsing due to the capillary effect. This phenomenon has yet to be explored experimentally. While this effect may be important at the microscopic scale for individual air bubbles, it is unlikely that would have a major effect on the macroscopic behavior of soils. Not all air bubbles would collapse at the same time, so the singularity will be smoothed out. Once the bubbles disappear, the difference between the air pressure and the water pressure no longer exists.

Based on Equation 4, the $P_c - S$ relationship at higher degrees of saturation on the wetting cycle is given by:

$$P_{c2}(S) = (u_{a0} - u_{w0})[\frac{(1 - S_0)}{(1 - S)}]^{1/3}, \, S \geq S_m \tag{5}$$

where $P_{c2}(S)$ is the relationship between P_c and S considering discontinuous air phase. It must be emphasized that whereas an increase in the capillary values decreases the saturation values in the existing models for continuous air phase, in the occluded zone the saturation increase is associated with an increase in the pressure difference between pore air and pore water. Therefore, the values of P_c at $S \geq S_m$ can not be replaced simply by decreasing P_c (Bicalho 1999).

2.2 Unsaturated hydraulic conductivity functions

Burdine (1953) derived a capillary model for prediction of the unsaturated hydraulic conductivity function from the basic law of fluid flow in porous media (Poiseuille's law). Burdine's equation for relative hydraulic conductivity, which is defined as the ratio of the effective hydraulic conductivity at a given saturation to the saturated hydraulic conductivity, is given by:

$$k_r(S) = = (\frac{T_1}{T_s}) \frac{\int_{S_{min}}^{S} \frac{dS}{P_c^2(S)}}{\int_{S_{min}}^{1} \frac{dS}{P_c^2(S)}} \tag{6}$$

where T_s, the tortuosity at saturation S, is defined as $(L_e/L)^2$, L_e is the length of the tortuous tube and L the direct distance between the ends of the tube, T_1 is the value of T_s at S = 1, and $P_c(S)$ is the capillary pressure at a particular saturation S.

The possibility of deriving a closed-form expression for the hydraulic conductivity-saturation function by substituting a capillary pressure-saturation relationship into Equation 6 makes the Burdine approach very attractive (Brooks & Corey 1964, Mualem 1976, van Genuchten 1980). However, the Burdine's equation is valid only for the drainage cycle, and the simulation of infiltration events with the main drying SWRC can be misleading. As shown in Equation 5, the increase in the saturation values increases the air-water pressure difference values in the occluded zone. Therefore, the capillary effects introduced by the Kelvin's equation into the Schuurman's model cannot be used as integral functions in Equation 6. It should also be stated that the Burdine's approach is based on the fact that the SWRC reflects the pore size distribution of the soil. It is then logical for the function to be used as the basis for predicting the hydraulic conductivity-saturation function. However, in the occluded zone the capillary values and pore water pressure values contain no information on the pore sizes or their distribution. The pore water pressure controls only the volume change of the air water mixture and as such should not be used as the basis for predicting the hydraulic conductivity function. Therefore, the assumption of the air pressure being equal to the atmospheric pressure in the occluded zone should be avoided and the statistical hydraulic conductivity functions derived from SWRC (drainage cycle) should not be used for modeling unsaturated flow (infiltration).

The author has proposed the following empirical expression for defining $k_r - S$ relationship at higher degrees of saturation on the wetting cycle (Bicalho 1999). The expression was determined by curve fitting to the experimental results and is given by:

$$k_r(S) = a + (1-a)[\frac{(S-b)}{(1-b)}]^n , S \geq S_m \tag{7}$$

where a, b and n are empirical parameters. Bicalho (1999) presents the effects of varying the three parameters a, b and n on the shape of the proposed hydraulic conductivity function. The experimental results suggest that k_r increases sharply when S increases from 0.9 to 1 and it has a value of 0.1 while $S \approx 0.8$ for a compacted silt. These results are consistent with the measured k_r values presented by Taibi (1993) for four different fine-grained soils. Bicalho et al. (2005a) show that hysteretic effect on the k-S relationship can be neglected for soils with discontinuous air phase.

3 NUMERICAL EVALUATION OF THE CONSTITUTIVE RELATIONSHIPS

A numerical simulation of the transient water flow measurements is obtained by following an approach that is analogous to Celia et al. (1990). Finite difference methods either explicit or implicit, belong to the most frequently used techniques in modeling unsaturated flow conditions (Remson et al. 1971). The advantage of the finite difference method is its simplicity and efficiency in treating the time derivatives. Therefore, a finite difference computer program was written to solve the Richards' equation and to model one-dimensional transient water flow through a compacted silt. The silt is from a sedimentary formation at the Bonny Dam Site, in eastern Colorado. The index properties for Bonny Silt are LL = 25%, PL = 21%, clay fraction 12%, Casagrande classification CL-ML, specific gravity 2.63, Proctor optimum water content 14.5% and the corresponding dry density 17.5 kN/m³. Relationships for θ_w (h_{pw}) and k_w (S) were calculated using Equation 4 and Equation 7 multiplied by the porosity and the saturated hydraulic conductivity, respectively. The one-dimensional unsaturated flow equation is solved assuming that the soil skeleton is rigid. Laboratory measurements were conducted to confirm that the assumption of no variation in the volume of void space was appropriate. The specimen's height was measured by using a dial gauge and a negligible change in the height of the sample was observed (i.e., $\Delta H < 0.02$ mm).

The input data for the program are the values of the sample's height, L_s, the initial and boundary conditions, and the soil parameters for the UHC functions. During testing of soil specimen B_1 ($L_s = 0.107$ m) the back pressure was increased in steps followed by hydraulic conductivity measurements for each step. Thus, each hydraulic conductivity test represents a separable boundary value problem with different initial conditions. The parameters of the SWRC function for this soil sample are: $S_0 = 84\%$, $u_{ao} = 80$ kPa (absolute), $u_{wo} = 75$ kPa (absolute), H = 0.02, $\theta_{sat} = 0.34$. The saturated hydraulic conductivity for this sample, k_{sat}, is 6.86×10^{-9} m/s. For each hydraulic conductivity test the transient pore pressure response was measured prior to reaching the steady state. The empirical parameters for the hydraulic conductivity function were adjusted until a good agreement between the experiment and the analyses was obtained. The obtained parameters for the hydraulic conductivity function for this soil specimen are a = 0.22, b = 0.7 and n = 5.7. Full details of the theoretical formulation and the numerical simulation are given in Bicalho (1999).

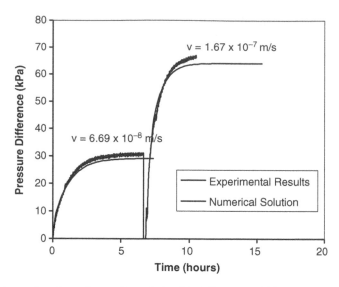

Figure 1. Experimental results and numerical solution (Back Pressure = 50 kPa).

The transient water flow measurements were accomplished by using a modified triaxial cell connected to a flow pump. The system includes pore-water back pressure facilities, a differential pressure transducer, a differential mercury manometer and a data acquisition system. The flow pump enables a precise control of water flow in and out of the soil sample. The flow pump forces water through the soil specimen at a constant predetermined rate and creates a pressure difference across the sample that is continuously recorded with a precision differential pressure transducer connected to a personal computer-based data acquisition system.

The saturation of the soil specimen was initially increased by applying back pressure and then by imposing upward flow rates at the bottom of the sample. Therefore, the analysis was performed in two steps. In the first step the saturation process due to the back pressure increase was simulated. The constant head boundary conditions were specified for the top and bottom boundaries in the numerical simulations. The output from the steady state analysis was then used as the initial conditions for the next transient analysis. The boundary conditions for the numerical simulation consisted of a constant head boundary, a Dirichlet type of boundary condition, at the top of the sample (i.e., applied back pressure) while constant flow boundary condition, a Neumann type of boundary condition, was specified at the bottom of the sample (i.e., flow rate imposed during the experiments).

Figures 1 through 3 show that good agreements between the predicted and measured curves were achieved by using the hydraulic conductivity function proposed for the tested soil specimen. Therefore, for this range of S the proposed hydraulic conductivity function gives an accurate prediction of the transient water flow and the steady state. While the second flow rate was imposed on the bottom of the soil specimen ($v = 1.67 \times 10^{-7}$ m/s), the experimental measurement was stopped before achieving the steady state, Figure 1. Therefore, the hydraulic conductivity function could not be investigated for this range of the degree of saturation. However, a tendency towards an agreement between the predicted and measured curves was observed. That was confirmed by the subsequent test, Figure 2. It can be seen that the same observations apply to Figures 2 and 3. The results of the simulation, in terms of the variation of moisture content, may be examined by consideration of saturation distributions at key times, as shown in Figure 4.

The predicted curves are in reasonably close agreement with the measured curves. The results achieved are viewed as particularly encouraging in terms of validation of the constitutive

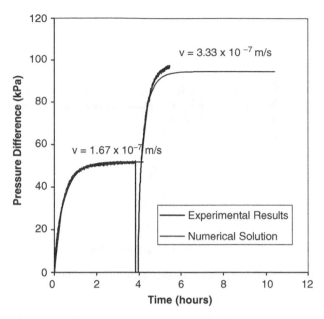

Figure 2. Experimental results and numerical solution (Back Pressure = 100 kPa).

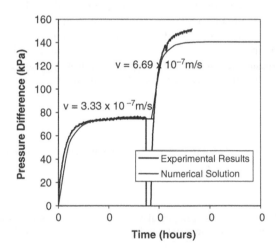

Figure 3. Experimental results and numerical solution (Back Pressure = 200 kPa).

relationships used in this study to describe the movement of the water in a porous medium containing discontinuous air phase.

4 CONCLUSIONS

This paper presents a consistent set of UHC functions for modeling infiltration process through compacted soils. A numerical solution for the Richards' equation for one-dimensional transient water flow was used to evaluate the proposed UHC functions. The comparisons between the numerical solutions and the measured transient water flow show that the constitutive relationships are

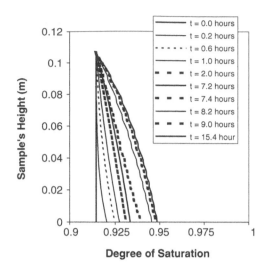

Figure 4.　Distribution of saturation within the Sample (Back Pressure = 50 kPa).

producing simulations of the measurements of the unsaturated flow that are both qualitatively and quantitatively realistic. Clearly, a good correlation has been achieved overall. However, for the routine application of the test methodology one needs to realize that reliable measurements of the transient portion of the experiments still pose a significant challenge.

The proposed UHC relationships (i.e., the soil water retention and the hydraulic conductivity functions) are based on the assumption that the overall volume of the soil remains constant. It is recommended to investigate the effect of the void ratio (or porosity) changes on the UHC functions for soils with discontinuous air phase. We leave it for future work.

ACKNOWLEDGMENTS

The author is grateful for the sponsorship by the Brazilian government agencies CNPq and CAPES.

REFERENCES

Arya, L.M. & Paris, J. F. 1981. A Physicoempirical Model to Predict the Soil Moisture Characteristic from Particle-Size Distribution and Bulk Density Data. Soil Sci. Soc. Am. J., 45, 1023–1030.
Bloomsburg, G.L. & Corey A. T. 1964. Diffusion of Entrapped Air from Porous Media. Colorado State University, Hydrology papers, 5, 1–27.
Bicalho, K. V. 1999. Modeling Water Flow in an Unsaturated Compacted Soil. Ph.D. dissertation, University of Colorado, Boulder, Colo, USA.
Bicalho, K. V., Znidarcic D. & Ko H.-Y. 2000. Air Entrapment Effects on Hydraulic Properties. *Geotechnical Special Publication,* ASCE, 99, 517–528.
Bicalho, K V., Znidarcic D. & Ko, H.-Y. 2005a. An experimental evaluation of unsaturated hydraulic conductivity functions for a quasi-saturated compacted soil. Advanced Experimental Unsaturated Soil Mechanics. Balkema. 325–329.
Bicalho, K V., Znidarcic D. & Ko, H.-Y. 2005b. Measurement of the soil-water characteristic curve of quasi-saturated soils, 16 International Conference on Soil Mechanics and Geotechnical Engineering, Rotherdam, Netherlands: Mill Press, 2, 1019–1022.
Brooks R. H. & Corey A. T. 1964. Hydraulic Properties of Porous Media. Colorado State University, Hydrology papers, 3, 1–27.
Burdine, N. T. 1953. Relative Permeability Calculation size distribution data. Transactions of the American Institute of Mining, Metallurgical and Petroleum Engineers, 198, 71–78.

Celia, M. A. Bouloutas, E. T. & Zarba, R. L. 1990. *A general Mass-Conservative Numerical Solution for the Unsaturated Flow Equation.* Water Resources Research, 26 (7), 1483–1496.

Chiu, T.-F. & Shackelford, C. D. 1998. Unsaturated Hydraulic Conductivity of Compacted Sand-Kaolin Mixtures. J. Geotech. Engrg, ASCE, 124(2), 160–170.

Corey, A.T. 1994. Mechanics of Immiscible Fluids in Porous Media. 3rd Ed., Water Resources Publications, Highlands Ranch, Colo.

Fredlund, D. G. & Xing, A. 1994. Equations for the Soil-water characteristic Curve. Can. Geotech. J., Ottawa, Canada, 31 (4), 521–532.

Fredlund, D. G., Xing, A. & Huang S. 1994. Predicting of the Permeability Function for Unsaturated Soil using the Soil-water characteristic Curve. Can.Geotech. J., Ottawa, Canada, 31(4), 533–546.

Hilf, J. W. 1956. An Investigation of Pore-Water Pressure in Compacted Cohesive Soils. Ph.D. dissertation, University of Colorado, Boulder, Colo., USA.

Kunze, R. J., Uehara, G., & Graham, K. 1968. Factors Important in the Calculation of Hydraulic Conductivity. Proc. Soil Sci. Soc. Amer., 32, 760–765.

Mualem, Y. 1976. A New Model for Predicting the Hydraulic Conductivity of Unsaturated Porous Media. Water Resources Research, 12(3), 513–522.

Remson, I., Hornberger G. M. & Molz F. J. 1971, Numerical Methods in Subsurface Hydrology, Wiley-Interscience, New York.

Richards, L. A. 1931. *Capillary conduction of liquids in porous media.* Physics 1: 318–333.

Schuurman, E. 1966. The Compressibility of an Air/Water Mixture. Geotechnique, 16 (4), 269–281.

Taibi, S. 1993. Comportement mécanique et hydraulique des sols soumis à une pression interstitielle négative – Etude expérimentale et modélisation, Thèse de Docteur-Ingénieur, Ecole Centrale Paris, France.

van Genuchten, M. Th. 1980. A closed-form Equation for Predicting the Hydraulic Conductivity of Unsaturated Soils. Soil Sci. Soc. Am. J., 44(5), 892–898.

Applications of Computational Mechanics in Geotechnical Engineering – Sousa,
Fernandes, Vargas Jr & Azevedo (eds)
© 2007 Taylor & Francis Group, London, ISBN 978-0-415-43789-9

Effects of post-peak brittleness on failure and overall deformational characteristics for heterogeneous rock specimen with initial random material imperfections

X.B. Wang

Department of Mechanics and Engineering Sciences, Liaoning Technical University, Fuxin, P R China

ABSTRACT: For heterogeneous rock with initial random material imperfections in uniaxial plane strain compression, the effects of post-peak brittleness on the failure and overall deformational characteristics are numerically modeled using FLAC. A FISH function is generated to prescribe the initial imperfections within the specimen with smooth ends using Matlab. For intact rock exhibiting linear strain-softening behavior beyond the occurrence of failure and then ideal plastic behavior, the failure criterion is a composite Mohr-Coulomb criterion with tension cut-off. Initial imperfections undergo ideal plastic behaviors beyond the occurrence of failure. As post-peak brittleness decreases, shear band thickness and peak volumetric strain increase; peak stress and corresponding values of axial and lateral strains increase; the stress-axial strain curve, stress-lateral strain curve, lateral strain-axial strain curve, calculated Poisson's ratio-axial strain curve and volumetric strain-axial strain curve become ductile at post-peak. The five kinds of the curves at different post-peak brittleness have separated prior to the peak strength. The calculated Poisson's ratio-axial strain curve can be divided into three stages. As axial strain increases, it always increases. In the second stage, the pre-peak Poisson's ratio linearly increases since material imperfections fail continuously so that the increase of the value of lateral strain exceeds the increase of axial strain. As post-peak brittleness increases, the precursors to failure tend to be less apparent and the failure becomes sudden.

1 INTRODUCTION

Rocks belong to heterogeneous materials. In compression, with an increase of axial strain, microcracking may occur in the vicinity of material imperfections, such as voids, pores and cracks. Further loading will result in the progressive coalescence of microcracks to eventually form the macroscopic fractures (Scholz 1968). As we know, using the fracture mechanics theory, the coalescence and propagation of cracks are difficult to describe since the dynamic microcracking events cause the boundary conditions to be changed complicatedly (Fang & Harrison 2002).

Recently, the failure process and mechanical behavior were numerically studied considering the heterogeneity of rocks (Tang & Kou 1998, Fang & Harrison 2002). Usually, the Weibull distribution function is used to describe the random variation of material strength for the elements. In these numerical simulations, the elastic-brittle constitutive relation is widely used. The relatively complex post-peak strain-softening constitutive relation is not considered. Moreover, all elements have different strength. Therefore, the failure process and macroscopic mechanical behavior of rock specimen with initial random material imperfections cannot be studied.

FLAC (Fast Lagrangian Analysis of Continua) is an explicit finite-difference code that can effectively model the behaviors of geomaterials, undergoing plastic flow as their yield limits are reached (Cundall 1989). One outstanding feature of FLAC is that it contains a powerful built-in programming language, FISH, which enables the user to define new variables, functions and constitutive models. The developed FISH functions by the user to suit the specific need extend FLAC usefulness or add user-defined features.

Recently, Wang wrote some FISH functions to calculate the overall deformational characteristics (such as average lateral strain, volumetric strain and calculated Poisson's ratio through axial and lateral strains) of rock specimen in plane strain compression. Using the FISH functions, the effects of dilation angle (Wang 2005d), softening modulus or post-peak brittleness (Wang 2006c), pore pressure (Wang 2005b, 2006b), initial cohesion and initial internal friction angle (Wang 2005c) on the failure process, final failure mode, precursors to failure and overall deformational characteristics were investigated for rock specimen with a material imperfection at left edge of the specimen. For jointed rock specimen, the effects of joint inclination on the failure process, anisotropic strength, final failure mode, precursors to failure and overall deformational characteristics (Wang 2005e, 2006d) were investigated using the FISH functions mentioned above and a FISH function written to automatically find the addresses of elements in the joint.

In the paper, for heterogeneous rock specimen with initial random material imperfections in uniaxial plane strain compression, the effects of post-peak brittleness on the failure process, final failure mode, overall deformational characteristics and precursors to failure were modeled using FLAC. Intact rock was assigned to obey linear strain-softening behavior once failure occurs and then ideal plastic behavior. A FISH function was generated to prescribe the initial imperfections within the specimen using Matlab. FISH functions written previously (Wang 2005c,d, 2006b,c,d) were used to calculate the axial, lateral and volumetric strains as well as the ratio (called the calculated Poisson's ratio) of negative lateral strain to axial strain of the specimen.

2 SCHEME AND CONSTITUTIVE RELATION

See Figure 1, the height and width of rock specimen with smooth ends are 0.1 m and 0.05 m, respectively. Black elements are material imperfections, while white elements are intact rock. The specimen deformed in small strain mode and plane strain condition is loaded at a constant velocity of $v_0 = 5 \times 10^{-10}$ m/timestep at the top. The specimen is divided into 3200 square elements with the same area.

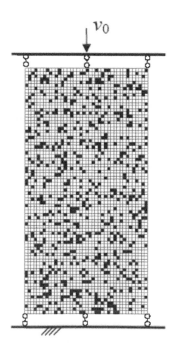

Figure 1. Model geometry and boundary conditions.

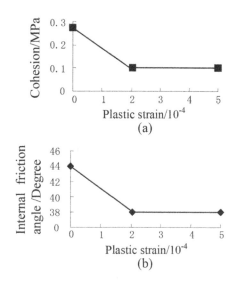

Figure 2. Post-peak constitutive relation of intact rock outside imperfections in scheme.

In linear elastic stage, the intact rock and the imperfections modeled by square elements have the same constitutive parameters. Shear modulus and bulk modulus are 11 GPa and 15 GPa, respectively.

Beyond the failure of the initial imperfection, the element or imperfection undergoes ideal plastic behavior. Cohesion and internal friction angle for the imperfection are 0.1 MPa and 38°, respectively.

The elements except for the imperfections are the intact rock. Once these elements yield, they undergo linear strain-softening behavior and then ideal plastic behavior. The adopted failure criterion for the intact rock is a composite Mohr-Coulomb criterion with tension cut-off and the tension strength is 2 MPa.

Five schemes are adopted in the present paper. For intact rock, each scheme has the same initial cohesion (0.275 MPa), initial internal friction angle (44°), residual cohesion (0.103 MPa) and residual internal friction angle (38°).

In scheme 1, the constitutive relation of intact rock in linear strain-softening stage is the most steep (Figure 2); in scheme 5, it is the least steep. In scheme 1, the absolute values of the slopes of cohesion-plastic strain relation and internal friction angle-plastic strain relation are $c_c = 0.86$ GPa and $c_f = 3.00 \times 10^4(°)$; in scheme 2, $c_c = 0.65$ GPa and $c_f = 2.28 \times 10^4(°)$; in scheme 3, $c_c = 0.47$ GPa and $c_f = 1.56 \times 10^4(°)$; in scheme 4, $c_c = 0.24$ GPa and $c_f = 0.84 \times 10^4(°)$; in scheme 5, $c_c = 0.03$ GPa and $c_f = 0.12 \times 10^4(°)$. The parameters c_c and c_f can be called cohesion-softening modulus and internal friction angle-softening modulus, respectively.

In scheme 5, c_c and c_f of intact rock are the lowest, reflecting the highest post-peak ductility; in scheme 1, c_c and c_f of intact rock are the highest, reflecting the highest post-peak brittleness.

3   FISH FUNCTIONS PRESCRIBING THE MATERIAL IMPERFECTIONS

Three steps are implemented in order to randomly prescribe the initial material imperfections within the heterogeneous specimen.

1. Choose random numbers from 1 to 3200 for 700 times using Matlab. The random numbers are chosen according to uniform distribution.
2. Remember the identification numbers equal to the chosen random numbers. In FLAC, every zone or element has an identification number. In the paper, the minimum and maximum identification numbers are 1 and 3200, respectively.
3. Give material properties of the ideal plastic constitutive relation to the elements remembered (the imperfections) and provide material properties of the linear strain-softening and ideal plastic constitutive relations to the other elements (the intact rock). The different linear strain-softening constitutive relations of intact rock are provided for the different schemes.

It is noted that we choose random numbers for 700 times in schemes 1 to 5. Therefore, the number of the imperfections can be less than 700 since a random number can be chosen more than one time. Using Matlab, a FISH function is generated to prescribe the initial random material imperfections within the specimen.

4   NUMERICAL RESULTS AND DISCUSSIONS

Figures 3–5(a–h) show the failure processes of schemes 1, 3 and 5, respectively. Figures 3–6(i) show the stress-axial strain curves of schemes 1, 3 and 5, respectively. Black elements mean that these elements have yielded, while white elements always remain elastic state.

In Figures 3–5(a), the timesteps t are 5000; Figures 3–5(b), 6000; Figures 3–5(c), 7000; Figures 3–5(d), 8000; Figures 3–5(e), 9000; Figures 3–5(f), 10000; Figures 3–5(g), 11000; Figures 3–5(h), 12000.

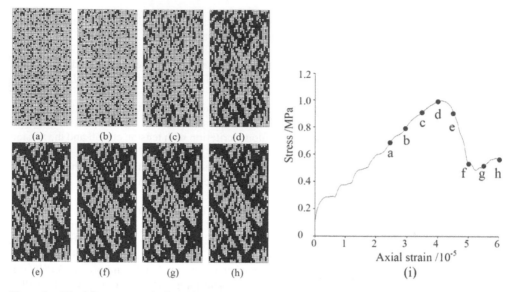

Figure 3.   The failure process (a–h) and macroscopic stress-axial strain curve (i) of scheme 1.

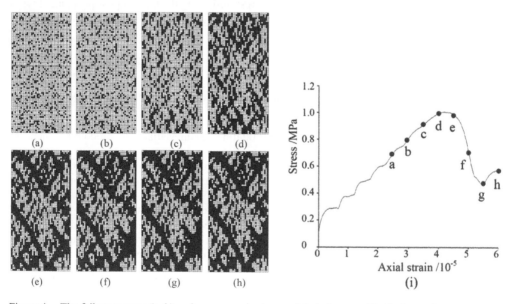

Figure 4.   The failure process (a–h) and macroscopic stress-axial strain curve (i) of scheme 3.

The point "a" in Figure 3(i) corresponds to Figure 3(a); the point "b" in Figure 3(i) corresponds to Figure 3(b), and so on.

The stress-axial strain curves and the stress-lateral strain curves of schemes 1 to 5 are shown in Figures 6 and 7, respectively. The axial strain ε_a is defined as (Wang 2005c, 2006b, c, d)

$$\varepsilon_a = \frac{v_0 t}{L} \tag{1}$$

where L is the length of the specimen; ε_a is positive for axial contraction.

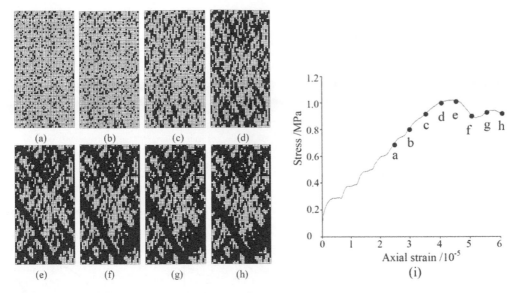

Figure 5. The failure process (a–h) and macroscopic stress-axial strain curve (i) of scheme 5.

According to the timesteps (t), the height (L) of the specimen and the velocity (v_0), we can calculate the axial strain (ε_a) of each picture in Figures 3–5. For example, in Figure 3(a), $\varepsilon_a = v_0 t/L = 5 \times 10^{-10} \times 5 \times 10^3/10^{-1} = 2.5 \times 10^{-5}$.

The average lateral strain ε_l is defined as (Wang 2005c, 2006b, c, d)

$$\varepsilon_l = \frac{1}{Bn} \sum_{i=1}^{n} (u_i - v_i) \tag{2}$$

where B is the width of the specimen; n is the number of nodes at one lateral edge of the plane specimen. Parameters u_i and v_i are values of horizontal displacements at the same height at left and right edges of the specimen, respectively. The difference $u_i - v_i$ is the value of relative horizontal displacement. Summing the value of the relative horizontal displacement leads to the value of total horizontal displacement.

Using the value of the total horizontal displacement and n, we can get the value of average horizontal displacement. The average lateral strain ε_l can be obtained if the value of the average horizontal displacement is divided by the width of the specimen. ε_l is negative for lateral expansion.

Volumetric strain ε_v is calculated as (Wang 2005c, 2006b, c, d)

$$\varepsilon_v = \varepsilon_a + \varepsilon_l = \frac{v_0 t}{L} + \frac{1}{Bn} \sum_{i=1}^{n} (u_i - v_i) \tag{3}$$

The calculated Poisson's ratio v in plane strain compression is expressed as (Wang 2005c, 2006b, c, d)

$$v = -\frac{\varepsilon_l}{\varepsilon_a} = -\frac{L}{Bnv_0 t} \sum_{i=1}^{n} (u_i - v_i) \tag{4}$$

Figure 8 shows the influence of post-peak brittleness on the lateral strain-axial strain curve. Figure 9 shows the influence of post-peak brittleness on the volumetric strain-axial strain curve. Figure 10 shows the influence of post-peak brittleness on the calculated Poisson's ratio-axial strain curve.

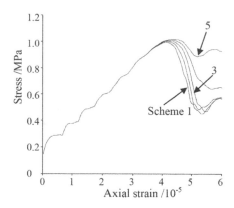

Figure 6. Stress-axial strain curves of five schemes.

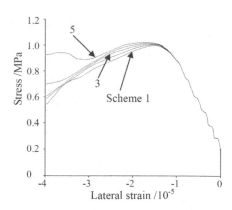

Figure 7. Stress-lateral strain curves of five schemes.

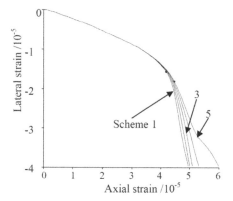

Figure 8. Lateral strain-axial strain curves of five schemes.

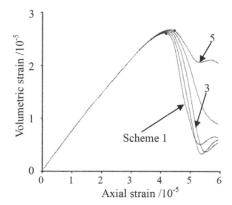

Figure 9. Volumetric strain-axial strain curves of five schemes.

The black point in Figures 8–10 corresponds to the onset of strain-softening behavior of stress-axial strain curve. Only the points in schemes 1 and 5 are marked in Figures 8–10.

4.1 *Progressive failure process for the specimen with initial random material imperfections*

The progressive failure of the specimen with initial random material imperfections is outlined as follows.

1. Firstly, imperfections fail (Figures 3–5(a)).
2. Secondly, a few imperfections finitely extend in the vertical direction (Figures 3–5(b)).
3. Then, some imperfections extend axially and the yielding elements compete each other in a disordered manner (Figures 3–5(c)).
4. Next, some yielded elements coalesce to form shorter shear fracturing bands (Figures 3–5(d)).
5. Finally, orderly longer shear fracturing bands are formed in the inclined direction through the linkage of shorter shear fracturing bands (Figures 3–5(e–h)).

4.2 *Effects of post-peak brittleness on failure mode*

Figures 3–5 show that shear bands become wider at lower post-peak brittleness. Moreover, in the later strain-softening stage, the number of the failed elements increases with a decrease of post-peak brittleness.

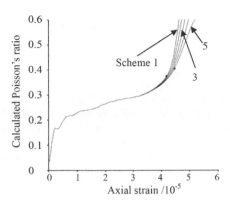

Figure 10. Calculated Poisson's ratio-axial strain curves of five schemes.

In classical elastoplastic theory, shear band with a certain thickness is simplified as slip line without thickness. Extensive experimental observations and theoretical analyses (Mühlhaus & Vardoulakis 1987, Vardoulakis & Aifantis 1991, Alshibli & Sture 1999) do not support this simplification.

It is believed that the thickness of shear band is influenced by the average grain diameter or the maximum aggregate diameter. Using gradient-dependent plasticity, the thickness w of shear band is $2\pi l$(De Borst & Mühlhaus 1992, Pamin & De Borst 1995), where l is an internal length parameter. However, this formula does not consider the effect of shear dilatancy. After the effect is taken into account, the thickness w of shear band is

$$w = 2\pi l(1 + \bar{\gamma}^{p}\sin\psi) \tag{5}$$

where $\bar{\gamma}^{p}$ is the average plastic shear strain in shear band and ψ is the dilation angle (Wang et al. 2004c).

For post-peak linear strain-softening rock material, $\bar{\gamma}^{p}$ is dependent on shear strength, shear stress acting on shear band and softening modulus c (Wang et al. 2004c).

The softening modulus c is equal to the absolute value of slope of the shear stress-average plastic shear strain in shear band curve (Wang et al., 2004b, Wang 2005a, 2006e). The softening modulus c is a constitutive parameter describing the post-peak brittleness. Higher softening modulus means that the post-peak brittleness of rock is higher.

The softening modulus c is different from cohesion-softening modulus c_{c} and internal friction angle-softening modulus c_{f}. However, the three parameters describe the post-peak brittleness.

The higher the post-peak brittleness is, the lower the average plastic shear strain is (Wang et al., 2004c). Thus, the thickness of shear band will be decreased as post-peak brittleness increases. The result is qualitatively consistent with the present numerical predictions. However, when $\psi=0$, the derived expression (Wang et al. 2004c) shows that the thickness of shear band is independent of post-peak brittleness. The present numerical results show that the thickness of shear band is also related to the post-peak brittleness when $\psi = 0$.

4.3 *Effects of post-peak brittleness on overall deformational characteristics*

4.3.1 *Stress-axial strain curves*
Figure 6 shows that lower post-peak brittleness leads to higher peak stress and corresponding axial strain. Moreover, the post-peak stress-axial strain curve tends to be less steep as post-peak brittleness decreases.

An analytical solution of the post-peak slope of stress-axial strain curve for rock specimen in uniaxial compression subjected to single shear failure in the form of shear band bisecting the specimen was derived (Wang and Pan 2003, Wang et al. 2004b, Wang 2005a, 2006e). The post-peak slope of the curve is influenced by the constitutive parameters of rock (elastic modulus, softening modulus c, shear band's thickness dependent on characteristic length of rock), the height of rock specimen and the orientation of shear band.

The analytical solution shows that higher softening modulus leads to steeper stress-axial strain curve in strain-softening stage (Wang and Pan 2003, Wang et al. 2004b, Wang 2005a, 2006e). The theoretical result is in qualitative agreement with the present numerical results.

As mentioned in Section 4.2, at lower post-peak brittleness, more failed elements are observed in the later strain-softening stage. The increased number of the failed elements will dissipate much energy. The dissipated energy can be provided if the post-peak stress-axial strain curve becomes less steep as post-peak brittleness decreases. The numerical results shown in Figure 6 support the analysis above.

4.3.2 *Stress-lateral strain curves*

Figure 7 shows that lower post-peak brittleness leads to higher peak stress and corresponding value of lateral strain. Moreover, the post-peak stress-lateral strain curve tends to be less steep as post-peak brittleness decreases.

An analytical solution of the post-peak slope of stress-lateral strain curve for rock specimen in uniaxial compression subjected to single shear failure in the form of shear band bisecting the specimen was derived (Wang et al. 2004a, Wang 2006e). Similar to the analytical solution of the post-peak slope of stress-axial strain curve (Wang and Pan 2003, Wang et al. 2004b, Wang 2005a, 2006e), the solution is also influenced by softening modulus c. Higher softening modulus results in steeper post-peak stress-lateral strain curve. The present numerical results are in qualitative agreement with the analytical solution.

4.3.3 *Lateral strain-axial strain curve*

Figure 8 shows that as post-peak brittleness decreases, the post-peak lateral strain-axial strain curve becomes less steep. Prior to the peak stress, the lateral strain-axial strain curve exhibits nonlinear characteristic.

The present predictions are different from the previous numerical results (Wang 2006c) showing that the pre-peak lateral strain-axial strain curve is linear. The reason for the nonlinear characteristic is that material imperfections fail progressively at pre-peak, leading to a rapid increase of lateral expansion.

4.3.4 *Volumetric strain-axial strain curve*

Figure 9 shows that as post-peak brittleness decreases, the post-peak volumetric strain-axial strain curve becomes less steep. The pre-peak volumetric strain-axial strain curve also demonstrates nonlinear behavior. This is due to the rapid lateral expansion of rock specimen.

As can be seen from Figure 9 that as post-peak brittleness decreases, the peak volumetric strain increases. This means that rock specimen composed of the material of lower brittleness at post-peak can reach the minimum volume in compression.

As post-peak brittleness decreases, the distance between the point of peak volumetric strain and the point of peak stress (black points) increases. This means that for more brittle material at post-peak, strain-softening behavior takes place once the volume dilates. However, for more ductile material at post-peak, strain-softening behavior occurs after volume dilates. The obtained numerical results suggest that the failure of more brittle material at post-peak is more sudden, as is in agreement with the common knowledge.

Figure 9 shows that though volume dilates beyond the peak volumetric strain, the volumetric strain is always positive. Phenomenon that the volume of deformed specimen is larger than the original volume is not found. The abnormal phenomenon can be observed in previously numerical results at higher dilation angles (Wang 2005d) and higher pore pressures (Wang 2005b, 2006b).

72

4.3.5 *Calculated Poisson's ratio-axial strain curve*

Figure 10 shows that calculated Poisson's ratio-axial strain curve can be divided into three stages. In the first stage, the pre-peak Poisson's ratio rapidly increases. The stage corresponds to the initially loading stage and the minimum value of the pre-peak Poisson's ratio is zero (Wang 2005c, 2006b, c, d).

In the second stage, calculated (pre-peak) Poisson's ratio linearly increases. The stage corresponds to the stage in which material imperfections fail progressively. It is noted that in the first and second stages, the pre-peak Poisson's ratio-axial strain curves at different post-peak brittleness overlap. This means that intact rock has not undergone post-peak linear strain-softening behavior.

In the third stage, the pre-peak Poisson's ratio rapidly increases. The calculated Poisson's ratio-axial strain curves at different post-peak brittleness are different, suggesting the occurrence of strain-softening behavior of intact rock.

It should be noted that each black point (peak stress beginning to decrease) falls into in the third stage of calculated Poisson's ratio-axial strain curve. Therefore, the third stage corresponds to the strain-softening stage and the strain-hardening stage.

Figure 10 shows that as post-peak brittleness decreases, the calculated Poisson's ratio-axial strain curve in the third stage becomes less steep.

It is found from Figure 10 that the calculated maximum value of the post-peak Poisson's ratio has exceeded 0.5. On the aspect of theoretical analysis, Wang (2006a) derived an expression for post-peak Poisson's ratio for rock specimen in uniaxial compression subjected to single shear fracture. For common specimen (height/width = 2), the predicted post-peak Poisson's ratio can reach 1.4 (Wang 2006a).

For homogenous rock specimen with a material imperfection closer to the base of the specimen, the numerical results show that in the second stage the pre-peak Poisson's ratio remains a constant (Wang 2005c, 2006b, c, d). The stage corresponds to the uniformly deformational stage. For heterogeneous rock specimen, the present numerical results show that in the second stage the pre-peak Poisson's ratio linearly increases. The phenomenon is due to the progressive failure of material imperfections, leading to the increase of the value of lateral strain exceeding the increase of axial strain.

4.4 *Effects of post-peak brittleness on failure precursors*

Figures 8–10 show that lateral strain-axial strain curves, volumetric strain-axial strain curves and calculated Poisson's ratio-axial strain curves at different post-peak brittleness have separated prior to the black points. When stresses just begin to decrease, the three kinds of the curves mentioned have apparently deviated the tangential lines of separate points. The higher the post-peak ductility is, the higher the deviation is. The deviation can be seen as a kind of precursors to unstable failure occurring in strain-softening stage. Therefore, higher post-peak ductility will result in more apparent precursors to failure. The present numerical predictions are different from the previously numerical results (Wang 2006c).

5 CONCLUSIONS

As post-peak brittleness decreases, shear band thickness and peak volumetric strain increase; peak stress and corresponding values of axial and lateral strains increase; the stress-axial strain curve, stress-lateral strain curve, lateral strain-axial strain curve, calculated Poisson's ratio-axial strain curve and volumetric strain-axial strain curve become ductile at post-peak.

The lateral strain-axial strain curves, volumetric strain-axial strain curves and calculated Poisson's ratio-axial strain curves at different post-peak brittleness have separated prior to the peak strength. When stress just begins to decrease, the lateral strain-axial strain curve, calculated Poisson's ratio-axial strain curve and volumetric strain-axial strain curve have apparently deviated the tangential lines of separate points. As post-peak brittleness increases, the precursors to failure tend to be less apparent and the failure becomes sudden.

73

The calculated Poisson's ratio-axial strain curve in uniaxial plane strain compression can be divided into three stages. As axial strain increases, it always increases. In the second stage, the pre-peak Poisson's ratio linearly increases since material imperfections fail continuously so that the increase of the value of lateral strain exceeds the increase of axial strain.

ACKNOWLEDGEMENTS

The study is funded by the National Natural Science Foundation of China (50309004).

REFERENCES

Alshibli, K.A. & Sture, S. 1999. Sand shear band thickness measurements by digital imaging techniques. *Journal of Computing in Civil Engineering, ASCE* 13(2): 103–109.
Cundall, P.A. 1989. Numerical experiments on localization in frictional material. *Ingenigeur-Archiv* 59(2): 148–159.
De Borst, R. & Mühlhaus, H.B. 1992. Gradient-dependent plasticity: formulation and algorithmic aspects. *International Journal for Numerical Methods in Engineering* 35(3): 521–539.
Fang, Z. & Harrison, J.P. 2002. Development of a local degradation approach to the modeling of brittle fracture in heterogeneous rocks. *International Journal of Rock Mechanics and Mining Sciences* 9(4): 443–457.
Mühlhaus, H.B. & Vardoulakis, I. 1987. The thickness of shear bands in granular materials. *Géotechnique* 37(3): 271–283.
Pamin, J. & De Borst, R. 1995. A gradient plasticity approach to finite element predictions of soil instability. *Archives of Mechanics* 47(2): 353–377.
Scholz, C.H. 1968. Microfracturing and the inelastic deformation of rock in compression. *Journal of Geophysical Research* 73(4): 1417–1432.
Tang, C.A. & Kou, S.Q. 1998. Crack propagation and coalescence in brittle materials under compression. *Engineering Fracture Mechanics* 61(3–4): 311–324.
Vardoulakis, I. & Aifantis, E. 1991. A gradient flow theory of plasticity for granular materials. *Acta Mechanica* 87(3–4): 197–217.
Wang, X.B. & Pan, Y.S. 2003. Effect of relative stress on post-peak uniaxial compression fracture energy of concrete. *Journal of Wuhan University of Technology-Materials Science Edition* 18(4): 89–92.
Wang, X.B., Liu, J., Wang, L. & Pan, Y.S. 2004a. Analysis of lateral deformation of rock specimen based on gradient-dependent plasticity (II): Size effect and snap-back. *Rock and Soil Mechanics* 25(7): 1127–1130. (in Chinese)
Wang, X.B., Ma, J., Liu, J. & Pan, Y.S. 2004b. Analysis of lateral deformation of rock specimen based on gradient-dependent plasticity (I)-basic theory and effect of constitutive parameters on lateral deformation. *Rock and Soil Mechanics* 25(6): 904–908. (in Chinese)
Wang, X.B., Tang, J.P., Zhang, Z.H. & Pan, Y.S. 2004c. Analysis of size effect, shear deformation and dilation in direct shear test based on gradient-dependent plasticity. *Chinese Journal of Rock Mechanics and Engineering* 23(7): 1095–1099.
Wang, X.B. 2005a. Analytical solution of complete stress-strain curve in uniaxial compression based on gradient-dependent plasticity. In P. Konečný (ed.), *Eurock 2005-Impact of Human Activity on the Geological Environment*: 661–667. Leiden: Balkema.
Wang, X.B. 2005b. Effect of pore pressure on entire deformational characteristics of rock specimen. *Journal of Shenyang Jianzhu University (Natural Science)* 21(6): 625–629. (in Chinese)
Wang, X.B. 2005c. Effects of initial cohesions and friction angles on entire deformational characteristics of rock specimen. *Journal of Shenyang Jianzhu University (Natural Science)* 21(5): 472–477. (in Chinese)
Wang, X.B. 2005d. Effects of shear dilatancy on entire deformational characteristics of rock specimen. *Journal of Sichuan University (Engineering Science Edition)* 37(5): 25–30. (in Chinese)
Wang, X.B. 2005e. Joint inclination effect on strength, stress-strain curve and strain localization of rock in plane strain compression. *Material Science Forum* 495–497: 69–74.
Wang, X.B. 2006a. Analysis of the post-peak Poisson's ratio of rock specimens in uniaxial compression. *Engineering Mechanics* 23(4): 99–103. (in Chinese)
Wang, X.B. 2006b. Effect of pore pressure on failure mode, axial, lateral and volumetric deformations of rock specimen in plane strain compression. In A. Van Cotthem et al. (eds), *Multiphysics Coupling and Long Term*

Behaviour in Rock Mechanics (Proceedings of ISRM Eurock 2006): 105–111. London: Taylor & Francis Group.

Wang, X.B. 2006c. Effect of softening modulus on entire deformational characteristics of rock specimen. *Chinese Journal of Geotechnical Engineering* 28(5): 600–605. (in Chinese)

Wang, X.B. 2006d. Entire deformational characteristics and strain localization of jointed rock specimen in plane strain compression. *Journal of Central South University of Technology* 13(3): 300–306.

Wang, X.B. 2006e. Shear deformation, failure and instability of rock specimen in uniaxial compression based on gradient-dependent plasticity. In R. Luna et al. (eds), *Proceedings of Sessions of GeoShanghai, Geotechnical Special Publication*: (150): 213–219.

Applications of Computational Mechanics in Geotechnical Engineering – Sousa,
Fernandes, Vargas Jr & Azevedo (eds)
© 2007 Taylor & Francis Group, London, ISBN 978-0-415-43789-9

Constitutive model calibration using an optimization technique

A.G. Guimarães, I.D. Azevedo & R.F. Azevedo
Federal Universitiy of Viçosa, MG, Brazil

ABSTRACT: Many problems in geotechnical engineering request the use of sophisticated and complex constitutive laws, which require sometimes the calibration of a great number of parameters of the involved materials. In this way, a more skilled approach than the traditional methodology for determination of those parameters becomes necessary. The problem of parameter identification can be treated as an optimization problem and solved by iterative algorithms that allow finding those parameters which turn minimal the difference between the measured values in the laboratory and those calculated by the model. Among the several methods to solve optimization problems, the quasi-Newton method of limited memory LBFGS was chosen due to its readiness and handling easiness. The technique was used to calibrate a constitutive non linear elastic model from triaxial experimental results, showing the capacity of the optimization procedure in correctly determining the model parameters. A sensivity analysis was also performed to explore the complexity found in the optimization of highly non linear functions.

1 INTRODUCTION

Numerical methods are frequently used to solve engineering problems. Load-deformation problems require the use of constitutive models to describe the stress-strain behavior of the involved materials. However, before a model can be used in a numerical procedure, it is necessary to calibrate it from laboratory test results. The objective of the calibration is to find appropriate parameters that produce the best answer of the model in relation to the available experimental results.

There is a great diversity of constitutive models for engineering materials, the simplest are based in the theory of linear elasticity and the calibration is reasonably simple. However, most geomaterials do not obey the theory of elasticity and demand the use of sophisticated and complex constitutive laws that involve a considerable number of parameters. The determination of those parameters becomes more difficult and, frequently, dependent of the user's experience. Therefore, it is a challenging task to calibrate a set of model parameters that best satisfy the available experimental data.

The traditional calibration methodologies use a limited number of laboratory test results. The use of optimization techniques in the determination of the best constitutive model parameters allows the use of several test types, making the calibration procedure more rational and objective.

2 OPTIMIZATION STRATEGY

The calibration of a constitutive model can be defined as the determination of appropriate values for the parameters in such a way that the behavior predicted by the constitutive model best reproduces the material behavior observed in the laboratory tests. Therefore, the calibration can be treated as an optimization problem in that an iterative algorithm is used to determine a set of model parameters that minimizes an objective function, F, of the differences between the model predictions and the experimental data. The selected model parameters assume the role of optimization variables.

Of the mathematical point of view the problem can be treated as:

Minimize $F(\mathbf{p})$ (1)

Submitted to: $\mathbf{m} \le \mathbf{p} \le \mathbf{n}$ (2)

where $\mathbf{p}$ is a vector of model parameters. $\mathbf{m}$ and $\mathbf{n}$ are the limit vectors, that restrict the search space for vector $\mathbf{p}$, for a domain defined by reasonable values of the parameters. The solution of the optimization problem is a vector $\mathbf{p}*$ that satisfies the condition:

$$\forall \mathbf{p}: \mathbf{m} \le \mathbf{p} \le \mathbf{n} \rightarrow F(\mathbf{p}*) \le F(\mathbf{p})$$ (3)

2.1 Objective function

The objective function, expressed by the sum of the normal values of the difference between the laboratory test results and the ones given by the numerical model for a certain number of points, n, supplies a scalar measure of the error between the experimental and numerical data.

The objective function defined by the minimum square method is expressed as:

$$F(\mathbf{p}) = \left[\mathbf{y}^* - \mathbf{y}(\mathbf{p})\right]^T \left[\mathbf{y}^* - \mathbf{y}(\mathbf{p})\right] = \sum_{i=1}^{nv} r_i^2(\mathbf{p})$$ (4)

where $\mathbf{y}*$ is the vector of the measured values, $\mathbf{y}(\mathbf{p})$ is the vector of calculated values in function of the parameters to determine and r is the residue, in other words, the difference between the measured and calculated values of the nv number of values.

The function $F(\mathbf{p})$ should include the soil stress-strain behavior in the several points involved in the optimization process and also be independent of the state variables. Therefore, known the measured values and computed results in discreet points, equation (4) can be represented by a sum of individual norms in such points, given as:

$$F(\mathbf{p}) = \sum_{i=1}^{npt} \left\| \varepsilon_{1i}^{exp} - \varepsilon_{1i}\left(\mathbf{p}, \sigma_1^{exp}\right) \right\|^2 + \left\| \varepsilon_{vi}^{exp} - \varepsilon_{vi}\left(\mathbf{p}, \sigma_1^{exp}\right) \right\|^2$$ (5)

where $\mathbf{p}$ is the vector of parameters and npt is the number of points of all tests. The values of axial and volumetric strains, calculated by the hyperbolic model, are given by $\varepsilon_{1i}(\mathbf{p}, \sigma_i^{exp})$ and $\varepsilon_{vi}(\mathbf{p}, \sigma_i^{exp})$, respectively; ε_{1i}^{exp} e ε_{vi}^{exp} are the experimental strains.

2.2 Iterative algorithm

The solution of the problem declared by the equations (1) and (2) requires the minimization of a function given by equation (5), which can be accomplished through a wide variety of iterative optimization algorithms. The algorithm is initialized by an estimate of the solution and in a sequence of operations, in that the object of each one is the precedent result, new solutions are generated until that the optimality conditions are satisfied.

The quasi-Newton methods combine the advantages of the steepest descent and the Newton's methods, seeking a commitment between the speed of convergence of Newton's method and the difficulty in determining the inverse of the Hessian matrix, by generating an approximation for the inverse matrix by a process that only uses first order derivatives.

The LBFGS algorithm, a free domain open code in FORTRAN, developed by Zhu et al. (1994), was used in this work. This algorithm does not require second order derivatives or the knowledge of the objective function structure. Further, it also makes use of matrixes compact representations that reduces the computational cost in each iteration.

In this algorithm, the method of projection of the gradient is used to determine the group of active restrictions in each iteration and the BFGS matrixes of limited memory to approach the objective function Hessian matrix, so that the demanded storage is linear.

More details about the LBFGS algorithm may be found in Liu & Nocedal (1989), Byrd et al. (1994) e Zhu et al. (1994).

3 CONSTITUTIVE MODEL

Soil stress-strain behavior depends on a great diversity of factors. Several authors have been working in the area of constitutive models seeking to find a stress-strain-strength relationship that appropriately represents the soil behavior and that takes into account the largest possible number of conditioning factors (Desai and Siriwardane, 2000).

The hyperbolic constitutive model, proposed by Duncan and Chang (1970), has been used in geotechnical engineering due to its simplicity and easiness in obtaining the parameters. This model can be used to represent the stress-strain behavior of cohesive or cohesionless, saturated or dry soils, in drained or undrained load conditions.

The model constitutive relations take into account soil characteristics such as non linearity and the influence of confining pressure. Soil features as dilatancy and the influence of intermediate principal stress are not considered in this model. Therefore, the model constitutive relationships present the same behavior for compression, tension and plane strain states.

The hyperbolic equation:

$$(\sigma_1 - \sigma_3) = \frac{\varepsilon_1}{a + b\varepsilon_1} \tag{6}$$

proposed by Konder (1963) to represent the relationship between the stress difference, $\sigma_1 - \sigma_3$, and the axial deformation, ε_1, describes the hyperbolic model, where a and b are constants associated, respectively, to the initial deformability modulus, E_i, and to the stress difference in the ultimate condition, $(\sigma_1 - \sigma_3)_{ult}$, as indicated in Figure 1.

The initial deformability modulus is defined as an exponential function of the confining stress (Jambu 1963):

$$E_i = K.pa.\left(\frac{\sigma_3}{pa}\right)^n \tag{7}$$

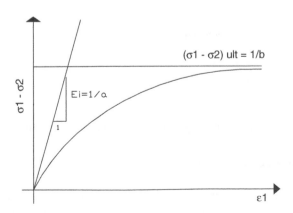

Figure 1. Hyperbolic stress-strain curve.

where K and n are constants obtained empirically from laboratory tests, and *pa* is the atmospheric pressure.

The stress difference in the ultimate condition is defined as the asymptotic value to the stress-strain curve and can be related with the shearing strength, $(\sigma_1 - \sigma_3)_f$, as:

$$(\sigma_1 - \sigma_3)_{ult} = \frac{(\sigma_1 - \sigma_3)_f}{R_f} \tag{8}$$

where R_f is a parameter known as failure ratio. For most soils, this factor varies from 0.7 to 1.1. It can be written as:

$$b = \frac{R_f}{(\sigma_1 - \sigma_3)_f} \tag{9}$$

The shearing strength given by Mohr-Coulomb criterion is defined as a function of the confining stress as:

$$(\sigma_1 - \sigma_3)_f = \frac{2c \cos\varphi + 2\sigma_3 \sin\varphi}{1 - \sin\varphi} \tag{10}$$

where c and φ are the material effective strength parameters, obtained experimentally.

The tangent elasticity modulus is determined by differentiating equation (1) in relation to ε_1 and, after some mathematical manipulations it is obtained:

$$E = K.pa.\left(\frac{\sigma_3}{pa}\right)^n \left[1 - \frac{R_f(1 - \sin\varphi)(\sigma_1 - \sigma_3)}{2c \cos\varphi + 2\sigma_3 \sin\varphi}\right] \tag{11}$$

Equation (10) represents the fundamental relationship of the hyperbolic constitutive model.

To represent the axial-volumetric strain relation, Duncan (1980) proposed an approach so that the non linear relationship between ε_1 and ε_v is represented through a constant value of the bulk modulus, B, which, in turn, is a function of the confining pressure, σ_3:

$$B = K_b.pa.\left(\frac{\sigma_3}{pa}\right)^m \tag{12}$$

where K_b and *m* are soil parameters. This way, the relationship between ε_1 and ε_v is expressed as:

$$\varepsilon_v = \frac{E}{3B}\varepsilon_1 \tag{13}$$

Based on Hooke's law, the incremental principal strains can be expressed as:

$$\begin{Bmatrix} d\varepsilon_1 \\ d\varepsilon_2 \\ d\varepsilon_3 \end{Bmatrix} = \frac{1}{E}\begin{bmatrix} 1 & -v & -v \\ -v & 1 & -v \\ -v & -v & 1 \end{bmatrix}\begin{Bmatrix} d\sigma_1 \\ d\sigma_2 \\ d\sigma_3 \end{Bmatrix} \tag{14}$$

where parameters B and E are defined by equations (7) and (8), respectively, and $v = \frac{3B-E}{6B}$.

4 SENSIVITY ANALYSYS

Besides obtaining an estimate of the parameters to adjust the experimental data, it is desirable that these parameters are reliable. A high uncertainty in the estimates may be involved due to low sensivity or high correlation between parameters.

The sensivity matrix is used to analyze the uncertainties associated to the estimated parameters. The sensivity matrix coefficients are defined as:

$$J_{ij} = \frac{\partial y_i(\mathbf{p})}{\partial p_j} \tag{15}$$

Sensivity coefficients are very important because they indicate the magnitude of change of the response $y_i(\mathbf{p})$ due to perturbations in the values of the parameters, p_j. They may also provide insight into the cases for which parameters can and cannot be estimated.

Parameters can be estimated if the sensivity coefficients over the range of the observations are not linearly dependent (Beck and Arnold, 1977). This is the criterion that is used in this work to determine if the parameters can be simultaneously estimated without ambiguity. If the parameters are linearly dependent, the variation of one can be compensated by the variation of other and the answer of the system will continue being the same.

A recommended practice is that the sensivity coefficients be plotted and carefully examined to see if linear dependence exists or even is approached. In order to compare the coefficients, it is convenient to transform them as:

$$J_{ij} = \frac{\partial y_i(\mathbf{p})}{\partial p_j} p_j \tag{16}$$

Another criterion to evaluate the results of the parameter identification problem is based on the parameter covariance matrix. For the least square method, the covariance matrix of the estimated parameters is given as:

$$\mathbf{C}_p = \sigma_0^2 \left[\mathbf{J}^T \mathbf{J}\right]^{-1} \tag{17}$$

The elements in the main diagonal, the variances, C_{pii}, are the square of the standard deviation, σ_{pi}, the coefficients C_{ij} are the covariances. The correlation coefficient is defined by

$$\rho_{ij} = \frac{C_{pij}}{\sqrt{C_{pii}C_{pjj}}} \tag{18}$$

Correlation coefficients assume values between -1 and $+1$. A correlation coefficient equal to zero indicates that there is no correlation between the parameters i and j, a larger value than 0.9 indicates high correlation, that is, the two parameters cannot be determined independently.

For non linear problems, the sensivity coefficients vary for different combination of parameters. Therefore, these analyses should be repeated for several hypotheses on the parameters.

The sensivity analysis for the hyperbolic model was accomplished with the parameters obtained using the traditional calibration methodology.

The hyperbolic model parameters, c (cohesion) and φ (friction angle), presented high correlation coefficients amongst themselves and also with parameter R_f (failure coefficient). As c and φ could not be estimated simultaneously, and since they can be easily determined from laboratory tests, these parameters were admitted to be constant during the optimization process.

Then, for the hyperbolic model the following vector of parameters was adopted:

$$\mathbf{p} = \{K, n, R_f, K_b, m\} \tag{19}$$

5 EXAMPLE

In order to show the applicability of the optimization strategy described previously, laboratory test results of three soils are used to determine the optimum parameters of the hyperbolic model.

Table 1. Traditional calibration parameters.

Example	K	n	Rf	Kb	m	c [kPa]	ϕ [°]	FO
A-04	318.13	−0.7925	0.8612	54.463	−0.5807	60	22	8.91×10^{-3}
A-05	143.85	−0.2075	0.8769	30.967	0.0688	19	30	7.43×10^{-2}
D-01	164.36	0.2492	0.8526	29.101	0.4246	47.4	20.6	2.13×10^{-1}

Table 2. Optimization strategy parameters

Example	K	n	Rf	Kb	m	c [kPa]	ϕ[°]	FO
A-04	318.13	−0.6179	0.8886	54.463	−0.5818	60	22	3.33×10^{-3}
A-05	143.85	−0.1683	0.9407	30.967	0.0527	19	30	5.71×10^{-2}
D-01	164.20	0.1215	0.7805	29.101	0.4246	47.4	20.6	5.11×10^{-2}

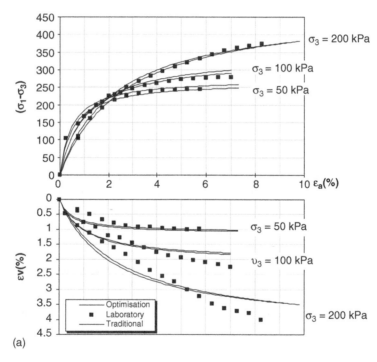

(a)

Figure 2. Comparison between experimental and numerical curves. (a) Example A-04; (b) Example A-05; (c) Example D-01

For each soil, results of conventional, drained, isotropically consolidated triaxial tests with three different confining pressures were used (Gouvêa, 2000 and Duarte, 2006).

Parameter values for the traditional calibration procedure are presented in Table 1. These values were used as initial estimatives for the optimization strategy. Final values of the model parameters are listed in Table 2. It can be seen that values of the objective function in Table 2 are smaller than the ones in Table 1, showing that the parameters were optimized.

Comparisons between experimental and numerical results obtained with these optimum parameters are presented in Figure 2. It can be observed that the numerical results adjusted well the experimental results.

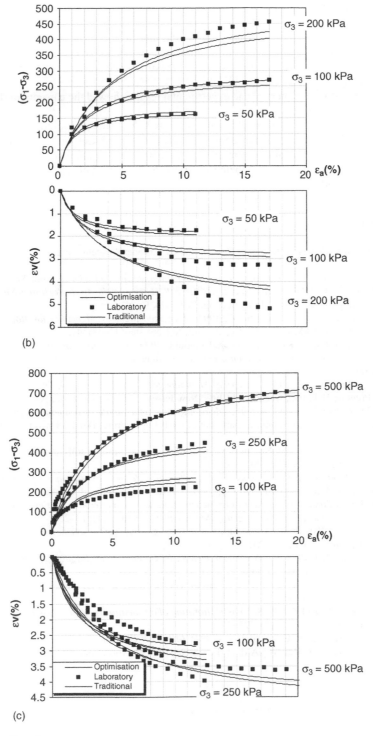

(b)

(c)

Figure 2. Continued

6 CONCLUSIONS

An optimization technique was used to calibrate the non linear elastic, hyperbolic model, using conventional triaxial test results. The optimization strategy took into consideration a set of parameters that minimize an objective function defined by the difference between experimental and results given numerically by the constitutive model.

Comparisons between numerical and experimental results showed the capacity of the procedure to obtain optimum parameters.

However, for the constitutive model and the soils used, the optimized solution was very close to the ones obtained with the parameters conventionally determined. Certainly, for other constitutive models and, or soils, the use of the optimization procedure may be more significative.

At the moment, the procedure is being extended to be used with elastic-plastic constitutive models and other soils.

REFERENCES

Arora, J.S. 2004. *Introduction to Optimum Design*. 2nd ed. Elsevier Academic Press. San Diego, California, USA.

Beck, J.V. and Arnold, K.J. 1977 *Parameter Estimation in engineering and science* , John Wiley & Sons, Inc.

Bicalho, K.V. 1992. *Modelagem das características e Resistências da Areia de Itaipú-Niterói/RJ*. MSc. Dissertation. Departament of Civil Engineering, PUC-Rio, Brazil.

Byrd, R.H. , Lu, P., Nocedal, J. e Zhu, C. 1994. *A Limited Memory Algorithm for Bound Constrained Optimization*. Report NAM-08. EECS Department, Northwestern University.

CeKerevac, C., Girardin, S., Klubertanz,G. and Laloui,L. 2006. *Calibration of an elasto-plastic constitutive model by a constrained optimization procedure*. Computers and Geotechnics (doi:10.1016/j.compeo 2006.07.009).

Dennis, J.E. e Schnabel, R.B. 1983. *Numerical Methodos for Unconsytrained Optimization and Nonlinear Equatios*. Prentice-Hall, Inc., Englewood Cliffs, New Jersey.

Duarte, L.N. 2006. *Estudo da relação tensão-deformação de uma sapata instrumentada, por meio de prova de carga direta*. MSc. Dissertation. Department of Civil Engineering, Federal University of Viçosa, MG, Brazil.

Duncan, M.J. e Chang, C.Y. 1970. Nonlinear analysis of stress and strain in soils. *Journal of Soil Mechanics and Foundation Division*, ASCE, SM5. 1629–1653.

Gouvêa, M.A.S. 2000. *Análise das Relações Carga-Recalque de uma Fundação em Verdadeira Grandeza* MSc. Dissertation. Department of Civil Engineering, Federal University of Viçosa, MG, Brazil.

Jambu, N. 1963. Soil Compressibility as determined by oedometer and triaxial tests. *European Conference on Soil Mechanics and Fundation Engeneering*, Wissbaden, Germany. 1(1). 19–25.

Kondner, R.L. 1983. Hyperbolic Stress-Strain response: coesive soils. *Journal of Soil Mechanics and Foundation Division*, ASCE, 89, SM1. 115–143.

Liu, D.C. and Polyak, B.T. 1989. *On the limited memory BFGS method for large scale optimization methods* Mathematical Programming. 45. 503–528.

Nogueira, C.L. 1998. *Análise não linear de Escavações e Aterros*, DSc. Thesis. Departament of Civil Engineering, PUC-Rio, Brazil.

Pal, S., Wathugala, G.W. and Kundu, S. 1996. *Calibration of a Constitutive Model Using Genetic Algorithms*, Computers and Geotechnics. 19(4). 325–348.

Zhu, C., Byrd, R.H., Lu, P. E Nocedal, J. (1994) *LBFGS-B:Fortran subroutines for large-scale bound constrained optimization,* Report NAM-11, EECS Department, Northwestern University.

Computational models

Applications of Computational Mechanics in Geotechnical Engineering – Sousa,
Fernandes, Vargas Jr & Azevedo (eds)
© *2007 Taylor & Francis Group, London, ISBN 978-0-415-43789-9*

Consideration of geological conditions in numerical simulation

G. Beer
Institute for Structural Analysis, Graz University of Technology, Austria

G. Exadaktylos
Technical University of Crete, Greece

ABSTRACT: The consideration of geological conditions in a numerical simulation is discussed. The main issue is how to go from laboratory tests and geological information to the numerical parameters and material laws for use in simulation models. Some novel concepts are presented.

1 INTRODUCTION

Numerical simulation programs are frequently used to assist in the selection of the best design and construction method of a tunnel. A numerical model allows the tunnel engineer to test various designs before construction commences and to assess with respect to economy and stability different alternative construction sequences or support designs. Simply put, the tunnel may be constructed in virtual space (i.e. the computer) prior to excavating it in reality.

The inclusion of geological conditions into a numerical simulation program, is an important and challenging task since geological and laboratory data have to be turned into digital information that is required by the model. The ideas presented here are currently being discussed in the European integrated project Technology Innovation in Underground Construction (TUNCONSTRUCT).

For numerical models based on continuum mechanics such as the Finite Element Method (FEM) or the Finite Difference Method (FDM) the input data required are:

- Elastic deformation modulus of rock mass, Poisson's ratio
- Failure condition (yield surface) with parameters describing it. The failure condition may be:
  - Isotropic (i.e. Hoek and Brown, Mohr-Coulomb etc)
  - Anisotropic (i.e. multi-laminate model)
- Flow law, describing stress-strain behavior at yield.
- Evolution law, describing the change in strength (deterioration/damage).

In addition knowledge of the virgin stress field (i.e. the stress that exists in the ground before excavation) is required.

Figure 1 shows a flow chart of how the information that may exist at the design phase of a tunnel may be turned into input data for a numerical model. In nearly all cases geological exploration is carried out prior to construction. This includes the borehole logs, the interpretation by a geologist of topological/site conditions and seismic logs. In addition there exist laboratory tests on small specimens usually taken from drill cores. However, in most cases – because of budget restraints – the information on geological conditions is patchy and a great degree of uncertainty exists.

To obtain a digital description of the material behaviour the geological information together with laboratory data is used to obtain "rock mass" data, which describe the response of the rock mass. This is sometimes known as "upscaling" and will be discussed in the following section.

Because of the uncertainty and patchiness of the information it is sometimes a good idea to include a sensitivity analysis, so that one can ascertain how sensitive the analysis results are to the

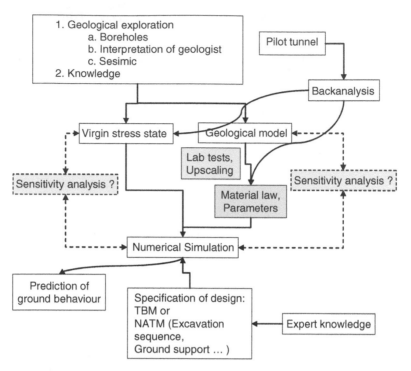

Figure 1.   Flow chart for determining the necessary input for numerical models at design stage.

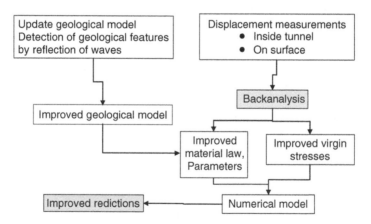

Figure 2.   Flow chart for updating material parameters and virgin stress information.

changing of parameters. In some cases pilot tunnels are constructed in order to get more detailed geological information for the final design. Here one can use back analysis procedures to obtain the rock mass properties from displacement measurements for example.

In tunnel construction based on the New Austrian Tunnelling Method (NATM) it is sometimes required to adjust the construction/support method according to updated information about geological conditions. In NATM tunnel sites extensive monitoring takes place and this allows to back analyse the virgin stress conditions and material parameters from displacement measurements. In addition innovative methods exist to detect major fault zones or changes in rock mass properties by the reflection of sound waves. Figure 2 shows a flow chart of a possible updating process during construction.

2 UPSCALING

Here we describe the proposed procedure for arriving at the input data required by the model. First we analyse the information that is usually available at the start of a project. Then we look at possible material models and finally at the parameter determination.

2.1 *Information available at design phase*

At the design phase of the project the following data is available:

- The constitutive behaviour of the rock at lab scale (i.e. of the order of 10 cm) in the frame of isotropic or anisotropic continuum elastoplasticity and elastoviscoplasticity theories, calibrated on several testing configurations (uniaxial and triaxial compression, indirect tension or direct tension if possible) on representative intact rock samples.
- The discontinuity mechanical properties at the lab scale, e.g. normal and shear stiffness and strength properties, respectively, that are derived from the laboratory shear tests on representative specimens collected in the field or from cores.
- From mapping on exposed surfaces, from (oriented) boreholes and from other exploration methods (i.e. seismic) the geometrical properties of the joints are collected and adequately described in a statistical context (probability density functions). These properties are the number of joint sets and their 3D orientation, persistence, frequency of joints and RQD, aperture, roughness, joint wall strength, water inflow etc. The guidelines for the collection of these joint data are provided by ISRM (Brown, 1981). This data may be subsequently used in rock mass classification criteria (i.e. RMR, Q or GCI etc) that assign properly weight factors to each joint set property measured in the field in order to derive some rating of the rock mass quality.

From this information a geological-discretized 3D model can be created (Figure 3). As it is illustrated in Figure 3 a given geological material can be assigned to each grid cell. In a subsequent step this material is then assigned a material law and parameters.

2.2 *Material models*

Material models for rock can be divided into elastic and elasto (visco-) plastic models. Elastic models do not consider irreversible deformation and failure (i.e. they do not consider any dissipation function). Elastic models can be isotropic or anisotropic. The first one only needs two parameters (E,v) and the second one (in the case of a stratified or transversely isotropic material) 5 parameters $(E_1, v_1, E_2, v_2, G_{12})$. The elastic deformation modulus of the rock mass is determined by "upscaling" from laboratory specimens considering the size effect and the geological conditions.

However elastic models are not very realistic for modelling the behaviour of the rock mass since failure and in-elastic deformations play a significant role. Basically two types of models can be used for describing the inelastic behaviour of rock masses: Elasto-plastic and visco-plastic models. The latter model describes the behaviour of rock masses in a more realistic way since it considers that the non-linear deformation process takes time. Indeed, in the NATM the timing of the installation of ground support is some importance, since in this way the amount of excavation loads that is carried by the rock mass itself through arching and by the supports can be controlled. The problem with using a visco-plastic material law is that it is not easy to determine the parameter that controls the time dependent behaviour (viscosity parameter).

However, it is not only important to consider the properties of the undisturbed rock mass but also the change in properties during the deformation process. For example the properties of joints can significantly change as asperities are sheared off after failure and during in-elastic deformation. These phenomena can be modelled by evolution laws (for example strain-hardening/softening behaviour) or using damage mechanics.

In most cases the behaviour of the rock mass is affected by the presence of joints and is therefore essentially anisotropic. In many cases, especially when there are a large number of joint sets,

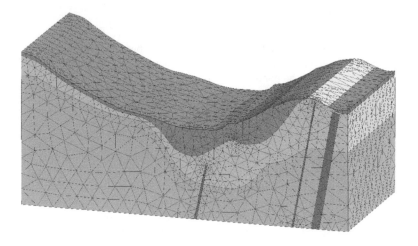

```
Flac3D>print zone information
        Zone Information ...
  ID   Type  Model        Group             Centroid
 ------ ----- ----------- ------  ----------- ----------- -----------
      1 Wedge Undefined    marble ( 1.510e+002, 5.372e+003, 9.203e+001)
      2 Brick Undefined    marble ( 1.510e+002, 5.372e+003, 9.208e+001)
      3 Brick Undefined    marble ( 1.510e+002, 5.371e+003, 9.213e+001)
      4 Brick Undefined    marble ( 1.510e+002, 5.370e+003, 9.218e+001)
      5 Brick Undefined    marble ( 1.510e+002, 5.369e+003, 9.224e+001)
      6 Brick Undefined    marble ( 1.510e+002, 5.368e+003, 9.229e+001)
```

Figure 3. Geological block model on a mesh with tetrahedral elements and data assigned to grid points (example of GID pre-processor ready for implementation into a finite difference or finite element code).

however the material behaviour is approximated as isotropic. The most popular failure conditions that are used in rock mechanics are: Mohr-Coulomb, Hoek and Brown and the multi-laminate model (Beer 2001). Only the last one considers the anisotropy of the rock mass.

The approach taken in the integrated project TUNCONSTRUCT is to develop a hierarchical model for the rock mass. This basically means that the model is chosen according to the amount of information available. For example if it is apparent that the rock mass is anisotropic but there is not enough information available on the properties of joint sets or there are too many joint sets to consider then an isotropic model is used. The number of parameters that are required range from 3 to 27.

Figure 4 shows the hierarchy of rock models. The simplest model is the Mohr-Coulomb model which requires only 3 parameters (cohesion, angle of internal friction, dilation angle). The most sophisticated model is the single surface model with hardening/softening behaviour and residual parameters and up to 3 families of discontinuities, requiring up to 27 parameters.

2.3 Determination of rock mass model parameters

After a model is selected it is necessary to determine the parameters for the model at the scale of the project (i.e. m, 10 m, 100 m and so on). Considering the information available at design stage outlined in section 2.1. we proposed the following approach.

2.3.1 Size effect, intact rock mass

Before considering the defects or discontinuities that are added to the rock mass as we move from the lab scale to the real scale of the problem, one must take into account the "size effect" exhibited by the brittle or quasi-brittle rocks. Size effects is a subject of increasing interest due to the fact that current applications in modern technology involve a variety of length scales ranging from a few

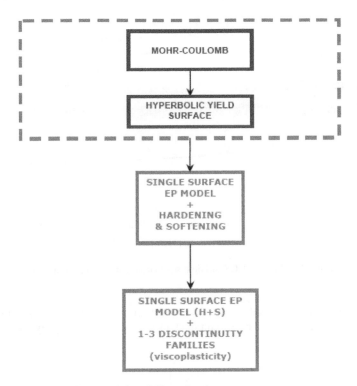

Figure 4.   Hirarchial concept for material modelling of rocks.

centimeters (sheet metal forming) down to few nanometers (thin film technology). This range of scales and the corresponding necessity for modeling and experiment have revealed that there is a strong connection between the various length scales involved and that the response may depend on size for otherwise dimensionally similar specimens. The size effect of intact rock strength is well known in the mechanics of brittle and quasi-brittle materials such as rocks and may be attributed to one of the following effects:

(a) the failure of rocks in uniaxial compression in the form of spalling cracks with subsequent buckling of the individual rock slender columns that are subsequently fail by buckling (Bazant et al., 1993)
(b) the transformation of stored volume deformation energy into fracture surface energy for the creation of new cracks (Bazant and Chen, 1997);
(c) the fractal character of crack surfaces produced at rock failure (Bazant and Chen, 1997);
(d) the stress or strain gradient effect (e.g. Aifantis, 1999; Exadaktylos and Vardoulakis, 2001).

In the simplest case of scale- or size-dependent response, one may envision the interaction between the geometric length of the specimen/externally applied load and an internal length/internal force associated with the underlying microstructure (e.g. grain, microcrack, pore etc.). The interaction between macroscopic and microscopic length scales in the constitutive response and the corresponding interpretation of the associated size effects may be modeled through the introduction of higher order stress or strain gradients in the respective constitutive equations. In turn, it is quite interesting that one may find a statistical interpretation of the stress or strain gradients based on the stochastic character of stresses and strains due to heterogeneity of the Representative Elementary Volume (REV) (Frantziskonis and Aifantis, 2002). The experimentally observed size effect exhibited by the Uniaxial Compression Strength (UCS) of intact marble and sandstone specimens with the same height:diameter ratio equal to 2, is illustrated in Figure 5.

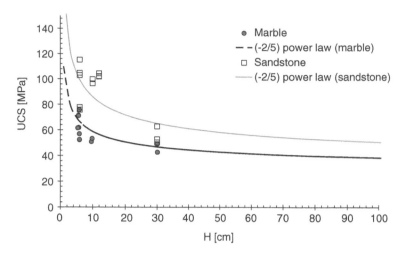

Figure 5. Size-effect exhibited by the UCS of Dionysos marble and Serena sandstone (Stavropoulou et al., 2006).

This size effect may be described by any of the models discussed above. For illustration purposes the mathematical model proposed by Bazant et al. (1993) – although slightly corrected – may be employed i.e.

$$UCS = UCS_d + C_1 H^{-2/5}; C_1 = \left(\frac{\sqrt{5}\left(2^{3/5}3^{4/5}\right)}{6} \right) \left(\pi^2 E' K_{IC}^4 \right)^{1/5} \tag{1}$$

wherein $E' = E/(1 - v^2)$, K_{IC} is the fracture toughness of the rock necessary for the formation of new mode-I cracks, H is the height of the specimen, and UCS_d is the ultimate UCS of the rock for sufficiently large height d of the rock (REV) with the geometrical similarity preserved (i.e. shape and height:diameter ratio). The above size effect law has been found to be valid for a marble and sandstone as it is displayed in Figure 5. Also there is experimental evidence that this inverse 2/5 relationship has been found to hold true in borehole breakouts (Papanastasiou, 2007).

By putting in the above equation (1) the representative elastic and fracture toughness properties of a given rock the constant C_1 may be calculated and then the mean UCS_d of the REV of this rock can be found from the respective representative value determined at the lab scale (i.e. most frequently on specimens of 10 cm height and with height to diameter ratio of 2:1), e.g.

$$UCS_d = UCS_{10} - C_1 10^{-2/5} \tag{2}$$

whereas UCS_{10} in MPa is the value of the UCS determined from the laboratory tests on specimens of height 10 cm. For example for marble with $E = 67\,GPa$, $v = 0.3$ and $K_{IC} = 0.6\,MPa \cdot \sqrt{m}$, $C_1 = 84.8\,MPa \cdot cm^{2/5}$ and for sandstone with $E = 30\,GPa$, $v = 0.1$ and $K_{IC} = 1.5\,MPa \cdot \sqrt{m}$, $C_1 = 147.8\,MPa \cdot cm^{2/5}$. From the above data it may be observed that the ratio of UCS_d/UCS_{10} varies between $0.3 \div 0.5$. Hence, in the absence of lab tests at various scales in the lab the lower value of $UCS_d/UCS_{10} \cong 0.3$ may be employed to be representative for the intact rock UCS for design purposes. It is expected that the Uniaxial Tensile Strength (UTS) exhibits also a similar or even more pronounced size effect. However, as the confining pressure increases the size effect should be depressed and this phenomenon must somehow considered.

2.3.2 Size effects, discontinuities
Next, we turn to the problem of the effect of discontinuities in the rock mass. First, it may be noted that fractal laws have been proven successful when the geologic mechanisms causing the studied

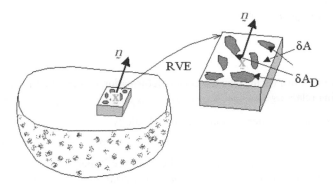

Figure 6. Representative Elementary Volume (REV) of damaged rock due to discontinuities.

characteristics were similar in both scales. When the large-scale discontinuities were formed by different mechanisms, e.g. thermal stresses locally and tectonic globally, the fractal approach has found limited success.

The starting hypothesis we make here is that the various forms of defects present in the rock mass alter the mechanical properties of the intact rock. These defects in general may be classified to cracks and inclusions (including cavities without any filling material). Therefore the most appropriate apparatuses one could employ to derive a model at the scale of the project are Fracture Mechanics and Continuum Damage Mechanics. Although it may be shown that the two theories are equivalent, the most appropriate to start with for practical purposes, is the latter, since the former is based on not easily comprehensible parameters such as the Stress Intensity Factor and so on.

So, a possible upscaling methodology may be based on two hypotheses, namely:

Hypothesis A: In a first approximation upscaling may be based on fundamental Damage Mechanics Principles through the use of the scalar damage parameter D or tensorial damage parameter $\mathbf{D} = D \cdot \mathbf{n}$ *for the general anisotropic case of joint induced anisotropy of the rock mass (* $\mathbf{n}$ *is the unit normal vector).*

It is noted that discontinuities may also alter the constitutive law of the rock during the transition from the lab scale to the full scale, i.e. they may induce anisotropy or they induce nonlinearity. Let us assume for the sake of simplicity that there are more than three joint systems in the rock mass so that it behaves like an isotropic material (the case of anisotropic geomaterial may be easily considered through appropriate tensorial analysis). If the area δA with outward unit normal n_j of the representative Elementary Volume (REV) with position vector x_i of Figure 6 is loaded by a force δF_i the usual apparent traction vector $\sigma_i = \sigma_{ij} n_j$ is

$$\sigma_i = \lim_{\delta A \to S} \frac{\delta F_i}{\delta A}, \quad i = 1,2,3 \tag{3}$$

where S is the representative area of the REV. The value of the dimensionless scalar damage quantity $D(n_i, x_i)$ (function of orientation and position of the surface) may be defined as follows

$$D = \frac{\delta A_D}{\delta A} \tag{4}$$

where δA_D is the Area of the discontinuities.

At this point we may introduce an "effective traction vector" $\sigma_i^{(e)}$ that is related to the surface that effectively resists the load, namely

$$\sigma_i^{(e)} = \lim_{\delta A \to S} \frac{\delta F_i}{\delta A - \delta A_D}, \quad i = 1,2,3 \tag{5}$$

From relations (4) and (5) it follows that

$$\sigma_i^{(e)} = \frac{\sigma_i}{1-D}, \quad i = 1,2,3 \tag{6}$$

According to the above definitions the elastic deformation of the intact rock can be described with the following relations

- The relation $\sigma_{ij}^{(e)} \rightarrow \varepsilon_{ij}^{(el)}$ which is obtained from elasticity
- The relation $\sigma_i \rightarrow \sigma_i^{(e)}$ which is obtained by employing the concept of damage (Lemaitre, 1992).

The above considerations lead to the "Strain Equivalence Principle" (Lemaitre, 1992), namely: *"Any strain constitutive equation for a damaged geomaterial may be derived in the same way for an intact geomaterial except that the usual stress is replaced by the effective stress"*.

The above principle may be applied on a linear isotropic elastic – perfectly plastic geomaterial obeying the *hyperbolic* Mohr-Coulomb yield criterion (which is a Griffith-type failure criterion accounting for the usual fact that the tensile strength of rocks is much lower than that predicted by the Mohr-Coulomb theory). For this type of geomaterial the relevant parameters will be given by the relations

$$\tilde{E} = E \cdot (1-D), \quad \frac{\tilde{v}}{\tilde{E}} = \frac{v}{E},$$

$$F\left(p, \sqrt{J_2}, \ \theta, \phi, (1-D) \cdot c_d, (1-D) \cdot p_{Td}, ...\right) = 0 \tag{7}$$

where the curly overbar indicates *equivalent* mechanical properties of the rock mass, the second of the relationships is simply the application of Castiliagno's theorem, F denotes the yield function, $p, \sqrt{J_2}, \theta$ are the 1st, 2nd deviatoric stress invariants and Lode's angle ($\theta = 0$ for triaxial extension and $\theta = \pi/3$ for triaxial compression conditions), respectively, ϕ is the internal friction angle, c_d and p_T are the cohesion and the tension limit (strength of the rock under hydrostatic extension), respectively at the large scale taking into account the size effect, E is the Young's modulus and v the Poisson's ratio of the intact rock. Figure 7 shows how the failure line is affected by the damage of the rock mass.

It is important to notice that both (1) the nature (constitutive behavior and parameters of the intact rock) and (2) damage state as well as possible change of the mechanical behavior due to the topology and properties of the inherited discontinuities of every geological material are considered in this way by the numerical model; the first one through the assignment to every element of the lithology through the combined geological-numerical model (Figure 3), and the second through the damage state variable D (Figure 7).

In Rock Engineering the effect of discontinuities present in rock mass is usually taken into account in a quantitative way by Classification Systems. The three most widely used rock mass classifications are the Rock Mass Rating (*RMR*) (Bieniawski, 1976; 1989), the *Q system* (Barton et al., 1974) and the GCI system (Hoek, 1994 and Hoek et al., 1995). The first two methods incorporate geological, geometric and design or engineering parameters in arriving at a quantitative value of their rock mass quality. The similarities between *RMR* and *Q* stem from the use of identical, or very similar, parameters in calculating the final rock mass quality rating. The differences between the systems lie in the different weightings given to similar parameters and in the use of distinct parameters in one or the other scheme. Bieniawski (1976) published the details of a rock mass classification called the Rock Mass Rating (*RMR*) system. Over the years, this system has been successively refined as more case records have been examined and Bieniawski has made significant changes in the ratings assigned to different parameters. The discussion which follows is based upon the 1989 version of the classification (Bieniawski, 1989). The following six parameters are used to classify a rock mass using the *RMR* system:

1. Uniaxial compressive strength of rock material.

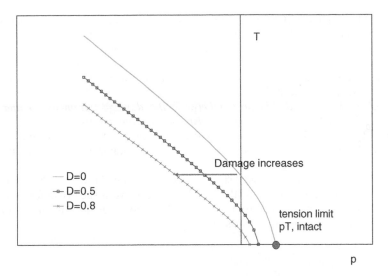

Figure 7. Schematic representation of the hyperbolic Mohr-Coulomb criterion applied to damaged geoma-terial in the T-p plane ($T = \sqrt{J_2}, \theta = \pi/3$) (compressive stresses are taken as negative quantities here).

2. Rock Quality Designation (RQD).
3. Spacing of discontinuities.
4. Condition of discontinuities.
5. Groundwater conditions.
6. Orientation of discontinuities.

On the basis of an evaluation of a large number of case histories of underground excavations, Barton et al., (1974) of the Norwegian Geotechnical Institute proposed a Tunnelling Quality Index (Q) for the determination of rock mass characteristics and tunnel support requirements. The numerical value of the index Q varies on a logarithmic scale from 0.001 to a maximum of 1000 and is defined by

$$Q = \frac{RQD}{J_n} \cdot \frac{J_r}{J_a} \cdot \frac{J_w}{SRF}$$

where J_n is the joint set number, J_r is the joint roughness number, J_a is the joint alteration number, J_w is the joint water reduction factor and SRF is the stress reduction factor. In explaining the meaning of the parameters used to determine the value of Q, Barton et al., (1974) offer the following comments: The first quotient (RQD/J_n), is a rough measure of the block or particle size. The second quotient (J_r/J_a) represents the roughness and frictional characteristics of the joint walls or filling materials. The third quotient (J_w/SRF) consists of two stress parameters, namely SRF *that* is a measure of: 1) loosening load in the case of an excavation through shear zones and clay bearing rocks, 2) rock stress in competent rock, and 3) squeezing loads in plastic incompetent rocks. It can be regarded as a total stress parameter. The parameter J_w is a measure of water pressure, which has an adverse effect on the shear strength of joints due to a reduction in effective normal stress. It appears that the rock tunnelling quality Q can be considered to be a function of only three parameters which are crude measures of:

1. Block size (RQD/J_n)
2. Inter-block shear strength (J_r/J_a)
3. Active stress (J_w/SRF)

Geological Strength Index (GSI) was proposed as a replacement for Bieniawski's RMR (Hoek, 1994 and Hoek et al., 1995). It had become increasingly obvious that Bieniawski's RMR is difficult to apply to very poor quality rock masses and also that the relationship between RMR and *m* and *s* *parameters employed in the Hoek and Brown failure criterion* is no longer linear in these very low ranges. It was also felt that a system based more heavily on fundamental geological observations and less on "numbers" was needed.

Hypothesis B: A universal relationship between the damage parameter D and RMR (or equivalently with Q or GCI or may be other) exists for all geomaterials.

This hypothesis is based on the fact that RMR or Q or GCI do take into account explicitly the discontinuities of the rock mass that deteriorate the rock mass parameters, hence they may be linked with the damage state parameter D.

Such a function must have a sigmoidal shape resembling a cumulative probability density function giving D in the range of 0 to 1 for RMR or GCI varying between 100 to 0 or for Q varying from 1000 to 0.001, respectively. On the grounds of the above considerations the normal (Gaussian) cumulative probability function is proposed, namely

$$D = 1 - \frac{1}{\sigma\sqrt{2\pi}} \int_0^{RMR./100} \exp\left(-\frac{1}{2}\left[\frac{x-\mu}{\sigma}\right]^2\right) dx \tag{8}$$

where μ, σ are the mean value and the standard deviation of the normal p.d.f., respectively, and the RMR does not include the correction term due to unfavorable tunnel orientation with respect to joints and the grounbd water (hence RMR considers only the joints, i.e. RMR_{89}). The first parameter depicts the value of RMR that corresponds to damage D = 0.5 and the second the spread of the function around D = 0.5. The above integral is the well-known "error function" that is built in Excel for example, and many other spreadsheets, so it is easy to be calculated. As it is illustrated in Figure 8 the two unknown parameters of this function were found from published data by Hoek and Brown (1997) and Ramamurthy and Arora (1994) on $UCS_m/UCS_d = 1 - D$ with UCS_m being the rock mass UCS. Based on these data $\mu \cong 0.6$ and $\sigma \cong 0.2$. It may be observed from this figure that for RMR = 100 i.e. for intact rock the "integrity parameter" which is equal to 1-D, is not equal to 1 but to 0.3, which is the ratio of the large scale UCS_d over the lab scale UCS_{10} (i.e. UCS_d/UCS_{10}).

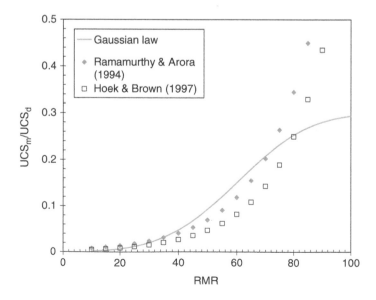

Figure 8. Calibration of the Gaussian damage law on existing large scale testing data.

To compare the above upscaling theory we consider the data presented by Cai et al. (2004). In that paper the GSI system is applied to characterize the jointed rock masses at two underground powerhouse cavern sites in Japan. GSI values are obtained from *block volume* and *joint condition factor*, which in turn are determined from site construction documents and field mapping data. Based on GSI values and intact rock strength properties, equivalent Mohr–Coulomb strength parameters and elastic modulus of the jointed rock mass are calculated and compared to in situ test results. The GSI system has been developed and evolved over many years based on practical experience and field observations. GSI is estimated based on geological descriptions of the rock mass involving two factors, rock structure or block size and joint or block surface conditions. Although careful consideration has been given to the precise wording for each category and to the relative weights assigned to each combination of structural and surface conditions, the use of the GSI table/chart involves some subjectivity. Hence, long-term experiences and sound judgment is required to successfully apply the GSI system. To overcome these difficulties, a different approach building on the concept of block size and conditions, namely, the idea of *block volume* and *joint condition factor* is proposed in the aforementioned paper.

Block size, which is determined from the joint spacing, joint orientation, number of joint sets and joint persistence, is an extremely important indicator of rock mass quality. Block size is a volumetric expression of joint density. In the cases that three or more joint sets are present and joints are persistent, the volume size can be calculated as

$$V_b = \frac{s_1 s_2 s_3}{\sin \gamma_1 \sin \gamma_2 \sin \gamma_3} \tag{9}$$

where s_i and γ_i (i = 1,2,3) are the joint spacing and the angle between joint sets, respectively. The above factor is appropriately also modified to take into account the persistence of each joint set. The joint condition factor, J_C is also defined as

$$J_c = \frac{J_W}{J_S J_A} \tag{10}$$

where J_W and J_S are the large-scale waviness (in meters from 1 to 10 m) (Palmstrom, 1995) and small-scale smoothness (in centimeters from 1 to 20 cm) (Palmstrom, 1995) and J_A is the joint alteration factor. From the above factors and the GCI chart the GCI estimation may be easily performed.

The authors present lab and in situ test data as well as GSI data for two projects. The first one is the Kannagawa pumped hydropower project located in Gumma Prefecture in Japan. The rock mass at the site consists of conglomerate, sandstone, and mudstone. The percentage of conglomerate in CG1 and CG2 rock mass zones are about 93% and 62%, respectively, and FSI, MI stand for sandstone and mudstone formations. The second one is the Kazunogawa power station in which the rock mass consists of sandstone and composite rock of sandstone and mudstone, described as two groups (C_H and C_M) of rock mass types. Table 1 presents the data for these two projects. UCS has been estimated from laboratory tests, m_i is the parameter used in the Hoek-Brown failure criterion and has been estimated from lab triaxial compression tests. The equivalent Mohr-Coulomb cohesion and internal friction angle of each rock type has been estimated and presented in Table 1 below. GCI for each rock formation has been also estimated through the use of the two indices V_b and J_C defined above. Unfortunately the authors did not present any data for the Young's modulus from laboratory tests. For this purpose we employ the following empirical relationship between Young's modulus and UCS derived from regression analysis of experimental data from a variety of rocks

$$E_d \cong 300 \cdot UCS \tag{11}$$

Also, from the empirical relationship $RMR_{89} = GCI + 5$ the damage parameter D can be estimated for each geological formation from the GCI value and formula (8). The estimations of the

Table 1. Characterization of the rock masses at the Kannagawa and Kazunogawa test sites using the GSI system and the damage theory.

	Geological formations											
	CG1	Test data	CG2	Test data	FSI	Test data	MI	Test data	C_H	Test data	C_M	Test data
UCS [MPa]	111.0		162.0		126.0		48.0		108.0		108.0	
m_i	22.0		19.0		19.0		9.0		19.0		19.0	
c [MPa]	19.6		29.1		22.6		9.9		19.4		19.4	
ϕ [deg]	50.8		49.5		49.5		41.8		49.5		49.5	
E [GPa][1]	33.5		48.9		38.0		14.5		32.6		32.6	
GCI	74.0		65.0		65.0		54.0		60.0		46.0	
D	0.2		0.3		0.3		0.5		0.4		0.7	
c_m [MPa][4]	5.0	5.2[2]	6.0	3.4	4.7	3.4	1.5	1.9	3.5	1.5	1.8	0.8
ϕ_m [deg][4]	50.8	57.0[2]	49.5	57.0	49.5	57.0	41.8	40.0	49.5	58.0	49.5	55.0
E [GPa][4]	28.2	45.3 (6.2)[3]	33.8	33 (3.6)	26.3	24.4 (2.5)	7.2	11.7 (1.7)	19.5	12.9 (2.84)	10.0	7.9 (1.2)

[1] Estimated from formula (11)

[2] Estimated from in situ block shear tests.

[3] Estimated from in situ plate bearing tests. Number in parenthesis indicated the standard deviation.

above Damage Mechanics theory are illustrated in Table 1 in order to be compared with in situ measurements performed on large scale specimens from these geological formations. From the same table it may be seen that the predicted values are in reasonable agreement with the large scale measurements.

3 SENSITIVITY AND BACKANALYSIS

Given the uncertainty of the material parameters and the virgin stress field sensitivity analysis and back analysis play an important role. In the sensitivity analysis it is determined what effect the different parameters have on the results (predicted deformations/stresses). The result of this allows to concentrate on the determination of parameters that are important. If displacement measurements are available form a tunnel under construction or a pilot tunnel that these can be used to back analyze the material properties and the virgin stress field. In the case of a pilot tunnel a much greater confidence can be obtained in this way on the material parameters before the construction of the main tunnel commences. In the case of measurements obtained during construction one may continually update the parameters of the simulation model in order to obtain predictions for the next stages of excavation. Details of the procedures are outlined in the public deliverable (Grešovnik, 2007) available from the TUNCONSTRUCT web site.

4 INCLUSION OF GEOLOGY INTO A NUMERICAL MODEL

With respect to the inclusion of geological features into a numerical model we distinguish between continuum models and discontinuum models. In the case of continuum models there are methods

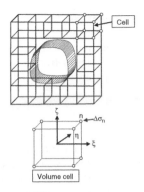

Figure 9. Definition of volume cells for the BEM.

based on domain discretisation such as Finite Element (FEM) and Finite Difference (FDM) Methods and those based on boundary discretisation such as the Boundary Element Method (BEM). We will discuss here only the inclusion into continuum models further. For the theoretical background of the BEM please consult Beer, 2001.

The definition of material parameters into an FEM or FDM model is fairly straight forward as each Element (or indeed each Gauss Point inside an Element) can be assigned different material properties. Since the BEM has a boundary discretisation only and no elements exist in the rock mass the inclusion of geological conditions requires further elaboration. As outlined in more detail in Beer & Dünser, 2007 the idea is to use internal cells for this purpose (see Figure 9). These cells actually look like finite elements but are significantly different because no additional degrees of freedom are introduced as the cells are only used to integrate the volume terms that arise. The internal cell mesh has the advantage that it can be unstructured (i.e. cells do not have to connect at nodal points as they have to do in the FEM) so that cell meshes are very easy to construct. The proposed procedure is to provide the "upscaling software" with coordinates of points and to receive the material law and parameters. Further details can be found in the public deliverable of TUNCONSTRUCT (Exadaktylos, 2006).

5   CONCLUSIONS

In this paper we have presented a possible concept on how information that is available about the rock mass can be converted into data that are suitable for a numerical model. The ideas are currently being discussed within the integrated EC project Technology Innovation in Underground Construction (TUNCONSTRUCT, www.tunconstruct.org).

Linking the geological conceptual model assembled from geological mapping, core drilling investigation and other measurements (e.g. geophysical data, hydrogeology data etc.), with the numerical model in the 3D space is essential for more realistic modeling of underground excavations in rock masses.

The above approximate theory for the estimation of model parameters is considered very useful in rock engineering, since:

(i) It is obvious that the "engineering judgment" plays a role in the process of assessing the damage of the rock mass.
(ii) Rock mass characterization systems are placed in a "mechanics of materials" context since they are linked with the damage parameter which is an *internal state variable* and central in Damage Mechanics Theory.
(iii) The damage parameter that is planned to be incorporated in the constitutive rock models in the BEM/FEM numerical simulation code is linked with empirical approaches, which

are faster although much less accurate, based on rock mass classification systems for rock support measures.

(iv) RMR, GCI or Q or other discontinuity or defect parameter that quantifies the rock mass structure and sampled from the boreholes or from rock mass exposed surfaces, such as for example the joint factor proposed by Ramamurthy and Arora (1994) defined as

$$J_f = \frac{FF}{n \cdot J_s},$$

where FF is the fracture frequency, n is a joint inclination factor depending on the orientation of the joint, and J_s is a joint strength factor, respectively, may be interpolated or extrapolated in 3D space by virtue of estimations from boreholes or rock exposures and the block Kriging technique (Stavropoulou et al., 2006). Hence, damage parameter spatial distribution is also evaluated based on relationship (8) that links Damage with RMR.

(v) The correlation of such rock mass damage measures with seismic P or S-wave velocities with geophysical methods may be also exploited for large scale invasive geophysical methods, such as for example the correlation proposed by Barton (2002), $V_P \approx 3.5 + \log_{10} Q$.

(vi) The drawback of rock mass classification systems approach, namely that complex properties of a rock mass cannot be satisfactorily described by a single number, is surpassed with this approach.

(vii) Also the drawback that the same rating can be achieved by various combinations of classification parameters, even though the rock mass behavior could be different, is also annihilated because we may choose different consitutive law based on experimental evidence.

(viii) The other drawback that the user is led directly from the geological characterization of the rock mass to a recommended ground support without the consideration of possible failure modes is also overcome.

Benchmark tunnel sites will be considered for the application of this type of analysis. The work performed in the project will lead to realistic results and knowledge on the applicability of the proposed methodology on certain problems. These benchmark cases will indicate also the applicability of the proposed upscaling relations for the deformability and strength of rock masses based on already existing rock mass indices (RMR, Q or other) and the Damage Mechanics theory.

ACKNOWLEDGEMENTS

The authors would like to thank the financial support from the EC 6th Framework Project TUN-CONSTRUCT (Technology Innovation in Underground Construction) with Contract Number: NMP2-CT-2005-011817.

REFERENCES

Aifantis, E.C. (1999): Strain gradient interpretation of size effects. International Journal of Fracture 95, 299–314.

Barton N, Lien R, Lunde J. Engineering classification of rock masses for the design of tunnel support. Rock Mech 1974;6(4):189–236.

Barton N. (2002): Some new Q-value correlations to assist in site characterization and tunnel design. Int. J. Rock Mech. & Min. Sci. 39, 185–216.

Bazant, Z. P. and Chen, E-P. (1997): Scaling of structural failure. *Appl. Mech. Rev.* 50 (10), 593–627.

Bazant, Z. P., Lin, F-B., and Lippmann H. (1993): Fracture energy release and size effect in borehole breakout, *Int. J. for Numer. & Analyt. Meth. In Geomechanics* 17, 1–14.

Beer, G. (2001): Programming the boundary element method, J. Wiley.

Beer, G. and Duenser Ch. (2007): Advanced numerical simulation of the tunnel excavation/construction process with the boundary element method, ISRM 2007 Congress, Lisbon.

Bieniawski, Z.T. 1976. Rock mass classification in rock engineering. In *Exploration for Rock Engineering, Proc. Of the Symp.*, (ed. Z.T. Bieniawski), 1, 97–106. Cape Town, Balkema.

Bieniawski Z.T. (1989): Engineering rock mass classifications. Wiley, New York.

Brown E.T. (ed.), Rock characterization, testing and monitoring: ISRM suggested methods, Pergamon Press, Oxford, 1981.

Cai, M., P.K. Kaiser, H. Uno, Y. Tasaka, M. Minami, Estimation of rock mass deformation modulus and strength of jointed hard rock masses using the GSI system, International Journal of Rock Mechanics & Mining Sciences 41 (2004) 3–19.

Exadaktylos G. E. and Vardoulakis I. (2001): Microstructure in Linear Elasticity and Scale Effects: A Reconsideration of Basic Rock Mechanics and Rock Fracture Mechanics. *Tectonophysics*, 335, Nos. 1–2, 81–110.

Exadaktylos G. E. (2006) Specifications for interface between geostatistical block model, UCIS and simulation models. Public deliverable TUNCONSTRUCT.

Frantziskonis, G., Aifantis, E.C. (2002): On the stochastic interpretation of gradient-dependent constitutive equations. European Journal of Mechanics A/Solids 21, 589–596.

Grešovnik I., (2006): Specifications for software to determine sensitivies for optimization of the design of underground construction as part of IOPT, Public deliverable TUNCONSTRUCT.

Hoek, E. 1994. Strength of rock and rock masses, *ISRM News Journal,* **2**(2), 4–16.

Hoek, E., Kaiser, P.K. and Bawden. W.F. 1995. *Support of underground excavations in hard rock.* Rotterdam: Balkema

Hoek E, Brown ET. Empirical strength criteria for rock masses. J Geotech Engng 1980;106:1013 ± 35.

Hoek E, Brown ET. Practical estimates of rock mass strength. Int J Rock Mech Min Sci 1997;34(8):1165–86.

Lemaitre J., 1992, *A Course on Damage Mechanics.* Springer-Verlag, Berlin.

Palmstrom, A. RMi—a rock mass characterization system for rock engineering purposes. PhD thesis, University of Oslo, Norway, 1995.

Papanastasiou, P. 2007, Interpretation of the scale effect in perforation failure Bifurcations, Instabilities, Degradation in Geomechanics, George E. Exadaktylos & Ioannis G. Vardoulakis (Eds.), Springer, in press.

Ramamurthy T. and Arora V.K., Strength predictions for jointed rocks in confined and unconfined states, *International Journal of Rock Mechanics and Mining Sciences & Geomech. Abstr., Vol. 31, No. 1, pp. 9–22, 1994.*

Stavropoulou M., G. Exadaktylos, and G. Saratsis, A Combined Three-Dimensional Geological-Geostatistical-Numerical Model of Underground Excavations in Rock, Rock Mech. Rock Engng. (2006).

Applications of Computational Mechanics in Geotechnical Engineering – Sousa,
Fernandes, Vargas Jr & Azevedo (eds)
© 2007 Taylor & Francis Group, London, ISBN 978-0-415-43789-9

Numerics for geotechnics and structures. Recent developments in ZSoil.PC

Th. Zimmermann[1,2], J.-L. Sarf[1,3], A. Truty[2,4] & K. Podles[2,4]

[1] *LSC-EPFL, Lausanne, Switzerland*
[2] *Zace Services Ltd, Lausanne, Switzerland*
[3] *BET J.L.Sarf, Aubonne,Switzerland*
[4] *Cracow University of Technology, Cracow, Poland*

ABSTRACT: Civil engineers are more and more often confronted with constructions in urban environment. It is then essential to be able to accurately assess the initial state, and the evolution of the safety during construction stages. While traditional geotechnical engineering has been using a family of simplified approaches for different problems, in association with two-dimensional modeling, the approach adopted herein is integrated, coupling all aspects of the problem, and 3-dimensional, thus offering a significant potential for optimization of constructions. Over the past 20 years, the authors have been developing numerical tools for static and dynamic analysis of soil, rock, structures and flow with the aim of providing to civil engineers a tool which unifies computational geomechanics, structural mechanics and soil-structure interaction within a single theoretical framework, while remaining user-friendly and transparent for the user. Recent developments in Z_Soil.PC are presented in this paper, along with three-dimensional simulations which illustrate the potential of the tool.

1   INTRODUCTION

Figure 1 shows a three-dimensional view of a model for the simulation of the excavation of a subway tunnel, corresponding to an early stage of design. The image illustrates the excavation procedure, lateral tunnels with concrete liners excavated first, followed by the excavation of the roof. The presence of a protection umbrella composed of injected pipes is visible and the presence of a water table is indicated . The tunnel is located in urban area so that an accurate initial state evaluation is essential and no surface deflection can be tolerated. Structural components interact with the soil continuum, and underground flow must be taken into account. This situation is fairly typical of the construction of a subway tunnel in urban environment.

Constructing a computational tool which unifies computational geomechanics, in partially saturated media, and structures, in order to analyze problems like the one just described, presents a number of difficulties which must be overcome in a coherent manner. Among these, locking phenomena due incompressible behavior, pressure oscillations in 2-phase media, implicit seepage boundary conditions depending on the solution, compatibility of structural elements with continuum elements, etc.

We present in this paper some aspects of program (ZSOIL 1985–2007) related to two-phase geomechanics and to urban underground construction.

The theory is summarized first. Difficulties related to numerical modeling and proposed remedies are presented next. Finally test cases of a slope instability due to rain, and a consolidation test are presented, followed by a full size case study.

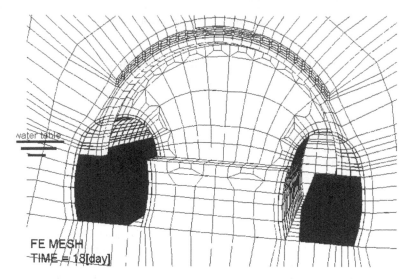

Figure 1. View of a 3D model.

2 GOVERNING EQUATIONS

2.1 Equations of two phase saturated media (ZSOIL 1985–2007, Truty & et al. 2006)

The overall equilibrium equation for the solid and fluid phases written in terms of the total stress, .

$$\sigma_{ij,j}^{tot} + b_i = 0 \tag{1}$$

$$b_i = \left(\gamma_{dry} + nS\gamma^F\right)\bar{b}_i \tag{2}$$

where n is the porosity and $S(p)$ the saturation ratio, 'γ_{dry} the unit weight of soil sample in dry state', γ^F the specific weight of interstitial fluid and $\bar{b}_i$ a cosin of gravity direction.

The extended effective stress concept after Bishop,

$$\sigma_{ij}^{tot} = \sigma_{ij} + \delta_{ij} S p \tag{3}$$

where $S p$ is the interstitial (partially) saturated fluid pressure.

The fluid flow continuity equation including partial saturation,

$$S \varepsilon_{kk} + \upsilon_{k,k}^F = c(p) \tag{4}$$

with $c(p)$ the rate of storage

$$c(p) = \left(n\frac{S}{K^F} + n\frac{\partial S}{\partial p}\right)\dot{p} \tag{5}$$

and K^F the fluid bulk modulus.

The linearized strain-displacement relations

$$\varepsilon_{ij} = \frac{1}{2}\left(u_{i,j} + u_{j,i}\right) \tag{6}$$

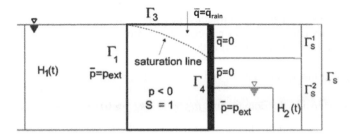

Figure 2. Boundary conditions of flow problem shown on soil sample elevation.

A nonlinear elasto-plastic constitutive relation for the solid phase, expressed in total integrated form,

$$\sigma_{ij} = D^e_{ijkl}\left(\varepsilon_{kl} - \varepsilon^P_{kl}\right) \tag{7}$$

An elastic perfectly plastic Mohr-Coulomb plasticity model is assumed.
The extended Darcy's law,

$$\upsilon^F_i = -k_r(S)k_{ij}\left(-\frac{p}{\gamma^F} + z\right)_{,j} \tag{8}$$

where k_r is a reversible function of S.

Constitutive equations for saturation ratio S, after (van Genuchten 1980), and relative permeability coefficient $k_r(S)$ (Aubry & et al. 1988) are defined as

$$S(p) = S_r + (1 - S_r)\left[1 + \left(\alpha\frac{p}{\gamma^F}\right)^2\right]^{-\frac{1}{2}} \tag{9}$$

$$k_r(S) = \left(\frac{S - S_r}{1 - S_r}\right)^3 \tag{10}$$

where (9) governs the evolution of saturation as a function of suction and (10) governs the influence of saturation on k_r and the permeability. S_r is the residual (minimum) saturation ratio, and α is a material parameter which controls the decrease of the saturation ratio with increasing pressure suction.

Boundary conditions to be satisfied at any time t in [0, T] are (see Figure 2)

$$\sigma^{tot}_{ij}n_j = \bar{t}_i \ on \ \Gamma_{t,j} \tag{11}$$

$$\upsilon^F_i n_i = \bar{q} \ on \ \Gamma_q = \Gamma_3 \tag{12}$$

$$u_i = \bar{u}_i \ on \ \Gamma_u \tag{13}$$

$$p = \bar{p} \ on \ \Gamma_p = \Gamma_1 \tag{14}$$

$\Gamma_t, \Gamma_q, \Gamma_u, \Gamma_p$ are parts of the boundary where the tractions, fluid fluxes, displacements and pore pressures are prescribed. A special treatment is required for so-called "seepage surfaces". As the point where the free surface intersects with the domain boundary is unknown a priori, none of the above boundary conditions can be set in an explicit form. This problem is illustrated in Figure. 2, in which a simple model of a permeable soil elevation is shown. On part Γ^1_S of the boundary Γ_S one should set a flux boundary condition $\bar{q} = 0$, while on part Γ^2_S a pressure boundary condition $\bar{p} = 0$

or $\bar{p} = p_{ext}$ should be set, depending on whether the point is above or below downstream water table (W.T.). This boundary condition can be enforced transiently through a penalty approach expressed $\bar{q} = -k_\upsilon(p - p_{ext})$ (Aubry & et al. 1988) with

$$\bar{q} = q_{imposed} \text{ if } p \geq 0 \text{ inside the domain and } \bar{p}_{ext} = 0 \quad (k_\upsilon = 0) \tag{15}$$

$$\bar{q} = k_\upsilon p \quad \text{if } p < 0 \text{ inside the domain and } \bar{p}_{ext} = 0 \quad (k_\upsilon \gg 0) \tag{16}$$

$$\bar{q} = k_\upsilon (p - p_{ext}) \quad \text{for any } p \text{ inside the domain if } \bar{p}_{ext} < 0 \quad (k_\upsilon \gg 0) \tag{17}$$

where k_υ is a penalty parameter
Initial conditions

$$u_i(t = t_0) = u_{i0} \tag{18}$$

$$p(t = t_0) = p_0 \tag{19}$$

2.2 Stabilized formulations for 2phase-partially saturated media

Most low order finite elements, applied to single-phase problems, exhibit severe locking phenomena when incompressibility or strong dilatancy constraints appear during plastic flow. The inappropriate handling of these constraints results in overestimated bearing capacities and strong stress oscillations. This defect can be eliminated by means of mixed formulations. A comprehensive review of these techniques can be found in textbooks by (Zienkiewicz & et al. 2000), (Hughes 1987), e.g.

In medium and large scale computations, especially in 3D, low order finite elements are very attractive due to their simplicity and numerical efficiency. In this class of elements BBAR and EAS (Enhanced Assumed Strain) four-node quadrilaterals and eight-node bricks seem to be the most robust ones to overcome locking phenomena related to incompressibility or dilatancy, typical of some soils. However, straightforward application of these elements to the problem of consolidation in two-phase fully or partially saturated media yields new deficiencies. It can be shown that finite elements, with the same interpolation order for both displacement and pore pressure fields may exhibit spurious spatial pressure oscillations. This was a reason for seeking new approaches.

Stabilized methods (Franca 1987), (Hughes 1989) have been found to be a powerful tool to circumvent spurious oscillation. The approach consists in adding to the standard Galerkin formulation least-squares term(s) based on the residual(s) of the Euler-Lagrange equation(s) of the problem at hand, integrated over each element, in order to enhance the stability of the resulting scheme.

The usefulness of such methods for a wide range of problems in computational mechanics has been established since the beginning: Stokes flow and incompressible elasticity in (Franca 1987), Timoshenko beam, fluid dynamics in (Hughes 1989) and a unifying multiscale framework generalizing the concept has been proposed by Hughes (1995).

Different approaches are commonly used. The first group consists of direct stabilization methods, based on the perturbation of the fluid mass conservation equation, following the general idea proposed by Brezzi & et al. (1984), Brooks & et al. (1982), Franca (1987), Hughes & et al. (1989), Hafez and Soliman (after Pastor & et al. 1997) and Zienkiewicz & et al. (1994). This is the approach adopted herein and described in details in (Truty & et al. 2006).

The effectiveness of the approach will be illustrated in the example of a one-dimensional consolidation test.

3 APPLICATIONS

3.1 One dimensional consolidation test

A one-dimensional consolidation test of a partially saturated soil column, of depth $H = 10$ m, is performed to verify the proposed formulations. The mesh, corresponding boundary conditions, load

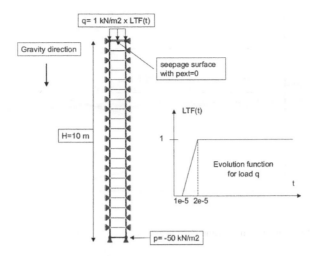

Figure 3.   One-dimensional test problem.

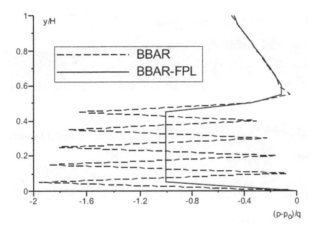

Figure 4.   Pressure difference profile a time $t = 2^{-5}$ year.

density and associated load time function *LTF(t)* are shown in Figure 3. The computation consists of two major steps. The initial state is generated first as a composition of two solutions, a steady state fluid flow problem $(div(v^F) = 0)$, followed then by a mechanical analysis, in which the computed pore pressures are treated as an explicit input. In the second step, the consolidation process, starting at $t = 10^{-5}$ [year], is activated and driven by the vertical load q. The following set of material properties is used: $E = E_{oed} = 10^4$ kPa, $v = 0$, $k_x = k_y = 10^{-4}$ m/year, $K^F = 2 \cdot 10^5$ kPa, $e_0 = 0.01$, $\gamma^F = 10\,kN/m^3$, $S_r = 0$, $\alpha = 2\,m^{-1}$. The initial time step is taken as $\Delta t_0 = 10^{-6}$ [year], which is far below the critical time step value defined by Eq. (20). In the range 10^{-5} [year] $\leq \Delta t \leq 2 \cdot 10^{-5}$ [year] a constant time step $\Delta t = 10^{-6}$ [year] was used and then a variable, based on the recurrence formula $\Delta t_{n+1} = 1.1\,\Delta t_n$ was used.

Standard formulation's are subject to a critical lower bound time step (Vermeer & et al. 1981):

$$\Delta t_{crit} = \frac{\gamma^F h^2}{E_{oed}\,\theta k_r\,k}\left[\frac{S^2}{4} + \frac{1}{6}E_{oed}\,c\right] \tag{20}$$

The profile of the pressure difference between the current pressure, at time $t = 2 \cdot 10^{-5}$ day, and the one in the initial state, is shown in Figure 4. One may notice that, in the zone of partial

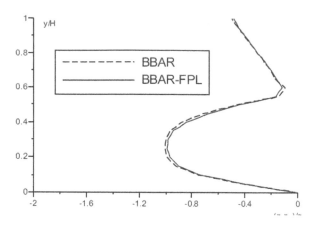

Figure 5. Pressure difference profile at time t = 4 years.

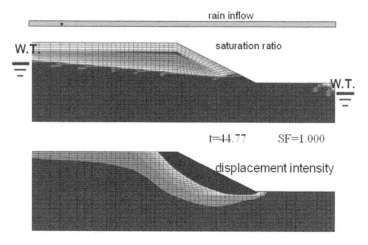

Figure 6. Failure of slope due to rain.

saturation, pressure oscillations do not appear for the standard Galerkin formulation and neither for the stabelized FPL form. This is so, because the effect of pressure oscillations, in the case of a compressible fluid, are eliminated not only by the time step size, but also through the spatial discretization. In the zone of full saturation spatial pressure oscillations are successfully eliminated by the stabilized FPL formulation. For larger times (see Figure 5) the differences between SG and FPL formulations are negligible (Truty & et al. 2006).

3.2 Slope instability due to rain

The simple case of a slope stability analysis under rain is presented next. The goal of this test is to illustrate flow and instability detection in ZSoil. The slope's characteristics are: friction angle (18°), cohesion (12 kN/m2), Darcy coefficient (0.1 m/h). Rain is characterized by an water inflow which is maintained until instability is detected. This inflow must be compatible with the Darcy flow coefficient, otherwise immediate surface saturation will occur.

Saturation of the domain will then progressively increase, starting at the wet surface, and the level of the water table will rise until an instability is detected.

Stability of the slope is evaluated after each time step and decreases with time. The safety factor is computed using a (C, φ) reduction algorithm. The advantage of this approach is that the failure

108

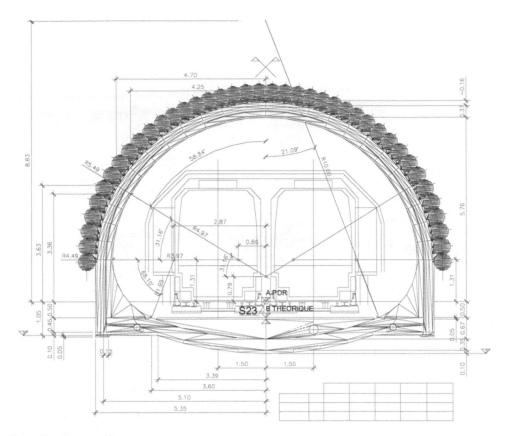

Figure 7. Cross-section.

surface is defined automatically by the program when localization of strain occurs; there is no need to make any assumption on the failure surface. For soils characterized by more complex plasticity theories, more general algorithms are available in ZSoil, which also identify the failure surface automatically.

The upper figure illustrates the rain inflow, the soil saturation ratio increases starting at the free surface, the lower figure illustrates displacement intensities, at the onset of instability, with no other loads than gravity and rain inflow. The figure gives a clear identification of the slope instability, automatically identified by the algorithm, here after 44.77 days of rain.

3.3 *Case study: the M2 subway extension in Lausanne*

The project of extension of the subway system in Lausanne led to extensive numerical simulations, most of them done with ZSOIL.PC. A segment of tunnel of about 75 m is analyzed here. The goal of the analysis is to assess underground flow, stress resultants in structural components, front stability, and critical displacements.

The main assumptions are: elasto-plastic soil behavior according to Mohr-Coulomb criterion, elasto-plastic behavior of the arch-pipe umbrella, elastic behavior of structures and concrete liner.

The geometry of the cross-section is given in figure 7

Figure 8 shows the main components of the simulation; symmetry with respect to a vertical plane through the tunnel axis is assumed.

Excavation steps are simulated according to the true excavation schedule, starting with an initial state evaluation (corresponding to a stressed undeformed state).

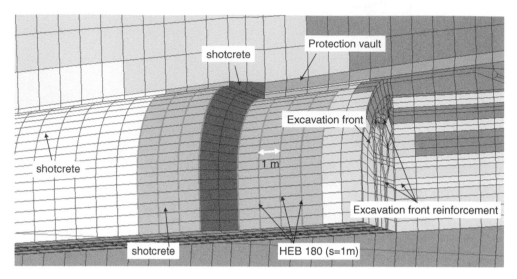

Figure 8.   Main components of the problem.

Table 1.   Material parameters.

n°	Legand	E Mpa		MN/m³	c MPa	°	kx = ky = kz m/s
1	Fissured rock	1000	0.3	0.025	elastic		impermeable
2	Marls	200	0.3	0.024	0.4	25	impermeable
3	Weathered rock	40	0.3	0.022	0.04	31	1.00E−07
4	Weathered rock	35	0.3	0.022	0.02	31	1.00E−07
5	Sand	28	0.3	0.021	0.005	31	1.00E−05
6	Sand	28	0.3	0.021	0.005	31	1.00E−06
7	Protection vault	1000	0.2	0.023	1.5	30	1.00E−08
8	Shortcrete	20,000	0.2	0.024			impermeable
9	Steel girders	210,000	0.2	0.078			
10	Shotcrete excavation front	5,000	0.2	0.024			

Material parameters are specified in Table 1, corresponding to the geometrical distribution given in figure 9.

The water table is located approximately 25 m. above the bottom slab of the tunnel. As a consequence, a drainage effect results when the tunnel penetrates into permeable soil, as illustrated in figure 10. Consequences on front stability are of course possible, and actually an incident occurred, but in another segment of the tunnel.

The front stability is improved by the use of reinforcement columns and a thin concrete liner, which is neglected in the safety assessment.

Figure 11 shows plastified zones at the tunnel front. At this point of the excavation time (t = 8), a safety factor of 1.18 is estimated at the front. The observed instability is however localized next to the roof and can be stabilized.

After this local stabilization, the front has a larger safety factor value of 1.62 at time 8, illustrated in figure 12.

Maximum deflections at the tunnel roof are plotted in figure 13, and a maximum value of 2.9 cm is observed, corresponding to a maximum surface deflection of 7 mm. Plotted displacements corresponding to increasing safety factor values (SF = 1.2 to 1.6) do not correspond to a physical state but they are indicators of where failure will initiate.

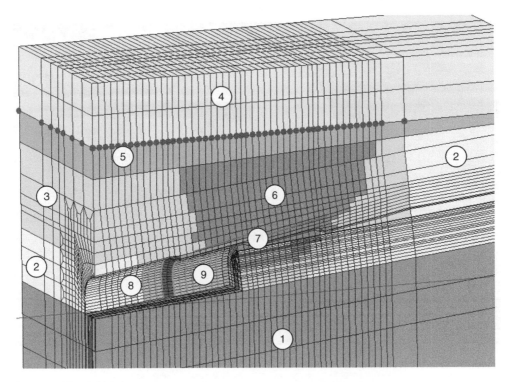

Figure 9.   Material labels.

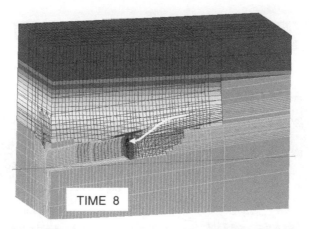

Figure 10.   Pore pressure at TIME 8.

Stress resultants in steel girders are finally shown in figure 14 (maximum stress of 110 MPa in HEB profiles, maximum moment 11.1 kNm, maximum axial force of 548 kN).

4   CONCLUSIONS

Selected aspects of a unified approach to modeling soils, structures and soil-structure interaction are presented in this paper. Emphasis is placed on two-phase media and related difficulties.

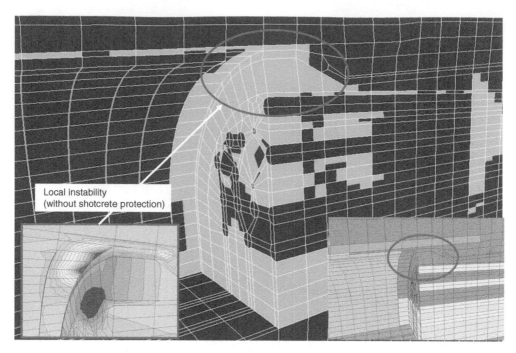

Figure 11.   Plastified zones.

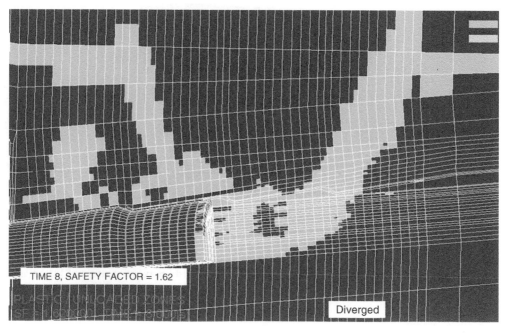

Figure 12.   Plastified zones at TIME 8 and a safety factor 1.62.

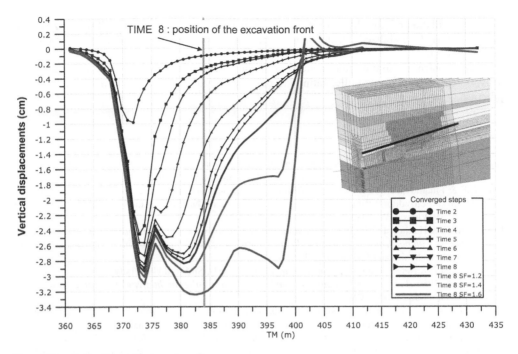

Figure 13.   Deflection at the tunnel roof.

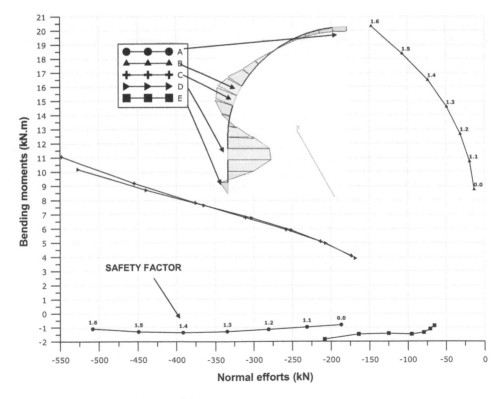

Figure 14.   Stress resultants in steel girders.

The formulation of partially saturated media and associated boundary conditions in ZSoil.PC is recalled first. Stabilized formulations necessary to achieve non-oscillatory solutions are described next.

A test problem illustrating rain induced instability, a consolidation test, and a case study of a subway tunnel are presented, illustrating the potential of such type of modeling in geotechnical practice.

REFERENCES

Aubry, D. & Ozanam, O. 1988. *Free-surface tracking through non-saturated models*. In Swoboda, editor, Numerical Met hods in Geomechanies, pages 757–763, Innsbruck, Balkema.

Brezzi, F. & Pitkàranta, J. 1984, *On the stabilization of finite element approximations of the Stokes problem*. In W. Hackbusch, editor, Efficient solutions of elliptic problems, Notes on Numerical Fluid Mechanics,1O, pages 11–19, Vieweg, Wiesbaden.

Brooks, A. N. & Hughes, T. J. R. 1982. *Streamline upwind/Petrov-Galerkin formulations for convection dominated fiows with particular emphasis on the incompressible Navier-Stokes equations*. Computer Methods in Applied Mechanies and Engineering, 32:199259.

Franca, L. P. 1987. *New mixed finite element methods*. Ph.d thesis, Stanford University.

Genuchten, V. 1980. *A closed form equation for predicting the hydraulic conductivity of unsaturated soils*. Sou Sciences Am. Soc., 44:892–898.

Hughes, T. J. R. 1987. *The Jinite element method. Linear Static and Dynamic Finite Element Analysis*. Prentice Hall, Inc. A Division of Simon & Schuster, Englewood Cliifs, New Jersey 07632.

Hughes, T. J. R. & Franca, L. P. & Balestra, M. 1989. *A new finite element formulation for computational fluid dynamics*: VIII. The Galerkin/Least-squares method for advective-diffusive equations. Computer Methods in Applied Mechanics and Engineering, 73:173–189.

Hughes, T. J. R. 1995. *Multiscale phenomena: Green's functions, the Dirichletto-Neumann formulation*, subgrid scale models, bubbles and the origins of stabilized methods. Computer Methods in Applied Mechanics and Engineering, 127:387–4001.

Pastor, M. & Quecedo, M. & Zienkiewicz, O.C. 1997. *A mixed displacement-pressure formulation for numerical analysis of plastic failure*. Computers & Structures, 62(1):13–23.

Truty, A. & Zimmermann, Th., 2006. *Stabilized mixed finite element formulations for materially nonlinear partially saturated two-phase media*, Computer Methods in Applied Mechanies and Engineering 195, 1517–1546.

Vermeer, P. A. & Verruijt, A. 1981. An accuracy condition for consolidation by finite elements. *International Journal for Numerical and Analytical Methods in Geomechanics*, 5:1–14.

Zienkiewicz, O. C. & Huang, M. &Pastor, M. 1994. *Computational soil dynamicsA new algorithm for drained and undrained conditions*. In Siriwardane and Zaman, editors, Computer Methods and Advances in Geomechanics, pages 47–59, Balkema, Rotterdam.

Zienkiewicz, O. C. & Taylor, R. L. 2000. *The Finite Element Method. Fifth edition*. Butterworth-Heinemann.

ZSOIL 1985–2007. User *manual*, Elmepress & Zace Services. 1985-2003. Limited, Lausanne, Switzerland.

Applications of Computational Mechanics in Geotechnical Engineering – Sousa,
Fernandes, Vargas Jr & Azevedo (eds)
© 2007 Taylor & Francis Group, London, ISBN 978-0-415-43789-9

Identification of parameters: their importance, attention to be paid, cares to be taken

J.C. André & A.C. Furukawa
Escola Politécnica da Universidade de São Paulo, São Paulo, Brasil

ABSTRACT: The problem of identifying parameters is formulated, considering measured and calculated data. Such data are expressed regarding parameters to be determined for a certain mathematical model. Thence one should determine the values of these parameters involved in the mathematical model which represent the observed physical process in the best possible way, either if these parameter values are related to the geometry of the structures, to the properties of the materials or to the load characterization. There should be obtained the parameter values which, when they are introduced into the equations which rule the problem being studied, lead to results as close as possible to the corresponding measured values. This paper presents an analysis of the application of a procedure developed to identify parameters, through the finite element method, in a reference problem with known results. Herein the influences of several factors involved are studied, such as the number and the position of measurements. The simulated problem deals with the digging of a tunnel, to which it was applied the procedure developed for the determination of physical parameters: modulus of elasticity and lateral pressure ratio. It should be pointed out that the conclusions established for this reference problem are not immediately extended to other problems, but their objective is to discuss a series of issues regarding parameter identification.

1 BACK ANALYSIS PROCESS

1.1 Introduction

Back analysis techniques allow the identification of models or of parameters of a model which best represent an observed physical process. In cases where there is a good definition of the model which rules the problem, there is the issue of identifying the model parameters which best adapt its measurements and forecasts, whether they be related to the geometry of the structure, to the properties of the materials, to the laws ruling the model, or to the characterization of the actions.

The process of parameter identification corresponds to a mathematical problem, usually a problem of minimizing a given function, whose solution are the values for these parameters which render minimal the difference between measured and calculated values. Therefore, the process can be defined by two main stages. During the first one, a criterion is used in order to define the objective function, the function of the measured values and of the calculated ones. In the second stage, through a minimization algorithm, the objective function is minimized, and thus the parameters are obtained.

1.2 Identification criteria

Identification criteria are basically differentiated by the degree of previous information available about the problem. Main criteria are: the criterion of maximum likelihood with previous information, the criterion of maximum likelihood, and the least squares criterion.

When the quality of the results from the observations is known, and the previous information on the parameter values to be estimated are available – based on acquired experience in a length of time through study and observation of the physical process, the criterion of maximum likelihood with previous information may be used. The objective function defined with this criterion takes the form:

$$F(\mathbf{p}) = [\mathbf{y}^* - \mathbf{y}(\mathbf{p})]^T \mathbf{C}_y^{-1} [\mathbf{y}^* - \mathbf{y}(\mathbf{p})] + [\mathbf{p} - \mathbf{p}^0]^T \mathbf{C}_{p^0}^{-1} [\mathbf{p} - \mathbf{p}^0]$$ (1)

where $\mathbf{y}^*$ is the vector of the measured values; $\mathbf{y}(\mathbf{p})$ is the vector of the values calculated regarding the parameters to be determined $(\mathbf{p})$; $\mathbf{C}_y$ is the covariance matrix of measurements; and $\mathbf{C}_{p^0}$ is the covariance matrix of the a priori estimated parameters $(\mathbf{p}^0)$.

If measurements are independent, matrix $\mathbf{C}_y$ is a diagonal one, in which elements are given through the variance of measurements (σ_y^2), that is, the square of standard deviation (σ_y).

In case when there is only available information on measurement precision, the most recommended criterion is the maximum likelihood one, which has the objective function defined by:

$$F(\mathbf{p}) = [\mathbf{y}^* - \mathbf{y}(\mathbf{p})]^T \mathbf{C}_y^{-1} [\mathbf{y}^* - \mathbf{y}(\mathbf{p})]$$ (2)

If measurements are independent and, moreover, present the same precision, matrix $\mathbf{C}_y$ is an unitary matrix multiplied by a scalar which is given by the variance of measurements (σ_y^2). In this case, the objective function is defined only by the sum of the squares of the differences between measured values and the ones calculated by multiplying by the above mentioned scalar. Since this is a minimization problem, this positive scalar does not modify the result, and the criterion of maximum likelihood is reduced to the least squares one. The objective function may be represented as:

$$F(\mathbf{p}) = [\mathbf{y}^* - \mathbf{y}(\mathbf{p})]^T [\mathbf{y}^* - \mathbf{y}(\mathbf{p})]$$ (3)

The least squares criterion is also applied when there is no other kind of additional information available about parameters or measurements. This paper employs the least squares criterion in order to define the objective function.

1.3 *Minimization algorithms*

The authors have studied the use of the Newton-Raphson, Gauss-Newton and Levenberg-Marquardt methods in parameter identification. These methods are characterized by an iterative process such as:

$$\mathbf{p}\big|_{r+1} = \mathbf{p}\big|_r + \Delta\mathbf{p}\big|_r$$ (4)

with

$$\Delta\mathbf{p}\big|_r = \mathbf{H}^{-1}\big|_r \mathbf{A}^T\big|_r \Delta\mathbf{y}\big|_r$$ (5)

where vector $\Delta\mathbf{y}$, of order $m \times 1$, is the vector of the differences between the m values measured and calculated m values given by:

$$\Delta\mathbf{y} = \begin{Bmatrix} \Delta y_1 \\ \vdots \\ \Delta y_m \end{Bmatrix} = \begin{Bmatrix} y_1^* - y_1(\mathbf{p}) \\ \vdots \\ y_m^* - y_m(\mathbf{p}) \end{Bmatrix}$$ (6)

Matrix $\mathbf{A}$, of order $m \times n$,

$$\mathbf{A} = \begin{bmatrix} \dfrac{\partial y_1(\mathbf{p})}{\partial p_1} & \cdots & \dfrac{\partial y_1(\mathbf{p})}{\partial p_n} \\ \vdots & \ddots & \vdots \\ \dfrac{\partial y_m(\mathbf{p})}{\partial p_1} & \cdots & \dfrac{\partial y_m(\mathbf{p})}{\partial p_n} \end{bmatrix}$$ (7)

is the sensitivity matrix formed by the partial derivatives of values calculated regarding parameters to be estimated.

Matrix **H**, of order $n \times n$, is the Hessian matrix of function $F(\mathbf{p})$ and differs in each of the methods. A more detailed discussion of these methods and additional references on them may be found in Ledesma (1987) and Castro (1997).

Bearing in mind the greatest simplicity in determining the Hessian matrix and the quality of the results it obtains, this paper uses the Gauss-Newton method. In the Gauss-Newton method, the Hessian matrix is approximated by:

$$\mathbf{H} \cong \mathbf{A}^T \mathbf{A} \tag{8}$$

1.4 Use of the finite element method

The determination of the sensitivity matrix is extremely important in the process of parameter identification. The establishment of an explicit expression for **A** demands an also explicit equilibrium equation. In this paper, it was used the equilibrium equation corresponding to the linear elastic model obtained through the finite element method.

The equilibrium equation established through the finite element method for a linear elastic model has this form:

$$\mathbf{K}\mathbf{U} = \mathbf{R} \tag{9}$$

in which **K** is the stiffness matrix of the structure, **U** is the vector of nodal displacements and **R** is the vector of nodal forces.

Deriving (9) in relation to the parameters, one has:

$$\mathbf{K}\frac{\partial \mathbf{U}}{\partial \mathbf{p}} = \frac{\partial \mathbf{R}}{\partial \mathbf{p}} - \frac{\partial \mathbf{K}}{\partial \mathbf{p}}\mathbf{U} \tag{10}$$

Vector $\frac{\partial \mathbf{U}}{\partial \mathbf{p}}$ represents the sensitivity matrix when a formulation through the finite element method is used for a linear elastic problem where the measured values are displacements.

1.5 Developed procedure

Based on what has been related above, a procedure is developed for identification of parameters using the finite element method, and considering that the measured values are displacements. For this it is used the finite element software Adina plus a routine developed in Matlab. Figure 1 illustrates schematically the developed procedure.

1.6 Result analysis

After the parameters are identified, it is important to run an analysis of the results, since there are several factors which may influence their quality.

The condition number of a system is an indicator of the conditioning of this system. The larger the condition number is, the greater is the probability of errors in the solution. In the iterative procedures applied to the parameter identification, the condition number of the Hessian matrix provides a measurement of the difficulty of convergence in the iterative process; the larger the condition number is in the Hessian matrix, the larger the likelihood of errors in the solution.

The sensitivity matrix contains information on the behavior of the system in regard to the parameters and the measured values. The elements of the sensitivity matrix, equation 7, are the sensitivity coefficients, defined by:

$$A_{ij} = \frac{\partial y_i(\mathbf{p})}{\partial p_j} \tag{11}$$

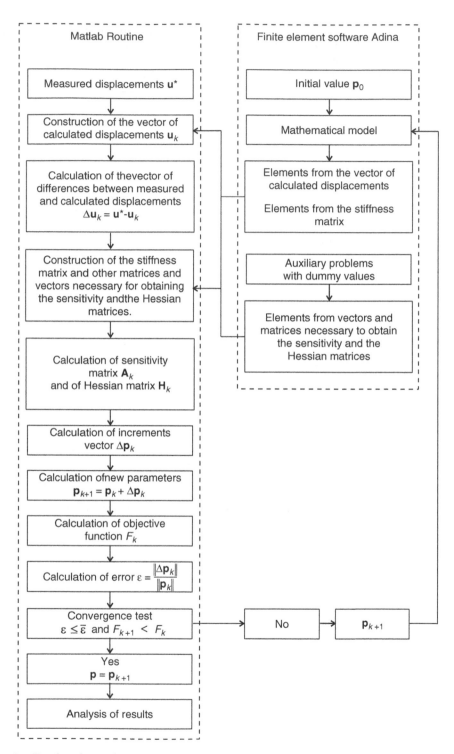

Figure 1.   Developed procedure.

These coefficients show the impact of a small variation of parameter p_j on the calculated value y_i. Therefore, the analysis of the sensitivity coefficients allows the determination of the parameters with the most influence on the system answer and the best measurement points.

The analysis of the elements in the covariance matrix delivers information about the scattering of parameters p_i and the linear correlation between such parameters. According to Finsterle & Persoff (1997) and Finsterle & Pruess (1995), for the least squares criterion the covariance matrix is estimated by:

$$\mathbf{C}_p = s_0^2 \left[\mathbf{A}^T \mathbf{A}\right]^{-1} \tag{12}$$

in which $\mathbf{A}$ is the sensitivity matrix calculated with the obtained parameters, and s_0^2 the final error variance, estimated through

$$s_0^2 = \frac{F_{minimo}}{m-n} \tag{13}$$

The elements in the diagonal of $\mathbf{C}_p$ contain the variances of the estimated parameters ($C_{p_{ii}}$) which allow the evaluation of absolute scatterings of the parameters by the standard deviations ($\sqrt{C_{p_{ii}}}$) and of the relative scatterings by the variation coefficients, defined by:

$$r_{ii} = \frac{\sqrt{C_{p_{ii}}}}{\overline{p}_i} \tag{14}$$

The variances of the estimated parameters represent a degree of the parameter uncertainty.

Elements outside the diagonal of $\mathbf{C}_p$ are the covariances which allow, by the adimensional coefficients of correlation given by:

$$r_{ij} = \frac{C_{p_{ij}}}{\sqrt{C_{p_{ii}} C_{p_{jj}}}} \qquad -1 \leq r_{ij} \leq 1 \tag{15}$$

the establishment of linear correlations between parameters.

At the end of the process, it is also possible to determine the confidence interval of the parameters and the final residues. The confidence interval of the parameters may be found through:

$$\left(p_i - \sqrt{C_{p_{ii}}} \, t_{(\alpha/2, m-n)}\right) \quad \text{to} \quad \left(p_i + \sqrt{C_{p_{ii}}} \, t_{(\alpha/2, m-n)}\right) \tag{16}$$

In which $t_{(\alpha/2, m-n)}$ represents Student's distribution value t for the confidence level $(1 - \alpha)100\%$ with $m - n$ degrees of freedom. It may also be observed that this interval is not defined when the measurement number is equal to the number of parameters.

Final residues are the difference between measured and calculated values with identified parameters, which are simply the objective function calculated at its minimum. Residues represent the sum of all the errors in the process.

2 REFERENCE PROBLEM

The reference problem corresponds to a simulated digging of a tunnel. It deals with the determination of physical parameters, of the elasticity model (E) and of the lateral pressure ratio (K_0) in a uniform massif, characterized by an isotropic and linear elastic half-plane, with a field of initial stresses and under the action of an instantaneous opening of a circular tunnel with the geometric characteristics presented in Figure 2.

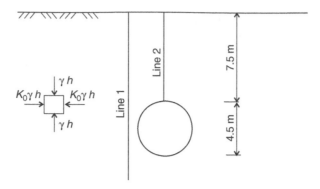

Figure 2.   Geometric characteristics of the reference problem.

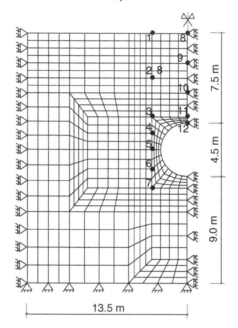

Figure 3.   Finite element model, showing locations of considered measurements.

The influences of the minimization methods, of initial values, of mesh refinement, of errors in the mesh, of the number of measurements, of the position of measurements and of the model were studied. This paper only presents a study of the influence of the number of measurements and of the measurement positions.

Measurements are considered to be horizontal displacements at some points of line 1 and vertical displacements at some points of line 2. In order to obtain the values of these displacements, the finite element method was used, considering $E = 10.000$ MPa and $K_0 = 1.000$. Plane strain elements of 9 nodes are employed and loading is characterized by efforts due to the relief provoked by the instantaneous opening of the tunnel. Figure 3 presents the model thus defined, with the mesh composed by 2081 nodes and 4007 degrees of freedom and boundary conditions, considering the symmetry.

In order to consider measurement errors, to the displacement values obtained from the analysis of this model it is added a random component with normal distribution and a null average, and a given standard deviation. Four groups of measurements are considered:

- 24 measurements, 15 of them horizontal and 9 vertical ones, located between the ones shown in Figure 3;

120

Table 1. Obtained results.

Description	Estimated parameters	F_{min} and (s_0^2)	Relative errors and number of iterations
24 measurements	$E = 9.986$ MPa	5.513×10^{-5}	0.39%
	$K_0 = 1.017$	5.506×10^{-6}	4
12 measurements	$E = 9.914$ MPa	2.889×10^{-5}	0.37%
	$K_0 = 0.961$	2.889×10^{-6}	4
6 measurements	$E = 10.041$ MPa	1.146×10^{-5}	0.40%
	$K_0 = 0.977$	2.864×10^{-6}	4
2 measurements	$E = 10.216$ MPa	1.053×10^{-11}	0.47%
	$K_0 = 1.004$	–	4

- 12 measurements, 7 of them horizontal (1, 2, 3, 4, 5, 6 and 7) and 5 vertical ones (8, 9, 10, 11 and 12);
- 6 measurements, 4 of them horizontal (4, 5, 6 and 7) and 2 vertical ones (11 and 12);
- 2 measurements, 1 of them horizontal (5) and 1 vertical one (12).

The choice of these measurements was based on a previous analysis of the sensitivity coefficients, which was done using the initial values of the parameters.

In order to obtain the calculated displacements, the same model is used with a less refined mesh, composed by 545 nodes and 1011 degrees of freedom.

With the developed procedure, the Gauss-Newton method is used to minimize the objective function defined with the least squares criterion, starting the process with $\mathbf{p}_0 = \begin{bmatrix} 5\,\text{MPa} \\ 0.5 \end{bmatrix}$.

Table 1 presents the values of estimated parameters (E and K_0), of the objective function calculated at its minimum (F_{min}), of final error variance (s_0^2), of relative error and of the number of iterations of the iterative process from the different studies.

The symbol "-" which appears in the table when two measurements are considered is due to the fact that the final error variance value (s_0^2) cannot be estimated by equation (13), since the number of measurements is equal to the number of parameters. This symbol will be used whenever a magnitude cannot be calculated.

The number of measurements has little influence on the results, and it is observed that only two measurements are enough to identify the parameters. By intuition, it might be expected that a larger number of measurements would render easier the iterative process; this, however, does not occur in this study: in the three cases there is the same number of iterations. Moreover, the larger the number of measurements is, the larger the value of the objective function calculated at the minimum is.

Figure 4 presents the sensitivity coefficients calculated at the minimum of the objective function defined with 12 measurements. It may be noticed that measurements with the largest sensitivity coefficients are the horizontal measurement 5 and the vertical measurement 12, thus confirming the analysis previously run with the initial values of the parameters. As presented in Table 1, only these two measurements are enough to identify parameters E and K_0.

Contour lines of the objective function defined with 24 measurements are represented in Figure 5, as well as in Figure 6 considering 12 measurements, in Figure 7 for 6 measurements and in Figure 8 for 2 measurements.

The analysis of the results is made with the analysis of the condition number, of the standard deviations of E and K_0, of the variation coefficients of E and of K_0 and of the coefficient of correlation between these parameters. Table 2 presents an analysis of the results for the conditions under study.

It should also be born in mind that, when two measurements are considered, one uses the variance value of the final residues in order to obtain the covariance matrix of the estimated parameters, that is, $\mathbf{C}_p = \sigma_r^2 [\mathbf{A}^T \mathbf{A}]^{-1}$. This variance value is found in Table 3.

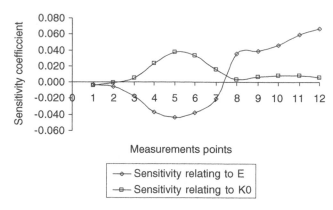

Measurements points

Figure 4. Sensitivity coefficients.

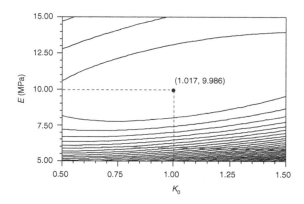

Figure 5. Contour lines of the objective function defined with 24 measurements.

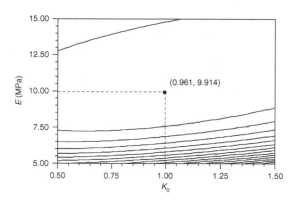

Figure 6. Contour lines of the objective function defined with 12 measurements.

The low values obtained from the condition number indicate the good convergence of the procedures. Analysing the values of the correlation coefficients, it is noticed the low correlation between parameters.

The analysis of the results shows that two measurements are not only enough to identify the parameters, but they also provide the smallest values of standard deviation and of variation coefficient of the estimated parameters.

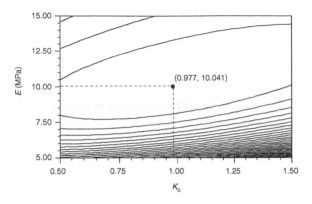

Figure 7. Contour lines of the objective function defined with 6 measurements.

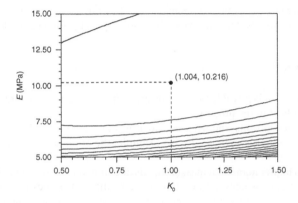

Figure 8. Contour lines of the objective function defined with 2 measurements.

Table 2. Analysis of obtained results.

Description of	Condition number	Standard deviation	Coefficient of variation	Coefficient correlation
24 measurements	25.628	$\sigma_E = 6.406 \times 10^{-2}$ MPa $\sigma_{K0} = 1.522 \times 10^{-2}$	$r_E = 0.641\%$ $r_{K0} = 1.496\%$	0.534
12 measurements	22.917	$\sigma_E = 9.023 \times 10^{-2}$ MPa $\sigma_{K0} = 2.225 \times 10^{-2}$	$r_E = 0.910\%$ $r_{K0} = 2.316\%$	0.508
6 measurements	36.279	$\sigma_E = 1.159 \times 10^{-1}$ MPa $\sigma_{K0} = 2.437 \times 10^{-2}$	$r_E = 1.155\%$ $r_{K0} = 2.494\%$	0.597
2 measurements	31.766	$\sigma_E = 3.105 \times 10^{-4}$ MPa $\sigma_{K0} = 6.830 \times 10^{-5}$	$r_E = 0.003\%$ $r_{K0} = 0.007\%$	0.573

In order to complete the analysis of the results, Table 3 shows the confidence intervals of the estimated parameters, calculated by considering a confidence level of 95%. In order to verify final residues, this Table also shows the variance of the final residues and average residue.

The smallest values in the analysis of final residues are also obtained by considering only two measurements. It is noticed that the quality of the results obtained considering the two measurements

123

Table 3. Confidence interval of the estimated parameters and analysis of final residues.

Description	Confidence interval	Average residue	Variance of final residues
24 measurements	$E = 9.986 \pm 0.133\,\text{MPa}$ $K_0 = 1.017 \pm 0.032$	$r = -7.425 \times 10^{-5}$	$\sigma_r^2 = 2.391 \times 10^{-6}$
12 measurements	$E = 9.914 \pm 0.201\,\text{MPa}$ $K_0 = 0.961 \pm 0.050$	$r = 6.235 \times 10^{-5}$	$\sigma_r^2 = 2.622 \times 10^{-6}$
6 measurements	$E = 10.041 \pm 0.322\,\text{MPa}$ $K_0 = 0.977 \pm 0.068$	$r = 1.954 \times 10^{-4}$	$\sigma_r^2 = 2.246 \times 10^{-6}$
2 measurements	– –	$r = -2.994 \times 10^{-7}$	$\sigma_r^2 = 1.035 \times 10^{-11}$

is related to the choice of those measurements, since they are the ones which present the highest sensitivity coefficients, that is, the ones which have the highest influence on the process.

3 FINAL COMMENTS

In this reference problem it is possible to study the procedure developed for parameter identification and some issues involved in the process.

The developed procedure may be used considering the Newton-Raphson, the Gauss-Newton and the Levenberg-Marquardt methods. In this example, the application studied is the one for the Gauss-Newton method.

As verified here, a higher number of measurements does not necessarily entail better results, since the number of measurements does not interfere in the results: what does interfere is the influence of the positions of measurements in the process. It is pointed out the importance of analysis of the sensitivity coefficients, identifying the measurements with the highest sensitivity, which, therefore, bear the highest influence on the process. In the case under study, the two measurements with the highest sensitivity coefficient by themselves suffice to identify the parameters.

It stands out that the conclusions obtained are valid for this problem and should not be taken as absolute ones, but as important references in real cases of parameter identification

REFERENCES

Castro, A.T. 1997. Métodos de retroanálise na interpretação do comportamento de barragens de betão. *Tese (Doutorado), Instituto Superior Técnico, Universidade Técnica de Lisboa*. Lisboa. 244p.
Costa, A. 2006. Identificação de parâmetros em obras civis. *Tese de Doutorado, Escola Politécnica da Universidade de São Paulo*. São Paulo. 240p.
Costa, A. & André, J.C. & Goulart, M. L. S. 2004. Back analysis for identification of parameters using the finite element method: iterative methods. In *Iberian Latin American Congress on Computational Methods in Engineering, 25, Recife, 2004* . Recife.
Finsterle, S. & Persoff, P. 1997. Determining permeability of tight rock samples using inverse modeling. *Water Resour. Res* 33(8): 1803–1811.
Finsterle, S. & Pruess, K. 1995. Solving the estimation-identification problem in two-phase flow modeling. *Water Resour. Res* 31(4): 913–923
Ledesma, A. 1987. Identificación de parâmetros em geotecnia: aplicación a la excavasión de túneles. *Tesis(Doctorado), Escuela Técnica Superior de Ingenieros de Caminos, Canales y Puertos de Barcelona, Universitat Politécnica de Catalunya*. Barcelona. 331p.

Applications of Computational Mechanics in Geotechnical Engineering – Sousa,
Fernandes, Vargas Jr & Azevedo (eds)
© 2007 Taylor & Francis Group, London, ISBN 978-0-415-43789-9

Inverse analysis on two geotechnical works: a tunnel and a cavern

S. Eclaircy-Caudron, D. Dias & R. Kastner
LGCIE laboratory, Site Coulomb 3, Villeurbanne, France

T. Miranda & A. Gomes Correia
University of Minho, Department of Civil Engineering, Guimarães, Portugal

L. Ribeiro e Sousa
University of Porto, Department of Civil Engineering, Porto, Portugal

ABSTRACT: One of the major difficulties for geotechnical engineers during project phase is to estimate in a reliable way the mechanical parameters values of the adopted constitutive model. In project phase, they can be evaluated by laboratory and in situ tests. But, these tests lead to uncertainties due to the soil reworking and to local character of the test which is not representative of the soil mass. Moreover for in situ tests interpretation difficulties exist due to the non homogeneous character of the strain and stress fields applied to the soil mass. In order to reduce these uncertainties, geotechnical engineers can use inverse analysis processes during construction. This article shows the application of two of these processes (a deterministic and a probabilistic method) on convergence leveling measurements realized during the excavation of the Bois de Peu tunnel (France). Moreover, these two processes are also applied on displacements measured by inclinometers during the excavation of the hydroelectric powerhouse cavern Venda Nova II (Portugal). The two inverse analysis methods are coupled with two geotechnical software (CESAR-LCPC and FLAC^{3D}) to identify soil parameters. Numerical and experimental results are compared.

1 INTRODUCTION

The purpose of an inverse analysis process is the identification of parameters using tests results or/and experimental measurement carried out during works. Various methods exist. Hicher (2002) distinguishes three kinds: analytical methods, correlation and optimization methods. Optimization methods are applied when model parameters are not appropriate to be used in a direct approach by graphic construction or in an analytical approach. By this method, the inverse problem is solved using an algorithm which minimizes a function depending on all parameters. This function is generally called "cost function" and corresponds to the difference between numerical and experimental measurements. The experimental results can come from various origins: laboratory tests, in situ tests or work measurement data. Two types of optimization methods are distinguished: deterministic and probabilistic methods. The deterministic methods include gradient, Newton, Gauss-Newton or Liebenberg Marquardt optimization algorithms. Several researches aiming at parameter identifications based on deterministic methods were carried out in the last years. Zentar (2001) tried to identify some mechanical parameters of the Saint-Herblain clay using the results of pressuremeter tests and the optimization software SiDoLo (Pilvin 1983). He considered an elastoplastic model with or without hardening. In spite of the identification methods development, a few of them were applied to real cases like tunnels or deep excavations (Jeon et al. 2004, Finno et al. 2005).

The deterministic methods present some advantages. The iteration number required to achieve the optimization process is relatively low. But, if the cost function presents several local minima, the deterministic methods can converge towards the first found minimum. This major drawback

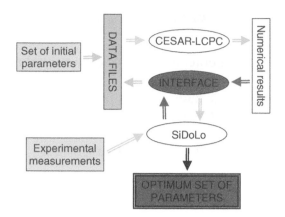

Figure 1.   Principle of the coupling.

explains why these methods are seldom used for complex problems. For such problems probabilistic methods are preferred. Probabilistic methods include evolutionary algorithms such as genetic algorithms and evolution strategies. Evolutionary algorithms reproduce the natural evolution of the species in biological systems and they can be used as a robust global optimization tool. The major principles of genetic algorithms were developed by Goldberg (1991) and renders (1995). Evolution strategies (Schwefel 1985) start searching from an initial population (a set of points) and transition rules between generations are deterministic. The search of new points is based on mutation and recombination operators. Recent researches applying genetic algorithms to soil parameters identification for a constitutive model have been undertaken in geotechnic (Levasseur et al. 2005, Samarajiva et al. 2005). Evolution strategies were also recently applied to problems in many domains (Costa et al. 2001).

First, this paper presents briefly the used optimization software, SiDoLo and the used evolutionary algorithm. Then the numerical modeling of the two geotechnical works and the followed approach are detailed. Finally, results obtained by the two processes are compared and influence of some data in the evolutionary algorithm is highlighted.

2   PRESENTATION OF THE OPTIMIZATION METHOD USED

2.1   *The optimization software, SiDoLo*

SiDoLo is coupled with the geotechnical software CESAR-LCPC (Itech 2002) or FLAC^{3D} (Itasca 2005). The coupling principle is illustrated in figure 1.

Through this coupling, SiDoLo compares numerical results with experimental measurements, in order to calculate the cost function. When the cost function is lower than a fixed value, then the optimization process stops. The cost function L(A) is expressed by the following finite sum:

$$L(A) = \sum_{n=1}^{N} \frac{1}{M_n} \sum_{i=1}^{M_n} {}^{T}[Z_s(A,t_i) - Z_s^*(t_i)].D_n.[Z_s(A,t_i) - Z_s^*(t_i)] \qquad (1)$$

where A are the model parameters, N is the number of experimental measurements; $[Z_s(A,t_i) - Z_s^*(t_i)]$ is the difference between numerical and experimental results evaluated only at Mn observation steps t_i and D_n is the weighting matrix of the nth test. Measurements accuracy can be taken into account by weighting coefficients.

SiDoLo uses a hybrid optimization algorithm, which combines two typical minimization methods: the gradient method and a variant of the Lavenberg-Marquardt method to accelerate the convergence when the solution is close. More details on the SiDoLo approach can be found in Eclaircy-Caudron et al. (2006).

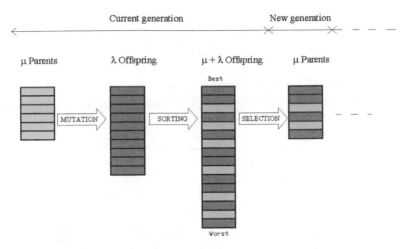

Figure 2. Principle of the $(\mu/\rho + \lambda)$ evolution strategies (Costa et al. 2001).

2.2 *The algorithm based on evolution strategies*

The algorithm used was developed by Costa et al. (2001). This algorithm was adapted to the problems studied in this paper, and as for SiDoLo, requires creating an interface with computer codes. The population entities are vectors of real coded decision variables that are potential optimal solutions. An initial population is generated and then each following generation, λ offspring are generated from μ progenitors by mutation and recombination. Then the best entities are selected for next generation among the $\mu + \lambda$ members according to their cost function value. Finally, the μ best of all the $\mu + \lambda$ members become the parents of the next generations. Important features of evolution strategies are the self adaptation of step sizes for mutation during the search and the recombination of entities that is performed between ρ individuals. This algorithm, named generally $(\mu/\rho + \lambda)$ algorithm, is illustrated in figure 2.

The algorithm stops when one of the following conditions is verified:

- The maximum number of generations is reached
- The difference between the two extreme values of the cost functions is lower than 10^{-5}
- This difference divided by the average of the cost functions values is lower than 10^{-5}

Various error functions are considered. First, the same function than in SiDoLo is introduced in the algorithm. Then, the following error function is considered.

$$L_\varepsilon(A) = \sqrt{\frac{1}{N} \sum_{i=1}^{N} \frac{[Z_s(A,t_i) - Z_s^*(t_i)]^2}{[\varepsilon + \alpha.Z_s^*(t_i)]^2}} \qquad (2)$$

The ε and α variables represent respectively the absolute and relative error of measurements. Different values of these two variables are tested.

3 THE FOLLOWED APPROACH

3.1 *Case 1: Application to the Bois de Peu tunnel*

3.1.1 *Presentation of the tunnel*

The Bois de Peu tunnel is situated near Besançon in France. It is composed of two tubes. The excavation length is about 520 meters per tube for a cover height which varies between 8 meters and 140 meters. Four kinds of supports were foreseen in the project phase for four materials

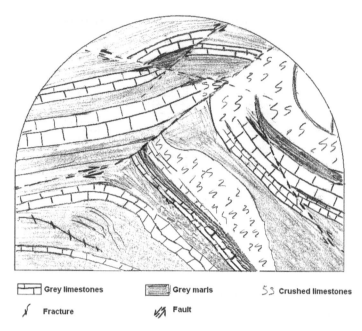

	Grey limestones		Grey marls	$\mathcal{S}\mathcal{S}$ Crushed limestones
$\int$	Fracture		Fault	

Figure 3.   Face leveling.

Table 1.   Design characteristics.

Parameters	γ kN/m3	E MPa	C MPa	ϕ °	ν –
Probable	24.8	1600	0.70	40	0.3
Exceptional	24.0	750	0.21	36	0.3

types: limestones, marls, interbeding of marls and limestones and clays. The studied section is located near one of the two tunnel portals where interbeding of marls and limestones were expected. A face levelling of this section is showed in figure 3. The lining support set up in this area is composed of a mixed tunnel support by shotcrete with a thickness of 0.2 m and steel ribs.

The tunnel is dug with a constant step of 1.5 m. The cover height is equal to twenty six meters. From the site investigations, two kinds of design characteristics were defined: probable and exceptional. They are resumed in table 1.

3.1.2   *Numerical model*

Due to time consuming of three dimensional calculations, tunnel excavations are often taken into account by two dimensional modelling. We adopted a transverse plane strain model, taking into account the real geometry, the excavation and tunnel support being modelled by an unconfinement ratio (Panet, 1995). Only a half section is represented due to symmetry. The grid extents are 75 m (=6D) in all directions in order to avoid the influence of the boundaries. The grid (figure 4) contains around 3900 nodes and 1850 elements with quadratic interpolation (six nodes triangles and 8 nodes quadrangles). The initial stress field is anisotropic. An earth pressure ratio of 0.7 is adopted. This value was obtained from in situ tests. Steel ribs and sprayed concrete are simulated by a homogeneous support with equivalent characteristics. This support is assumed to have a linear elastic behaviour. A linear elastic perfectly plastic model with a Mohr Coulomb failure criterion and non associate flow rule is considered to represent the soil mass.

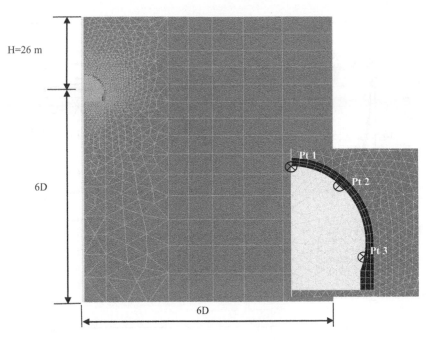

Figure 4. The grid and its dimensions.

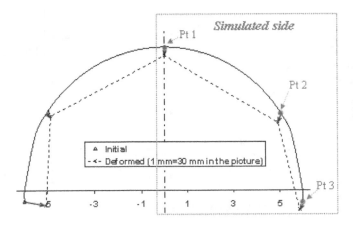

Figure 5. The observations points and deformed section.

Three computing phases are defined. First the initial stress field is applied. After, the tunnel is excavated. In this phase, the unconfinement ratio at the time of the support installation is applied. Finally the homogeneous support with 0.24 m width is activated. A total unconfinement is applied.

3.1.3 Experimental results

Figures 4 and 5 locate the three experimental measurements points used for optimization. Five measurements are observed for these three points: levelling for these three points and horizontal displacements for the points 2 and 3. They correspond to convergence and levelling measurements of the tunnel wall in a given cross section (point 1 to 3 in figures 4 and 5).

The convergence and levelling measurements of this section are reported in figures 6 and 7. These measurements permit to deduce the horizontal and vertical displacements of each point.

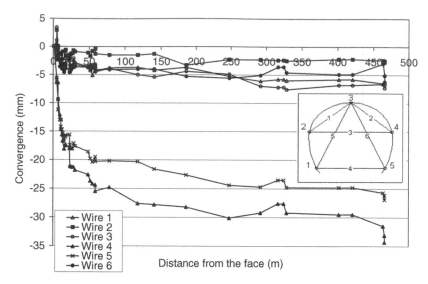

Figure 6.　Convergence measurements.

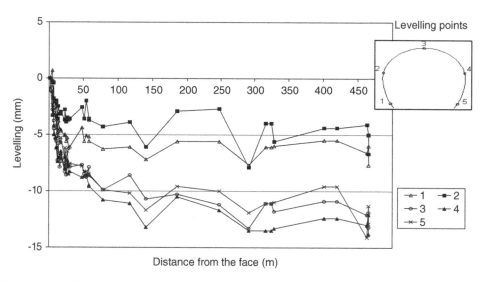

Figure 7.　Levelling measurements.

The convergence is maximal for wires 4 and 5. It reaches 30 mm for wire 4. Moreover leveling is more important for targets 3, 4 and 5 than for the others. It reaches 13 mm. So, these measurements showed a dissymmetrical deformation of the section. In optimization process average displacements measured at equilibrium are used. The dissymmetry is confirmed by the section deformation presented in figure 5 where the crown displacement is assumed purely vertical. The experimental results considered in the optimization process correspond to the average displacements observed at the two sides.

3.1.4 *Unknown parameters*
In plane strain calculation, the introduced unconfinement ratio λ at the time of support installation might be calculated by the convergence-confinement method (Panet 1995) if the constitutive

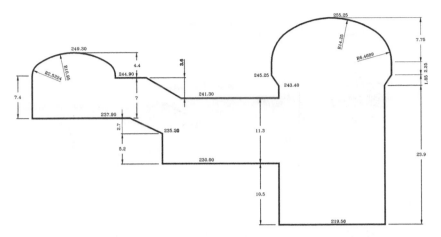

Figure 8. The powerhouse cavern geometry.

model parameters were well known. But in this study, it is considered as an additional parameter to identify. Before using inverse analysis, a sensibility study was performed. The Poisson ratio and the dilatancy angle are parameters which have low influence on the observed variables. So, these parameters could not be identified with these measurements. Only parameters λ, C, E and ϕ, might be evaluated.

Validation studies (Eclaircy-Caudron et al. 2006) were performed with SiDoLo on this numerical model in order to define which parameters might be identified according to the available experimental measurements. These studies showed that with convergence and levelling measurements no more than two parameters could be identified. But SiDoLo always succeeded to identify ϕ and λ even if they were unknown because they have more influence on the convergence and levelling measurements than the others. In order to verify the provided solution an evolution strategies algorithm is used in complement of the software SiDoLo.

3.2 Case 2: Application to the Venda Nova II hydroelectric powerhouse cavern

3.2.1 Presentation of the powerhouse cavern

The Venda Nova II powerhouse cavern is located in the North of Portugal, about 55 km of the town Braga. The powerhouse complex includes two caverns interconnected by two galleries. The caverns have respectively the dimensions as follows: 19.00 m × 60.50 m and 14.10 m × 39.80 m. Their axes are spaced by 45.00 m. They are situated at a 350 m depth. Both caverns have vertical walls and arch roofs. The roof of both caverns is situated at different levels. Moreover, the main cavern has various floors while the other has a single floor. The geometry is illustrated in figure 8. The powerhouse complex is situated in a rock mass characterized by medium-size grain granite of a porphyritic trend with quartz and/or pegmatitic veins and beds, which are, sometimes, rose. The rock mass also includes embedment of fairly quartzitic mica-schist. In order to investigate the geotechnical characteristics of the rock mass in depth, four deep and subvertical boreholes were performed. Laboratory tests were done on rock samples resulting from boreholes. These tests permitted to identify three geological and geotechnical zones. The zone where the caverns are located presents a Young modulus of 54 GPa and a Poisson ratio of 0.17. Moreover, in situ tests are carried out before the beginning of the main work and after completing the access tunnel to the powerhouse. In fact, an exploration gallery was excavated from the top of the access tunnel and parallel to the caverns axis. In situ tests confirm the presence of four main families of discontinuities and led to values ranging from 33 to 40 GPA for the deformability modulus. The in situ stress state of the rock mass was also characterized. The vertical and horizontal stresses aligned with the longitudinal axes of the caverns correspond to the earth weight. The horizontal stress normal to the longitudinal axes of the caverns is 2 to 3 times higher than the other stresses.

131

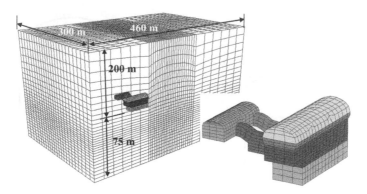

Figure 9.   The 3D mesh.

Stage 1	Stage 2	Stage 3
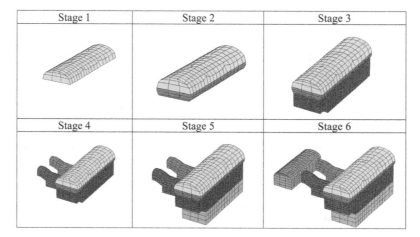		
Stage 4	Stage 5	Stage 6

Figure 10.   The excavation stages.

3.2.2   *Numerical model*

– The caverns excavation was modelled in 3D with the finite difference computer code FLAC^{3D} in order to simulate the real complex geometry of them and the excavation stages. Figure 9 shows the mesh performed with the hexahedral-Meshing Pre-processor, 3DShop. The grid dimensions are 300 m in the transversal direction, 460 m in the longitudinal direction and its height is 275 m. A layer of 75 m is modelled under the caverns. A cover height of 350 m is simulated. The mesh contains around 47000 grid-points, 44000 zones and 1100 structural elements. The simulated excavation stages are lightly different from the real stages but the rock mass staying in elasticity, final displacements are the same. Figure 10 shows the excavation sequences modelled.

The excavation steps are the following:

– 1: Total excavation of the main cavern arch. Set up of fibre sprayed concrete (thickness: 25 cm) on the arch.
– 2: Excavation of the main cavern arch until 1.5 m below the base level of the support beams of the rail tracks and set up of fibre sprayed concrete on the walls cavern (thickness: 25 cm).
– 3: Excavation of the main cavern until the level of the interconnecting galleries.
– 4: Excavation of the interconnecting galleries and set up of fibre sprayed concrete on the galleries arch (thickness: 25 cm).
– 5: Achievement of the main cavern excavation.

132

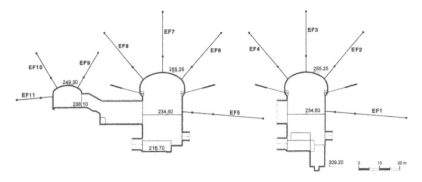

Figure 11. Location of the extensometers in the two cross sections.

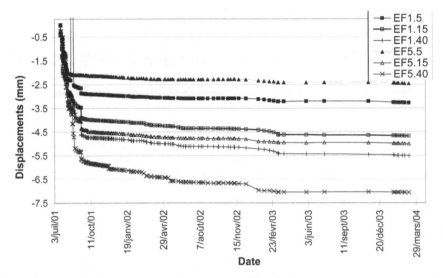

Figure 12. Evolution of the measured displacement in extensometers 1 and 5.

– 6: Excavation of the transformer cavern and set up of fibre sprayed concrete (thickness: 25 cm) on the arch.

In the numerical modelling, an elastic perfectly plastic constitutive model with a Mohr Coulomb failure criterion is assumed to represent the rock behaviour. The sprayed concrete is simulated by shell elements with a linear elastic and isotropic constitutive model, with a Young modulus of 15 GPa and a Poisson ratio of 0.2. The rock bolts are simulated by cable elements, with two nodes and one axial free degree. These elements can yield a tensile stress.

3.2.3 *Experimental results*
A monitoring program was established in order to evaluate the behaviour of the rock mass and the support system and to observe displacements in the rock during and after construction. Convergence targets were installed in several sections. Besides the convergence measurements, 11 extensometers were installed in two sections along the caverns axis (figure 11). In the powerhouse cavern extensometers were installed in two sections, while only one in the other cavern was installed. Almost all the extensometers are double. Just the ones installed in the wall on the main caver are triple and of larger length (EF1 and EF5 in Figure 11). Figure 12 shows the evolution of displacements in extensometers 1 and 5. Figure 13 gives the displacement value measured at the last stage for each extensometer.

133

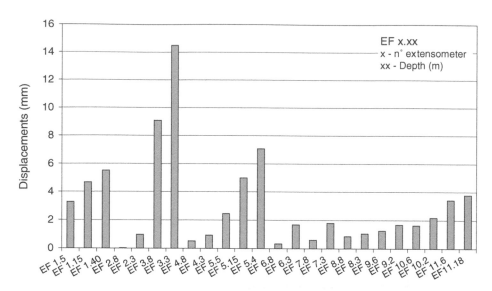

Figure 13.   Displacements value measured in each extensometer at the last excavation stage.

In optimization process, displacements measured at the last excavation stage are used. In validation study, all extensometers are considered. So, 24 displacements are available in the optimization process. In the case of the application on the in situ measurements, only 20 displacements are used in the optimization process because some values are not considered due to measurement errors (EF 3.8, 3.3, 4.8 and 4.3).

3.2.4 *Unknown parameters*

Previously studies permitted to obtain mechanical properties of the rock by using Artificial Intelligence techniques (Miranda et al. 2005). These are the following: 45 GPa for E, 54° for ϕ and 4 MPa for C. Due to the good strength of the rock, a few plastic zones appear. So the value of the two parameters C and ϕ has low influence on the numerical results. A value of 0.2 is adopted for the Poisson ratio and the dilatancy angle is taken equal to 0° . These two parameters have also low influence on the numerical result. Then, only the Young Modulus might be identified by inverse analysis. Besides, many uncertainties reside on the in situ stresses ratio value R. So, as this parameter influences the numerical results, it might be also identified by inverses analysis.

Validation studies were performed with the two optimization processes on this numerical model. These studies show that it is possible to identify the Young modulus and the in situ stresses ratio just from the displacements measured by extensometers. So, this paper presents the identification results performed on these two parameters and compares the results provided between the two processes.

4 RESULTS

4.1 *Case 1: Bois de Peu tunnel (France)*

4.1.1 *Comparisons between the two methods*

Several identification attempts were performed with SiDoLo, confirming that SiDoLo provides approximately the same values of ϕ and λ in all cases. They also showed that the friction angle was lower than the project value (between 10 and 20°).

One example of identification is presented here, where three vertical and two horizontal displacements are used in the optimization process. For each identification several initial values (referred as

Table 2.   The identification results realized with SiDoLo.

Set	E MPa	C MPa
Range	100–2000	0.10–1
a	217	0.13
b	100	0.19
c	102	0.19

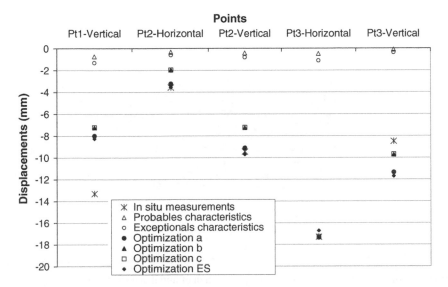

Figure 14.   Comparisons between the measured and computed displacements after optimization by SiDoLo.

a, b and c) of the unknown parameters are tested in order to show the influence of the initial value. Only the Young modulus E and the cohesion C are identified. λ and φ are fixed respectively to the average of values provided by identifications of the four parameters (E, C, φ, λ). A friction angle of 14° and an unconfinement ratio of 0.7 are adopted. The enabled ranges in SiDoLo are resumed in table 2. Table 2 gives also the obtained values. Several solutions are found by the optimization process according to the introduced initial values. SiDoLo does not succeed to find the best couple. It provides local minima. Displacements computed after optimization are showed in figure 14.

For points 2 and 3 numerical results are close to the measured displacements. The vertical displacements computed at point 1 are lower than the one measured. The vertical displacement measured at the tunnel crown appears difficult to reproduce in the numerical model. Although the face leveling showed in figure 3 highlights a disturbed geology with folds and faults, an homogeneous medium is considered in the model. A more complex model seems to be required to simulate the real behavior of the tunnel crown.

In order to avoid local minima, the evolution strategies algorithm is used. The same identification is realized. This identification allows verifying the solutions found by SiDoLo. The same ranges are enabled for parameters. The number of generations is limited to 50 in order to keep acceptable computation times. The parent and recombination population size are 10 and the offspring population size is 20. The evolution strategies algorithm stopped when the maximum number of generations is reached, the cost function being still relatively important and the stop criteria on this function cannot be reached. The design characteristics, the best member provided by the Evolution Strategies algorithm (ES) and solutions provided by SiDoLo are resumed in table 3.

Table 3. Design characteristics and optimized values.

Parameters	E MPa	C MPa	φ °	λ −
Probable	1600	0.7	40	–
Exceptional	750	0.21	36	–
SiDoLo	100–217	0.13–0.19	14*	0.7*
ES	242.7	0.12	14*	0.7*

* Fixed values

Error (×10²)

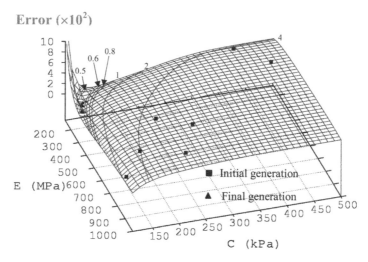

Figure 15. Evolution of the cost function and representation of initial and final generation's parents.

Figure 15 presents the evolution of the cost function. The cost function is not convex and shows a valley. A decrease of parameter E is counterbalanced by an increase of C in a small variation domain. So it is difficult to identify these two parameters with only convergence and leveling measurements. Moreover the value of the cost function is important, confirming that the numerical model should be improved. Figure 15 shows also the parents at initial and final generations. At initial generation parents are dispersed in all research space. At later generation they are located in the valley and are close to each other.

Figure 16 locates the three couples found by SiDoLo and the best member found by the evolution strategies algorithm after identification. The error function value is also reported. Solutions found by SiDoLo are situated in the valley and correspond to local minima. The evolution strategies algorithm provides a global minimum.

4.1.2 *Influence of the error function*
The two error functions presented in equations 1 and 2 are compared. For the second error function, different values of ε and α are tested. All the error functions tested present a valley but the valley extend depends on the error function and on the ε and α values. Figure 17 showed the evolution following E of the different cost functions for a fixed value of C close to the global minimum. The error value reported in figure 17 is not the real value. It corresponds to the real value minus the minimum value in order to set all curves at the origin. The minimum error is reported on each curve. In this figure the different valleys are visible. For an absolute value of 1 mm and a relative value of 0 the valley is more condensed than for the others. So it is easier to find the global minimum with this error function. The algorithm convergence is similar for all the error function.

136

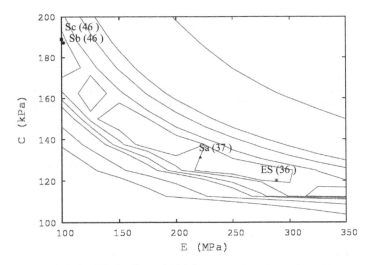

Figure 16.   Location of the solutions found by SiDoLo (Sa, Sb and Sc) and the best member found by the evolution strategies algorithm (ES).

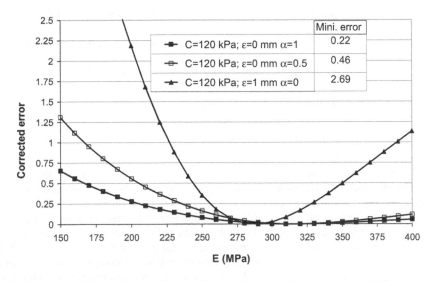

Figure 17.   Evolution of the $L_E(A)$ error function for several couples of parameters (ε, α) following the Young modulus for a fixed cohesion value and their minimum value.

The solution found by the evolutionary algorithm varies according to the values of absolute and relative errors. Table 4 resumes the found solutions and the computation time. The computation time with SiDoLo is also given. The same solution is found if just a relative error or an absolute error is considered but this solution changes according to the considered type of error. The calculation time is similar for all the error function. But it is very important compared to the computation time required by SiDoLo.

4.1.3   Influence of the population size
Table 5 gives the solutions provided by the algorithm and the calculation time in two cases. In the first case a parent and recombination population size equal to 10 and a offspring population size of

Table 4. Found solutions according to the error function and calculation time.

Error	E MPa	C MPa	Error	Time Hours
$\varepsilon = 0; \alpha = 1$	332	0.118	0.22	56
$\varepsilon = 0; \alpha = 0.5$	332	0.118	0.46	55
$\varepsilon = 1$ mm; $\alpha = 0$	289	0.12	2.69	53
SiDoLo	100–217	0.13–0.19	37–46	1,5

Table 5. Found solutions according to the population size and calculation time.

Cases	E MPa	C MPa	Error	Time Hours
Case 1	288	0.120	2.69	53
Case 2	252	0.125	2.69	84
Difference	12%	4%	0%	50 %

20 is adopted. In the second case, values of 15 and 30 are respectively adopted for these population sizes.

For each calculation, the stop criteria (10^{-5}) cannot be reached; then it is the fixed maximum number of generations which stopped the calculation. For the second case the time is greater about 50%. With a difference of 12% on the Young modulus and 4% on the cohesion, these two calculations give the same error function value. So, due to the computation, it does not seem necessary to increase the population size in our studied case.

4.2 Case 2: hydroelectric powerhouse cavern (Portugal)

4.2.1 Identifications with SiDoLo

The initial values introduced in SiDoLo are the following: 45 GPa for E and 2 for the in situ stress ratio R. First, the twentieth values of displacements are considered in the optimization process. Figure 18 compares the measured displacement to the computed displacements with the initial values and with the optimized values of the two parameters. It also gives the respective error function values. Displacements computed with the optimized values are closer to the experimental measurements than those computed with initial values excluded for 1.5, 1.15, 5.5, 5.15, 9.2, 10 and 11 (so for 9 observations points). Despite of that, the error function computed with the optimised value is lower than with initial values. So, the using of the optimization software SiDoLo permits to improve the fitting of the numerical values on experimental values. Extensometer 1 and 5 are the same location in the two main cavern cross sections. Extensometers 9, 10 and 11 are located in the small cavern. It could be interesting to do identification only on the main cavern monitoring and in two steps: first, on the monitoring of the first cross section (EF 1 to 4) and after on the other section (EF 5 to 8).

The difference $[Z_s(A,t_i)- Z_s^*(t_i)]$ between the computed value $Z_s(A,t_i)$ and the experimental value $Z_s^*(t_i)$ is always of the same sign excluded for values EF 1.15 and EF 5.4. In fact, for these observations points, the computed value before optimization is more important than the measured values while after optimization, this value is lower than the measurement. In a second time, these two displacements will not be considered in the optimization process. So, only 18 displacements will be used for the identification.

Figure 19 shows the error function value for each displacement and in the two cases: before and after optimization. With initial values, maximal error functions are obtained for EF 1.4 and 2.3 while they are obtained for EF 1.5 and 11.6 with optimized values.

138

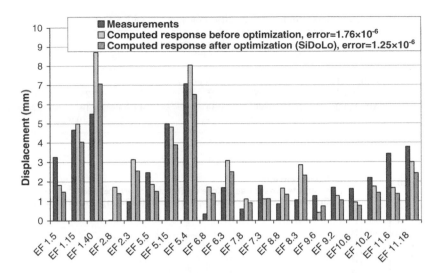

Figure 18.   Comparisons between experimental and numerical measurements after the first identification attempt with SiDoLo.

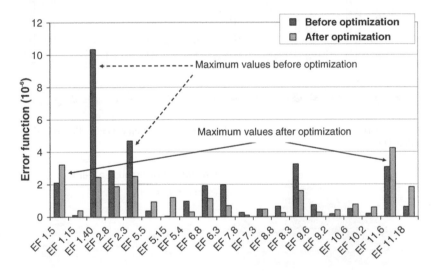

Figure 19.   The error function values during the first identification attempt with SiDoLo.

In a second time, identifications are performed only on 18 displacements. Table 6 resumes the parameters values provided by SiDoLo at the end of the two identification attempts, the respective error function and the iteration number required to achieve the identification process. For the second attempt the iteration number is not so important than for the first attempt. But, the error function is more important than for the first identification attempt. So, the fitting of the numerical values on experimental values is not improved.

Figure 20 compares the measured displacement to the computed displacements with the initial values and with the optimized values of the two parameters. It also gives the respective error function values. Similar observations to those obtained for the first identification attempt can be made. The error function computed with the optimised value is lower than with initial values. The difference $[Z_s(A,t_i) - Z_s^*(t_i)]$ between the computed value $Z_s(A,t_i)$ and the experimental value $Z_s^*(t_i)$ is always of the same sign.

Table 6. Results of the two identification attempts with SiDoLo.

Cases	E GPa	R –	Error function × 10⁻⁶	Iteration	Time Hour
Initial values	45.0	2	1.76 or 1.7	–	–
Attempt 1	55.0	1.98	1.25	25	27
Attempt 2	56.7	1.90	1.34	12	14
Difference	3%	−4.2%	7.2%	−108%	−93%

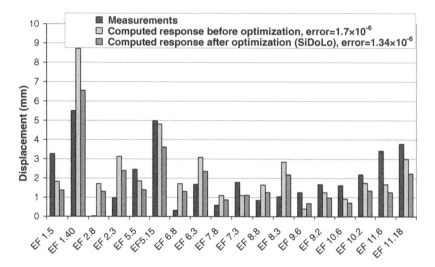

Figure 20. Comparisons between experimental and numerical measurements after the second identification attempt with SiDoLo.

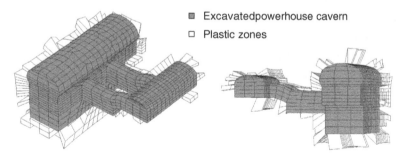

Figure 21. Plastic zones at the last excavation stage in the case of the second identification attempt.

A similar error function value for each displacement to the one computed in the first identification attempt is obtained.

Figure 21 shows the plastic zones computed at the last excavation stage with the optimized values obtained for the second identification attempt (E=56.7 GPa and R=1.9). Plastic zones are very limited with the two optimized parameters sets. So, it could be very difficult to identify C and φ. These parameters have no influence on the results. Maybe, it might be interesting to adopt an elastic linear constitutive model to represent the rock behaviour and to compare the computed displacements. Besides, the computation time will be shorter.

Table 7. Comparisons between the two optimization methods.

Cases	E GPa	R –	Error function $\times 10^{-6}$	Time Hour
SiDoLo	56.7	1.90	1.34	14
ES	52.1	1.72	1.37	49
Difference	−8.1%	−9.4%	2.2%	250%

4.2.2 *Identifications with the evolution strategies algorithm and comparisons between the two optimizations methods*

In order to verify that SiDoLo does not provide a local minimum, a probabilistic optimization method is performed in a second step. An evolution strategies algorithm is used to find the optimized values of the two parameters and obtained results are compared between the two methods. Only 18 displacements are used in the optimization process.

The computed displacements with the optimized values provided by the two methods are close to each other. Table 7 compares the parameters value provided by SiDoLo at the end of the second identification attempt to the values provided by the evolution strategies algorithm (ES), the respective error functions and the computation time required to achieve the identification process.

The error functions are quite similar. A difference of 8.1% is observed on the Young modulus value provided by the two methods and for the in stresses ratio a difference of 9.4% is obtained. The time required by the evolution strategies algorithm is more than 3 times higher in comparison with the time required for the other method. Only one generation is required in the evolution strategies algorithm to achieve the optimization process due to that for the parents of the first generation the stop criterion is verified. In order to verify that the solution provides by the evolution strategies algorithm corresponds to the global minimum it is necessary to decrease the tolerance criteria in order to obtain error function lower than 10^{-7} and to launch a new identification. But, the computation time will be important.

5 CONCLUSION

The use of inverse analysis on in situ measurements carried out during the construction of a real work is rare. This might be due to the difficulty met to apply inverse analysis (measurements quality, complexity of the geometry and of the excavation stages, ...). In an underground works case, the numerical model required many simplifications in order to obtain acceptable computation times. The adopted constitutive model should not be too sophisticated with many parameters but should properly simulate the soil behaviour. For the studied cases, a linear elastic perfectly plastic model with a Mohr Coulomb failure criterion is considered to represent the soil mass.

This article shows the coupling of two optimization methods with 2D and 3D numerical modelling of two geotechnical works: a tunnel and underground powerhouse caverns. The use of a deterministic method may provide local minima. The probabilistic method based on evolution strategies permits to find global minimum. But this method required more important computation time. In the tunnel case, with two unknown parameters, the studied cost functions are not convex and present a valley. Moreover the error values are important for all the studied cost function (> 20 %). Thus the problem is difficult to solve. The solution found by the evolutionary algorithm varies according to the studied error function. For the second used function $L_E(A)$ it depends on the values of absolute and relative errors. But if just a relative error or an absolute error is considered the same solution is found. The solution changes according to the considered type of error. The calculation time is similar for all the error functions. This article shows also that the parent, combination and offspring population size influence the results. With an increase of the population size, the same error function value is obtained at the same number of generations. The provided value sets are different of about 10%

and the calculation time can increase of 50%. In order to improve results a non linear elasticity and a Young modulus which varies with depth should be introduced. Moreover, maybe the evaluation of the earth pressure ratio by means of in situ tests needs to be more investigated. This ratio might be considered as a complementary parameter to identify.

In the complex powerhouse caverns case, a linear elastic constitutive model could be considered to represent the rock behaviour because a few plastic zones appear. And, this model should permit to reduce computation time. Moreover, in order to verify that the solution provides by the evolution strategies algorithm corresponds to the global minimum in the case of the complex powerhouse caverns, it is necessary to decrease the tolerance criteria and launch some new identifications.

ACKNOWLEDGMENTS

The authors wish to express their acknowledge to Professor Lino Costa for providing the evolution strategy algorithm.

REFERENCES

Costa, L. & Oliveira, P. 2001. Evolutionary algorithms approach to the solution of mixed integer non linear programming problems. *Computers and Chemical Engineering* 25: 257–266.

Eclaircy-Caudron, S., Dias, D., Kastner, R. & Chantron, L. 2006. Identification des paramètres du sol rencontré lors du creusement d'un tunnel par analyse inverse. *Proc. JNGG 2006*. Lyon, France.

Finno, R.J. & Calvello, M. 2005. Supported Excavations: the Observational Method and Inverse Modeling. *Journal of Geotechnical and Geoenvironmental Engineering* 131(7): 826–836.

Goldberg, D. 1991. *Algorithmes génétiques : exploration, optimisation et apprentissage automatique*. Adisson-Wesley Edition.

Hicher, P.Y. & Shao, J.F. 2002. *Modèles de comportement des sols et des roches 2: lois incrémentales, viscoplasticité, endommagement*. Hermès Science Publications.

Itasca Consulting group 2005. *FLAC^{3D} user's manual*.

Itech 2002. *Cleo^{2D} user's manual*.

Jeon, Y.S. & Yang, H.S. 2004. Development of a back analysis algorithm using FLAC. *International Journal of Rock Mechanics and Mining Sciences* 41(1): 447–453.

Levasseur, S., Malecot, Y., Boulon, M. & Flavigny, E. 2005. Analyse inverse par algorithme génétique en géotechnique : application à un problème d'excavation. *17th Congrés Français de Mécanique*. Troyes, France.

Miranda, T., Gomes Correira, A., Ribeiro e Sousa, L. & Lima, C. 2005. Numerical modelling of a hydroelectric underground station using geomechanical parameters obtained by artificial intelligence techniques. *4th Portuguese-Mozambican of Engineering, Maputo, 30 August-1 September 2005:807–816*.

Panet, M. 1995. *Le calcul des tunnels par la méthode convergence-confinement*. Presse de l'école nationale des Ponts et Chaussées.

Pilvin, P. 1983. Modélisation du comportement d'assemblages des structures à barres. *Ph. D. thesis*. Université Paris VI. Paris, France.

Renders, J.M. 1995. *Algorithmes génétiques et réseaux de neurones*. Hermès Science Publications.

Samarajiva, P. 2005. Genetic algorithms for the calibration of constitutive models for soils. *International Journal of Geomechanics. ASCE*. September 2005: 206–217.

Schwefel, H.-P. 1985. *Evolution and optimal seeking*. John Wiley and Sons.

SiDoLo version2.4495. 2003. *Notice d'utilisation*. Laboratoire Génie Mécanique et Matériaux de l'Université de Bretagne-Sud, Lorient.

Zentar, R. 1999. Analyse inverse des essais pressiométriques : Application à l'argile de Saint-Herblain. *Ph. D. thesis*. Ecole Centrale. Nantes, France.

Applications of Computational Mechanics in Geotechnical Engineering – Sousa,
Fernandes, Vargas Jr & Azevedo (eds)
© 2007 Taylor & Francis Group, London, ISBN 978-0-415-43789-9

The influence of the permeability model in unconfined seepage problems

J.M.M.C. Marques
Universidade do Porto, Faculdade de Engenharia, Departamento de Engenharia Civil, Porto, Portugal

J.M.P. Marado
Instituto Politécnico de Viseu, Escola Superior de Tecnologia, Departamento de Engenharia Civil,
Viseu, Portugal

ABSTRACT: The paper reviews two distinct techniques that can be used for the finite element solution of unconfined seepage problems with a phreatic surface. In the first technique the flow is considered to be circumscribed to the saturated region of the domain, with recourse to either a variable or a constant mesh approach. The second technique considers the phreatic surface as the boundary between the fully saturated and the unsaturated regions of the domain. This involves the adoption of a function for each soil type that accurately conveys the dependence of permeability on pore pressure, making this approach much smoother in numerical terms and physically more realistic. The numerical examples presented will illustrate the pros and cons of the various techniques.

1 INTRODUCTION

Unconfined seepage flow occurs in cases such as the homogeneous earth dam of Figure 1, when there is a free water boundary BC, known as the phreatic surface that is exposed to subsurface air. The phreatic surface intersects at C the downstream slope of the dam, with water seeping out from the so called seepage or wet face CD. AB and DE are equipotential lines and AE is an impervious boundary. It should be noted that both the shape of the phreatic surface BC and the location of the exit point C on the seepage face are unknown *a priori*. This geometric indetermination makes unconfined seepage problems more demanding.

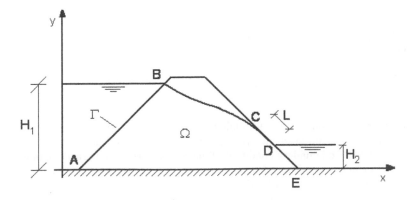

Figure 1. Unconfined seepage in homogeneous earth dam.

2   GOVERNING EQUATIONS

Admitting the validity of Darcy's law and the absence of sources or sinks, the steady seepage flow through a saturated porous medium is governed by the so called quasi-harmonic equation:

$$\frac{\partial}{\partial x_i}(k_{ij}\frac{\partial \phi}{\partial x_j}) = 0 \qquad (i,j = 1,2,3) \tag{1}$$

where x_i are Cartesian coordinates, k_{ij} is the permeability tensor, ϕ is the total head and summation over repeated indices is implied (Zienkiewicz, Taylor & Zhu(2005).

If the domain Ω is homogeneous ($\partial k_{ij}/\partial x_m = 0$) and isotropic ($k_{ij} = k\delta_{ij}$) Equation 1 reduces to the well known Laplace equation

$$\frac{\partial^2 \phi}{\partial x_i \partial x_j}\delta_{ij} = 0 \tag{2}$$

with δ_{ij} being the Krönecker delta.

The boundary Γ of the flow domain, in cases such as depicted in Figure 1, is divided into three zones in what concerns the nature of the boundary conditions:

(i) The head is prescribed along AB, DE and CD ($\phi = H_1$ on AB, $\phi = H_2$ on DE and $\phi = y$ on CD).
(ii) The flux is assumed to be zero across the dam-foundation interface. The velocity vector is therefore tangential to the flow line AE, and its normal component v_n is zero:

$$v_n = k_{ij}\frac{\partial \phi}{\partial x_j}n_i = 0 \tag{3}$$

(iii) On the phreatic surface BC both the head ($\phi = y$) and the flux ($v_n = 0$) are prescribed but its location and shape are unknown. BC is a flow line whose geometrical configuration has yet to be determined.

3   FINITE ELEMENT SOLUTION

Applying the finite element method to the solution of Equation 1 an algebraic equation system is obtained:

$$K_{ij}\phi_j = 0 \qquad (i,j = 1,...,N) \tag{4}$$

where N is the number of nodes of the finite element mesh, ϕ_j is the total head at node j and K_{ij} is a global matrix coefficient given by

$$K_{ij} = \int_{\Omega} k_{mn}\frac{\partial N_i}{\partial x_m}\frac{\partial N_j}{\partial x_n}d\Omega \quad (i,j=1,...,N)\ (m,n=1,2,3) \tag{5}$$

where N_i is the interpolation function for node i and Ω is the flow domain.

The flux boundary condition of the type $v_n = 0$ is automatically incorporated in Equation 4 while the prescribed head boundary condition of the type $\phi = H_1$ can be easily applied using, for example, the procedure employed in structural problems for imposing support settlements.

The solution of the system given by Equation 4 becomes immediate if the flow domain Ω is defined beforehand. However in unconfined seepage problems the flow domain Ω is not fully known *a priori* due to the geometrical indetermination of the phreatic surface.

4 SOLUTION TECHNIQUES FOR UNCONFINED SEEPAGE PROBLEMS

For the finite element solution of unconfined seepage problems there are essentially two alternative approaches: the variable mesh (Taylor & Brown 1967, Finn 1967, Neuman & Witherspoon 1970) and the fixed mesh techniques (Desai 1976, Bathe & Khoshgoftaar 1979, Cividini & Gioda 1984, Lacy & Prevost 1987).

4.1 Variable mesh techniques

An initial guess is made for the phreatic surface location and either the $\phi = y$ head condition or the $v_n = 0$ flux condition are imposed on the corresponding nodes. The system given by Equation 4 is solved for the flow domain Ω thus defined and the phreatic surface nodes are checked for compliance with the other boundary condition. The nodal coordinates are then adjusted taking into account the detected error magnitude and a new mesh is in this way obtained. The process is iterated until the errors on the phreatic surface nodes are deemed small enough.

This technique leads to the reformulation of the global coefficient matrix K_{ij} and consequently to the solution of a new equation system at each iteration. The method is cumbersome and requires a good initial guess, otherwise the mesh may get too distorted during the iteration process. If in addition to the seepage analysis a stress analysis is to be performed on the dam of Figure 1, the convenience of working with the same finite element mesh for both analyses is lost with the variable mesh technique.

4.2 Fixed mesh techniques

In this case the mesh remains constant while the permeability is iteratively adjusted at Gauss point level to reflect its dependence on the pore pressure value.

Two major numerical alternatives exist at this level in correspondence with two distinct physical assumptions about unconfined seepage phenomena.

In the first assumption (which is also implicit in the variable mesh techniques described above) the flow is considered to be circumscribed to the saturated region of the domain. In terms of fixed mesh technique this corresponds to the so called variable permeability method, whereby the permeability for those Gauss points found to have negative pore pressure is drastically and somewhat arbitrarily reduced to a fraction (typically 0.1%) of the initial value (Bathe & Khoshgoftaar 1979).

In the second approach the phreatic surface is taken as the boundary between fully saturated and unsaturated flow regions and recourse is made to a permeability function for accurately modelling its dependence on pore pressure taking into account the soil properties. This is physically more realistic and leads to better numerical results, as shown by the following examples (Marado 1994).

5 NUMERICAL EXAMPLES

The homogeneous dam of Figure 2 has been analysed with the fixed mesh technique considering two permeability scenarios:

(i) isotropy, with $k_x = k_y = 8.467\text{e-}05$ m/s;
(ii) anisotropy, with $k_x = 3k_y = 2.540\text{e-}04$ m/s.

5.1 Results obtained with the variable permeability method

The variable permeability method has been applied adopting the curve of Figure 3.

The flow nets obtained for the isotropic and anisotropic permeability cases are shown in Figures 4 and 5, respectively.

Note that in both cases the top flow line corresponding to the phreatic surface intersects the upstream slope at point P somewhat below the reservoir level.

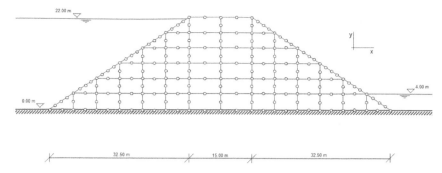

Figure 2. Geometry and finite element mesh for homogeneous earth dam.

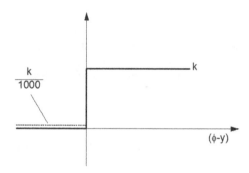

Figure 3. Permeability vs. pressure head for the variable permeability method.

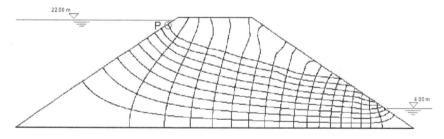

Figure 4. Flow net for the isotropic case obtained with the variable permeability method.

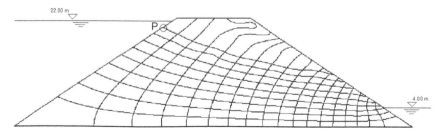

Figure 5. Flow net for the anisotropic case obtained with the variable permeability method.

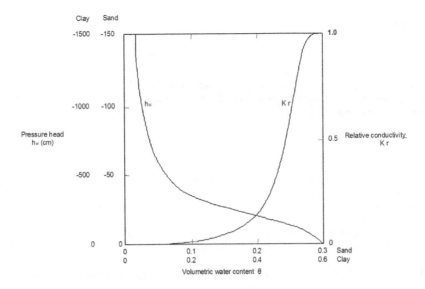

Figure 6.   Permeability function.

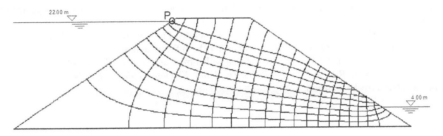

Figure 7.   Flow net for the isotropic case obtained with the permeability function method.

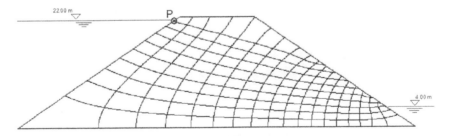

Figure 8.   Flow net for the anisotropic case obtained with the permeability function method.

5.2   *Results obtained with the permeability function method*

The permeability (or hydraulic conductivity) is multiplied by the relative conductivity parameter K_r of Figure 6 (Neuman 1975) when the pressure head becomes negative. This gives a very smooth transition between the fully saturated and the unsaturated zones which is also reflected in the better quality of the flow nets obtained for the isotropic and anisotropic permeability cases, as shown in Figures 7 and 8, respectively. Note that point P, at the intersection of the phreatic surface and the upstream slope, now coincides with the reservoir level.

6 CONCLUSIONS

Finite element solution strategies for unconfined seepage problems have been reviewed. The drawbacks of the variable mesh technique were outlined and are clearly overcome by the fixed mesh procedures. In this context the numerical examples presented show the convenience of using a permeability function to accurately model the effect of negative pore pressure, instead of adopting a step function type reduction.

REFERENCES

Bathe, K.J. & Khoshgoftaar, M.R. 1979. Finite element free surface seepage analysis without mesh iteration. *Int. J. Num. Anal. Meth. Geomechanics*, 3: 13–22.
Cividini, A. & Gioda, G. 1984. An approximate f.e. analysis of seepage with a free surface, *Int. J. Num. Anal. Meth. Geomechanics*, 8: 549–566.
Desai, C.S. 1976. Finite element residual schemes for unconfined flow, *IJNME*, 10: 1415–1418.
Finn, W.L.D. 1967. Finite element analysis of seepage through dams. *J. Soil Mech. Found. Div. ASCE*, 93: 41–48.
Lacy, S.J. & Prevost, J.H. 1987. Flow through porous media: a procedure for locating the free surface. *Int. J. Num. Anal. Meth. Geomechanics*, 11: 585–601.
Marado, J.M.P. 1994. *Analysis of Seepage Problems by the Finite Element Method*. MSc thesis (in Portuguese), Porto: FEUP.
Neuman, S.P. & Witherspoon, P.A. 1970. Finite element method of analyzing steady seepage with a free surface. *Water Res. Research*, 6: 889–897.
Neuman, S.P. 1975. Galerkin Approach to Saturated-Unsaturated Flow in Porous Media. In R.H Gallagher et al. (eds.), *Finite Elements in Fluids (Vol.1)*, Chichester: John Wiley and Sons.
Taylor, R.L. & Brown, C.B. 1967. Darcy flow solution with a free surface. *J. Hydraulic Div. ASCE*, 93: 25–33.
Zienkiewicz, O.C., Taylor, R.L. & Zhu, J.Z. 2005. *The Finite Element Method: Its Basis and Fundamentals*, Oxford: Butterworth-Heinemann.

Artificial intelligence

Applications of Computational Mechanics in Geotechnical Engineering – Sousa,
Fernandes, Vargas Jr & Azevedo (eds)
© 2007 Taylor & Francis Group, London, ISBN 978-0-415-43789-9

Alternative models for the calculation of the RMR and Q indexes for granite rock masses

T. Miranda, A. Gomes Correia & I. Nogueira
University of Minho, Department of Civil Engineering, Guimarães, Portugal

M. F. Santos & P. Cortez
University of Minho, Department of Information Systems, Guimarães, Portugal

L. Ribeiro e Sousa
University of Porto, Department of Civil Engineering, Porto, Portugal

ABSTRACT: Empirical classification systems like the RMR and Q are often used in current practice of geotechnical structures design built in rock masses. They allow obtaining an overall description of the rock mass and the calculation, through analytical solutions, of strength and deformability parameters which are determinant in design. To be applied these systems need a set of geomechanical information that may not be available or can be difficult to obtain. In this work it is intended to develop new alternative regression models for the calculation of the RMR and Q indexes using less data than the original formulations and keeping a high accuracy level. It is also intended to have an insight of which parameters are the most important for the prediction of the indexes and in the rock masses behaviour. This study was carried out applying Data Mining techniques to a database of the empirical classification systems applications in a granite rock mass. Data Mining is a relatively new area of computer science which concerns with automatically find, simplify and summarize patterns and relationships within large databases. The used Data Mining techniques were the multiple regression and artificial neural networks. The developed models are able to predict the two geomechanical indexes using less information that in the original formulations with a good predictive capacity.

1 INTRODUCTION

Rock mass characterization is normally carried out through the application of empirical classification systems which use a set of geomechanical data and provide an overall description of the rock. Moreover, they allow obtaining other important information like support needs, stand-up time, geomechanical parameters among others.

The different classification systems have some well known drawbacks and limitations due mainly to their empirical base (Miranda, 2003). However, they are still very useful in practice therefore there is a need to improve their efficiency.

Two of the most used classification systems are the RMR-Rock Mass Rating (Bieniawski, 1989) and the Q-system (Barton et al., 1974). The RMR system is based on the consideration of six geological/geotechnical parameters. To each parameter is assigned a relative weight related to the rock mass characteristics and the final RMR value is the sum of these weights and can vary from 0 to 100. The parameters considered by this system are the following: P_1 – uniaxial compressive strength; P_2 – Rock Quality Designation (RQD); P_3 – Discontinuities spacing; P_4 – Discontinuities conditions; P_5 – Underground water conditions; and P_6 – Discontinuities orientation.

The Q-system also uses six parameters to which values have to be assigned depending on the rock mass characteristics. The final Q value is then obtained through the following expression:

$$Q = \frac{RQD}{J_n} \cdot \frac{J_r}{J_a} \cdot \frac{SRF}{J_w}$$
(1)

where the different parameters are related with: J_n – number of discontinuities sets, J_r – discontinuities rugosity, J_a – discontinuities alteration, SRF – stress state and J_w – underground water.

Most of the times there are some difficulties to apply these classifications systems. Some of the data required to their application may not be available, can lack of reliability or may be difficult/expensive to obtain. Also, the considered parameters may have different importance depending on the type of rock mass being analyzed.

In this work it is intended to develop new alternative models to calculate the RMR and Q indexes for the particular case of granite rock masses which are very important in the North of Portugal. They are intended to use only the most important parameters in the behavior of granite rock masses with a good predictive accuracy.

This study was carried out using a large database of the empirical systems application in an important underground structure built in a granite rock mass. On this database Data Mining techniques were applied to obtain the new models. Multiple regression techniques and artificial neural networks (ANN) were used. The first are simpler to use and analyze and allow having an insight of which parameters are the most important in the indexes prediction while the latter are more complex and suitable for highly non-linear problems.

2 KNOWLEDGE DISCOVERY IN DATABASES AND DATA MINING

Currently, there is a great expansion of information that needs to be stored. It is important to use computational tools to explore this data which often presents high complexity and can hold valuable information such as trends or patterns that can be very useful (Goebel & Gruenwald, 1999).

In the past, two major approaches have been used for this goal: classical statistics and knowledge from experts. However, the number of human experts is limited and they may overlook important details, while classical statistic analysis does not give adequate answer when large amounts of complex data are available. The alternative is to use automated discovery tools to analyze the raw data and extract new and useful knowledge (Hand et al., 2001).

Due to the awareness of the great potential of this subject there has been an increasing interest in the Knowledge Discovery from Databases (KDD) and Data Mining (DM) fields. These terms are often confused. KDD denotes the overall process of transforming raw data into knowledge and DM is just one step of the KDD process, aiming at the extraction of useful patterns from the observed data. The knowledge derived through DM is often referred to as models or patterns and it is very important that this knowledge is both novel and understandable.

The KDD process consists in the following steps (Figure 1):

• Data selection: the application domain is studied and relevant data is collected.

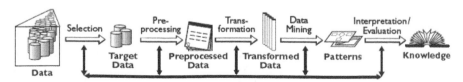

Figure 1.   Phases of the KDD process (Fayyad et al., 1996).

- Pre-processing or data preparation: noise or irrelevant data is removed (data cleaning) and multiple data sources may be combined (data integration). In this step appropriate prior knowledge can be also incorporated.
- Transformation: data is transformed in appropriate forms for the Data Mining process.
- Data Mining: intelligent methods are applied in order to extract models or patterns.
- Interpretation: results from the previous step are studied and evaluated.

DM is a relatively new area of computer science that lies at the intersection of statistics, machine learning, data management, pattern recognition, artificial intelligence and others. DM is thus emerging as a class of analytical techniques that go beyond statistics and concerns with automatically find, simplify and summarize patterns and relationships within large data sets.

There are several DM techniques, each one with its own purposes and capabilities. Examples of these techniques include Multiple Regression Analysis, Decision Trees and Rule Induction, Neural and Bayesian Networks, Learning Classifier Systems and Instance-Based algorithms (Lee & Siau, 2001; Berthold & Hand, 2003).

3 MATERIALS AND METHODS

3.1 *The database*

The data for these models was assembled from the Venda Nova II powerhouse complex which is an important underground work recently built in the North of Portugal. The interested rock mass is a granite formation so the conclusions drawn in this study are only applicable to formations with similar characteristics. The overall process was carried out in the following steps:

- collect geotechnical data from Venda Nova II powerhouse complex;
- build and organize a database with the collected data;
- explore the data using DM techniques to induce the models.

The collected data was composed by applications of the empirical RMR and Q systems. After some data cleaning work it was then organized and structured in a database composed of 1230 examples and 21 attributes which are described in Table 1.

Other attributes were added to the database in order to check their possible influence on the models. Globally, 9 new attributes were added and are presented in Table 2.

The histograms of some variables presented skewed distributions (Figure 2). This fact can influence the quality of the models specially those based on neural networks since this kind of algorithm can learn better the behaviour of variables with normal distributions. This way, and after some preliminary trial calculations, it was decided to proceed to the transformation of some variables in order to maximize their normality.

The data is based on the results obtained in a granite rock mass with a good overall quality. More specifically, the main limitations that should be considered are high uniaxial compressive strength (>100 MPa), RQD values over 65% and slightly wet to dry rock mass. The models developed in this work should only be applied to rock masses with similar characteristics.

3.2 *Modelling and Evaluation*

The SAS Enterprise Miner, registered trademark of the SAS Institute Inc., was used as modelling tool. It performs DM tasks and combines statistical analysis with graphical interfaces and delivers a wide range of predictive models.

In the SAS Enterprise Miner the DM tasks are carried out programming and connecting nodes in a graphical workspace, adjust settings, and run the constructed workflow. In Figure 3 the workflow used in this work is presented.

The algorithms used for the regression models were multiple regression and ANN. The applied ANN was a multilayer feed-forward network with one hidden layer of six neurons. Focus was

Table 1. Name and description of the attributes in the original database.

Name	Description
RQD	*Rock Quality Designation*
J_w	Factor related with the underground water
J_n	Factor related with the number of discontinuities sets
J_r	Factor related with discontinuities rugosity
J_a	Factor related with the weathering degree of discontinuities
SRF	Factor related with the stress state in the rock mass
Q	Rock mass quality index proposed by Barton et al. (1974)
Q'	Altered form of the Q index ($Q' = RQD/J_n * J_r/J_a$)
RCU	Uniaxial compressive strength
P_1	Weight related with the uniaxial compressive strength of the intact rock
P_2	Weight related with the RQD
P_3	Weight related with discontinuities spacing
P_4	Weight related with discontinuities conditions
P_5	Weight related with the underground water conditions
P_6	Weight related with discontinuities orientation
P_{41}	Discontinuities conditions – persistence
P_{42}	Discontinuities conditions – aperture
P_{43}	Discontinuities conditions – rugosity
P_{44}	Discontinuities conditions – filling
P_{45}	Discontinuities conditions – weathering
RMR	Rock Mass Rating proposed by Bieniawski (1978)

Table 2. List of attributes added to the original database.

Name	Description
RQD/J_n	Ratio which represents the compartimentation of the rock mass
J_r/J_a	Ratio which represents the shear strength of discontinuities
J_w/SRF	Ratio which represents an empirical factor named "active stress"
logQ	Base 10 logarithm of the Q value
logQ'	Base 10 logarithm of the Q' value
GSI	Geological Strength Index proposed by Hoek et al., 2002
N	Altered form of the Q index ($Q' = RQD/J_n * J_r/J_a * J_w$)
RCR	Altered form of the RMR index ($RCR = P_2 + P_3 + P_4 + P_5 + P_6$)
RCU	Uniaxial compressive strength

drawn to the multiple regression models because it was intended to obtain the explanatory physical knowledge behind the models. Moreover, these models are simpler to use and to implement. The neural network models were used mainly for comparison purposes and are an open issue for further research since it was not tried to optimize their behaviour.

In regression problems the goal is to induce the model which minimizes an error measurement e between real and predicted values considering N examples. The used error measures were the following:

$$\text{Mean Absolute Deviation: } MAD = \frac{\sum_{i=1}^{N} |e_i|}{N} \qquad (2)$$

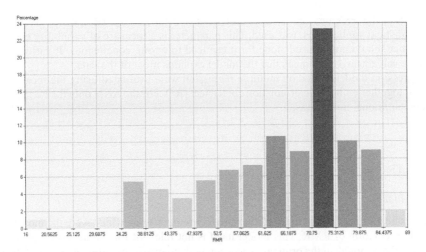

Figure 2. Histogram of the RMR variable where is possible to observe the skewness of the distribution.

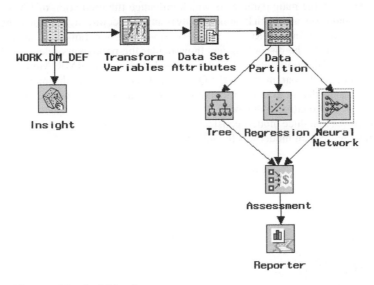

Figure 3. Workflow used for the DM tasks.

Root Mean Squared Error: $RMSE = \sqrt{\dfrac{\sum_{i=1}^{N} e_i^2}{N}}$ (3)

To validate and assess the models accuracy the holdout method was used. In this method data is randomly partitioned into two independent sets, a training set and a test set. In this case, 2/3 of data was used for training and 1/3 for testing. The training set is used to induce the model and its accuracy is estimated with the test set. For each model 10 runs were carried out randomizing the data within the training and testing sets. The mean and confidence intervals for the error measures were then computed considering the results of the 10 runs and a 95% confidence interval of a T-student distribution. These statistical measures define the range of expected errors for future predictions of the final model which is induced using all the data for training.

In addition to the error measures also the coefficient of determination (R^2), which is very common in many statistical applications, was used. This parameter is a measure of variability explained by the model but should not be used alone for it can lead to wrong conclusions. It varies between 0 and 1 and a value near 1 may mean that the model explains most of the data.

The regression models based on multiple regression were evaluated using the measures MAD and RMSE together with the determination coefficient (R^2). For the ANN only the RMSE was used due to computational limitations.

4 RESULTS

4.1 *RMR index*

This study started considering firstly all the variables to determine which were the most important ones in the prediction of RMR. This model itself is not relevant for prediction purposes since it uses more information than the original expression with no profit. In Figure 4, a plot of the relative importance of the main attributes for the RMR variable is presented.

As it was expected, the main parameters which influence the prediction of RMR are the ones related to its calculation even though P_1 appears only in an indirect way in the form of the unconfined compressive strength (defined as RCU in the plot). Among these parameters, the most important are the ones related with the discontinuities. In particular the parameters related with conditions (P_4) and orientation of discontinuities (P_6) are very good predictors of RMR. Moreover, in the scale of relative importance, the parameters of the Q system also related with discontinuities appear (J_n and J_r/J_a). This means that in granite formations data related to the discontinuities is a very good predictor of the overall quality of the rock.

The next step was to induce models considering only the most important parameters: P_3, P_4 and P_6. The obtained regression model was the following:

$$RMR = 34.77 + 0.065 \times P_3^2 + 1.369 \times P_4 + 0.977 \times P_6 \tag{4}$$

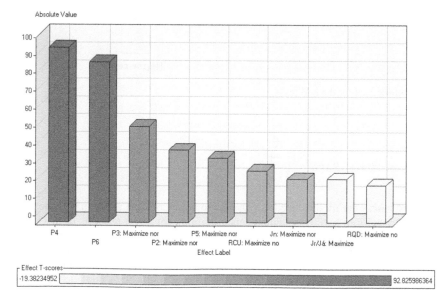

Figure 4.   Relative importance of the attributes for the prediction of the RMR variable.

In Table 3 the results for the regression and ANN models are presented in terms of average errors and determination coefficient and correspondent T-student 95% confidence intervals. The results for the models which use all the attributes are presented only for comparison matters.

The models which use all attributes are very accurate. The error measures are low and the determination coefficient is near 1. Using only the three main parameters, the error significantly increases. This is because only half of the parameters used in the original expression are applied. Nevertheless, the error can be considered low for engineering purposes. Analysing the MAD and RMSE values a prediction error around 3 is expected. This means that, for instance, if a rock mass has a "real" RMR value of 65, a value within [62; 68] will be predicted which is acceptable. This expression can be useful for preliminary stages of design or when only information about discontinuities is available or is reliable.

Considering the RMSE, the ANN slightly outperforms the regression models. Only for the ones with less attributes the difference can be considered significant. In this case the RMSE for the ANN is approximately 20% less than the correspondent value of the regression model. In Figure 5 the plot of real versus predicted RMR values is presented.

As it can be seen, the values lay near a 45 degree slope line which means that the prediction model shows a good accuracy. However, the deviations between real and predicted values increase with decreasing rock mass quality. For RMR values below 30–35 the prediction error increases and the model tends to overestimate the RMR. Since the model is based in the discontinuities characteristics this fact can be explained by the loss of importance of discontinuities for poorer rock masses.

Table 3. Results of the models for the RMR index.

All attributes				P_3, P_4 and P_6			
Regression			ANN	Regression			ANN
R^2	MAD	RMSE	RMSE	R^2	MAD	RMSE	RMSE
0.995 ± 0.001	0.650 ± 0.050	1.094 ± 0.073	1.070 ± 0.070	0.944 ± 0.005	2.565 ± 0.083	3.522 ± 0.169	2.857 ± 0.114

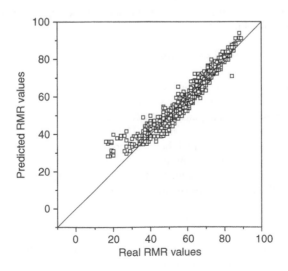

Figure 5. Real versus Predicted RMR values for regression model with parameters P_3, P_4 and P_6.

Table 4. Results for the multiple regression model considering parameters P_3, P_4 and P_6 and using RMR^2.

Regression		
R^2	MAD	RMSE
0.954 ± 0.004	2.179 ± 0.081	3.172 ± 0.119

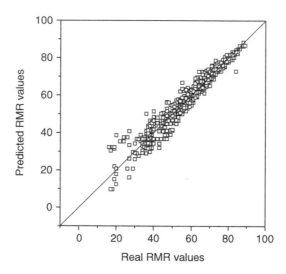

Figure 6. Real versus Predicted RMR values for regression model with parameters P_3, P_4 and P_6 and considering the transformation RMR.

The plot of Figure 5 shows a tail with an almost quadratic trend. In order to minimize this fact a transformation of the RMR variable was performed and calculations were repeated using the squared value of RMR. The obtained regression model is presented in equation 5 and the results are resumed by Table 4 and Figure 6.

$$RMR^2 = 1036.7 + 7.148 \times P_3^2 + 166.3 \times P_4 + 116.7 \times P_6 \qquad (5)$$

This transformation led to a slight reduction on the error measurements (approximately 0.4 for each) and a small increase on R^2. In Figure 6, a loss of accuracy for lower RMR values can still be observed. However, this only happens for RMR values below 30 and the overestimation trend is no longer observed has in the previous model. The points are almost equally distributed along the 45 degree slope line which means that the mean prediction error is close to 0.

As it was already referred, the Q system related parameters J_n and J_r/J_a are also important to the RMR prediction. These attributes were added to this model and calculations were again performed. However, only marginal increased performance was achieved.

4.2 Q index

The preliminary runs for the Q variable using all attributes indicated that the use of the base 10 logarithm of Q (logQ) led to a significant improvement of the results. Table 5 shows the results for the models which use all the attributes and the most important ones. As can be observed in Figure 7,

Table 5. Results of the models for the Q index.

All attributes				J_r/J_a, J_n, SRF			
Regression			ANN	Regression			ANN
R^2	MAD	RMSE	RMSE	R^2	MAD	RMSE	RMSE
0.997 ± 0.000	0.016 ± 0.001	0.031 ± 0.003	0.030 ± 0.003	0.989 ± 0.001	0.049 ± 0.002	0.075 ± 0.004	0.075 ± 0.005

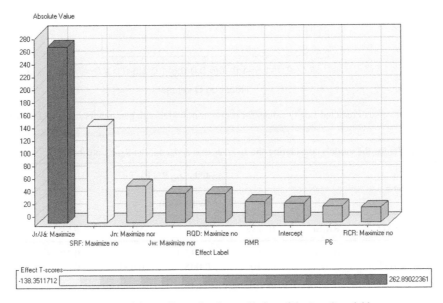

Figure 7. Relative importance of the attributes for the prediction of the Log Q variable.

the most important attributes are the J_r/J_a ratio and the SRF and J_n variables. This regression model is translated by the following expression:

$$\log Q = 2.00 + 0.47 \times \ln\left(\frac{J_r}{J_n \times J_a \times SRF^{1.07}}\right) \tag{6}$$

As it happened for the RMR, the parameters related with discontinuities have a significant effect on the prediction of this quality index together with the parameter related with the stress state. This point corroborates the previous conclusion that the discontinuities characteristics are good predictors of the overall rock mass quality. Analysing the values of R^2 in Table 5 it can be seen that the values are very high for both models. The error values are low considering that the target variable ranged approximately from −1.85 to 2.13. Figure 8 shows the plot of real against predicted values and a good relation can be observed.

Since it was concluded that the parameters related with the discontinuities are very much related with both studied indexes, two more sets of variables were tested: one using only the variables J_r/J_a and J_n and other using these variables together with the parameters related with the discontinuities of the RMR system (P_3, P_4 and P_6). The latter is justified since once it is possible to obtain information

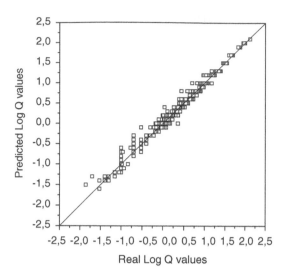

Figure 8. Real versus Predicted Log Q values for regression model with parameters J_r/J_a, SRF and J_n.

Table 6. Results for the models considering the J_r/J_a, J_n and J_r/J_a, J_n, P_3, P_4, P_6 attributes.

J_r/J_a, J_n				J_r/J_a, J_n, P_3, P_4, P_6			
Regression			ANN	Regression			ANN
R^2	MAD	RMSE	RMSE	R^2	MAD	RMSE	RMSE
0.908 ± 0.009	0.149 ± 0.007	0.214 ± 0.013	0.204 ± 0.010	0.933 ± 0.005	0.128 ± 0.004	0.184 ± 0.009	0.152 ± 0.007

for the J_r/J_a and J_n variables it is not difficult to deduce values for P_3, P_4 and P_6. The regression models are translated by equations 7 and 8 and the overall results are presented in Table 6 and in Figure 9.

$$\log Q = 2.17 + 0.57 \times \ln\left(\frac{J_r}{J_n^{1.03} \times J_a}\right) \tag{7}$$

$$\log Q = 1.27 + 0.43 \times \ln\left(\frac{J_r}{J_n^{0.95} \times J_a}\right) + 0.0015 \times P_3^2 + 0.015 \times P_4 + 0.0094 \times P_6 \tag{8}$$

Even though the R^2 value is still within acceptable values, for the simpler model the errors significantly increase. This is especially true again for poorer rock mass conditions. Figure 9 shows high dispersion for logQ values approximately below −0.5 (Q < 0.3 or RMR < 35). This is also due to the loss of discontinuities importance for rock masses with low geomechanical characteristics as discussed before and shows the importance of the stress parameter consideration. The results also show that the behaviour of the models is significantly enhanced with the inclusion of the discontinuities parameters of the RMR system resulting in reduced dispersion and error values. A thorough discretization about the discontinuities minimizes the lack of information about the stress state parameter.

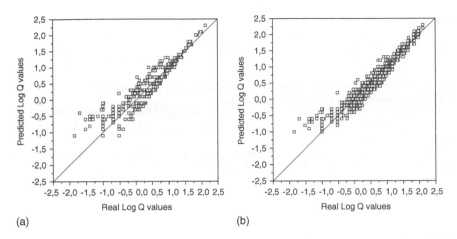

(a)   (b)

Figure 9. Real versus Predicted Log Q values for regression models with: (a) parameters J_r/J_a and (b) J_n and J_r/J_a, J_n, P_3, P_4 and P_6.

5 CONCLUSIONS

The empirical geomechanical classification systems like the RMR and Q are still very much used in practice. In this paper new alternative models for the calculation of these indexes were developed using data collected during the Venda Nova II hydroelectric scheme construction. The interested rock formation was a granite with good overall geomechanical quality so the induced models should only be used in rock masses with similar characteristics.

The models were developed applying DM techniques to the data previously organized and structured. DM is a recent are in computer science which uses a set of different tools from areas like machine learning and artificial intelligence among others to automatically find new and relevant knowledge from raw data.

Regression models for the RMR and Q indexes were developed using multiple regression and ANN. In all cases it was possible to induce accurate and reliable models that can be useful for practitioners and researchers using different sets of parameters. They have the advantage of using less information than the original formulations maintaining high accuracy levels. Moreover, they allowed drawing some conclusions about the physical aspects and main phenomena behind the behaviour of granite rock masses.

One interesting issue was the fact that the most important parameters for prediction were the ones related with the discontinuities. This means that in good quality granite formations this data is a very good predictor of the overall quality of the rock masses. The prediction models loose accuracy for rock formations which lay in the border between hard-soil and soft rock due to the loss of discontinuities importance.

ACKNOWLEDGMENTS

The authors wish to express their acknowledge to EDP Produção EM for authorization and making available the necessary data. This work was financed by the Foundation for Science and Technology (FCT) in the framework of the research project POCI/ECM/57495/2004, entitled *Geotechnical Risk in Tunnels for High Speed Trains*.

REFERENCES

Barton, N., Lien, R., & Lunde, J. 1974. Engineering Classification of Rock Masses for the Design of Tunnel Support. *Rock Mechanics, Springer-Verlag*, Vol. 6, pp. 189–236.

Berthold, M. & Hand, D. 2003. Intelligent Data Analysis: An Introduction. *Springer*, Second Edition.

Bieniawski, Z. T. 1989. Engineering Rock Mass Classifications. *John Wiley & Sons*, 251p.

Fayyad, U.; Piatesky-Shapiro, G.; Smyth, P. 1996. From Data Mining to Knowledge Discovery: an overview. In Fayyad et al. (eds) *Advances in Knowledge Discovery and Data Mining*. AAAI Press/The MIT Press, Cambridge MA, pp 471–493.

Goebel, M. & Gruenwald, L. 1999. A survey of Data Mining and Knowledge Discovery Software Tools. *SIGKDD Explorations*, 1 (1): 20–33, ACM SIGKDD.

Hand, D.; Mannila, H.; Smyth, P. 2001. Principles of Data Mining. *MIT Press*, Cambridge, MA.

Lee, S. & Siau, K. 2001. A review of data mining techniques. *Industrial Management & Data Systems*, 101 (1): 41–46, MCB University

Miranda, T. 2003. Contribution to the calculation of geomechanical parameters for underground structures modelling in granite formations. *MCs thesis*, UM, Guimarães, 186p (in Portuguese).

Applications of Computational Mechanics in Geotechnical Engineering – Sousa,
Fernandes, Vargas Jr & Azevedo (eds)
© *2007 Taylor & Francis Group, London, ISBN 978-0-415-43789-9*

The role of intelligent systems in the safety control
of civil engineering works

E.A. Portela
University of Madeira, Funchal, Portugal

ABSTRACT: The safety control of civil engineering works is an extremely important matter of public safety and economics. The evaluation of the structure safety condition relies in field measurements. Measurement data is very often not fully analysed and exploited by user and valuable information is very often not extracted from the measurements and properly reported. The increased computational power of personal computers together with availability of techniques provided by various disciplines, such as intelligent decision support systems, has improved and simplified our ability to effectively cope with the large volumes of data collected during long term civil engineering works monitoring programs. In the future, engineers will rely much more on intelligent tools and procedures than today for data storage, data retrieval and for processing, analysis and reporting. This paper discusses the application of intelligent systems in the domain of the safety control of civil engineering works and presents an approach applied to the field of dam safety control.

1 INTRODUCTION

A successful monitoring program for evaluating the performance of any civil engineering structure is one that provides the right type and amount of data, at the right time, with an acceptable level of accuracy and in a form that can be readily processed and interpreted. To set up a monitoring system it is necessary to have a good knowledge of the engineering problem being studied and to know and fully understand the objectives of the monitoring program. This provides the basis for deciding on what to monitor, when to make observations, for how long, and at how many points. All parameters of rock, soil, and concrete used in the structure design are measured and tested in a small scale. Thus, the location and the number of measurement points must be selected carefully. This is normally done on the basis of engineering judgment derived from past experience, aided by numerical modelling to determine beforehand the most beneficial locations to place instruments.

For the structure safety control the collection of unnecessary data should be avoided so that the users attention can be focused on the critical parameters that need to be monitored and which will provide early indication of a developing safety problem. Real-time monitoring of selected instruments that track the "vital signs" of the structure can be used as a primary detection network. This close monitoring provides a reasonable basis for reading the designated secondary instruments on a less frequent basis. However, care should be taken in selecting the primary detection instruments to assure that adequate coverage is achieved so that a developing condition will be detected.

The ultimate measure of success of monitoring programs lies in the quality and completeness of the data obtained from instruments and the ability to efficiently process and evaluate data collected from extensive instrumentation arrays – particularly delivered by the automatic data acquisition systems.

The main issue that is facing the structural health monitoring community is not the lack of measurements per se, but rather how to take advantage of the emerging technology in the fields of informatics and electronics, which provide the means to refurbish and improve measurement equipment and the acquisition systems, but above all, improvements regarding the way data is processed and analysed in order to extract useful information from the massive amount of data that is currently coming on-line.

This paper presents an overview of the use of intelligent systems in the safety control of civil engineering works, emphasizing the role of information technology to enhance the structure safety assessment and its great potential for new developments in the near future.

2 A FRAMEWORK FOR STRUCTURE MONITORING

Internet technologies are increasingly facilitating real-time monitoring. The advances in wireless communications are allowing practical deployment for large extended systems. Sensor data, including video signals, can be used for short and long-term condition assessment, emergency response and safety assessment applications. Computer-based automated signal-analysis algorithms routinely process the incoming data and determine anomalies based on pre-defined response thresholds. Upon authentication, appropriate action may be authorized for a strategy, data from thousands of sensors can be analysed with real-time and long-term assessment and decision making implications.

The framework for the monitoring of civil engineering works should network and integrate online real-time heterogeneous sensor data, database and archiving systems, computer vision information, data analysis and interpretation strategies, numerical simulation of complex structural systems, visualisation, probabilistic risk analysis, historical case strategies, and rational statistical decision-making procedures.

Figure 1 depicts a framework for civil engineering works safety assessment. The objective is to improve usability of information technology applications and services and access the knowledge they embody in order to encourage their wider adoption and faster deployment. The focus is on technologies to support the process of acquiring and modelling, navigating and retrieving, representing and visualising, interpreting and sharing knowledge.

The benefits of using new technology may include: (1) increase data reliability; (2) reduction in the labor required to collect and evaluate data; (3) timely results allowing efforts to be focused on the evaluation of the data rather than the process of data collection/reduction; (4) allow *real-time*[1] monitoring of the structure performance, and (5) a better understanding of the structure's performance so that changes can be identified in a timely manner and corrective actions taken.

2.1 *Data and information storage capabilities*

Dam monitoring systems typically consist of a dense array of heterogeneous sensors. Some current trends in sensor development and measurement technology include intelligent instruments or *smart sensors*, which are becoming more common and their level of intelligence is constantly being elevated. Some important advantages of smart sensors are: built-in capability for self-checking and automatic warning of malfunctions, automatic compensation for nonlinearity or systematic errors due to temperature drift, etc., built-in signal processing with internal memory, and a networking capability thereby allowing a number of instruments to be connected to the same instrument cable, thus reducing the amount of cable that has to be installed.

Performance assessment relies not only in the "seeing eyes" of the monitoring system, the sensors, but rather in a complexity of data sources, such as photos and video inspection data, geodetic survey data, design history (material properties, hydrology and hydraulics, seismicity, geology and foundations), construction history (foundation preparation, materials of construction, seepage control, drainage control), operational history, including extreme events (earthquakes, extreme flooding, etc.) and regulatory requirements.

Currently, instrumentation on remote areas of the world can be accessed and transmitted to offices in other continents. The data can instantly be reduced by a desktop computer and plotted in

---

[1] Real-time monitoring is the ability to gather and process instrumentation data rapidly and in a clear and organized manner so that decisions can be made regarding the safety of the structure during critical periods.

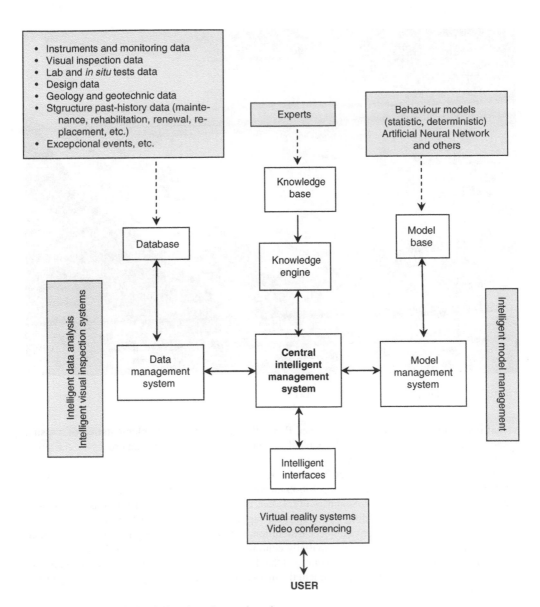

Figure 1.   Framework for civil engineering works safety assessment.

a format which immediately allows engineers to identify deviations from the expected performance of the structure.

Thus, the ultimate goal for all surveillance programs is to have high-performance multimedia management databases containing not only data from automated performance monitoring systems but all information relative to the soil, rock or structure.

2.2   *Visual inspection*

Up to now visual inspections have been performed in a rather conventional way and without the support of sophisticated technology. Combined laser scanning and digital image technologies are recent fields of research and the use of such technologies will certainly enable the automation of

Figure 2. Visual inspection of a dam using laser scanning technology.

visual inspection data collection. Moreover, this calls for an effective implementation of a visual inspection support system which will enable the identification of deterioration processes in a rather accurate, complete and faster way.

During the last decade Combined Terrestrial Imaging Systems (CTIS) have been matured and marketed, Figure 2. This multi-sensor solution integrates laser scanning and digital photo-imagery technologies. Terrestrial Laser Scanners (TLS) can get the co-ordinates of millions of points in reflecting surfaces thus providing new means for rapid and precise geometric, discrete but very dense, electronic representation of objects. In addition, digital photo-imagery acquired by calibrated photographic cameras can obtain the spatially continuous radiometric record of the same objects. The corresponding data fusion has encouraged the development of a new methodology to register and to codify into an electronic environment the main deficiencies typically surveyed during visual inspections (Berberan, 2007).

2.3 *Enhanced analysis capabilities*

Modelling structural behaviour is a complex process in which one should consider the nonlinear behaviour of materials, interaction between the structure and its foundation, influence of water load on the structure and its foundation bedrock and environmental actions. However, despite the complexity of the model, they encompass a number of simplifications of real world, uncertainties of model parameters are always present. Thus, careful monitoring of the structures and its surroundings are required in order to verify and enhance the model. Once the model is set up it is possible to determine if the structure still behaves as expected by comparing results of monitoring measurements with a prediction model.

A number of different techniques is being used for structural modelling. It is not always clear whether one technique gives better results than another. Some rules of thumb are emerging regarding the selection and choice of method. Innovative ways of combining the different techniques especially

166

when data is limited or incomplete, or when there is complementary additional knowledge available, is opening up a number of different areas for research.

For civil engineering works, the set up of the monitored quantities expected range is very often based on statistical and/or deterministic models, also referred as behaviour models. New attractive areas of research include techniques such as data-mining for complex query processing to provide query support to the structural analyst and neural network for prediction of monitored quantities. Neural networks do not require information concerning the phenomenological nature of the system being investigated, which makes them a robust means for representing model-unknown systems encountered in specific engineering problems. In such cases, the data-driven model is dependent entirely on the data. There are real dangers here in that there is strong temptation to focus the modelling on the available data and to ignore the physics situation. In fact it could be stated that data driven modelling cannot be guaranteed to give safe and reliable results unless there is proper attention given to the physics of the problem (Price, 2001).

2.4 *Knowledge-base capabilities*

Instrumentation and measurement systems provide data for evaluating the structure performance. If the readings fall outside the expected range, reflect undesirable trends, or show increasing rates of change, then the instrumentation and measurement system will provide early warning that something is occurring that might affect the integrity of the dam. At this time the expert team is called. The resultant action might be that remedial actions can be taken to ensure the integrity of the civil engineering work or to prepare for a significantly hazardous situation. An example of such an activity is the activation of an Emergency Action Plan.

Skill and judgment are required to properly evaluate the structure performance. The value of using experienced engineering judgment cannot be underestimated. However, a number of factors are common to the performance of a group of structures (dams, tunnels, bridges, etc.). Moreover, there very often exist interrelationships among the common factors and other performance factors. An understanding of these factors is important in making the transition from data to knowledge and in unraveling the mysteries of the structure performance (ASCE, 2000).

Highly increasing computing power and technology could make possible the use complex intelligent systems architectures for diagnosis, decision-making, modelling and analysis, taking advantage of more than one intelligent technique, not in a competitive, but rather in a collaborative sense.

2.5 *Intelligent interfaces*

Intelligent interfaces can provide more effective ways of accessing ubiquitous information and hide the complexity of technology by supporting a seamless human interaction between humans, humans and devices, virtual and physical objects and the knowledge embedded in everyday environments. This includes research on virtual and augmented reality.

A well-designed, intuitive, and appealing user interface is crucial to the success of the structure monitoring and assessment system. Ambient intelligence[2] will become an important part of the framework puzzle and it will become unthinkable to demand from users to become acquainted with hundreds of systems; rather, these systems themselves have to be usable in an efficient and appealing way, must be adaptive and intuitive, context-sensitive and, often, personalized. Therefore, new approaches to automatic, on-the-fly creation of user oriented, ergonomic interfaces must be pursued.

The internet is already being use to transfer information from almost any place to almost any-where. The internet, or future generations of it, will undoubtedly play a key role in future safety

---

[2] The term Ambient Intelligence is defined by the Advisory Group to the European Community's Information Society Technology Program (ISTAG) as "the convergence of ubiquitous computing, ubiquitous communication, and interfaces adapting to the user".

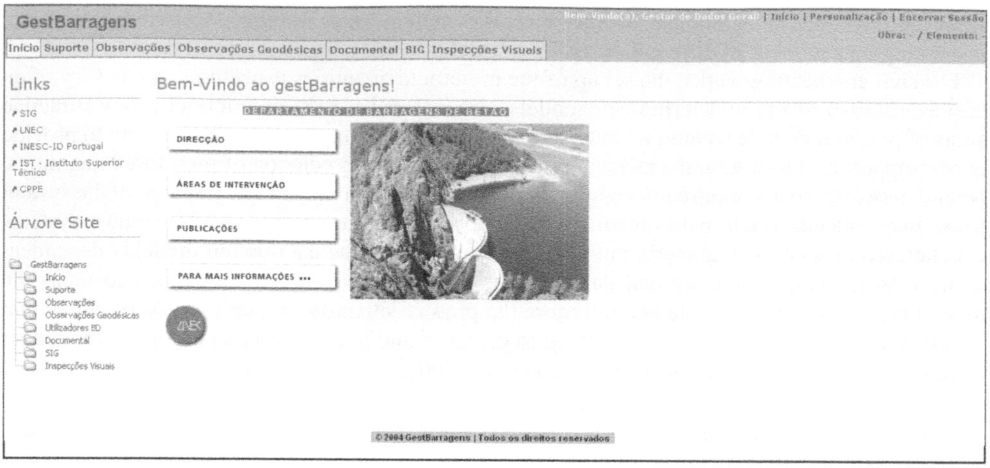

Figure 3.   GestBarragens main interface.

control systems. The internet also provide a forum where specialists can communicate and exchange information world-wide, for instance via video-conferencing.

3   THE PROJECT GESTBARRAGENS

The envisioned integrated monitoring framework is encapsulated in an evolving project named gestBarragens (Figure 3). The system is applied to the field of dam engineering and depicts key elements of the safety control of those structures (Portela, 2005; Silva, 2005). gestBarragenss results of a close collaboration of three Portuguese institutions (LNEC[3], INESC-ID[4] and EDP[5]). The overall project is currently deployed in the organizational context of LNEC and EDP and it is the result of a successful partnership between dam experts and information system professionals.

GestBarragens is now being implemented for the most important large concrete dams in Portugal. Nevertheless, one must be fully aware that the installation of a completely new information system brings additional constraints and that one of the main factors for success is the training of people in charge. Furthermore, populating such a complex information system is a major task and a team of competent and motivated specialists is essential to the success of the system otherwise, even the best performing system will be useless.

3.1   *System background*

The system was designed and developed to replace the existing information system, named SIOBE ("Sistema de Informação para Observação de Barragens de Betão"). SIOBE was written in Fortran, in the 70's, and the underlying information is stored in binary and ASCII files. Through the years, new requirements have arisen, concerning the type and variety of information stored as well as the functionalities offered to handle and visualize it.

Data maintained by the old system SIOBE had to be migrated into the relational database that stores the data maintained by the new gestBarragens. Besides this legacy information mainly

---

[3] LNEC – Laboratório Nacional de Engenharia Civil, Lisbon, Portugal.
[4] INESC-ID – Instituto de Engenharia de Sistemas e Computadores, Investigação e Desenvolvimento, Lisbon, Portugal.
[5] EDP – Electricidade de Portugal, Porto, Portugal.

stored in text files, a collection of Excel data files containing measurements of different types had to be transformed in order to populate the gB database. A specific data migration application, named LegacyData2GB has been developed for that purpose. This application had to deal with the heterogeneity of the data formats present in excel data files, as well as ensuring the quality of the new data produced.

Thus, the primary goal of gestBarragens system was to import and store data from previous data repositories into an integrated information system, ensuring the security continuity and follow up of previous databases. In addition, the system requirement included, among others:

- multi-user and remote access for information sharing;
- security by using predefined personal profile and personal privileges;
- acquisition and storage of manual, semi-automatic and automatic measurements as well as of the comments made during measuring or the ones related to all events concerning the structure and its instrumentation, a "structure biography" has to be stored;
- software to be generic within the type of dam addressing and embracing all types of instruments used in the structure framework;
- immediate and on site checking of the plausibility of the measurements results (based on data history or simple models);
- ease transfer of data to graphic interpreter like Excel or AutoCad;
- remote access to the database for storing or/and exporting operations either by modem, cable net or through the Web;
- clear, automatic and unique system of identification of each measuring points;
- checking and validation of the data and the notes by engineer in charge of the work. After this check, the data cannot be modified by other users. They are set and frozen;
- easy to expand, and
- friendly and well designed user interface.

3.2 The underlying technology of gestBarragens

The integrated system gestBarragens is a multi-user web-based decision support system and it was built on the basis of a modular approach. GestBarragens was developed in ".net", the application is accessible through a Web browser (e.g., Internet Explorer or Mozilla) in an Intranet or Internet context by a proper user access authentication Data is stored and managed in Oracle 9i, reports are designed and implemented using Microsoft Reporting Services, and data is graphically and spatially visualized using the ESRI technology (GIS).

3.3 gestBarragens modules

The main modules of gestBarragens system are:

- gB-Support – integrates tasks such as users management (users with different roles or functions in the safety control process), dams related data.
- gB-Observations – all data collected from existing monitoring systems are stored in this module, including data from manual, semi-automatic or automatic acquisition systems. Data validation and check for anomalies is carried out before storage. The management of instruments and measurement devices can be done in this module (location, installation date, calibration details, etc.). It is also possible to prepare a field inspection file which will support the field personnel during an inspection through the use of portable devices or manually. This module has a special feature to support geodetic inspections. The module has basic functionalities such as data capture, storage and recall, presentation and report facilities.
- gB-Visual Inspection – this module allows the storage of all information collected by the field personnel during routine visual inspections.

- gB-Tests and Analysis – this module stores information on material properties provided by specific tests and analysis.
- gB-Models – this module supports the dam modelling process.
- gB-Documental – stores historical documents, reports, photos drawings, etc. Any document can be uploaded to the system and linked to the dam and, if it is the case, to a specific inspection and become available to system users upon a proper user authentication.
- gB-SIG – the visualization component of gestBarragens integrates geo-referenced information acting as a graphical user interface and allows database inquires. It provides drawing of the dam and its elements, key cross-sections, key location of instruments, basic graphics functionalities such as time history plots, scatter/correlation plots, position plots (data from multiple instruments at a particular cross section and time), attachment of comments to data points, etc.

Due to modularity and expandability of gestBarragens system new modules will be added to the system in the future.

3.4 *Data reporting*

Reports for all modules are available on user request in the Web interface or they can be generated for use by MsWord, MsExcel, Adobe Acrobat (pdf). They include text, tables and graphs; pre-defined report sets of tables and graphs; data maximum and minimum values; correlation reports for model pre-defined values and observed values; alarm thresholds; instrument configuration report and instrument history report; inspections and observations reports; photos, videos and documents reports, among others.

Special attention is dedicated to data analysis with spatial visualization of the dam. Furthermore, on the top of the structure, it is possible to show different layers that represent specific types of instruments. Moreover, graphical representations of data captured by the instruments are required to provide easy and high-level views to better understand the dam behaviour. ESRI products have been selected to support this kind of visualization. The Web interface provides mechanisms that allow users to easily navigate and explore physical elements of the dam. They are able to monitor mouse click events, for instance to select a dam or an instrument type; or activate common GIS features such as zoom, pan, undo, redo , measures, printing, spatial objects identification and information. Figure 4 shows an example of spatial graphics delivered by gestBarragens GIS module (Silva, 2006).

3.5 *Benefits*

There is no doubt that for the engineers in charge assessing the performance of their many dam projects and preparing the annual monitoring reports is a very labor-intensive and unpopular task. The implementation of intelligent information systems can significantly reduce the labour effort required and improve the quality of the information. The end result is that engineering personnel will be able to utilize their time evaluating data in a timely manner to continually monitor the performance of the many dams that they are responsible for. Checking of long term monitored quantities is extremely easy and data are permanently at the disposal of the engineers and the experts in charge of the dams.

4 CONCLUSIONS

Increased awareness of safety issues in modern society calls for an increased level of sophistication of means and methods of control. Sound decisions must be based on relevant and reliable information that is readily accessible through an affordable medium. The evolution, or some may say revolution, of intelligent information systems has been a major contributing factor to the

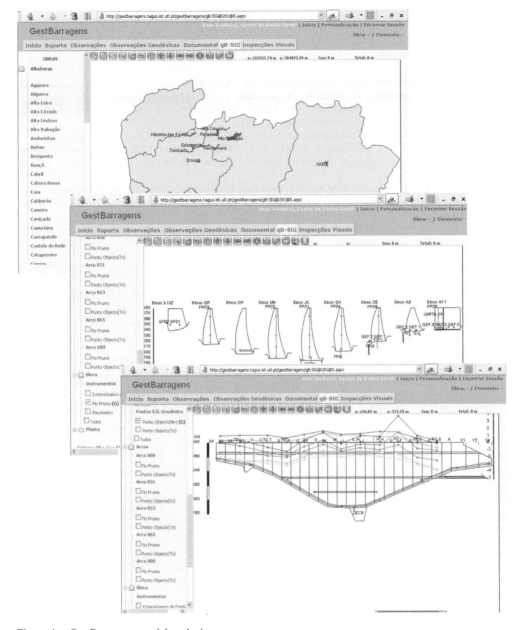

Figure 4. GestBarragens spatial analysis.

development, implementation and maturation of computerized systems to support safety control activities of civil engineering works.

This paper reports the application of an integrated framework system to support the safety control activities of concrete dams. The technologies and processes discussed in the context of dam engineering can be extrapolated with benefits to other areas of civil engineering.

REFERENCES

ASCE Task Group Committee on Instrumentation and Monitoring Dam Performance, 2000. Guidelines *for Instrumentation and Measurements for Monitoring Dam Performance, ASCE, USA.*

Berberan, A., Portela, E.A., Boavida, J. 2007. Enhancing on-site dams visual inspections, *5th International Conference on Dam Engineering, Lisbon, February 2007* A.T. 1980.

Portela, E.A., Pina, C., Silva, A.R., Galhardas, H., Barateiro, J. 2005. *A Modernização dos Sistemas de Informação de Barragens: o Sistema gestBarragens, Seminário Barragens: Tecnologia, Segurança e Interacção com a Sociedade, Lisbon, Outubro 2005* (in portuguese).

Price, R. 2001. *State of the Art Paper, Seminar on Hydroinformatics in Portugal, Lisbon, Portugal.*

Silva, A.R., Galhardas, H., Barateiro, J., Portela, E.A. 2005. *O Sistema de Informação gestBarragens, Seminário Barragens: Tecnologia, Segurança e Interacção com a Sociedade, Lisbon, Outubro 2005* (in portuguese).

Silva, A., Galhardas, H., Barateiro, J., Matos, H., Gonçalves, J., Portela, E. 2006. *Data Analysis Features of the gestBarragens System, Rio de Janeiro, International Conference on Innovative Views of .NET Technologies (IVNET'06).*

Underground structures

Applications of Computational Mechanics in Geotechnical Engineering – Sousa,
Fernandes, Vargas Jr & Azevedo (eds)
© 2007 Taylor & Francis Group, London, ISBN 978-0-415-43789-9

Prediction of safe performance of TBM tunnels on the basis of engineering rock mass classification

M. Zertsalov & A. Deyneko

Moscow State University of Civil Engineering, Moscow, Russian Federation

ABSTRACT: The paper presents engineering rock mass classification (ERMC), developed for engineering design of circular tunnels, constructed by tunnel boring machines (TBM). The ERMC is based on the numerical modeling of interaction of tunnels with rock mass. The parameters of the well known Q-system are used for rock mass description. The numerical calculations were performed in accordance with the theory of experiment plan. As a result the parametrical equations have been obtained where strength safety coefficients of both unsupported excavation and tunnel concrete lining were taken as response functions. The solutions of parametric equations are presented in the form of nomograms, which provide the values of strength safety coefficients. The parameters of Q-system can not be used directly in numerical modeling as they contain the information, concerned with mechanical properties of rock mass in uncertain form. The technique for formalization of this information is considered.

1 INTRODUCTION

In the modern design practice of underground structures engineering rock mass classifications (ERMC) are used. Generally ERMC are considered to be the procedures, providing the prediction of interaction of rock mass and underground structure on the basis of rock general characteristics, not undertaking direct analysis of stress state.

The available ERMC, in spite of different approaches to their development, have the following-common practical shortcoming: the offered classifying indices (parameters), being estimation of mass various properties, are purely empirical and are mostly based on ERMC author's intuition and engineering experience. Therefore, the estimation may be somewhat subjective and the reliability of the results obtained appears difficult to be estimated quantitatively.

Nevertheless, ERMC proves to be a very useful engineering tool makes it possible at the preliminary investigation stage to have information on rock mass quality and to plan engineering measures, providing safety during underground structure construction and performance. So there is no doubt in the necessity of ERMC further improvement.

The paper deals with the classification system, developed for engineering design of circular tunnels in rock mass, constructed by tunnel boring machines (TBM). It may be used for designing long tunnels such as transportation tunnels, water pressure tunnels and so on.

The system offered is based on the use of initial parameters of the well known Q-system (Barton et al. 1974) in combination with numerical analysis of rock mass interaction with underground structure. Here in the process of numerical modeling combinations of parameters, characterizing rock mass and the structure are being selected with the use of the theory of experiment plan. The latter provides the predicted reliability of the obtained results, which are represented in the form of classifying criteria.

Generalized safety coefficients of unsupported rock mass and tunnel lining as quite definite in engineering respect criteria of static interaction of underground structure and rock mass were selected as classifying criteria.

2 FORMALIZATION OF ROCK MASS MECHANICAL CHARACTERISTICS

Parameters of Q-system make it possible to describe wide range of engineering and geological conditions. However they cannot be used directly for numerical modeling, as they contain the information on mechanical properties of rock mass in uncertain form. The paper considers the method for formalization of that information and its representation as values of specific mechanical characteristics of rock mass.

In recent three decades Q-system has been improved and developed significantly (Barton 2000, 2002). As a result there can be found two types of expressions for determination of quality index for rock mass Q and Q_c:

$$\left. \begin{aligned} Q &= \frac{RQD}{J_n} \cdot \frac{J_r}{J_a} \cdot \frac{J_w}{SRF} \\ Q_c &= \frac{RQD}{J_n} \cdot \frac{J_r}{J_a} \cdot \frac{J_w}{SRF} \cdot \frac{\sigma_{ci}}{100} \end{aligned} \right\} \tag{1}$$

where RQD = rock quality designation; J_n = joint set number; J_r = joint roughness number; J_a = joint alteration number; J_w = joint water reduction factor; SRF = stress reduction factor; σ_{ci} = rock sample uniaxial compression strength, MPa.

The first relationship in the (Eq. 1) characterizes the rock mass structure; the second one represents shear strength of joints; the third one is empirical parameter, accounting for active stress; and the last is empirical coefficient of correction.

Correlation between Q-value and rock mechanical characteristics is known (Barton 2000, 2002). Using this correlation may be obtained both the modulus of rock mass deformation E and compressive strength of rock mass σ_{cm-Q}:

$$E = 10 \cdot \sqrt[3]{Q_c} = E(\sigma_{ci}), \text{GPa} \tag{2}$$

$$\sigma_{cm-Q} = 5\gamma \cdot \sqrt[3]{Q_c} = \sigma_{cm-Q}(\sigma_{ci}), \text{MPa} \tag{3}$$

where γ = rock mass density, t/m³.

Other strength mechanical characteristics, the information of which is contained in parameters of Q-system in uncertain form, are defined by the technique, proposed earlier in references (Deyneko 2004, Zertsalov & Deyneko 2005a, b) on the basis of combined application of Q-system and Hoek-Brown method (Hoek 2000, Hoek et al. 2002) and the use of the theory of fuzzy sets concepts.

Along with (Eq. 2) rock mass modulus E may be calculated as follows (Hoek et al. 2002):

$$E = \left. \begin{cases} (1-D/2)\sqrt{\sigma_{ci}/100} \cdot 10^{(GSI-10)/40}, & \sigma_{ci} \le 100 \text{ MPa} \\ (1-D/2) \cdot 10^{(GSI-10)/40}, & \sigma_{ci} > 100 \text{ MPa} \end{cases} \right\} = E(\sigma_{ci}, GSI), \text{GPa} \tag{4}$$

where GSI = geological strength index; $D = 0$ for TBM excavations.

Equating (Eq. 2 and 4) the following formula is obtained:

$$GSI = \left. \begin{cases} 10 + 40\log\left(\dfrac{E}{\sqrt{\sigma_{ci}/100}}\right), & \sigma_{ci} \le 100 \text{ MPa} \\ 10 + 40\log E, & \sigma_{ci} > 100 \text{ MPa} \end{cases} \right\} = GSI(\sigma_{ci}) \tag{5}$$

Index GSI makes it possible to determine Poison's ratio for rock mass. Also geological strength index being known, rock mass compression strength can be also obtained (Hoek et al. 2002):

$$\left. \begin{aligned} \sigma_{cm-HB} &= \sigma_{ci} \frac{\left[m_b + 4s - a(m_b - 8s)\right](m_b/4 + s)^{a-1}}{2(1+a)(2+a)} = \sigma_{cm-HB}(\sigma_{ci}, m_i, GSI), \text{MPa} \\ m_b &= m_i \cdot e^{(GSI-100)/(28-14D)}, \quad s = e^{(GSI-100)/(9-3D)}, \quad a = 0{,}5 + \left(e^{-GSI/15} - e^{-20/3}\right)/6 \end{aligned} \right\} \tag{6}$$

where m_i & m_b = Hoek-Brown's constants for undisturbed rock and rock mass; s & a = Hoek-Brown's parameters, defining rock structure.

Equating (Eq. 3 and 6) the following equation is obtained:

$$\sigma_{cm-Q}\left(\sigma_{ci}\right)=\sigma_{cm-HB}\left(\sigma_{ci},m_i\right) \tag{7}$$

The (Eq. 7) has a lot of solutions. Taking into account the geomechanical sense of parameters, the interval of possible values of σ_{ci} for the given rock may be determined. Let us introduce an additional parameter characterizing the strength level for undisturbed rock within possible values. Thereafter the (Eq. 7) being solved, the value of m_i is calculated. Then, basing on parameters σ_{ci}, m_i and GSI, all rock characteristics, required for numerical modeling, are obtained.

3 IDENTIFICATION OF ROCKS ON THE USING THE THEORY OF FUZZY SETS

As was stated above, in the presented engineering rock mass classification of the input data for determination of mechanical properties of rock mass, used for numerical analysis, are considered to be combinations of Q-system parameters (thereafter factors), obtained in accordance with the rules of the theory of experiment plan. Under such approach the combination of various level factors between each other takes place: for example, maximal value of one factor is considered with maximal, mean and minimal value of the other one. Some combination may appear to be impossible in reality and then the result of numerical experiment is incorrect. So before to conduct each calculation it is necessary to evaluate to what degree the combination of factors under consideration corresponds to the actual rock. Special procedure, based on application of the theory of fuzzy sets, has been developed for such evaluation.

In the theory of fuzzy sets (Kophman 1982, Levner et al. 1998) the membership to set of elements A is characterized by membership function μ_A. The value of membership function is usually interpreted as subjective evaluation of membership grade of element to the set A: $\mu_A(x) = 0.8$ means that x is a member of set A for 80% (expert's evaluation as a rule). For assignment of numerical parameters, which may take more or less representative values at some interval, fuzzy numbers are used.

Let us take practical example for explanation (Deyneko 2004, Zertsalov & Deyneko 2005a, b). There exist reference values of m_i for various rocks, for instance $m_i = 15 \pm 5$ for diabase.

The following interpretation is used (diabase as an example): value = 15 is most representative; values = 10 and = 20 are representative not much, but possible; values $m_i \leq 9$ and $m_i \geq 21$ are not representative absolutely. Fuzzy numbers for diabase ($m_i = 15 \pm 5$), as well as for dolerite ($m_i = 16 \pm 5$), and sandstone ($m_i = 17 \pm 4$), are illustrated on Figure 1.

Parameters m_i and σ_{ci} characterize mechanical properties of rock as a whole, so consideration of the set of those parameters proves to be quite enough for its identification. Let us demonstrate this procedure considering for example the case when $m_i = 16.5$ and $\sigma_{ci} = 135$ MPa.

Figure 1 illustrates that the specified value $m_i = 16.5$ is representative for dolerite with membership function of = 0.917; for sandstone of = 0.900; and for diabase of = 0.750.

If to assume that the strength of undisturbed rocks (diabase, dolerite and sandstone) σ_{ci} is also represented in the form of fuzzy numbers, then $\sigma_{ci} = 135$ MPa has the following values for membership function, respectively: for diabase = 0.897; dolerite = 0.845; sandstone = 0.633.

Character of combination of m_i and σ_{ci} for each of considered rocks will be defined by arithmetic mean of non zero values of membership functions of both those parameters. In our example the indicated arithmetic mean values are as follows: for dolerite = 0.881; for diabase = 0.824; for sandstone = 0.766. As a result the combination of m_i and σ_{ci} in the case under consideration proves to be most representative for dolerite. Generally, the combination representative to the actual rock, thus this combination can be used in numerical modeling.

Similarly there have been checked all the combinations of factors, defined by the plan of numerical experiments. The combination of factors was included into the experiment if the corresponding

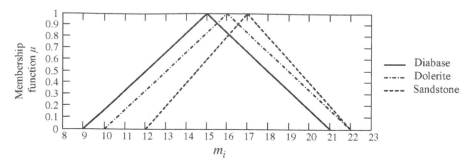

Figure 1. Constants m_i of rocks as fuzzy numbers.

Table 1. Factors.

Factors	Factors characteristic
X_1	Ratio RQD/J_n in Q-system
X_2	Ratio J_r/J_a in Q-system
X_3	Parameter J_w in Q-system
X_4	Relative level of strength in undisturbed rock mass
X_5	Underground structure span
X_6	Underground structure deepening

Table 2. Intervals of variation of initial factors.

Variation level	X_1	X_2	X_3	X_4	X_5, m	X_6, m
+1	90	5.33	1	0.8	15	250
0	45.33	2.70	0.665	0.5	10	150
−1	0.667	0.0625	0.33	0.2	5	50

values of m_i and σ_{ci} were representative for at least single type of rock. The main advantage of the stated approach is preservation of equal objectivity when taking decisions on the base of fuzzy information independently of the number of solutions.

4 DEVELOPMENT OF THE ENGINEERING ROCK MASS CLASSIFICATION

Summarizing the above stated the conclusion can be drawn that initial parameters of Q-system (factors), used for numeric simulation, shall meet the requirements as follows: factors should show the features of real rock mass state; the information which they contain might be formalized in the form of mechanical characteristics; combinations of factors shall comply with basic concepts of the theory of experiment plan.

The selected factors and intervals of their variation are illustrated in Tables 1 and 2. Algorithm for tacking into consideration of strength and reinforcement of concrete is discussed below in the paper.

When using the theory of experiment plan, the basic task is to detect on the basis of optimal set of experiments the response functions. The latter connects the experiment results with variable factors and makes it possible to find out response criterion (resulting data of experiment) for arbitrary combination of initial factors within the intervals of their variation.

Maximally informative parameter shall be taken as response criterion, which characterizes interaction of underground structure and rock mass. The criterion shall contain the information, which would provide for engineering solutions at the initial design stages, without conducting direct calculations of stress state. Two such criteria were selected for the procedure offered.

Generalized safety coefficient of concrete lining strength m_{lin} was selected as the first criterion for compression and tension conditions, and generalized safety coefficient of unsupported rock mass m_{rk} was selected as the second criterion:

$$\left. \begin{array}{l} m_{lin} = \min\left(R_b/\sigma_{bc}, R_{bt}/\sigma_{bt}\right) \\ m_{rk} = \sigma_{cm}/\sigma_c \end{array} \right\} \tag{8}$$

where R_b & R_{bt} = design compressive and tensile strength of lining concrete; σ_{bc} & σ_{bt} = the highest compressive and tensile strength of lining concrete; σ_c = the highest compressive stresses within unsupported excavation.

For obtaining optimal combinations of variable factors Box-Benkan experiment plan was selected (Markitantov 2003). More than 200 numerical experiments for analysis of stress state in concrete lining were performed with the use of geotechnical software Z_Soil.PC (Switzerland).

Quadratic polynomial was selected as a basic response function. The final response functions for generalized safety coefficients were found out in nonlinear composite form with the use of exponential transformation function:

$$\left. \begin{array}{l} m_{lin} = \exp\left(m_{lin}'^{\,5/4} - t\right) \\ m_{rk} = \exp\left(m_{rk}'\right) \end{array} \right\} \tag{9}$$

where m' component of functions is quadratic polynomial determined by applying the theory of experiment plan technique on a basis of experimental response data analysis; t = empirical coefficient of correction.

With six variable factors the quadratic polynomial contains 28 members:

$$m'\left(\overline{X}\right) = b_0 + \sum_{i=1}^{6} b_i X_i + \sum_{i=1}^{6} b_{ii} X_i^2 + \sum_{i=1}^{6}\sum_{j=1}^{6} b_{ij} X_i X_j \tag{10}$$

where b_0, b_i, b_{ii} and b_{ij} = estimation of unknown regression coefficients by least-squares technique; X_i, X_j = the factors.

5 NOMOGRAPHIC REPRESENTATION FOR THE ENGINEERING ROCK MASS CLASSIFICATION

Nonlinear composite form functions (Eq. 9–10) are seemed cumbersome. That is why nomograms were developed for practical determination of generalized safety coefficients. Figure 2 illustrates as an example the nomogram for determination of coefficient m_{lin}.

Algorithm for determination of coefficients m_{lin} by the nomogram is as follows. First, the point of intersection of lines, corresponding to prescribed values X_1 and X_3, shall be determined at the left top binary field. Then in a similar way the point of intersection shall be obtained at the right top binary field to prescribed values X_2 and X_3. As the values of X_3 are equal in both cases, the points obtained lie on one horizontal straight line.

Afterwards the distance between the points determined shall be measured (Step 1) and the obtained segment plotted horizontally to the right from the point of left lower binary field (Step 2), corresponding to prescribed X_4 and X_6. The segment end will indicate the value of the coefficient m_{lin} being determined at the right lower binary field.

Let us consider practical application of the nomogram for obtaining concrete numerical result for the following initial factors: $X_1 = 45.33$; $X_2 = 2.70$; $X_3 = 0.66$; $X_4 = 0.5$; $X_5 = 10$ m; $X_6 = 150$ m. The points, corresponding to the prescribed pairs of factors $X_1 - X_3$, $X_2 - X_3$ and $X_4 - X_6$, are recorded on the nomogram binary fields. Having measured the distance between the points determined on top binary fields and having plotted the segment obtained to the right from the determined point on the left lower binary field (factors $X_4 - X_6$), we get the point on the right lower binary field, which corresponds to the value $m_{lin} = 3.4$.

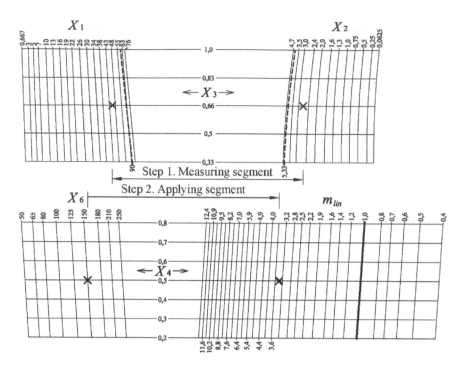

Figure 2. Nomogram for determination of m_{lin} (lining $0.15r$).

Algorithm of determination of generalized safety coefficient m_{lin} discussed above is valid for the linings with strength of concrete of constant grade. Thus, it is necessary to take into consideration the actual strength and reinforcement of concrete.

It has been defined that with increasing lining concrete strength generalized safety coefficient m_{lin} of both concrete and reinforced concrete linings is also increased according to the linear law. For numerical assignment of this increase generalized transition coefficient m_b, equal to the relationship between generalized safety coefficients for linings from concrete of two constant grades (B30 and B15 in Russian Standard notation) was introduced.

A response functions and the nomogram presented on Figure 3 was developed for the generalized transition coefficient m_b. It has been also defined that the level of generalized transition coefficient shows the possibility of predominance of tensile stress in the tunnel lining. The correspond range of m_b values highlighted on the nomogram on the right lower binary field.

For practical calculation transition coefficient M_b, equal to the relationship between generalized safety coefficients for linings from prescribed concrete and concrete of constant grade B15 in the form of chart was introduced and presented on Figure 4.

Change of m_{lin} with increasing reinforcement area and concrete grade of lining has been analyzed on the base of reinforced concrete theory and the same chart was plotted for reinforced concrete with coefficient of reinforcement = 3%. The chart is presented on Figure 5.

Thus, final m_{lin} of lining of prescribed strength and reinforcement is obtained as a product of nominal m_{lin} (defined by Figure 2) and transition coefficient M_b (defined by Figures 4 and 5 on a basis of generalized transition coefficient m_b, the latter defined by Figure 3) as follows:

$$m_{lin}^{final} = m_{lin} M_b \tag{11}$$

So, nomograms and charts allow to obtain generalized safety coefficient of lining for pre-scribed concrete grade and reinforcement percentage or vice versa to select concrete grade and reinforcement percentage for the required safety coefficient.

180

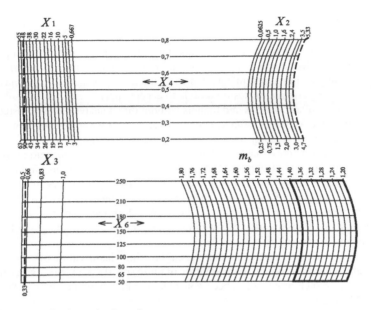

Figure 3. Nomogram for determination of m_b.

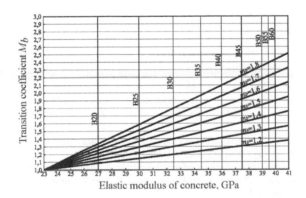

Figure 4. Chart of transition coefficients M_b for concrete linings.

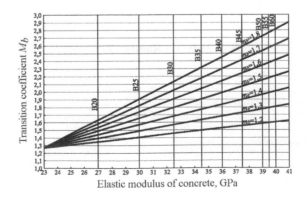

Figure 5. Chart of transition coefficients M_b for reinforced concrete linings (reinforcement = 3%).

181

6 CONCLUSION

In conclusion it shall be stated that the technique for prediction of safety of unsupported excavations and circular TBM tunnels under consideration in the paper, may be referred to ERMC. The technique, basing on initial parameters, which characterize the state of rock mass, and using the obtained classifying criteria, makes it possible to accept specific engineering solutions and to predict strength safety of excavations and tunnels without direct analysis of stress state.

At the same time the technique offered is considered to be further development of ERMC. It allows formalizing the information on rock mass properties, contained in initial parameters of Q-system, in the form of rock mass mechanical characteristics. The technique using the theory of experiment plan makes it possible to find out response functions, linking experiment responses (generalized safety coefficients) and initial variable parameters of Q-system on the basis of numerical modeling of static interaction of rock mass and underground structure.

Nomographic representation of functions is proposed. Both the problems with respect to the need in strengthening of unsupported excavations and selection of concrete grade for linings and their reinforcement can be solved without delay.

In conclusion it is worth to note again that the engineering classification offered shall be taken as a guide (as any other one) first of all for solution of those specific problems, which it is designed for. In the above case such problem has been considered to be the prediction of safe performance of circular TBM tunnels.

At the same time the basic principles of the given paper and the methods for its realization may be used for the development of engineering rock mass classification, designed for other specific underground structures.

REFERENCES

Barton N., Lien, R. & Lunde J. 1974. Engineering classification of rock masses for the design of tunnel support. In *Rock Mechanics*, 1974, Vol. 6/4, 189–236.

Barton N. 2000. General report concerning some 20th century lessons and 21st century challenges in applied rock mechanics, safety and control of the environment. In *Proc. of 9th ISRM Congress*, Paris, 2000, 1659–1679.

Barton N. 2002. Some new Q-value correlations to assist in site characterization and tunnel design. In *Int. J. of Rock Mechanics & Mining Science*, 2002, Vol. 39, No. 2, 185–216.

Deyneko A. 2004. (In Russian). Identification of rock masses by the group of parameters with the use of theory of fuzzy sets. In *Proc. of the All the Russian Exhibition on Scientific and Technical Creative Works of Youth NTTM-2004*, Moscow, 2004, 127–129.

Hoek E. 2000. *Practical rock engineering*. Balkema.

Hoek E., Carranza-Torres C. & Corkum B. 2002. The Hoek-Brown failure criterion. 2002 edition. In *Proc. of the 5th North American Rock Mass Mechanical Symp. and 17th Tunneling Association of Canada Conf*, Toronto, July 2002, 267–271.

Kophman A. 1982. (In Russian). *Introduction into fuzzy arithmetic (Transl. from French)*. Moscow: Radio and Communication.

Levner E., Ptuskin A. & Phridman A. 1998. (In Russian). *Fuzzy sets and their application*. Moscow: TsAMI RAN.

Markitantov I. 2003. (In Russian). *Investigation of organizational and technological processes on the basis of experiment planning methods with the use of Box-Benkan's three-level plans (chocolate production as an example). Procedure*. Saint-Petersburg: Saint-Petersburg University of Engineers and Economists.

Zertsalov M. & Deyneko A. 2005a. Formalization of rock mass mechanical characteristics and rock type identification on the basis of the Q-system. In *Proc. of the 10th ACUUS Int. Conf. & of the ISRM Regional Symp*, Moscow, January 2005, 277–280.

Zertsalov M. & Deyneko A. 2005b. The engineering rock mass classification on the basis of numerical analysis of static interaction of underground openings and rock masses (applied for TBM tunnels). In *Proc. of the 40th US Symposium on Rock Mechanics (USRMS)*, Anchorage, June 2005.

*Applications of Computational Mechanics in Geotechnical Engineering – Sousa,
Fernandes, Vargas Jr & Azevedo (eds)
© 2007 Taylor & Francis Group, London, ISBN 978-0-415-43789-9*

Modelling approaches used in innovative station designs for Metro do Porto

C. Maia
Babendererde Engineers LLC, Kent, Washington, USA

K. Glab
Babendererde Ingenieure GmbH, Bad Schwartau, Germany

ABSTRACT: The underground stations of Marquês and Salgueiros are part of the Metro do Porto light rail system. They are unique in their design, which avoids the traditional box-like geometries typically used in "cut and cover" solutions. During the run of the project, these stations were changed from conventional "cut and cover" box stations into large diameter elliptical shafts. This paper describes the most significant station design and construction issues, as well as the numerical modelling approaches used in the design of the temporary shotcrete lining for each station.

1 INTRODUCTION

Traditionally, underground metro or light rail stations have been conceived either as a rectangular "cut and cover" box, or as large diameter tunnels connected to smaller "cut and cover", near surface structures for passenger access and ventilation. In these types of rail projects, circular or elliptical shafts are typically used to house ventilation and pumping equipment, as well as to provide emergency escape ways to the surface.

The main reason why designers avoid using circular (or elliptical) shafts to contain underground metro stations is their limitation in size. In spite of the fact that a closed ring is known to have structural advantages over other shapes, its use has been limited to diameters usually less than 15 m.

In Oporto, Portugal, a recently commissioned light rail system has made use of innovative design to accommodate underground stations inside large diameter elliptical shafts. The technical challenges and advantages of this solution, as well as their numerical modelling are discussed in this publication.

2 LARGE DIAMETER SHAFTS CAN ACCOMMODATE A METRO STATION

It is commonly accepted that small diameter, circular shafts behave like a closed ring. As such, nearly all hydrostatic and lateral earth pressures acting on the shaft wall turn into compression loading of the shaft lining. Because commonly used shaft lining materials, such as concrete and shotcrete, handle compression well and primarily because a circular shaft does not require anchors or struts, this combination is ideal for the support of excavation works.

Designers have made extensive use of circular geometries for shafts up to 15 m in diameter. Beyond this, whenever a large excavation footprint is required, designers choose traditional box-like geometries which, for stability reasons, require anchors, struts, or both. The reasons for this seem to be founded on the notion that, as the diameter increases the loading regime on the shaft wall shifts from mostly compressional to mostly flexural, and becomes similar to that of a slurry wall, typically requiring support by anchors or struts. Also, openings in the shaft wall create design difficulties only solved by 3D modeling. In Oporto, it was possible to show that these problems can be overcome and that a light rail underground station can be accommodated inside a large diameter shaft, while having all of its functions preserved.

This important project was built under a design-build contract, involving contractors and design offices from Portugal and several other countries. Two underground stations named Marquês and Salgueiros were commissioned to Brazilian design offices; CJC Engenharia and Figueiredo Ferraz Consultoria e Engenharia de Projeto. The innovative design, involving the use of large shafts, was followed by on site technical construction management carried out by the principal designer (CJC), and supervised by the main author, acting at the time as the Project Director for the civil group.

3 MARQUÊS STATION

3.1 *General design and construction aspects*

The Marquês station is made up of a central elliptical shaft and two NATM tunnels. The shaft is 27 m deep and its elliptical footprint is 48 m along the major axis, and 40 m along the minor axis. The tunnels are 18 m long and have a cross section of 180 m².

The heterogeneous nature of the Oporto granite was evident at the station site. A sharp sub-vertical fault, oblique to the shaft main axis, separated weathered soil-like granite from moderately to slightly weathered good granite rock. This strong heterogeneous character was one of the main reasons for avoiding slurry walls at Marquês.

A shotcrete lined, large diameter shaft allowed the excavation of most of the station volume in an open cut, requiring no anchors or struts. This contributed a great deal to a fast paced excavation, advancing at a rate of 4.5 m each month. Shaft excavation was completed in six months, between June and November 2002. An added difficulty was the requirement to preserve the century old maple trees that embellish the Plaza. Only six trees, in the center of the Plaza, were allowed to be relocated. This was taken as a design input and contributed to the final shape of the shaft.

According to the sequential excavation design guidelines, the Marquês shaft was excavated in panels. Panels were 1.8 m high, with horizontal lengths varying between 4 m and 12 m. The exposed granite on each panel surface was immediately protected by three layers of shotcrete and welded wire mesh. The shotcrete wall thickness varied with depth between 0.3 m and 0.6 m. A final lining was provided by a cast-in-place concrete wall.

During excavation, the rock mass was dewatered by horizontal drainage holes, installed systematically on the shaft wall (4 m long, spaced at 1.8 m). This allowed the temporary shotcrete lining to remain drained and depressurized. Inside the shaft, deep vertical relief wells were also employed to reduce water uplift forces and to minimize the risk of hydraulic failure of the bottom.

Figure 1. Footprint and aerial view of Marquês Station. Note tunnel alignment oblique to the plaza and the surrounding maple trees.

The challenge with large diameter shafts is maintaining the stability of the shaft walls around large openings. At Marquês Station, in order to accommodate the light rail platforms with a length of 70 m, two NATM tunnels 18 m wide and 18 m long were constructed from the shaft. This produced large openings on the shaft walls and therefore required that a reinforced concrete (RC) frame be built prior to excavating the tunnels.

3.2 *Modelling Approach*

The loading imposed on the shaft by creating these two large openings is best modelled with three dimensional numerical models. 3D numerical analyses were carried out with STRAP (finite elements) to determine bending moments, axial and shear forces acting on the shaft wall and frame, as well as their deformation, before and after the excavation of the openings. The cast-in-place reinforced concrete frame, built prior to the openings, was incorporated in the model. It is essential for the stability of the shaft that the RC frame be stiff enough and capable of maintaining, as much as possible, the vicinity of the opening in compression. The basic dimensions of the RC concrete frame are shown in Figure 2.

The NATM tunnels attached to the openings were analyzed with a two dimensional finite difference model named FLAC. The Table 1 gives the main aspects regarding the modelling approach used in the design of the Marquês shaft.

A distinction is made between the short term, when the only acting structure is the shotcrete lining and the long term, when the final cast in situ concrete lining is in place.

Convergence movement in the shaft was estimated with FLAC 2D and STRAP 3D. Surface settlement due to shaft excavation was estimated with the Bowles approach, using wall deformation from the numerical runs as input for computations.

3.3 *Modelling results and comparison with construction behavior*

The results of the simulations show that the large Marquês shaft behaves like a closed ring and that the major component of the loading is compression. Table 2, gives moments, normal and shear forces actiong on the shaft lining at several depths prior to the openings. The reinforcement calculated from these values resulted in the minimum allowable steel area.

During construction, a geotechnical monitoring program was put in place to observe the performance of the structure, as well as nearby settlements and ground water level variations. Whilst it was not possible to install stress measuring devices, the Marquês shaft had an elaborate settlement and convergence monitoring program. Figure 3 gives convergence (-) and divergence (+) measurements taken during shaft excavation. Considering that a convergence cord measurement is equal

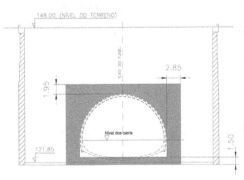

Figure 2. Schematic and photographic view of the reinforced concrete frame and adjacent platform tunnel at Marquês Station.

185

Table 1. Main aspects of the numerical modelling for Marquês Station.

Modelling Aspect	Short term	Long term
Numerical Tools	STRAP (by Atir). A three dimensional elastic finite element scheme.	Same.
Structural Conditions	Shotcrete lining and RC frame simulated by plate elements. Ground simulated by Winkler springs.	Shotcrete lining disregarded. Final cast in situ concrete lining spans reduced by several slab levels. Lining connection at slab levels is rotular.
Water	No water loading. Dewatering and depressurization provided by horizontal drainage holes and vertical wells.	Full hydrostatic loading. Temporary drains are plugged prior to execution of the final concrete lining.
Ground and Seismic Loads	Only static ground loading and surface surcharge.	In addition to the static ground loading, a full hydrostatic loading is added, as well as dynamic ground and water loads.

Table 2. Some results of the three dimensional modelling for Marquês Station.

Depth	Mmax (kNxm)	Nmax (kN)	Vmax (kN)	Displacement (cm)
2 m	9.5	289	12	0.82
10 m	16.1	1481	41	0.98
18 m	35.4	2492	49	0.92

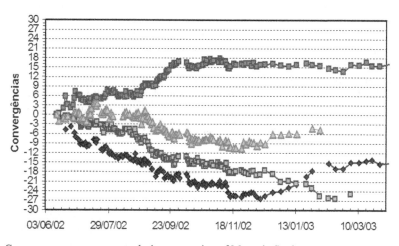

Figure 3. Convergence measurements during excavation of Marquês Station.

to double the model displacement of Table 2, a close agreement can be seen. Surface settlement predictions were also comparable to actual values.

The RC frame was introduced in the STRAP model to simulate the tunnel openings. This stiff frame, being 0.7 m thick, was placed into a properly designed notch in the shotcrete lining. Maximum moments and normal force, resulting from the tunnel openings, were 730 kNm and 9100 kN, respectively. The presence of the frame allowed the remaining shaft perimeter to stay heavily compressed.

3.4 Advantages of the shaft plus tunnel lay-out

From a construction point of view, some important advantages can be associated with a combination of large diameter shafts plus short tunnels. During excavation, a large, strut-free volume is available. Because no anchors are required, no time is lost in waiting for the anchor subcontractor to finish a level before excavation is allowed to proceed to the next level. Also, as the shaft is excavated in sequential panels, if ground treatment is required due to weaker than expected geology, ground improvement techniques may be applied at the desired location, while the rest of the shaft perimeter may continue with routine panel excavation or shotcreting. At Marquês, some weak granite exposed on the shaft wall required treatment by means of jet grouting. This was applied in about 20% of the shaft perimeter and the jet grouting columns had about half the shaft depth.

During construction of the station's internal structures, another major advantage of the "large shaft plus tunnel" arrangement becomes evident. The volume in the shaft is large enough to house all station technical rooms, commercial spaces and public areas. The tunnels are used for platforms only. Again, as the shaft is free of struts, the unobstructed bottom-up construction of internal walls, pillars, slabs and beams, becomes similar to any RC building on the surface. Construction of all station internal structures consumed eight months only.

4 SALGUEIROS STATION

4.1 General design and construction aspects

The double shaft design of Salgueiros Station is unique in Europe. The geometry is made up of two incomplete ellipses. Combined, they produce an open cut more than 80 m long and about 40 m wide. The excavation depth is 24 m.

The geology at the station site is predominantly made up soft, highly weathered granite. Medium to hard granite is found only at excavation bottom. The engineering solution for the excavation works is bolder than at Marquês.

As the two large ellipses are not excavated as closed rings, a very stiff, reinforced concrete portal type frame, located where the ellipses touch each other, was required in order to provide stability. The RC frame was constructed before station excavation. Two small diameter circular shafts (3.3 m) were excavated and filled with a rebar cage and concrete, to form the pillars of the portal frame. At the surface, a large RC beam, cast on surface ground, provided the connection between the two pillars. The beam is 30 m long, 2 m high and 1.6 m wide.

Once the pillars and the beam acquired enough strength, the large ellipses were excavated and supported with a temporary shotcrete lining. In order to maintain both ellipses with uniform ground loading on the shotcrete shell, excavation occurred on both sides simultaneously.

The entire station volume was excavated without struts or anchors. The ground was dewatered by means of vertical wells prior to excavation. Excavation panels were 1.8 m high, with horizontal lengths varying between 4 m and 12 m. The thickness of the lining increased with depth, going from 0.35 m near the surface, to 0.60 m near the bottom.

4.2 Modelling Approach

The general modelling approach is similar to the one described in Table 1. In the 3D STRAP model, the shotcrete lining was modeled with plate elements. The large beam and pillars that make up the RC frame were modeled as beam elements having properties of the geometric sections (area, moment of inertia, center of gravity, etc) equivalent to the actual structures.

Similarly to Marquês, loading scenarios were quite different in the construction stage from the final operational stage. During construction, as the water table had been drawn down, the loads acting on the temporary structure were due to lateral earth pressure and surface surcharge only. In the final stage, with the internal reinforced concrete structure in place, hydrostatic pressure and seismic loads were included. In the final station configuration, intermediate slabs at concourse and

Figure 4. (a) Top Left:Placing rebar cage into for portal frame pillar into shaft; (b) Top Right: Final stages of concreting the portal frame beam; (c) Bottom Left: Early excavation stage; (d) Bottom Right: Full excavation depth at Salgueiros Station.

mezzanine levels provided a strutting action for the final RC lining which has a design life of one hundred years. For the long term, the temporary shotcrete lining was disregarded.

4.3 Critical elements for excavation stability

The large circular pillars and the surface beam that make up the central RC frame are critical for station stability during construction. In the excavation stage shown in Figure 4d, corresponding to maximum excavation depth of 24 m, the circular pillars are subject to large bending moments. These moments result from the lateral earth pressure acting on the shotcrete lining. The pillars were designed for moments in the range of 48000 kNm for the excavation stage. High moment values are located at excavation midheight.

The pillars transfer these high loads to the surface beam and also to the ground, through an embedment lenght of 6 m. The surface beam was designed to withstand the span of the station width. There are two different loading situations during construction. As excavation begins, and the ground is removed from below the beam, the loading is by its own weight. As the excavation progresses, the circular pillars are subject to forces transmitted by the shotcrete lining. With increasing excavation depth, more lateral earth pressure is transferred to the pillars which, in turn, transfer axial loading to the surface beam.

The maximum axial loading on the beam was high, around 22000 kN which demanded reinforcement to be placed on the sided of the beam, in addition to usual bending reinforcement at the bottom of the cross section. In addition to a careful design of the pillar and beam elements, it was extremely important that the connection between the pillar and the shaft wall was reliable, both in shotcrete and in final concrete. Several construction details were supervised by the designer on site, and this connection was one of particular concern. Figure 5 gives details of this connection.

a)   b)   c)

Figure 5.   (a) Removing styrofoam used to block out connection rebars; (b) Connection couplers for final lining; (c) Concreting near the pillar.

ACKNOWLEDGEMENTS

The authors would like to acknowledge and thank the following entities: Metro do Porto, Normetro A.C.E., Transmetro A.C.E, CJC Engenharia Ltda and Figueiredo Ferraz Cons. Eng. Proj. Ltda.

REFERENCES

Andrade, J. C. et al., 2004. Estações suberrâneas em poços e túneis no Metro do Porto: Aspectos gerais de projecto e acompanhamento técnico da obra- ATO. Conference on Geotechnics, Aveiro, Portugal.
Franco, S. G., et. al., 2004. Estação do Marquês em poço no Metro do Porto. Modelação e Segurança. Conference on Geotechnics, Aveiro, Portugal.
França, P.T. et. al., 2004. Estação de Salgueiros em poço no Metro do Porto. Modelação e Segurança. Conference on Geotechnics, Aveiro, Portugal.

Applications of Computational Mechanics in Geotechnical Engineering – Sousa,
Fernandes, Vargas Jr & Azevedo (eds)
© *2007 Taylor & Francis Group, London, ISBN 978-0-415-43789-9*

Mining very large adjacent stopes using an integrated rock mechanics approach

L. Meira de Castro
Golder Associates Ltd., Mississauga, Ontario, Canada

ABSTRACT: Simple, 3D elastic numerical analyses are commonly applied in the mining industry to estimate the best mining sequence, design the support system and stope dimensions and evaluate the monitoring performance. This paper presents the applications of numerical analyses for the excavation of four large open stopes at the 1500 m Level of the underground Beaufor Mine in Canada. This excavation required the application of an integrated rock mechanics approach from the planning stage through to final extraction in order to control dilution and reduce the potential for loss of reserves. The excavation of these stopes was a success, as more than 190,000 tonnes of ore were extracted using open stopes with no backfill. Dilution was between 5% and 10% and access drifts, located as close as 8 m from the hangingwall, were fully operational throughout this excavation. By gaining experience with the local rock mass conditions, mine personnel gained the confidence necessary for implementing a more aggressive extraction ratio by excavating stopes of up to 40 m long × 26 m wide × 75 m high at a depth of 450 m

1 GENERAL INSTRUCTIONS

Two mineralized zones, labelled B and C zones were planned to be mined between the 1500 and 1200 m levels, using both longitudinal and transverse stopes. Mining was to be carried out by creating 20 m long (strike length), 20 m to 24 m wide stopes in about 15 m increments of height to a maximum height of up to 75 m. Mining included the development of four sublevels, spaced up to 15 m along dip, and labelled 610, 625, 637 and 652. Along strike, mining was planned to be carried out from section 1230 E to 1300 E , leaving a 10 m wide vertical "rib pillar" from 1270E to 1280 E (Fig. 1).

More than 190,000 tonnes were mined from the B and C series of stopes (Figs 1 & 2). The B zone was located in the hangingwall (HW) and in close proximity to the C zone, resulting in a narrow pillar, as listed in Table 1 for stopes B-1250E to C-1250E.

Stability of this pillar and the HW of the B stopes were crucial for the successful mining of these ore zones. Selection of the best mining sequence, design and evaluation of the ground support systems and development of a monitoring program, required the implementation of an integrated rock mechanics approach, which included several 3D numerical analyses.

The integrated rock mechanics approach was used from the planning stage through to final extraction in order to mine the stopes located between the 1500 and 1200 levels (Elev. 2600 m and Elev. 2675 m corresponding to depths of 450 to 375 m). This integrated approach consisted of (Castro & Beaudoin, 2006):

- Assessment of the rock mass quality;
- Development of 3D numerical models;
- Evaluation of the mining sequence;
- Application of ground support using cable bolts;

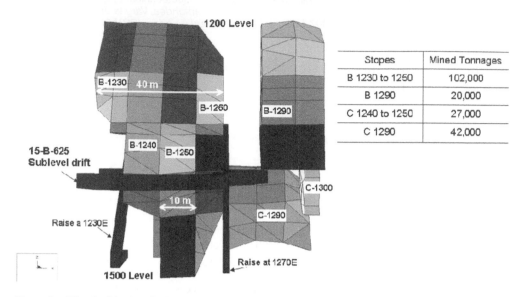

Stopes	Mined Tonnages
B 1230 to 1250	102,000
B 1290	20,000
C 1240 to 1250	27,000
C 1290	42,000

Figure 1.   View looking North showing the stope layout.

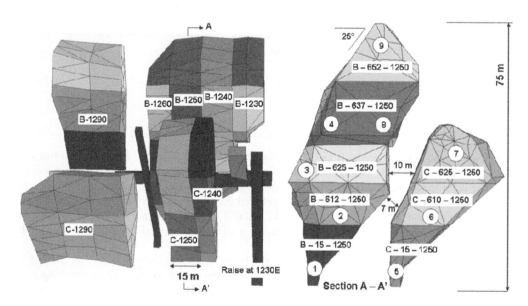

Figure 2.   View looking South to Southeast showing the 3D geometries and the mining sequence.

Table 1.   Summary of the stopes in the 1250 E section.

Stope number B Zone	Tonnes mined	Stope number C Zone	Tonnes mined	Minimum distance between stopes (m)
B-652-1250E	14699	–	–	–
B-637-1250E	13441	–	–	–
B-625-1250E	13245	C-625-1250E	15171	10
B-612-1250E	7848	C-610-1250E	9754	7.2
B-15-1250E	2790	C-15-1250E	2457	14

192

- Monitoring of HW stability using Smart-Cables and CMS surveys;
- Application of control blasting with pre-shearing and electronic detonators in the underground mine;
- Calibration of the 3D numerical model; and
- Re-evaluation of the mining sequence and the stability of the access drifts and drawpoints.

The challenges associated with mining these open stopes were related to their large size and close proximity, with the potential for high dilution and loss of access drifts, located as close as 8 m from the HW.

This paper describes this integrated approach and discusses the 3D numerical analysis results and the monitoring program.

2 BACKGROUND

2.1 *Stereographic projection*

Structural mapping was carried out in 1996, 2000 and 2002 at different depths and locations within the B and C zones. For the 1200 to 1500 levels, 159 measurements were processed to prepare the stereographic projections. Figure 3 presents the stereographic projection of these poles and suggests the fabric of rock mass can be represented by three major discontinuity sets (labelled V1, J1, and J2) with possibly several minor discontinuity sets.

The major vein set V1 (mean dip/dip direction = 44°/211°) is sub-parallel to the mineralized zones. The joint set J1 (57°/175°) is parallel to the east-west trending faults (South and Perron faults). Joint set J2 (51°/319°) strikes in the SW direction and dips to the NW (Fig. 3). The minor set j3 (53°/103°) strikes almost North-South and dips to East. Minor sets j4 and j5 strike in the NW direction and dip to NE.

Kinematic analysis of the HW indicated the potential for planar instability along minor sets j4 and j5 and wedge instability formed by the intersection of sets J2 and the minor sets j3 to j5 (Fig. 3). This means that the HW stability was relatively favourably oriented in relation to the major discontinuity sets.

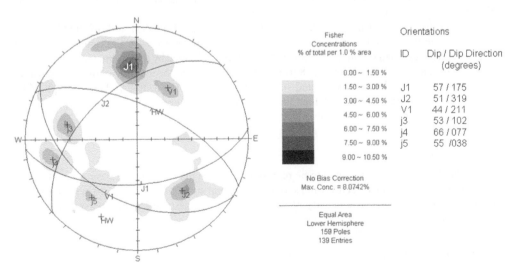

Figure 3. Summary of the major (upper case) and minor (lower case) discontinuity sets, based on previous structural data collected at the mine.

Table 2. Laboratory intact rock test results.

Test results	Description	Rock type	
		Fresh granodiorite*	Altered Granodiorite*
Uniaxial Compressive Strength (UCS)	Average (MPa)	151.4	128.3
Tensile Strength	Average (MPa)	−8.5	−13
Young's Modulus	Average (GPa)	36.9	28.8
Poisson's ratio	Average	0.14	0.13
Hoek-Brown (1988) parameters for intact rock	mi	15	16
	s	1	1

* Altered Granodiorite located in the immediate HW (say less than 5 m) and Fresh Granodiorite located farther inside the HW of the B zone.

2.2 Intact Rock Strength

The Beaufor property is within the Val d'Or mining district, located in the SE part of the Abitibi "greenstone" belt, in Quebec, Canada. The host rock to the ore zones essentially consists of the granodiorite rock unit, which varies in composition from altered to fresh. The alteration is characterized mainly by chloritization and carbonitization with local sericitization. The Fresh Granodiorite occurs intercalated within the Altered Granodiorite in the HW and FW of the B and C zones. Results of laboratory rock testing are presented in Table 2.

2.3 Rock Mass Properties

Based on RMR_{1976} rock mass classification (Bieniawski, 1976), the Fresh Granodiorite can be represented by RMR > 75 (or Geological Strength Index, GSI = 75) and the Altered Granodiorite by RMR (or GSI) in the range of 68 to 75.

Rock mass elastic properties were assumed to be represented by a Young's Modulus = 30 GPa and Poisson's ratio = 0.14 and rock mass strength Hoek and Brown (1988, 2002) strength parameters, UCS = 130 MPa, m = 5.5 and s = 0.035, in the numerical models. Using these parameters yields a tensile strength for the rock mass of only −0.8 MPa.

Based on the empirical method, developed by Mathews et al. (1981) and updated by Potvin (1988), the modified stability number N' was initially estimated to be in the range of 15 to 25, which would suggest a maximum hydraulic radius, HR (area divided by the perimeter of the exposed stope surface analysed) in the range of 8 to 10, when considering the average within the transition zone between stable with no support and stable with support in the Mathews/Potvin graph. If the stope surface is properly supported, then the HR values would increase to values of up to 11 and 12.3 for N' of 15 and 25, respectively.

2.4 Far-field Stresses

The far-field stresses were assumed to be represented by a major horizontal stress, $\sigma_{H1} = 1.9 \times \sigma_v$ (in the true North to South direction) and a minor horizontal stress, $\sigma_{h1} = 1.3 \times \sigma_v$ (in the East-West direction) (Arjang, 1994). The density of the rock mass was assumed to be on average equal to $0.026\,MN/m^3$.

2.5 Criteria for Defining the Overstressed Zones

Both the deviatoric stress and the Hoek-Brown (1988) strength approaches were used for assessing the potential for stress controlled rock mass failure.

The deviatoric stress approach was applied for assessing zones with high (induced) compressive stresses. Where the induced deviatoric stress (i.e. $\sigma_1 - \sigma_3$) exceeds the rock mass system strength (Castro & McCreath, 1997), then potential for stress induced failure is considered high. This rock mass system strength is approximately 0.5 to 0.6 of the uniaxial compressive strength (UCS) of the intact rock.

In order to assess the potential for rock mass relaxation, due to the reduction of the induced stresses that are parallel to the excavation walls, the minor principal stress contours (σ_3) were prepared, representing the confining conditions. For these contours, zones with nil or negative minor principal stress indicate areas under low confinement where there is reduced clamping stresses for holding any potential rock blocks or wedges. These tensile zones would also correspond to areas with factors of safety <1, when using the Hoek-Brown strength approach.

3 NUMERICAL ANALYSIS

Three-dimensional numerical analyses were carried out using the boundary element program Map3D© to evaluate the planned mining sequence; estimate the potential and extent of over-stress zones; assess the stability of the pillar located between the B and C zones and the HW; and assess the potential impact of mining the 15 B stopes (i.e. B-123, 124 and 125) on the HW drifts located at the sub-levels and at the 1200 level.

3.1 Analysis results

Based on the numerical analysis results, plots of normalized deviatoric stress ($\sigma_1 - \sigma_3/100\,\mathrm{MPa}$), minor principal stress (σ_3), and factors of safety (using the Hoek-Brown strength parameters) were initially prepared. In addition, it became evident from these initial results that the potential for rock mass failure would likely be due to rock mass relaxation rather than by high compressive induced stresses. As a result, for this technical paper, only minor principal stress contours are presented.

A summary of main observations from the numerical results are summarized here:

3.1.1 Assessment of overall stability

1. The numerical analysis results indicated that the confining pressure around the stopes would be reduced, and even become negative (tensile), as the stopes are mined upwards to the 1200 level. The acceptable tensile strength for the rock mass became a critical variable. In general, mines consider that the rock mass would be working near the ultimate strength or even after peak. The results suggested that the induced minor principal stress would reduce from $-1.5\,\mathrm{MPa}$ up to $-3\,\mathrm{MPa}$, which is about three times higher than the estimated rock mass tensile strength and approximately 20% to 30% of the intact rock tensile strength.
2. Usually, due to the potential for rock mass relaxation, potential instabilities are considered to likely be structurally controlled. Consequently, the orientation and surface conditions of the major joint sets was considered to control the potential for dilution and instability of the walls and back of the stopes. However, based on the kinematic analysis presented on Figure 4, the potential for planar or wedge failure would involve one major joint set (J2) and minor discontinuity sets (e.g. j3 to j5). Based on the kinematic analysis results, which indicated that the HW was favourably oriented in relation to the major discontinuity sets, should any ground support be recommended for the HW? For this particular mine, cable bolt drifts were relatively easy to implement; however, in other places, it might not be feasible to excavate them, increasing the pressure for not using ground support in the HW.
3. Recommendations were provided to continuously carry out structural mapping of the existing and newly excavated sublevel drifts; and to install cable bolts in the HW of the stopes from dedicated cable bolt and/or planned access drifts to improve stability conditions for the

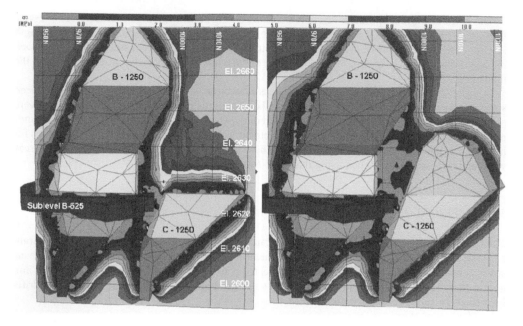

Figure 4. Minor principal stress (σ3) contours after completely mining the B-1250 stopes and then the C-1250 stopes to Elev. 2625 m and Elev. 2640 on a vertical grid plane passing at 1255E. Note the large tensile zone (dark grey shaded area) in the HW of the C-1250 stope.

HW. The question that the rock mechanics practitioner is usually faced with is "Do we really need to install these cable bolts?" For this mine, the recommendation was based on the estimated HR which would be about 13 and the risks and costs involved in losing the ore reserves in the C zone and the drawpoints for both the B and C zones compared to the cable-bolts costs.

Based on these observations, the mine decided to install birdcage cable-bolts in the HW of both the B and C zones, in order to protect the drawpoints and access drifts. Single cable bolts installed with no face plates and using a fan type of distribution were installed from cable bolt drifts for the 15 B and C – 1250 stopes.

3.1.2 Mine sequence

1. There was a potential for rock mass failure (by relaxation) of the vertical pillar that would separate the B-1250 & 1260 and C-1250 & 1260 stopes, when considering the initially proposed mining sequence, which consisted of completely mining the B zone and then the C zone (Fig. 4).
2. From a rock mechanics viewpoint, it was considered important to increase the horizontal confinement of the pillar in order to maximize the chances of reducing this potential rock mass failure. An alternative mining sequence was proposed which consisted of delaying excavation of the upper portion of the B-1250 stope until after mining stope C-1250 (Fig. 5). A comparison of the tensile zones (dark grey areas) on Figures 4 (initially planned mining sequence) and 5 (suggested mining sequence) shows an increase in the horizontal confining pressures within the pillar.
3. Based on this alternative mining sequence, the following recommendations were provided during the planning stage: a) delay the excavation of the B-1250 stope above Elev. 2657 m (B-650 sublevel); b) install cable bolts in the HW of the C-1240 and 1250 stopes, from the sublevels 625 and 637; c) excavate the entire C-1250 and C-1240 stopes with blasting control; d) complete excavation of B-1250 and e) excavate stopes B-1240 and B-1260.

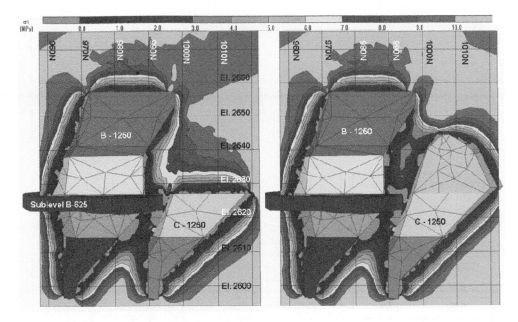

Figure 5. $\sigma 3$ contours as mining of the C-1250 stope advances upwards, while the upper part of B1250 stope is kept at the same elevation. Compare the results with those from Figure 4.

4. In the section where the vertical pillar was narrow (i.e. around Elev. 2627 m or B-625 sublevel), recommendations were provided for the installation of cable bolts with a high density (say, every 1.5 m to 2 m) and 3 m long Swellex bolts in the HW of the C-1240 and 1250 stopes.

Based on the numerical analysis results and the discussion presented above, the mine decided to install cable-bolts as suggested and to alter the initially planned mine sequence to follow that recommended in item 3.

3.2 *Instrumentation*

Recommendations were provided which included the installation of Smart-cables to monitor the performance of the cable bolts used to control the potential for HW dilution. A total of five instrumented Smart-Cables (supplied by MDT, www.mdt.ca), each with six anchors, was installed in the HW of the stopes (Fig. 6). It should be noted that the number of cable bolts and Smart cables were increased after their initial performance and field observations made by the mine personnel, confirming the ground behaviour. Figure 6 also shows the Cavity Monitoring Surveys (CMS) carried out after each stope lift was excavated.

For exemplification purposes, the monitoring results of the Smart cable located at the 612 sublevel will be discussed here. By April 2, 2004, the anchor located at a distance of approximately 3.5 m from the HW of stope B-15-1250E was showing a load of 20 tons. A linear extrapolation of the loading curve to where it would intercept an assumed maximum load of 25 tonnes, suggests that even with no additional excavation, there would be the potential for failure of the HW cable bolt(s) by March 2005 (Fig. 7). As a result, and recognizing that it was essential to protect the two drawpoints, recommendations were provided to immediately install additional support on the HW of stope B-1250 between the 1500 Level and the 612 sublevel. Reinforcement was calculated to hold the estimated 3.5 m wide rock slab with a minimum safety factor of 1.3. Additional cable-bolts (double cables) and another Smart-cable were installed at this area (Fig. 6).

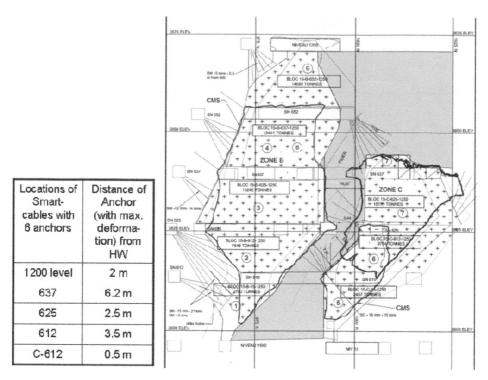

Locations of Smart-cables with 6 anchors	Distance of Anchor (with max. deformation) from HW
1200 level	2 m
637	6.2 m
625	2.5 m
612	3.5 m
C-612	0.5 m

Figure 6. Section 1250E showing the five locations with Smart cables and the final pillar thickness between the B and C zones (based on the CMS surveys).

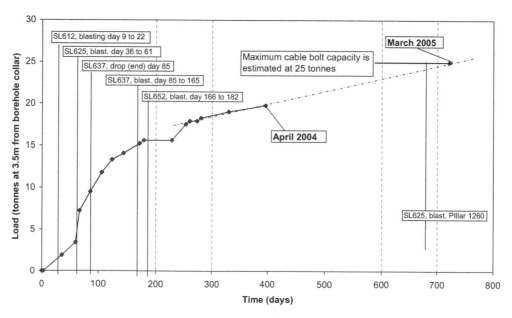

Figure 7. Loading curve of the Smart-Cable located at the HW of the stope B-1250E from the sublevel 612 (as prepared by Beaufor Mine) after 400 days after installation. It also includes a projected linear extrapolation of the loading curve to estimate the potential for failure at that location.

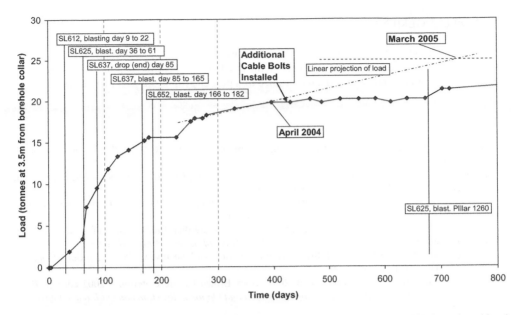

Figure 8. Actual loading curve of the Smart-Cable located at the HW of the stope B-1250E from the sublevel 612 (as prepared by Beaufor Mine) after 720 days.

At approximately 700 days after mining of the B zone began, the HW was still in place and the anchor located at 3.5 m from the HW had shown a maximum load of 21 tons (Fig. 8).

4 FINAL REMARKS

Mining the 15 B and C stopes was successfully completed in December 2005. The thin pillar between the B and C zone is still in place (although the mining at this location is now closed). Some minor dilution was observed on the northern wall of stope C-1250E, where it was transected by the Beaufor Fault. The back and walls of stope B-1250 do not show visible signs of stope dilution or deterioration.

Although not described in detail in this paper, control blasting was adopted for excavation of the final stope walls. A Instantel Minimate Plus blast vibration monitor was used to establish an appropriate load of explosives per delay. Testing established an optimum load of 120 kg/delay resulting in peak particle velocities (PPV's) of 60 mm/sec at 20 m distance from the charge. Electronic delays were selected over non-el to assure minimum PPV. Pre-shearing blasting techniques were used along the hangingwall of the stopes to minimize over-break.

In summary, the excavation of these stopes was a success, as more than 190,000 tonnes of ore were extracted using closely spaced open stopes with no backfill. The numerical analyses were a useful tool and the use of cable bolts and Smart cables proved to be important in controlling dilution, which was calculated as being between 5% and 10%. The access drifts, located as close as 8 m from the HW, were fully operational throughout this excavation.

Full contribution from the team, including consultants, was essential for this integrated rock mechanics approach, right from mining planning through to the final excavation. Constant communication between the team was also important for continuous updating on the understanding of the ground behaviour. By gaining experience with the local rock mass conditions, mine personnel gained the confidence necessary for implementing a more aggressive extraction ratio by excavating stopes of up to 40 m long × 26 m wide × 75 m high at a depth of 450 m.

ACKNOWLEDGEMENTS

The author wishes to acknowledge Mines Richmont – Beaufor Mine for authorizing the publication of this paper and the mine personnel involved in the day-by-day operations of these stopes for their cooperation and understanding of the importance of a proper ground support installation.

REFERENCES

Arjang, B. 1994. An analysis of in situ stresses in the Abitibi mining district. *CANMET Division Report MRL 94-069 (TR)*.

Bieniawski, Z.T. 1976. *Rock Mass Classification in Rock Engineering Exploration for Rock Engineering*, ed. Z.T. Bieniawski, A.A. Balkema, Johannesburg, 97–106.

Castro, L.A.M. & McCreath, D.R. 1997. How to Enhance the Geomechanical Design of Deep Openings. *Proc. of 99th Annual General Meeting – CIM '97*, Vancouver, Canada.

Castro, L.A.M. & Beaudoin, M. 2006. Successful Extraction of more than 190,000 tones of ore through very large adjacent Open Stopes – Beaufor Mine. *Val d'Or – 21st Colloque - Contrôle de Terrain*.

Hoek, E. & Brown, E.T. 1988. The Hoek-Brown Failure Criterion – a 1988 update. *Proc. 15th Canadian Rock Mechanics*, Univ. of Toronto: 31–38.

Hoek, E.; Carranza-Torres, C. & Corkum, B. 2002. Hoek-Brown Failure Criterion – 2002 edition. *Proc. NARMS-TAC 2002 – Mining and Tunnelling Innovation and Opportunity. Hammah et al. (eds)*, University of Toronto.

Mathews, K. E., Hoek, E., Wyllie, D.C. and Stewart, S.B.V. 1981. Prediction of Stable Excavations for Mining at Depth below 1000 metres in Hard Rock. *CANMET Report DSS Serial No.OSQ80-0081, DSS File No. 17SQ.23440-0-9020, Ottawa: Dept. Energy, Mines and Resources*.

Potvin, Y. 1988. Empirical Open Stope Design in Canada. *Ph.D. thesis, Dept. Mining and Mineral Processing, Univ. of B. Columbia*.

Applications of Computational Mechanics in Geotechnical Engineering – Sousa,
Fernandes, Vargas Jr & Azevedo (eds)
© 2007 Taylor & Francis Group, London, ISBN 978-0-415-43789-9

Numerical analysis of the Venda Nova II powerhouse complex

T. Miranda & A. Gomes Correia
University of Minho, Department of Civil Engineering, Guimarães, Portugal

S. Eclaircy-Caudron & D. Dias
INSA, Lyon, France

C. Lima
EDP-Produção, Porto, Portugal

L. Ribeiro e Sousa
University of Porto, Department of Civil Engineering, Porto, Portugal

ABSTRACT: In the North of Portugal a hydroelectric scheme called Venda Nova II was recently built in order to optimize the resources of the reservoirs created by Venda Nova and Salamonde dams. The scheme, almost fully composed by underground structures and built in a predominantly granite rock mass, include several tunnels with a total length of about 7.5 km, inclined and verti- cal shafts with a total length of 750 m and two caverns which compose the powerhouse complex. The complex consists of two caverns interconnected by two galleries at a dept of about 350 m. For this complex, 2D and 3D numerical models were developed considering the different con- struction stages. The geomechanical parameters of the granite formation for the numerical models were obtained using GEOPAT. This software is a knowledge based system which allows obtaining geomechanical parameters for underground structures modelling in granite formations. The 2D model was developed in the Phases2 software while the 3D model in FLAC3D. In this paper results of these models are analysed. Some comparisons are carried out between the models results and the monitored data. The numerical results show in general a good agreement with the monitored ones.

1 INTRODUCTION

In the 90's the CPPE (Portuguese Company of Electricity Production) decided to reinforce the power of Venda Nova hydroelectric scheme by building a new one, named Venda Nova II, that took advantage of the high existing head – about 420 m – between two reservoirs (Lima et al., 2002; Plasencia, 2003). Venda Nova II is equipped with two reversible units in order to optimize the use of the water resources for energy production. It was built in a predominantly good quality granite rock mass and involved the construction of important geotechnical underground works of which the following can be mentioned:

- the access tunnel to the caverns, with about 1.5 km, 10.9% slope and 58 m² cross-section;
- the hydraulic circuit with a 2.8 km headrace tunnel with 14.8% slope and a 1.4 km tailrace tunnel and 2.1% slope, with a 6.3 m diameter modified circular section;
- the powerhouse complex located at about 350 m depth with two caverns, for the powerhouse and transforming units, connected by two galleries;
- an upper surge chamber with a 5.0 m diameter and 415 m height shaft and a lower surge chamber with the same diameter and 60 m height.

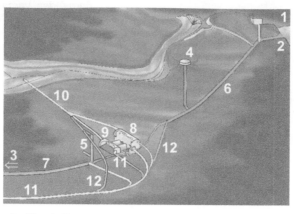

1 – Venda Nova reservoir	7 – Tailrace tunnel;
2 – Upper intake	8 – Powerhouse cavern
3 – Lower intake	9 – Transformer cavern
4 – Upper surge chamber	10 – Ventilation galleries
5 – Lower surge chamber	11 – Access tunnel
6 – Headrace tunnel	12 – Auxiliary tunnels

Figure 1. General perspective of the power reinforcement scheme (Lima et al., 2002).

Figure 1 shows a general perspective of the power reinforcement scheme.

2 THE UNDERGROUND POWERHOUSE COMPLEX

2.1 Description

The powerhouse complex, located in a intermediate position of the hydraulic circuit, was built at a depth of approximately 350 m. It is composed by two caverns interconnected by two galleries (Figure 2). The dimensions of the main cavern are, in plan, 19.0 × 60.5 m while for the transforming units cavern are 14.1 × 39.8 m. The distance between their axes is 45.0 m.

The caverns are located in an area where the existence of two subvertical discontinuities was detected as shown in Figure 3. These discontinuities are identified as E and F. Based on the interpretation of the geotechnical survey results, it was decided to move the caverns in the NE direction.

A coarse porphyritic, both biotitic and moscovitic, granite prevails in the region. The rock mass on which the hydroelectric complex is installed is characterized by medium-size grain granite of a porphyritic trend with quartz and/or pegmatitic veins and beds, which are occasionally, rose. The rock mass also presents embedment of fairly quartzitic mica-schist.

2.2 Geotechnical survey and monitoring plan

The complex was built in a granite rock mass with good geomechanical quality. To characterize the rock mass in the area of the caverns, four vertical boreholes with continued sample recovery were performed (Plasencia, 2003). The lengths of these boreholes varied between 271.0 m and 381.6 m and their positioning controlled each 50 m. A total of 98 samples were collected. The laboratory tests performed in these samples allowed the zoning of the rock mass as presented in Table 1. Caverns are located in the ZG1C zone. Lugeon permeability tests were also executed.

Between boreholes, seismic tests using longitudinal waves (P waves) were performed in order to obtain tomographies of the rock mass and to detect important geological structures (LNEC, 1997;

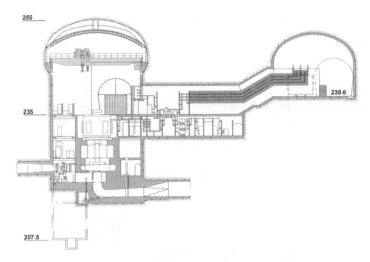

Figure 2.   Powerhouse complex.

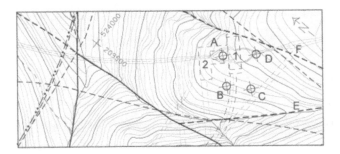

Figure 3.   Implantation of the powerhouse complex.

Table 1.   Geological-geotechnical zoning of the rock mass.

	Weathering	Disc.	RQD	Perm.	I_r (MPa)	UCS (MPa)	E_r (GPa)
ZG3C	W3/W4–5	F3/F4-5	0–90	>10 UL	3.8	57.7	42.0
ZG2C	W1–2/W3	F1–2/F3	50–90	0–8 UL	6.3	96.9	51.0
ZG1C	W1/W2	F1/F2	90–100	<2 UL	7.0	110.1	54.9

UL – Lugeon units; E_r – deformability modulus of the intact rock; I_r – Point load index; UCS – Uniaxial compressive strength

Plasencia et al., 2000). These tests were executed at depths varying between 95 and 370 m and the results confirmed the previous zoning. The area where the caverns are located was characterized with P waves velocities between 5250 and 6000 m/s, sometimes 4750 to 5250 m/s. These values confirmed the good geomechanical characteristics of the rock mass.

After the construction of the access tunnel to the caverns a gallery was excavated in order to characterize the rock mass and confirm the previous geomechanical characterization and to measure the *in situ* state of stress. This gallery was excavated from the top of the access tunnel and parallel to the caverns axis. Large Flat Jack tests were used to obtain the deformability modulus of the rock mass. The values ranged from 33 to 40 GPa. Strain Tensor Tube and Small Flat Jack tests were carried out for the stress state determination. The results showed that the vertical and horizontal

Table 2. Characteristics of the four main families of discontinuities (Plasencia, 2003).

Family	1	2	3	4
Direction	N81°E	N47W	N8E	N50E
Inclination	77NW	12NE	83NW	80NW
Continuity	1 to 3 m	1 to 10 m	3 to 10 m	3 m
Alteration	W1–2, occasionally W3	W1–2	W1–2, occasionally W4	W1–2
Opening	closed at 0.5 mm	closed at 0.5 mm	closed at 0.5 mm, sometimes 2.5 mm	closed
Thickness	none at 0.5 mm	none at 0.5 mm	none, sometimes 2.5 mm	none
Roughness	undulating poorly rough to rough	Undulating poorly rough, sometimes rough stepped	rough plane, sometimes polished	undulating poorly rough
Seepage	Dry	Dry	Dry, occasionally with continuous water flow	Dry
Spacing	2 to 3 m, sometimes 1 or 4 m	2 to 3 m, sometimes 1 m	1 to 2 m	5 to 6 m

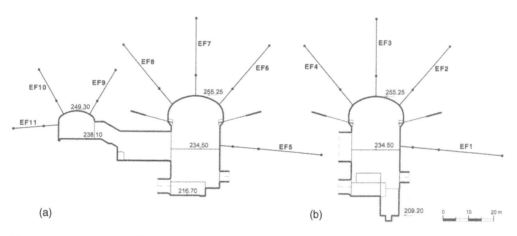

Figure 4. Cross-sections of the monitoring plan.

stress parallel to the caverns axis have the same magnitude and correspond to the overburden dead load. In the perpendicular direction the stress values are 2 to 3 times higher (K_0 value between 2 and 3). From the litological characterization it was possible to identify four main discontinuities sets. In Table 2 their main characteristics according to the ISRM (1978) criteria are summarized.

To evaluate the displacements in the rock mass surrounding the caverns, a monitoring plan using extensometers and convergence marks was established. The extensometers, in a total number of eleven, were placed in two sections and have lengths varying from 5 m to 40 m (Figure 4). The convergence marks were installed in several sections (5 to 7 each section). The three-dimensional convergence measurements readings were carried out using an optical system based on the total station technology. Six load cells were also installed for the anchors. Figure 5 shows the evolution of the measured displacements in the extensometers EF5 and EF11.

For the numerical models developed in this work, the geomechanical parameters were obtained using GEOPAT which is a Knowledge Based System especially developed at University of Minho for this purpose (Miranda, 2003). It uses well organized and structured knowledge from experts together with artificial intelligence techniques for decision support in the geomechanical parameters calculation and has been used with success in different applications. Using the gathered

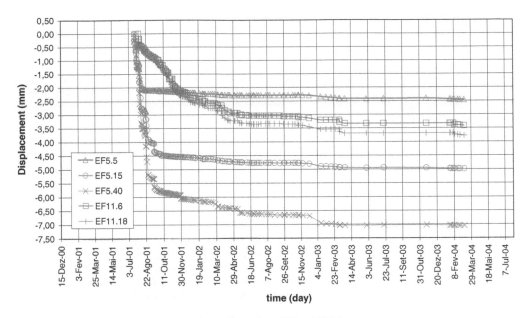

Figure 5. Evolution of displacements in extensometers EF5 and EF11.

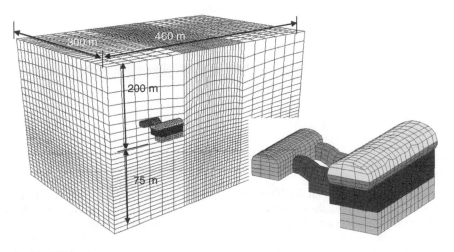

Figure 6. The 3D mesh.

geotechnical information together with GEOPAT the following geomechanical parameters were obtained: deformability modulus E = 45 GPa, friction angle $\phi' = 54°$ and cohesion c' = 4 MPa.

3 NUMERICAL MODELLING

The 3D model was developed using the finite difference software FLAC3D to simulate the complex geometry of the powerhouse complex and its construction sequence. It is composed by 43930 zones, 46715 grid-points and 1100 structural elements (Figure 6). Since the filed stress around the caverns was constant it was possible to simplify the mesh in order to be computationally more efficient. This way instead of the real 350 m depth of the cavern axis only 200 m was modelled.

Table 3. Adopted construction stages for the 3D numerical model.

Stage	Model	Description
1		Excavation of the upper part of the main cavern arch. Application of 25 cm of fiber sprayed concrete on the arch and 6 m length and 25 mm diameter rockbolts in a 2 × 2 m mesh.
2		Excavation of the remaining main cavern arch.
3		Excavation of the main cavern until the base level of the interconnecting galleries and the transforming units caverns. Application of 25 cm of fiber sprayed concrete on the arch of the second cavern and 6 m length and 25 mm diameter rockbolts in a 2 × 2 m mesh.
4		Excavation of the two interconnecting galleries and application of 25 cm of fiber sprayed concrete in the roof of the galleries.
5		Completion of the main cavern excavation.

The section analysed through the 2D numerical modelling was section a) referred in Figure 4. When comparisons between the two models are performed they are always referred to the results obtained for this cross-section where reliable monitoring values are available.

The sprayed concrete was simulated by shell elements with a linear elastic and isotropic constitutive model, with a Young modulus of 15 GPa and a Poisson ratio of 0.2. The rock bolts were simulated by cable elements which can yield tensile strength with two nodes and one axial degree of freedom.

For the numerical modelling, the construction sequence was simplified relatively to those defined in design. Therefore, the adopted stages are the ones presented in Table 3.

The construction sequence adopted for the 2D modelling was very similar. The only difference was the way the two interconnecting galleries were simulated. Three different approaches were carried out in a preliminary analysis: i) considering the total excavation of the galleries; ii) non considering the effect of the galleries excavation due to their small influence in the global behaviour of the structure; iii) replacing the material in the area of the galleries for another with lower equivalent geomechanical properties. The first approach led to unrealistic results with multiple shearing zones and high displacement levels which were not observed in the field. Since the model was developed considering plain strain conditions this consideration was too unfavourable. The remaining two approaches showed very similar results. The differences were insignificant

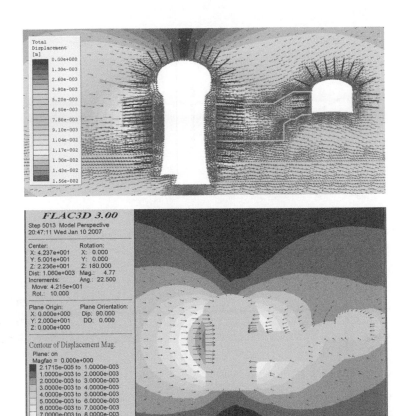

Figure 7. Displacement contours and vectors for the 2D (upper image) and 3D models.

therefore it was chosen not to consider the effect of the interconnecting galleries excavation in the following analysis.

The *in situ* tests pointed out for a K_0 coefficient between 2 and 3. For this analysis a starting value of 2 was considered. Due to the high K_0 ratio and the span of the main cavern vertical wall, the higher displacements are expected to take place in that area. Figure 7 presents the displacements contours along with the corresponding vectors for the two models in the referred cross-section. It is possible to observe the same qualitative displacement patterns in both models. The displacement vectors show the strong influence of the high horizontal stress translated by higher displacements in the vertical walls of the main cavern.

For a more thorough analysis Figure 8 shows the computed displacements along lines coinciding with extensometers 5 and 7 (near the wall and roof of the main cavern, respectively). The displacements of the 2D calculation along the sub-horizontal line are much higher than for the 3D model which was expected due to the plain strain consideration. For the 3D model the maximum displacement along this line is approximately 10 mm while for the 2D model is almost 50% higher. The displacements near the ceiling of the main cavern are small for the two models. In this zone the gravity loads, which would cause a downward movement, are almost compensated with the high horizontal stress which pushes the arch upwards causing a near-zero displacement.

Due to the good overall quality of the rock mass the displacements magnitude is small. The maximum computed displacements in the rock mass are 15 cm for the 2D model and 10.5 cm for the 3D case. Moreover, there are a small number of yielded zones which are confined to small areas near the arch and wall of the main cavern.

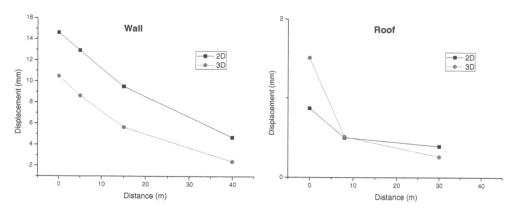

Figure 8. Computed displacement near the wall and roof of the main cavern.

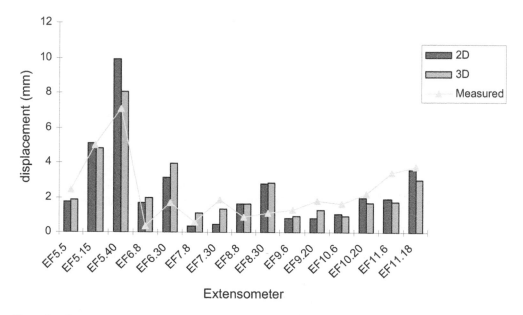

Figure 9. Comparison between computed and measured displacements.

The behaviour of the structure and surrounding rock mass is almost elastic. This means that the most important parameters for the behaviour prediction of the structure are E and K_0. Also, the maximum computed shear strains were low with values ranging from 0.02% and 0.1% for the 2D model and 0.015% and 0.04% for the 3D model. Once more lower values were obtained for the 3D model. These values are within the expected range considering the quality of the rock mass and the construction method which caused very low damage to the rock mass.

Figure 9 compares the results of the models with the measures of extensometers 5 to 11. The results of the 2D and 3D models are very similar for most of extensometers. Also, the computed values follow the same qualitative trend as the observed ones. The worst results are observed for the inclined extensometers of the main cavern (6 and 8) where the displacement values are clearly overestimated. In the remaining cases the 3D model is more accurate for the measurements of extensometers 5, 7 and 9 while the 2D model slightly outperforms the 3D model for extensometers 10 and 11. In a qualitative perspective it can be concluded that, excepting for extensometers 6 and 8, the results of the models are very acceptable.

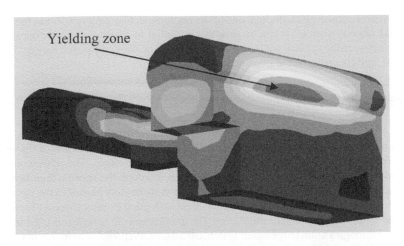

Figure 10a. 3D visualization of the shear strain contours for the last non-equilibrium state.

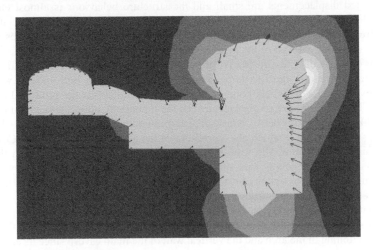

Figure 10b. 2D visualization of the shear strain contours and velocity vectors.

For a more thorough insight of the results, some statistical analysis was carried out. Tests were performed to the mean values of the measured and computed values in the extensometers and they can be considered statistically identical. The mean computed displacement is 2.47 mm (equal for both models) and the measured is 2.34 mm.

The Shapiro-Wilk normality test was performed to the error values of the models. It was verified that they follow a normal distribution for a 95% significance level. This fact suggests a good distribution of the errors with a mean value near 0 (≈ 0.14 mm) and points out for the good quality of the results. Also, a Smirnov test was performed and it was concluded that the observed and the computed values follow the same statistical distribution.

For the 3D model a calculation of the factor of safety was carried out. FLAC3D uses the method defined by Dawson et al. (1999) in which the strength parameters are consecutively reduced until significant plastic flow appears in some zone of the structure. The computed factor of safety was 4.63 which can be considered satisfactory in terms of security level. Figure 10 shows an image of the last non-equilibrium state produced by the methodology of strength reduction applied to calculate the factor of safety. The shear strain contours allow the visualization of the failure mode.

Plastic flow appears in the connection zone between the vertical wall and the beginning of the arch which is an area of stress concentration. This fact can be corroborated by the observation of

Figure 10 where a cutting plane through one of the interconnecting galleries shows the shear strain contours and velocity vectors. It can be seen that potential instability zones are located near the connections between the vertical walls of the main cavern and the ceiling arch mainly near the high span vertical wall (opposite to the interconnecting galleries).

4 CONCLUSIONS

The Venda Nova II hydroelectric scheme built in the North of Portugal includes a set of very important underground structures. In this work 2D and 3D numerical models of the powerhouse complex were developed considering the different construction stages. The powerhouse complex is composed by two caverns connected by two galleries. Through the models their behaviour was analysed and compared with the monitored values by extensometers placed in the caverns.

The scheme is located mainly in a granite rock mass with good geomechanical characteristics as shown by the results of the geotechnical survey. The geomechanical parameters used in the models were obtained through a knowledge based system called GEOPAT. The system was developed at the University of Minho with the purpose of calculate geomechanical parameters in granite formations.

The computed displacements are small and the structure behaviour is almost elastic due to the good quality rock mass. The displacements configuration is very much influenced by the high horizontal stress perpendicular to the caverns axis. The maximum displacement values are observed near the high span vertical wall of the main cavern.

The results of the models, in a qualitative perspective, are close to the observed values. Excepting for extensometers 6 and 8, the fit can be considered acceptable for both models which present similar values. Also in a statistical point of view it is concluded that the produced results are of good quality. The mean computed displacement by the numerical models was 2.47 mm against the measured 2.34 mm. The errors between measured and computed values follow a normal distribution with a mean value close to 0 which also corroborates the good quality of the fit. Concluding, in an engineering point of view, the results of the models fits very satisfactory to the observed displacements. This is especially true since the monitored displacements are very small which turns the fit more difficult to obtain (due to lack of precision in the readings, simplifications of the constructions sequence and constitutive models, etc.).

The computed factor of safety on the 3D model is 4.63 which translate an acceptable security level. The most probable failure mode taken from this calculation is plastic flow in the connection between the beginning of the arch and the vertical wall of the main cavern since its an area of stress concentration.

Backanalysis of the geomechanical parameters is being carried out using the measured and computed displacements of the 3D model. The preliminary results are presented in other work. Different techniques are being used namely the optimization software Sidolo which uses conventional search algorithms and an evolutionary algorithm. This is very innovative technique and is intended that it overcomes some limitations of conventional algorithms like the convergence to local minima.

ACKNOWLEDGMENTS

The authors wish to express their acknowledge to EDP Produção EM for authorization and making available the necessary data. This work was financed by the Foundation for Science and Technology (FCT) in the framework of the research project POCI/ECM/57495/2004, entitled *Geotechnical Risk in Tunnels for High Speed Trains*.

REFERENCES

Dawson, E., Roth, W., Drescher. 1999. Slope stability analysis with finite element and finite difference methods. *Géotechnique* 49(6), 835–840.

ISRM. 1978. Suggested methods for the quantitative description of discontinuities in rock masses. *International Journal of Rock Mining Sciences & Geomechanics* Abstracts, Vol. 15, n° 16, pp. 319–368.

Lima, C.; Resende, M.; Plasencia, N.; Esteves, C. 2002. Venda Nova II hydroelectric scheme powerhouse geotechnics and design. *ISRM News*, Vol. 7, no. 2, pp. 37–41.

LNEC. 1997. Seismic tomography between boreholes in the mass interesting the central cavern of the Venda Nova II scheme. Lisboa (in Portuguese).

Miranda, T. 2003. Contribution to the calculation of geomechanical parameters for underground structures modelling in granite formations. *MCs thesis*, UM, Guimarães, 186p (in Portuguese).

Plasencia, N. 2003. Underground Works – Aspects of the engineering geology contribution and design. *MSc thesis*, IST, Lisboa, 155p (in Portuguese).

Plasencia, N.; Coelho, M. J.; Lima, C., Fialho, L. 2000. Contribution of the seismic tomography for the characterization of the mass interesting the central cavern of the Venda Nova II scheme. *7th Geotechnical Portuguese Congress*. Porto, pp. 113–122 (in Portuguese).

Applications of Computational Mechanics in Geotechnical Engineering – Sousa,
Fernandes, Vargas Jr & Azevedo (eds)
© 2007 Taylor & Francis Group, London, ISBN 978-0-415-43789-9

Time dependent settlements of the ground due to shallow tunnelling

Sh. Jafarpisheh
Geotechnical Expert in consulting engineers, Tehran, Iran

M. Vafaien & B. Koosha
Lecturer in Civil Eng. Dept., IUT, Iran

ABSTRACT: A parametric study have been done for the purpose of evaluating the long term surface settlement of the soft ground in response to the excavating a shallow tunnel. The type of deformation around and above the tunnel is principally known as the settlement components which are the maximum settlement at the ground surface (S_{max}), the settlement at the tunnel roof (S_c) and are related to each other by a parameter defined as the settlement ratio (λ). In this study the time effect is due to the consolidation phenomena which depends on the permeability coefficient of the ground. The results indicate that the maximum long term settlement can be as high as the immediate settlement and the relative amounts depend on the ground properties and tunnel depth and diameter.

1 INTRODUCTION

Evaluating the ground settlement in long time due to tunneling is sometimes necessary for many cases of underground construction in urban areas. The theoretical analyses for this subject have been mostly developed within some assumed simplified circumstances in order to achieve some formulations which can be solved.

As the main cause of the time dependent deformations of the soil around an excavation is attributed to the effect of consolidation, so the main part of the analytical studies has been so far concentrated on the consolidation behavior of the soil.

Bowers et al. (1996) proposed a formula for computing the settlement at time t_b relative to time t_a. Carter and Booker (1982) have proposed an analytical procedure which resulted in the graphs and formulae for evaluating the deformation components at any given point of soil by some dimensionless terms for time and the ground properties considering a circular tunnel at a very deep conditions.

In addition to these types of theoretically based tools for estimating the amounts of time dependent deformations, there are some promising amounts of observational data obtained by monitoring the actual projects. These available empirical data are very much useful in evaluating the analytical formulae or even for comparisons between the reality and the results obtained from numerical analyses like the finite element programs.

In the present study the finite element program Plaxis 2D has been used for investigating the relative importance of different effective parameters on the long-term subsidence of the ground above and around the new excavated opening.

The present study mainly consist of the following topics:

(a) Estimating the long term amounts of ground subsidence corresponding to different defined conditions.
(b) Comparisons between the immediate settlement characteristics and the long term settlement for some assumed conditions.

(c) Illustrating the effect of tunnel lining on the subsidence components.
(d) Comparison between the present computed results and some available published experimental data.

2 MODEL IDEALIZATION AND COMPUTATION

The assumed model for the present computations consisted of an opening with diameter D at the depth of Z_o within a medium of width b and the physical properties of E, υ, and γ.

For the time dependent behavior, the permeability coefficient (k) is also necessary, based on which the consolidation settlement is computed. As is expected, the amount of this type of settlement increases with time ending to a final amount, and the settlement rate depends on the magnitude of k. Because in the consolidation procedure a drainage boundary is necessary, for the tunnel type problems the opening perimeter is presumed as the drain boundary.

The physico-geometrical properties assigned to an assumed model as the basic model in the present analyses are shown in Tables 1 and 2 and lining conditions are shown in Table 3 other chosen values are as follows:

D(tunnel diameter) = 3, 5, 8 and 10 m, γ_{sat} = 18, 20 and 23 kN/m^3; Z_0(tunnel depth, the distance from the surface to the tunnel center) = 15, 20 and 30 m; h_w(depth of water table from the surface) = 2, 6 and 9 m, l(thickness of soil layer below the tunnel base) = 2, 5 and 8 m, B(width of domain) = 30,60 and 80 m, k(soil permeability) = 1e-4, 1e-6 and 1e-2 m/days; k_0 = 0.5, 0.6 and 1; υ = 0, 0.2 and 0.33.

Because of symmetry of both geometrical and physical properties of the model assumed in the present analyses, the computations and displays can be made by only half space, except for some cases in which the full section is necessary.

Throughout this study, the effect of the most influential effective parameters have been examined and the corresponding results have been compared and discussed. As the principal aim of this study was to evaluate the ground settlements due to tunneling , so the main part of the results are shown in terms of s_{max}, (the maximum amount of surface settlement), s_c (maximum settlement at the tunnel crown) and $\lambda = s_{max}/s_c$ for each case.

As is expected for a parametric study for the present subject, the combination of many physico-geometrical parameters for evaluating the immediate settlement and then the same combinations

Table 1. Geometrical parameter of the basic model.

Z_0(m)	D(m)	h_w(m)	B(m)
20	8	−2	60

Table 2. Soil properties of the basic model.

γ_{dry}(kN/m^3)	γ_{sat}(kN/m^3)	E(MPa)	υ	K_x(m/day)	K_y(m/day)	K_0
17	20	50	0.33	4e-4	4e-4	1

Table 3. Lining condition of the basic model.

d(thickness of lining) (cm)	υ	EA(kN/m)	EI(kN/m^2)	W(kN/m)
20	0.15	4e6	1.33e4	4.8

for the time settlements can involve a wide area of several computations which results in several graphs and tables. For the sake of simplicity, in the present article, some of the results are shown by appropriate graphs while some others are mentioned by discussion.

3 COMPUTED RESULTS

3.1 *Unlined tunnel*

In this part, the results of the present computations are discussed firstly for an unlined tunnel with the properties indicated in Tables 1 and 2, and the results corresponding to different cases are compared.

In Figure 1, the half section of surface settlement trough for the immediate settlement compared with the time settlements for 5, 20, 100, 1000 days and the infinity (final long term). As this analysis and other similar cases show, the time of 1000 days can be considered almost as the final position with a small percent of approximation.

To examine the effect of width of the domain (B), three sizes of 30, 60 and 80 m have been tested for the basic case of $z_0 = 20$ m, and it was found that the width 30 m is long enough for the immediate settlement but for the time settlements, the width of 60 meter can be taken as the boundary value, the amounts more than that do not show any significant differences in the results.

As the computations show, corresponding to the crown settlements of 49 and 62 mm for the immediate settlement and the final settlement, the horizontal inward deformations at the tunnel spring line are 38 and 44 mm respectively. The computations reveals that the rate of increasing settlement is not constant in time, as the trend of spreading the settlement can be divided into two parts as shown in Figure 2: the first part is rather fast following a parabolic type, and the second part nearly on a constant slope. The effect of tunnel diameter is also seen in Figure 2.

The unit weight of soil affects directly on the settlements values because the heavier soil results in larger settlements for either immediate or time dependent settlements, but the relative settlement (λ) tends to decrease for the long time (but not much) with increasing the weight of soil.

The type of variations of settlement ratio (λ) with time (as shown in Fig. 3) is dependent upon all the factors involved (like E, υ, γ, k, D, z_0) ,but the corresponding results are not shown here for the sake of shortening the length of the text. Though the amounts of both settlements (at the ground surface and at the tunnel roof) increase with elapsing the time, but their ratio (λ) does not remain constant, and its values increase considerably with elapsing time. This increasing trend is more pronounced with decreasing the values of υ as shown in Fig. 3.

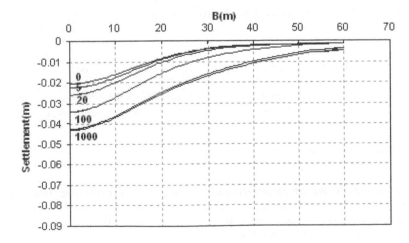

Figure 1. Settlement trough on the ground surface for the immediate and time settlements.

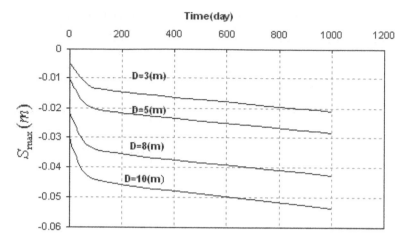

Figure 2.   Effect of tunnel diameter on the max amount of surface settlement.

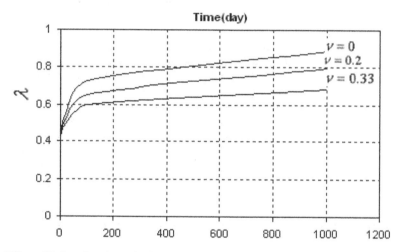

Figure 3.   Effect of Poisson's ratio on the time variations of relative settlements (λ).

The computed values of settlements in different cases are examined in the present study. It can be concluded from this examination that the long term settlements can reach to the very large values in comparison to the immediate settlements (as much as 10 times for example), though these final amounts are dependent on the different combinations of involved parameters. A simple formula as:

$$S_B = S_A + \left[\log(t_B / t_A) m\right] \text{ and } m = n/r^2 \tag{1}$$

relating two amounts of time settlements (S_A and S_B) corresponding to the times t_A and t_B has been proposed by Bowers et al. (1996) for the trial tunnel of Heathrow, in which the parameters m and n should be determined experimentally in any particular cases and r is the horizontal distance of any point to the vertical axis of the tunnel. Some of the results from the present computations reveal that the values of n can be around 2.65 to 4.5.

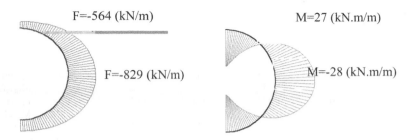

F=-564 (kN/m)   M=27 (kN.m/m)

F=-829 (kN/m)   M=-28 (kN.m/m)

Figure 4.   Typical shape of axial force and bending moment around the tunnel lining.

Table 4.   Computed results of settlements components for the gap value of zero.

Final settlement	1000 days	100 days	20 days	Immediate settlement	
63	61	18	6	+0.6	S_{max}(mm)
70	66	22	12	0.5	S_c(mm)
−61	−53	−37	−30	−28	M_{max}(kN.m/m)
−1247	−1236	−1210	−1197	−1410	F_{max}(kN/m)
58	51	36	29	24	M_c(kN.m/m)
−916	−925	−931	−935	−1142	F_c(kn/m)

3.2   Lined tunnel

Similar to the computations for unlined tunnels, some cases have been analyzed for the lined tunnels. The lining properties for three different thickness of lining, i.e.: 5, 20, and 35 cm is mentioned in Table 3.

As it is well known, for this type of analysis, it is necessary to define an amount for the so called gap value (which is the assumed or allowable space between the lining and the tunnel perimeter before the full contact of lining and the ground occurs), so in the present study three values of gap, i.e. 0, 1% and 2% have been assigned. Referring to the definition of gap, (i.e. $\Delta A/A*100$ in which A is the designed cross section area of tunnel and ΔA is the decreasing amount of area due to inward deformations of tunnel perimeter), it can be easily related to the settlement of the tunnel roof (S_c). Therefore, if the maximum inward deformation occurs at the roof as S_c and the minimum is zero at the floor, so: $\Delta A/A*100 = S_c/r*100$, and for the diameter of 8 meter a gap of 1% is equivalent to $S_c = 40$ mm.

Based on the parameters in Table 1, for an unlined tunnel, the amount of S_c is computed as 114 mm, which equals to a gap of 4.3%. However, for the gap values less than this amount the lining becomes under some pressures and the bending moment. If the gap value is taken equivalent to 4.3% (the same as free deformation of tunnel perimeter without lining), then no force or bending moment should occur within the lining.

Typical shape of the distribution of bending moments and axial forces around the tunnel perimeter inside the lining under symmetrical conditions are shown in Figure 4.

Tables 4 to 6 represent the computed results of immediate and long term settlements, the values of axial force and bending moments within the lining corresponding the gap values indicated as 0, 1 and 2 percent for the assumed lined tunnel, the tunnel characteristics are as shown in Table 3.

If the variations of internal force within the lining is drawn versus the corresponding assumed gap values, then the achieved curve which can be accepted as the ground- lining response curve is expected to be an straight line as shown in Figure 5.

To examine the effect of lining thickness, two other values of thickness, i.e. 5 cm and 35 cm (in comparison with 20 cm) have been used in the computations for the gap value of 2%. The corresponding results are shown in Tables 7 and 8.

Table 5. Computed results of settlements components for the gap value of 1%.

Final settlement	1000 days	100 days	20 days	Immediate settlement	
78	72	31	19	9	S_{max}(mm)
86	77	48	43	26	S_c(mm)
−63	−54	−38	−31	−28	M_{max}(kN.m/m)
−961	−950	−922	−912	−1120	F_{max}(kN/m)
60	52	37	30	24	M_c(kN.m/m)
−628	−636	−643	−647	−852	F_c(kn/m)

Table 6. Computed results of settlements components for the gap value of 2%.

Final settlement	1000 days	100 days	20 days	Immediate settlement	
99	83	44	29	19	S_{max}(mm)
154	100	70	64	52	S_c(mm)
−63	−54	−38	−31	−28	M_{max}(kN.m/m)
−669	−658	−630	−619	−835	F_{max}(kN/m)
62	54	39	32	27	M_c(kN.m/m)
−339	−348	−354	−357	−564	F_c(kN/m)

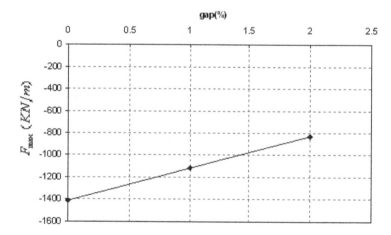

Figure 5. Ground-lining response.

Table 7. Computed results of settlements components for the thickness of 0.35.

Final settlement	1000 days	100 days	20 days	Immediate settlement	
94	80	42	28	18	S_{max}(mm)
149	96	69	60	50	S_c(mm)
−222	−192	−139	−117	−115	M_{max}(kN.m/m)
−712	−694	−661	−648	−866	F_{max}(kN/m)
218	190	139	116	109	M_c(kN.m/m)
−300	−314	−328	−335	−542	F_c(kN/m)

Table 8. Computed results of settlements components for the thickness of 0.05.

Final settlement	1000 days	100 days	20 days	Immediate settlement	
101	84	44	30	20	S_{max}(mm)
170	110	76	66	54	S_c(mm)
−1.2	−1	−0.7	−0.6	−1	M_{max}(kN.m/m)
−629	−620	−595	−586	−793	F_{max}(kN/m)
1.1	1	0.7	0.55	0.5	M_c(kN.m/m)
−886	−893	−896	−898	−548	F_c(kN/m)

Table 9. Ground properties and tunnel geometrical parameters.

Case	γ kNm^{-3}	E_u MPa	C_u kPa	k_0	K_x m/day	K_y m/day	ν	D m	Z_0 m	Ref.	h_w m
Bankok	17	2e4	15	1	4e-4	4e-4	0.33	2.67	18	1	−2.2
Belfast	15	4e3	10	0.7	0.0013	1e-6	0.33	2.74	4.85	1	−1.2
Heahtrow	15	3.5e4	160	1.15	1.3e-3	1e-6		8.5	19	2	
Singapoor	15–17		15–70	1	1e-7 1e-9	1e-5 1e-8	0.3	5.4	16.5	3	
Grimsby	15	4e4	12	0.6	5e-4	9.3e-4	0&0.33	3	6.5	4	−2
Wilington	15	8e4	12	0.6	1.4e-4	1e-4	0&0.33	4.25	13.37	4	−2

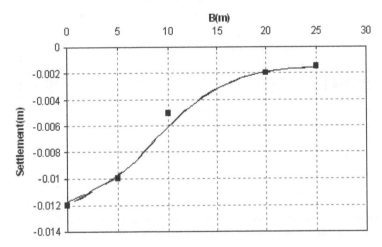

Figure 6. Comparison between the computed values of surface immediate settlements and the measured quantities for Bangkok tunnel (empirical data from Chou and Bobet, 2002).

3.3 Comparing the results with field data

At the moment there are ample amounts of data from different tunnels of many projects published in Journals and proceedings. The selected examples in the present computations, have not posses any specific privilege for this purpose, but the reason for this selection was their availability to the authors. The properties of the grounds in which the tunnels were excavated have been quoted from their original references which are shown in Table 9. Based on these properties the tunnels are modeled within the appropriate computations and the results are compared with the available observed data. The actual tunnel cases selected for the comparison in this study are the tunnels named: Bangkok, Belfast, Heathrow, Singapore, Grimsby, and Wilington.

Except for the first case for which the comparisons are for the immediate ground surface settlement (Figure 6) and second case for which comparisons are for immediate and long term surface

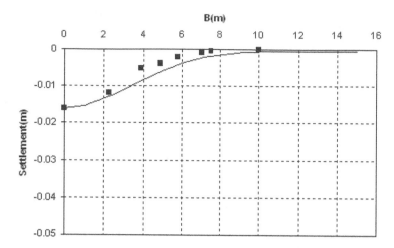

Figure 7. Comparison between the computed values of surface immediate settlements and the measured quantities for Belfast tunnel (empirical data from Chou and Bobet, 2002).

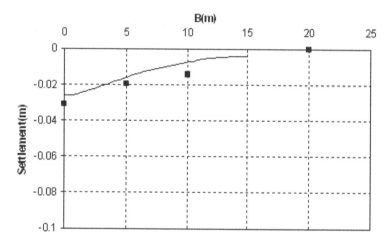

Figure 8. Comparison between the computed values of surface time settlements and the measured quantities for Belfast tunnel for 35 days (empirical data from Chou and Bobet, 2002).

settlement (Figures 7 and 8), for other cases the long time settlements are compared and the comparisons are shown in Figures 9 to 12. As these figures indicate, the corresponding comparisons are in excellent or acceptable agreement. Also, for the immediate settlements separate comparisons have already been made by Vafaeian and Mirmirani (2003).

4   CONCLUSIONS

A parametric study have been done for the purpose of evaluating the long term behavior of soft ground in response to the excavating a shallow tunnel. The behavior of ground in this study is mainly attributed to the deformation around and above the tunnel which is principally the settlement components which are the maximum settlement at the ground surface (s_{max}), at the tunnel roof (S_c) and the settlement ratio (λ).

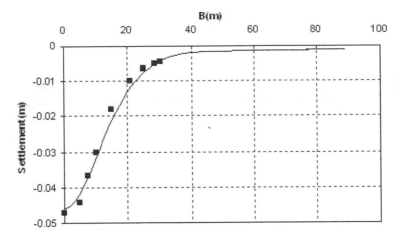

Figure 9.   Comparison between the computed values of surface time settlements and the measured Quantities for Heathrow tunnel for 390 days (empirical data from Browers et al., 1996).

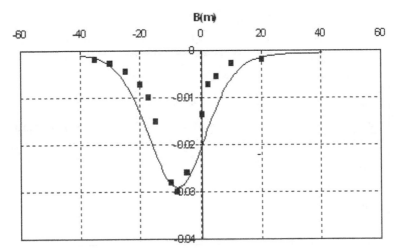

Figure 10.   Comparison between the computed values of surface time settlements and the measured quantities for Singapore tunnel for 14 days (empirical data from Shirlaw, 1994).

In the present article the effect of many influential parameters (i.e. geotechnical properties, physico-mechanical parameters, permeability and consolidation, and the structural properties of lining) have been investigated for both lined and unlined tunnels. Finally, some comparisons have been made between the available observational data and the computed results for the same cases.

Concluding remarks can be summarized in the following points:

(1) Plaxis 2D program is a suitable F. E. program for modeling the ground behavior for both immediate settlement and time settlement of the ground due to tunneling.
(2) The final amount of settlement (corresponding to the very long time) due to tunneling is clearly dependent upon almost all variables involved, so the ratio between the final settlement to the immediate amount can be varied in a wide range from 1.1 to large values, say about 10 or even more.
(3) The effect of lining stiffness and gap values are almost linear and proportional.

221

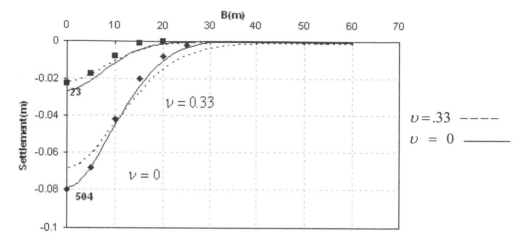

Figure 11. Comparison between the computed values of surface time settlements and the measured quantities for Wilington tunnel (empirical data from Attewell et al., 1986).

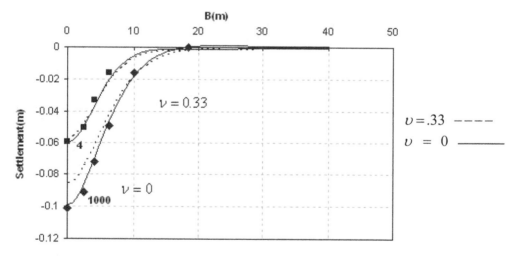

Figure 12. Comparison between the computed values of surface time settlements and the measured quantities for Grimsby tunnel (empirical data from Attewell et al., 1986).

REFERENCES

Attewell, P. B., Yeates, J. and Selby, A. R. 1986. Soil movement induced by tunneling and their effect on pipeline and structures. *Chapman and Hall.*

Browers, K. H., Hiller, D. M. and New, B. M. 1996. Ground movement over three years at Heathrow Express Trial Tunnel, *Geotechnical Aspects of Underground Construction in Soft Ground*, Mair & Tailor (eds.), pp. 632–647.

Chou, W. I. & Bobet, A. 2002. Predictions of ground deformations in shallow tunnels in clay. *Tunnelling and underground space technology*, Vol. 17. pp. 3–19.

Shirlaw, J. N. 1994. Observed and calculated pore pressure and deformation induced by an earth balance shield. *Canadian Geotechnical Journal*, Vol. 30, PP. 476–490.

Applications of Computational Mechanics in Geotechnical Engineering – Sousa,
Fernandes, Vargas Jr & Azevedo (eds)
© 2007 Taylor & Francis Group, London, ISBN 978-0-415-43789-9

Numerical analysis of a tunnel from the Brasilia metro using the finite element method

F. Marques & J. Almeida e Sousa
University of Coimbra, Coimbra, Portugal

A. Assis
University of Brasilia, Brasilia, Brazil

ABSTRACT: About 7.2 km of the Brasilia metro has been built in tunnel, excavated in a layer of porous clay, using sequential excavation methods, based on the NATM principles. During tunnelling, the observed settlements surpassed the values foreseen in the design. It was also observed an amplification of the displacements with the distance from the opening, which was unusual. In this paper, the results of numerical analyses, accomplished to evaluate the displacement and stress fields in the tunnel surrounding ground, are presented. Particular attention was given to the reproduction of the Brasilia porous clay behaviour. Two constitutive models were used: an elastic perfectly plastic model (Mohr-Coulomb) and the Lade elastic-plastic model. The results obtained from the numerical analyses carried out were compared to monitoring data in order to validate the numerical study.

1   INTRODUCTION

The Brasilia metro was designed in order to connect Brasilia to the main cities nearby. Despite a great extension has been built on the surface, almost 7.2 km has been built in tunnel in the south wing of Brasilia. The excavation of the tunnel developed using the NATM method and involved a layer of porous clay (Figure 1).

During tunnelling, it was observed a particular behaviour of the Brasilia porous clay (Ortigão 1996). As Figure 2 illustrates, the observed settlements were very high, surpassing the values foreseen in the design. It was also observed an amplification of the displacements with the distance from the opening, which was unusual.

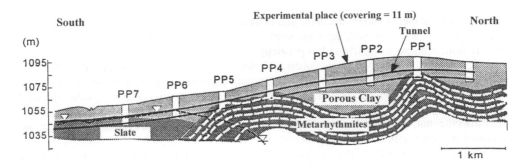

Figure 1.   Geotechnical profile of the Brasilia south wing (Ortigão 1994).

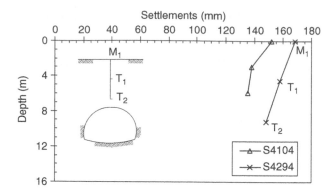

Figure 2. Evolution with depth of the maximum observed settlements in the tunnel axis.

Table 1. Identification characteristics.

Grain size			Atterberg limits		
sand %	silt %	clay %	w_L (%)	w_P (%)	I_P (%)
5–21	15–24	57–81	54–74	39–54	14–22

This particular behaviour has motivated the accomplishment of several studies (Almeida et al. 1996, Ruiz 1997, Moraes Júnior 1999). Recently a quite detailed study of the Brasilia porous clay behaviour due to tunnelling has been developed (Marques 2006), which demanded the accomplishment of a geotechnical characterization campaign. This characterization was made by some in situ tests and by an extensive programme of laboratory tests that includes oedometric, isotropic compression tests and triaxial drained compression tests, which allowed defining the constitutive models parameters used in the numerical simulation of the Brasilia porous clay behaviour.

2 IDENTIFICATION AND PHYSICAL PROPERTIES

The identification of the Brasilia porous clay layer was accomplished through grain size analysis and Atterberg limits tests. Table 1 shows, in terms of range, the grain size composition of the porous clay and the values of the Atterberg limits: liquidity limit, w_L, plasticity limit, w_P, and plasticity index, I_P.

The analysis of Table 1 shows that in the grain size distribution, the clay fraction is predominant, which allows to classify the soil, using the unified classification, as MH (elastic silt), because its liquidity limit is higher than 50% and its plasticity index is lower than 20%.

In Figure 3 it is represented the evolution with depth of the main physical indexes of the soil: void ratio, e, unit weight, γ, dry unit weight, γ_d, saturation degree, S_r, and soil particles density, G.

The analysis of Figure 3 allows verifying that the soil is a very porous material, with high void ratio values, higher than 2.0 near the surface and decreasing with depth, but never presenting values lower than the unit. As a consequence of those high void ratio values, the unit weight of the soil presents low values, especially in the first meters, and with tendency to increase with depth.

In relation to the saturation degree, the obtained values are lower than 90%, which evidence the unsaturated condition of the Brasilia porous clay. Finally, concerning the density of the solid particles, significant differences were not verified in depth.

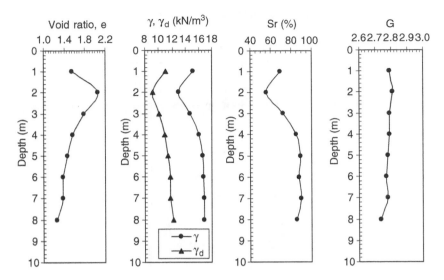

Figure 3. Main physical indexes of the Brasilia porous clay.

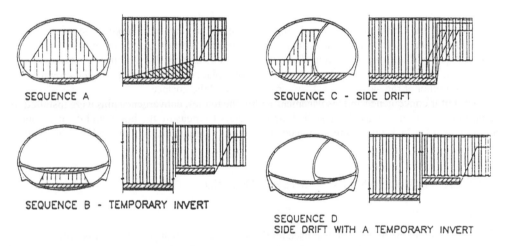

Figure 4. Construction sequences used in the south wing tunnel.

3 DESCRIPTION OF THE WORKS

In the south wing tunnel of the Brasilia metro it has been employed four different construction sequences, according to the NATM principles, each choice dependant of the geological and geotechnical conditions found near the excavation face. Among those sequences (Figure 4), the sequence A was the construction method employed in the extension between stations PP2 and PP3, studied in this paper.

According to sequence A the excavation starts by the roof and the sides, leaving a central core that functioned as a support for the excavation face. The central core and the bench were excavated 2.4 m behind the face. The invert was closed at a distance varying between 4.8 and 7.2 m behind the face.

The primary lining of the tunnel consists of lattice guiders (spaced between 0.6 and 1.0 m) and a layer of sprayed concrete (0.21 m). The final lining, 0.19 m thick, was done with sprayed concrete once the tunnel was completely excavated.

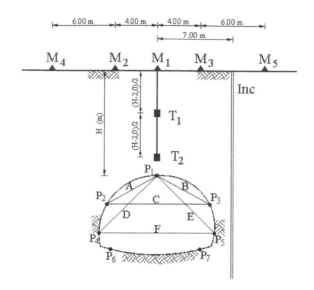

Figure 5. Main monitoring section.

To monitor the tunnel performance during construction, several main monitoring sections were installed (Figure 5). In those sections it was measured the vertical displacements in five marks placed on the surface (one on the tunnel axis and the others arranged symmetrically at distances of 4 and 10 m of the tunnel axis). One extensometer was placed around 2 m above the crown of the tunnel and other at the half distance between that one and the surface.

To monitor convergence and displacement within the tunnel, convergence pins were installed. In some of the sections it was also installed inclinometers to measure the horizontal displacements. Some Mini Flat Jack tests were also accomplished to evaluate the acting loads in the primary support.

4 DISCRETISATION AND PARAMETERS ADOPTED IN THE SIMULATIONS

4.1 F.E. programs

The finite element method (FEM) was used for the numerical simulations of the tunnel construction. Calculations were made with an automatic FEM programme developed in the Departments of Civil Engineering of the Universities of Oporto and Coimbra (Almeida e Sousa 1998). Besides permitting analysis of problems in conditions of plane states of strain and stress, and also axis-symmetric states, it enables the three-dimensional simulation, which develops at the excavation face of a tunnel.

The programme incorporates several types of finite elements capable of represent all the components of a geotechnical structure. Concerning the numerical simulation of the mechanical behaviour of the materials and their interfaces, different constitutive models can be used. Analyses may be performed in terms of total or effective stresses.

4.2 Finite element mesh

The finite element mesh used in the analyses, illustrating the step in which the face reaches the section of study, is presented in Figure 6. It has 24261 nodal points and 6272 isoparametric elements of 20 nodal points.

The definition of the finite element mesh was made to simulate adequately all construction process. As Figure 6 suggests, it was simulated in incremental stages, assuming that the crown, bench and invert moved forward simultaneously in steps of 2.4 m.

226

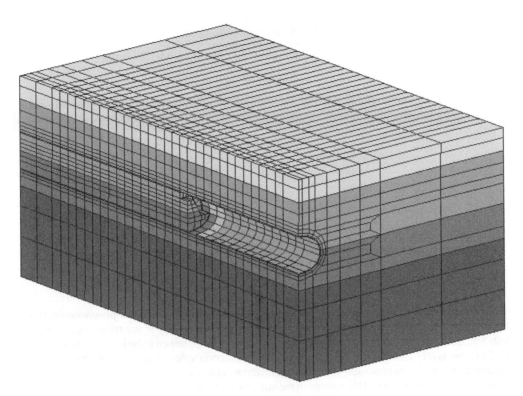

Figure 6.   3D Finite element mesh.

Table 2.   Adopted values of unit weigh and coefficient of earth pressure at rest.

	γ (kN/m³)	K_0
Layer 1	14.5	0.55
Layer 2	17.0	0.55
Layer 3	18.0	0.55
Layer 4	18.0	0.55

The total length of the mesh in the tunnel axis direction is 84.0 m. In the extreme sections of the mesh the longitudinal displacements are impeded. In the inferior border, located at 40 m of depth, all displacements are impeded, while in the lateral borders the horizontal displacements are impeded in respect to perpendicular direction.

In order to attribute different mechanical characteristics with depth, the porous clay ground surrounding the tunnel was divided in four layers. The first layer extends to the depth of 4.6 m and the second up to the tunnel horizontal axis, located at a depth of 15.0 m. The third and fourth layers are located below the tunnel horizontal axis, being the transition between them done at the depth of 22.0 m.

4.3   Initial state of stress and constitutive laws

The initial vertical stress was taken to be geostatic. The initial stresses were calculated based on the values of unit weight and the coefficient of earth pressure at rest for the different layers, shown in Table 2.

Table 3. Parameters of the Lade elastic-plastic model.

	Depth = 2.0 m (1st layer) Suction = 20 (kPa)	Depth = 6.0 m (2nd layer) Suction = 20 (kPa)
K	320.26	488.99
n	0.81	0.77
ν	0.25	0.25
p	0.0510	0.0300
η_1	19.94	32.84
m	0.106	0.326
s_1	0.361	0.328
s_2	−0.062	0.052
t_1	0.671	−0.153
t_2	−2.955	−4.067
P	0.274	0.276
l	1.045	1.086
α	0.780	0.947
β	−0.323	−0.081

The behaviour of the first two layers was reproduced through the Lade elastic-plastic model (Lade 1977, 1979). The calibration of the model was made from the results of the triaxial compression tests and the isotropic compression tests accomplished with undisturbed samples collected at depths of 2 m (1st layer) and 6 m (2nd layer). To take the suction effect into account the tests were accomplished with samples submitted to three different values of suction: 20, 60 and 100 kPa.

To evaluate the influence of the suction in the numerical results three calculations were performed. Those calculations differed only with respect to the Lade elastic-plastic model parameters used to simulate the mechanical behaviour of the first two soil layers. In each calculation it was used the parameters corresponding to each one of the suction values applied to the samples in the triaxial tests. The first conclusion obtained from the analysis of the calculation results is that the suction variation does not have great importance in the tunnel surrounding ground behaviour. The calculation results show that the cross-section settlements profiles at the surface obtained from the three calculations are identical, fitting well the monitoring results.

In this paper it is presented the results obtained from the calculation C1, which corresponds to the Lade elastic-plastic model parameters for the suction of 20 kPa. The calibration procedure adopted was described in a previous paper (Marques et al. 2006). The parameters obtained from the calibration procedure and used in the numerical calculation are shown in Table 3.

It was verified that the Brasilia porous clay presents some cementation among the particles, what is translated in the oedometric test results by a virtual pre-consolidation stress. The OCR values obtained from the oedometric tests were 1.38 (depth = 2 m) and 1.10 (depth = 6.0 m). In terms of the Lade elastic-plastic model, it is necessary to admit that the plasticity surfaces of the model (the contraction and the expansible surfaces) had already been previously activated, which means that the plastic deformations only happen when the stress variation tends to overstep those surfaces previously expanded. In other words, it happens when the cementation among the soil particles are broken. For the calculation of the initial stress levels (f_{c0} e f_{p0}) it was adopted, for each Gauss's point, the values of the major principal stress and of the minor principal stress corresponding to the pre-consolidation.

The importance of the initial plasticity surfaces activation is clearly evidenced in Figure 7. It shows the evolution of the superficial settlements with the distance to the tunnel axis of two calculations, which differ only with respect to the activation, or not, of those surfaces. The analysis of Figure 7 allows verifying that in the calculation without the consideration of the soil particle cementation leads to vertical displacements at the surface that overestimate the observed values. On the contrary, if the soil particle cementation is taken into account in the numerical calculation, the obtained vertical displacements fit quite well the observed values.

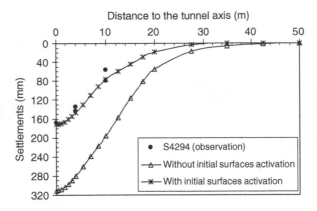

Figure 7. Cross-section settlements profile at surface – analysis of the influence of the initial plasticity surfaces activation.

Table 4. Parameters of the elastic perfectly plastic model (Mohr-Coulomb).

	c' (kPa)	ϕ' (°)	E (MPa)	v
Layer 3	15.0	30	28	0.25
Layer 4	15.0	30	100	0.25

In all calculations the third and fourth layers were described as linear isotropic elastic perfectly plastic media, with its strength defined by the Mohr-Coulomb criterion, because it was not possible to make the calibration of the Lade elastic-plastic model to those layers. The parameters adopted in the calculations are summarized in Table 4.

The primary support was modelled to have linear elastic and isotropic behaviour. It was attributed a value of 20 GPa to the deformability modulus of the sprayed concrete after closing the invert and a value of 10 GPa before that, to take into account the lower stiffness of the early-age concrete. The Poisson ratio was assumed equal to 0.2.

5 3D NUMERICAL ANALYSES RESULTS

Three-dimensional and two-dimensional analyses were accomplished, using the finite element method, to evaluate the displacement and stress fields in the tunnel surrounding ground. In this paper the results of one of those 3D numerical analyses are presented. In this calculation (C1) it was used the parameters corresponding to the suction value of 20 kPa applied to the samples in the triaxial tests.

The first conclusion obtained from the analysis of calculation C1 is that the cross-section settlement profile at the surface obtained from the calculation is identical to the monitoring results.

This good agreement at the surface between the numerical results and the monitoring ones is equally verified in Figure 9, where the distortions deduced from the Gauss curve and from the yield density curve that best fits the monitoring results are compared with the results obtained from the calculation C1.

In Figure 10 it is represented, jointly with the monitoring results, the evolution with the excavation face progress of the vertical displacements of points located on the tunnel axis: at the surface (Figure 10a), in the extensometer 1 (Figure 10b) and in the extensometer 2 (Figure 10c).

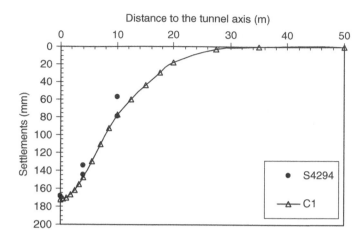

Figure 8.   Cross-section settlements profiles at surface.

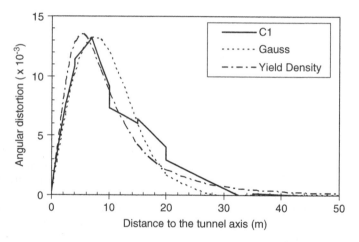

Figure 9.   Angular distortions at the surface.

The analysis of Figure 10 allows verifying that, despite in the numerical calculation the face excavation effects in the study section are noticed earlier than in the monitoring data, a quite reasonable displacement evolution with the face progress obtained from the numerical calculation to the observed one is verified.

Concerning vertical displacements in the tunnel crown (Figure 11), a good agreement is verified between the numerical results and the monitoring ones, especially in terms of the evolution with the face progress, as well in terms of the value of the final displacement.

Figure 12 illustrates the vertical displacements evolution with depth along the tunnel axis in the study section for different excavation face positions relatively to that, representing d, in the figure, the distance among both.

The analysis of Figure 12 allows verifying that before the face excavation reaches the study section, the calculated displacements are higher than the monitoring ones. That difference has, however, tendency to decrease as the excavation face moves forward. The analysis of Figure 12 still allows verifying that, similarity to what was already observed, the calculated displacements are higher at the surface. This aspect can also be observed in Figure 13, which shows the vertical displacements distribution in the tunnel surrounding ground in a step calculation near the end of the excavation.

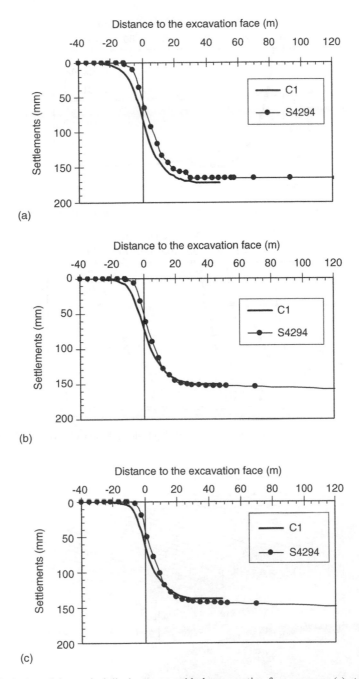

Figure 10. Evolution of the vertical displacements with the excavation face progress: (a) at the surface, (b) in the extensometer 1, (c) in the extensometer 2.

Figure 14 allows confronting the relative displacements obtained from the numerical calculation and the measured ones in the convergence base installed lightly above the tunnel axis (Base C). From the figure analysis it is verified, once installed the support, a removal between the convergence pins, being its evolution with the excavation progress very identical to the observed one. Also in quantitative terms, the comparison between the numerical and observed results is quite reasonable.

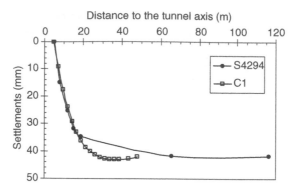

Figure 11. Evolution of the vertical displacements in the tunnel crown with the excavation face progress.

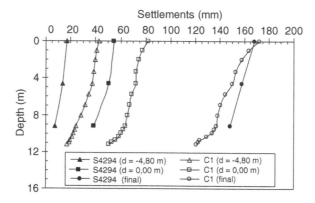

Figure 12. Evolution of the vertical displacements with depth along the tunnel axis for different positions of the excavation face.

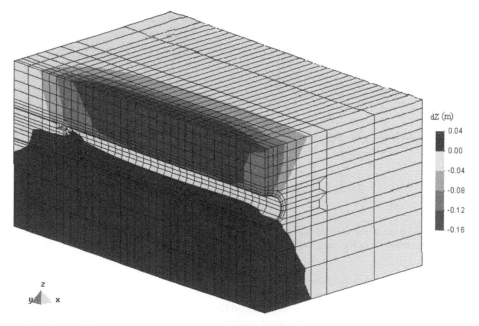

Figure 13. Vertical displacements in the tunnel surrounding ground near the end of the excavation.

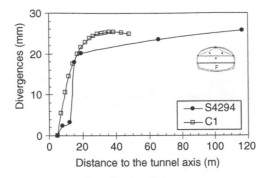

Figure 14. Evolution of the divergences inside the tunnel at Base C level.

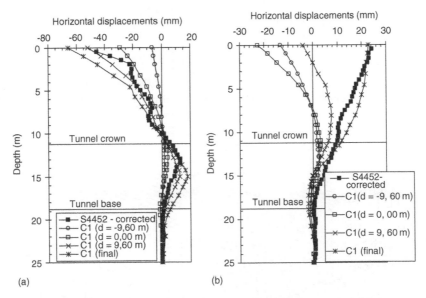

(a)   (b)

Figure 15. Evolution of the horizontal displacements with depth in the study section with the excavation face progress: (a) perpendicular to the tunnel axis, (b) in the direction of the longitudinal tunnel axis.

Figures 15a and 15b shows the evolution with depth of the final horizontal displacements in the study section, perpendicular to the tunnel axis and in the direction of the longitudinal tunnel axis, respectively, with the face excavation progress, in the points located at 7.0 m of the symmetry axis. It is also represented in those figures the corrected values of the measured horizontal displacements in the inclinometer, installed in the section S4452, in the last measurement campaign. The correction was made because in the section where the inclinometer was installed, the vertical displacements were substantially inferior to the observed in the study section (S4294).

The analysis of Figure 15a displays, in qualitative terms, that a good agreement exists among the calculated horizontal displacements and the observed ones. Above the crown both displacements are driven to the opening and in the area of the tunnel, the movements are driven from the interior to the exterior of the tunnel.

Figure 15b allows verifying that before the excavation face reaches the study section, the horizontal displacements in the direction of the longitudinal tunnel axis are driven to the opening. These displacements increase with the excavation face progress reaching their maximum when the face reaches the study section. After the excavation face passed the study section, the longitudinal

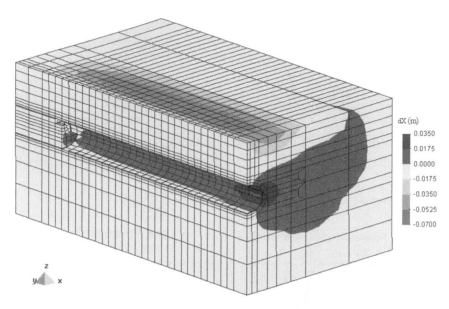

Figure 16.   Horizontal displacements, perpendicular to the tunnel axis, in the tunnel surrounding ground near the end of the excavation.

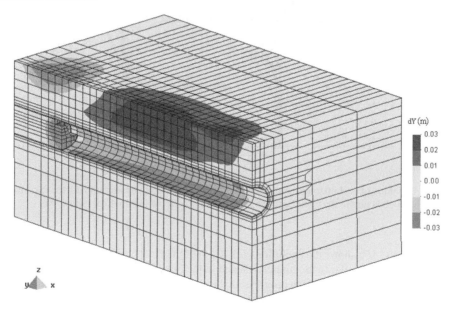

Figure 17.   Horizontal displacements, in the direction of the longitudinal tunnel axis, in the tunnel surrounding ground near the end of the excavation.

horizontal displacements decrease progressively until they are annulled when the face is at about 1.5 tunnel diameter from the study section. After that, the displacements are driven in the sense of the tunnel progress.

Figures 16 and 17 show the horizontal displacement distribution (perpendicular to the tunnel axis and in the direction of the longitudinal tunnel axis, respectively) in the tunnel surrounding ground, in a calculation step near the end of the excavation.

6 CONCLUSIONS

In this paper the results of the numerical simulation of a tunnel from the Brasilia metro, using the Finite Element Method, were presented. For the accomplishment of that numerical study there was a need of a geotechnical testing programme to complement the data available from other researches, in order to allow the calibration of advanced constitutive models to simulate the behaviour of the Brasilia porous clay.

The identification, classification and determination of the physical indexes tests, evidenced that it is a very porous material, characterized by high void ratio values and low unit weight values. It was still evident the unsaturated condition of the soil, due to the low values found to the saturation degree.

The 3D numerical analyses accomplished to evaluate the displacement and stress fields in the tunnel surrounding ground tried to reproduce the actual construction sequence. In those analyses two constitutive laws were use to reproduce the porous clay behaviour: a simpler constitutive model (elastic perfectly plastic model) and a more advanced model (Lade elastic-plastic model).

The calibration of the elastic perfectly plastic model was made based on the results of the laboratory tests of mechanical characterization (more superficial layers) and of the field rehearsals (deeper layers). For the Lade elastic-plastic model, just used to reproduce the behaviour of the most superficial layers of the porous clay, the determination of their parameters was done, for the suction levels of 20, 60 and 100 kPa, from the results of the laboratory tests of mechanical characterization.

The confrontation of the displacements obtained from three calculations, in which it was attributed to the soil layers, simulated using the Lade model, the parameters obtained for the three different suction values, allowed verifying that, for the considered values, the variation of the suction does not have great influence in the soil behaviour.

The comparison of two calculation results allowed verifying that the activation of the initial plasticity surfaces of the Lade model influences in a very important way the displacements induced in the soil by tunnelling. It was equally verified that the calculation accomplished with the activation of the initial plasticity surfaces reproduced in an appropriate way the monitoring results.

Concerning the Lade model, it could be verified that this model was adequate to the problem analysis, since the comparison between numerical and monitoring results was quite satisfactory, either in qualitative or quantitative terms.

ACKNOWLEDGEMENTS

The authors would like to express their gratitude to the Metrô/DF for their co-operation and for their permission to publish the results and to the Bureau of Projects and Consulting for making available the monitoring data. Thanks are also due to the Fundação para a Ciência e a Tecnologia (FCT) for the financial support given to this study through the scholarship SFRH/BD/4862/2001 and to CNPq (National Research Council of Brazil).

REFERENCES

Almeida, M. S. S.; Kuwajima, F. M. e Queiroz, P. I. B. (1996). Simulação da construção de um túnel por mecânica dos solos de estado crítico via elementos finitos. Infogeo, ABMS, SP, Vol. 1, 223–230.

Almeida e Sousa, J. (1998). Túneis em Maciços Terrosos. Comportamento e Modelação Numérica. DSc. Thesis in Civil Engineering. Faculty of Science and Technology, University of Coimbra, Portugal.

Lade, P. V. (1977). Elasto-plastic stress-strain theory for cohesionless soil with curved yield surfaces, International Journal of Solids and Structures, Vol. 13, p. 1019–1035.

Lade, P. V. (1979). Stress-strain theory for normally consolidated clay, Proceedings 3rd International Conference on Numerical Methods in Geomechanics, Aachen, Vol. 4, p. 1325–1337.

Marques (2006). Comportamento de túneis superficiais escavados em solos porosos – O caso do metro de Brasilia/DF. DSc. Thesis in Civil Engineering, Faculty of Science and Technology, University of Coimbra.

Marques, F.; Almeida e Sousa, J. e Assis, A. (2006). Simulação do comportamento tensão-deformação da argila porosa de Brasilia através do modelo de Lade. 10° Congresso Nacional de Geotecnia, Lisboa.

Moraes Júnior, A. H. V. (1999). Simulação numérica tridimensional de túneis escavados pelo NATM. MSc. Thesis in Geotechnics. Faculty of Technology, University of Brasilia, Brazil.

Ortigão, J. A. R. (1994). O túnel do Metrô de Brasilia. Propriedades geotécnicas e o comportamento da obra. Monograph presented for the contest for Soil Mechanics Professor. UFRJ, Brazil.

Ortigão, J. A. R.; Kochen, R.; Farias, M. M. e Assis, A. P. (1996). Tunnelling in Brasilia porous clay. Canadian Geotechnical Journal, Vol. 33, n° 4, 565–573.

Ruiz, A. P. T. (1997). Análise de Túneis Rasos em Solos Porosos – Mecanismos de Formação de Recalques. MSc. Thesis in Civil Engineering. Engineering School of São Carlos, USP, São Paulo, Brazil.

Applications of Computational Mechanics in Geotechnical Engineering – Sousa,
Fernandes, Vargas Jr & Azevedo (eds)
© *2007 Taylor & Francis Group, London, ISBN 978-0-415-43789-9*

The influence of unlined excavation length in face stability

A. Costa & A. Silva Cardoso
Faculty of Engineering of the University of Porto, Portugal

J. Almeida e Sousa
Faculty of Sciences and Technology of the University of Coimbra, Portugal

ABSTRACT: The understanding of the three-dimensional aspects involved in tunnel face stability can only be accomplished by 3D numerical analysis. This approach is, however, very complex and time consuming and consequently the influence of some parameters, such as the unsupported length at the face, has been neglected. As exceptions we can refer the experimental work of Casarin & Mair (1981) and of Kimura & Mair (1981) with cohesive materials, the experimental work of König et al. (1991) and Chambon & Corté (1994) with frictional materials and the numerical analyses of Antão (1997). Although not usually accounted for, the unlined excavation length is always present, since there is always a delay in closing the lining behind the face that must be considered in design. It is the author's intention to address this problem, using the commercial code FLAC3D. Several analyses have been conducted, both in drained and undrained conditions, for a wide range of unsupported lengths. Graphs of the results, comparing drained and undrained behaviour are presented and discussed.

1 INTRODUCTION

One of the main concerns in tunnel design is to adequately support the excavation face and thus minimize deformations and ensure safe working conditions. When the tunnel is not stable without internal support it becomes necessary to apply at the face an internal support pressure. This pressure is highly dependant on parameters such as the tunnel geometry (diameter, cover and unsupported length at the face), the loads (soil volumetric weight and surface charge) and the soil strength characteristics. The understanding of the three-dimensional aspects of face stability can only be accomplished by 3D numerical analysis, which is very complex and time consuming, and consequently very rarely performed.

Most existing studies address mainly unlined and fully lined tunnels, neglecting the importance of the unsupported excavation length behind the face. Nonetheless this parameter is of paramount importance since the referred situations are extremely rare in tunnel construction practice.

2 THE INFLUENCE OF THE UNSUPPORTED LENGTH PARAMETER

2.1 Definition of the unsupported length parameter

The unsupported length, P, is the distance between the excavation face and the closure of the lining (Fig. 1). Usually this parameter is divided by the tunnel diameter D, in order to obtain the non dimensional parameter P/D. When in the text a reference is made to the unsupported length, we will actually be referring to the "relative unsupported length" P/D.

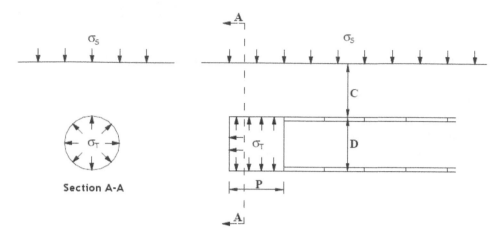

Figure 1. Definition of the parameter P/D.

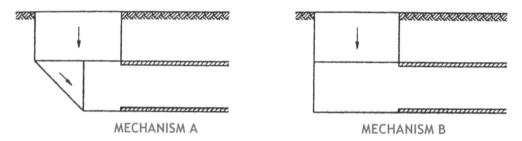

Figure 2. Observed failure mechanisms in undrained conditions (Casarin & Mair, 1981).

2.2 *Background*

Casarin & Mair (1981) have identified two distinct failure mechanisms, based on the results of tests conducted in centrifuge of the Cambridge University, with cohesive materials. For small values of P/D, we encounter what the authors defined as Mechanism A, with displacements on front and heading and the failure mechanism propagating from face to surface. For greater values of P/D a transition occurs from 3-D to 2-D behaviour, along with a decrease in stability, that is, an increase in the minimum support pressure. In this new mechanism, Mechanism B, the plastic deformation zone is almost confined to the region above the heading (Fig. 2). Kimura & Mair (1981), also working with cohesive materials, have concluded that for P/D > 2 the failure mechanism was similar to the two-dimensional mechanism, with longitudinal displacements much smaller than the vertical ones on top of the heading. Later, König et al. (1991) and Chambon & Corté (1994) came to similar conclusions, referring to friccional materials. Additionally, the above cited authors observed that the minimum support pressure remains unchanged for a small unsupported length behind front (the "small" unsupported length would de the P/D = 0.2 in the author's tests). More recently, Antão (1997) has identified the transition, in drained conditions, with a value of P/D larger than 0.3.

3 PARAMETRIC STUDY

3.1 *Objectives*

This study tries to evaluate how support pressures and plastic zones respond to increasing unsupported span.

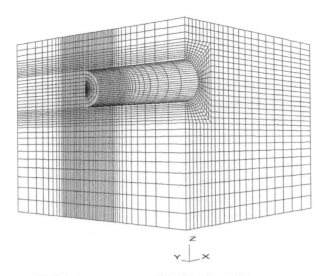

z
Y └─x

Figure 3.   Finite-difference mesh used in the analysis of the influence of the unsupported spam.

Concerning the support pressure σ_T, and since it attains much smaller values in drained conditions, we adopted a "normalized value" σ_T*, in order to become possible the comparison of this parameter evolution in drained and undrained conditions.

This normalized value is obtained from the minimum support pressure for a certain unsupported length (P), σ_T, the minimum support pressure for a fully lined tunnel, $\sigma_T(0)$ and the minimum support pressure for an unlined tunnel, $\sigma_T(\infty)$, in the following manner:

$$\sigma_T{}^* = \frac{\sigma_T - \sigma_T(0)}{\sigma_T(\infty) - \sigma_T(0)} \qquad (1)$$

Its value ranges, obviously, from 0 to 1.

The values for $\sigma_T(\infty)$ and $\sigma_T(0)$ were obtained from previous calculations performed by the author for an unlined tunnel and a fully lined tunnel.

3.2   Materials

The analysis performed addressed both drained and undrained conditions.

As to the undrained conditions, the cohesive material considered is characterized by an undrained shear resistance constant in depth and by a parameter $\gamma D/c_u$ equal to 2.6, common to the great majority of cohesive materials currently tested (Kimura & Mair (1981), Mair et al. (1984)).

For the friccional material, the characteristics of loose dry "Fontainebleau" sand were adopted, $\gamma = 15.3$ kN/m3, $\phi' = 35.2°$, $c' = 2.3$ kPa. An associated flow law was also considered ($\psi = 35.2°$).

3.3   Geometry

The adopted geometry consisted of circular tunnel with a 10 m diameter, with a 10 m cover (C/D = 1).

The finite-difference mesh comprises 56791 nodal points (Fig. 3).

As for boundary conditions, they were established according to the symmetry of the problem. The gridpoints along the left and right boundary planes are fixed in the xx direction. The gridpoints along the front and back boundary planes are fixed in the yy direction (corresponds to the tunnel

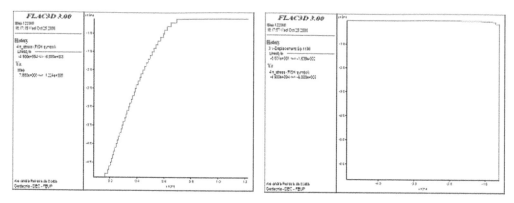

Figure 4.   Definition of failure adopted in the FLAC^{3D} code.

axis). The top boundary is free and the bottom boundary is fixed in the zz direction (corresponds to the gravity direction).

With the adopted geometry the maximum value available for P/D is 3.5. It will be shown that this value is however quite sufficient since both in drained and undrained conditions the results for such unlined length converge to the results obtained for a fully unlined tunnel.

3.4   Definition of failure

In the performed analyses the tunnel is fully excavated and a normal internal pressure is applied to the face and the unlined segment. This support pressure is reduced gradually in 1 kPa steps. The curves representing the pressure reduction with the calculation step and the evolution of the horizontal displacement at the face are similar throughout the different analyses. The pressure reduction curve attains a constant value when the pressure value reaches the minimum support pressure (Fig. 4). The displacement curve is characterized by a succession of small vertical jumps that mark the evolution of the deformations and horizontal platforms associated to a pressure reduction. As failure approaches the horizontal platforms tend to disappear and the vertical jumps become larger until the displacement keeps increasing for constant face pressure (Fig. 4). This was considered as the minimum support pressure required to maintain stability. This value is highly dependent on the "mechanical ratio". The mechanical ratio is the ratio of the maximum unbalanced force for all the gridpoints divided by the average applied gridpoint force. This value has been set as 1×10^{-6}, being sufficiently small not to interfere with the minimum pressure.

3.5   Results of the parametric analysis

In the present we have tried to analyse the influence of the relative unsupported length in face stability. We have addressed both the undrained and the drained response of the earth mass. All the calculations were conducted with the commercial code FLAC3D.

In Figure 5 we represent graphically the values obtained for the normalized support pressure, for the full series of eighteen analyses performed.

The most significant failure zones obtained for the drained conditions are also represented, from Figure 6 to Figure 8. Figure 9 to Figure 11 report the undrained behaviour.

The observation of the failure zones immediately points out a major difference between the drained and the undrained answer. The disturbed area is much smaller in drained conditions which might explain the small values we usually encounter for the minimum support pressure in model tests in friccional materials.

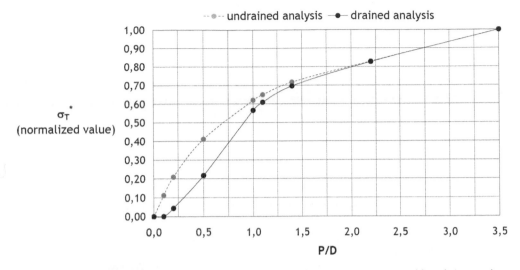

Figure 5. Dependence of the normalized support pressure with increasing unsupported length (comparison of drained and undrained responses).

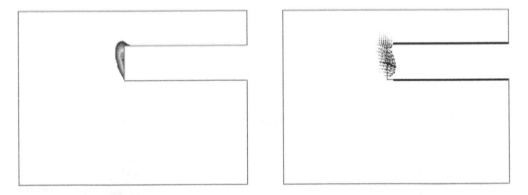

Figure 6. Magnitude (left) and direction (right) of displacements for $P/D = 0.2$ (drained conditions).

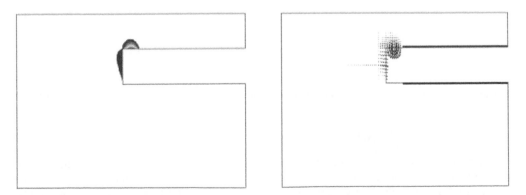

Figure 7. Magnitude (left) and direction (right) of displacements for $P/D = 0.5$ (drained conditions).

Figure 8.   Magnitude (left) and direction (right) of displacements for P/D = 1.0 (drained conditions).

Figure 9.   Magnitude (left) and direction (right) of displacements for P/D = 1.0 (undrained conditions).

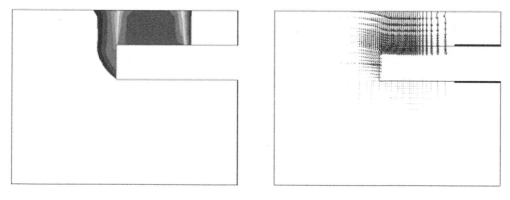

Figure 10.   Magnitude (left) and direction (right) of displacements for P/D = 2.2 (undrained conditions).

Several interesting observations arise from the obtained results:

– The transition between Mechanism A and Mechanism B, both identified by Casarin & Mair (1981), can be associated with a value of P/D close to 0.5 (Fig. 7) in drained conditions. In fact, for P/D = 0.2 (Fig. 6) we clearly have Mechanism A and for P/D = 1.0 (Fig. 8) the failure mechanism is unmistakably Mechanism B.

 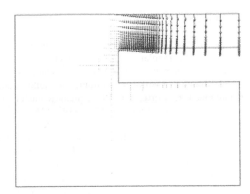

Figure 11. Magnitude (left) and direction (right) of displacements for P/D = 3.5 (undrained conditions).

- When the behaviour is undrained the transition is located, for reasons similar to those stated regarding drained conditions, around P/D = 2.2 (Fig. 9 to Fig. 10).
- Nonetheless the three – dimensional effects are still noticeable on Mechanism B, since, as can be seen on Figure 5 the normalized pressure only reaches the value 1.0 for P/D = 3.5. It should be remembered that a unitary value for P/D implies that the minimum support pressure is identical to the one obtained for an unlined tunnel.
- In drained conditions, the value of the minimum support pressure remains unchanged for unsupported lengths P/D up to 0.1(Fig. 5).
- From this value on, we can see a fast increasing of the support pressure up to P/D = 1.0.
- From P/D = 1.0 the increasing velocity diminishes to 1/3 of the initial value, until the pressure stabilizes for P/D = 3.5 ($\sigma_T{}^* = 1.0$).
- In undrained conditions any small increase of P/D is responsible for an immediate growth of minimum support pressure.
- As in the drained conditions, three main trends are noticeable in the evolution curve of the minimum support pressure:
    - Up to P/D = 0.2, the support pressure increases very fast.
    - From P/D = 1.0 on growth diminishes visibly, converging to the drained response form P/D = 2.2 on.

3.6 Conclusions

We have tried to evaluate how support pressures and plastic zones respond to increasing unsupported span.

We concluded that the transition between the three-dimensional and two-dimensional mechanism occurs much more abruptly and for significantly lesser values of P/D in drained conditions. The issue of lining closure is therefore of enormous importance in those conditions, since any delay could generate an accident. Nevertheless it was verified that for small values of P/D (up to 0.1) the support pressure remains unchanged. Such allows for the possibility of a small delay in lining closure without any significant increase in minimum support pressure.

The observations are consistent with the work of Kimura & Mair (1981) with cohesive materials and König et al. (1991) with friccional materials. The limiting value of P/D proposed by König et al. (1991) as 0.2 was however found to be 0.1.

ACKNOWLEDGEMENTS

The authors wish to express their sincere thanks to FCT for their financial support (Project poci/ecm/61934/2004).

REFERENCES

Antão, A. M. Sequeira Nunes 1997. Analyse de la stabilité des ouvrages souterrains par une méthode cinématique régularisée. Thèse de doctorat de L'École Nationale des Ponts et Chaussées.

Casarin, C., Mair, R. J. 1981. The assessment of tunnel stability in clay by model tests. In D. Resendiz and P. M. Romo (ed.). Soft-ground tunnelling, failures and displacements: 33–44.

Chambon, J. F., Corté, J. F. 1994. Shallow tunnels in cohesionless soil: stability of tunnel face. Journal of Geotechnical Engineering (120): 1148–1165.

Kimura, T., Mair, R. J. 1981. Centrifugal testing of model tunnels in soft clay. In Proc. 10th Int. Conf. on Mechanics and Foundation Engineering. Stockholm. (1): 319–322.

König, D., Gütter, U. and Jessberger, H. L. 1991. Stress redistributions during tunnel and shaft constructions. In Proc., Int. Conf. Centrifuge 1991, Balkema, Rotterdam, The Netherlands: 129–138.

Mair, R. J., Phillips, R., Schofield, A. N., Taylor, R. N. 1984. Application of centrifuge modelling to the design of tunnels and excavations in soft clay. In Proc. Symposium on Application of Centrifuge Modelling to Geotechnical Design. Manchester: 357–380.

Soil and rock excavations

Applications of Computational Mechanics in Geotechnical Engineering – Sousa,
Fernandes, Vargas Jr & Azevedo (eds)
© 2007 Taylor & Francis Group, London, ISBN 978-0-415-43789-9

Earth pressures of soils in undrained conditions. Application to the stability of flexible retaining walls

Armando Nunes Antão
UNIC, FCT, New University of Lisbon, Lisbon, Portugal

Nuno M. da C. Guerra
ICIST, IST, Technical University of Lisbon, Lisbon, Portugal

António Silva Cardoso & Manuel Matos Fernandes
FEUP, University of Porto, Porto, Portugal

ABSTRACT: A numerical implementation of the upper bound theorem with a tension truncated Tresca criterion is validated. The validation is performed with previously known results of limit loads in the same conditions. Values of active and passive earth pressures obtained through both classical Tresca and tension truncated Tresca criteria are presented. Some results are applied to the problem of the equilibrium of flexible earth retaining structures and the results issued from the application of the two criteria are presented.

1 INTRODUCTION

The literature reports a number of incidents and accidents of tied-back flexible retaining walls related with inadequate bearing resistance to vertical loads applied by the anchors (Broms and Stille, 1976; Ulrich, 1989; Clough and O'Rourke, 1990; Winter, 1990; Stocker, 1991; Gould et al., 1992).

Some of the incidents and accidents mentioned were also observed in soldier-pile walls. In a few other cases of direct knowledge of the authors, involving concrete soldier-pile walls, the deficient support of vertical loads was related with buckling of the soldier piles, which indicates that their design loads had been significantly underestimated.

Figure 1 illustrates the typical construction sequence of a concrete soldier-pile wall. Figure 2 shows the forces involved in vertical equilibrium of both wall and supported soil mass. It can be understood that the degree of mobilization of the shear stresses at the back wall face (note that in this type of wall there is contact with the soil through that face only) will considerably affect the magnitude of the load to be supported by the soldier-piles.

Studies on the vertical equilibrium of soldier-pile walls have been carried out by Guerra et al. (2001, 2004) and Cardoso et al. (2006), involving field monitoring, f. e. analyses and analytical and numerical limit analyses.

These studies showed that the role of the shear stresses mobilized at the soil-to-wall interface is not as straightforward as it could seem. In brief, the problem can be described as follows:

1. considering the equilibrium of the wall only, it would appear that higher mobilization of upward tangential stresses applied to the wall would lead to lower required pile resistance;
2. however, higher mobilized adhesion at the interface increases the total vertical downward force on the soil mass;
3. then, the equilibrium of this mass will demand a larger horizontal force applied by the anchors;

247

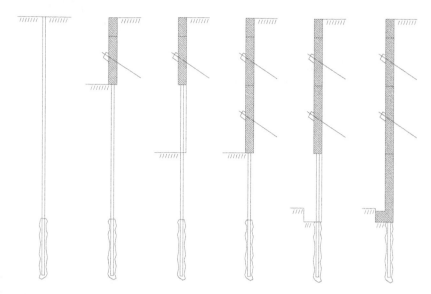

Figure 1. Construction stages of concrete soldier-pile walls (Cardoso et al., 2006).

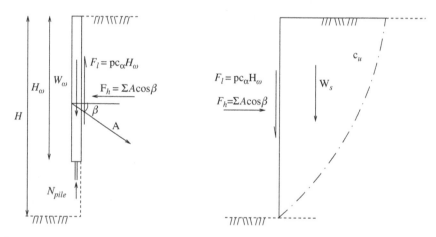

Figure 2. Forces involved in the equilibrium of a soldier-pile wall (left) and in the soil mass (right) (Cardoso et al., 2006).

4. since anchors are inclined downwards, this will lead to a greater vertical force on the wall;
5. so, in some circumstances, the mobilization of upward tangential stresses applied to the back of the wall by the soil may give rise to higher vertical loads on the soldier piles.

Vertical equilibrium of the wall requires that the following equation be verified (see the left side of Figure 2):

$$W_w + \sum A \sin \beta = N_{pile} + F_l \tag{1}$$

where:

- W_w is the weight of the wall per unit length;
- $\sum A \sin \beta$ is the vertical force applied by the anchors per unit length of the wall;

248

- N_{pile} is the load per unit length applied on the soldier piles;
- F_l is the shear force mobilized at the back soil-to-wall interface, per unit length of the wall and is positive if applied upwards to the wall interface.

The shear force, F_l, can be determined by:

$$F_l = pc_a H_w \tag{2}$$

where:

- c_a is the soil-to-wall interface adhesion;
- H_w is the current wall height;
- p is the fraction of mobilization of the soil-to-wall interface adhesion c_a, considered positive if F_l is applied upwards to the wall.

For a mobilized fraction p of the interface adhesion c_a, equation (1) can be written as:

$$\sum A \sin \beta = (N_{pile} - W_w) + pc_a H_w = (N_{pile} - W_w) + \frac{2\mu}{N_S}\chi_{am} \tag{3}$$

where:

- χ_{am} is given by

$$\chi_{am} = \frac{pc_a H_w}{c_u H} \tag{4}$$

- N_S is

$$N_S = \frac{\gamma H}{c_u} \tag{5}$$

- μ is

$$\mu = \frac{1}{2}\gamma H^2 \tag{6}$$

The right side of Figure 2 shows the external forces applied to the soil mass. The problem to be solved consists of determining the horizontal load $F_h = \sum A \cos \beta$ necessary to ensure stability of the soil mass of total height H, in undrained conditions, submitted to gravity (total weight of the unstable zone, W_s), and to a shear force applied at the cut face, F_l.

The problem described is, in fact, an earth pressure problem, in undrained conditions and with a shear force due to adhesion at the soil-to-wall interface.

Evaluation of these pressures has been performed using analytical (Guerra et al., 2004; Cardoso et al., 2006) and numerical methods (Antão et al., 2005; Cardoso et al., 2006).

The finite element numerical tool which implements the upper bound theorem of limit analysis used in the cited references allows the optimization of continuous displacement fields in order to obtain good approximations of limit loads without any *a priori* knowledge of the velocity field of plastic flow (Antão, 1997).

In previous works, the Classical Tresca (*CT*) criterion has been used to model soil strength. In this criterion any tension value is possible, as long as the Mohr circle radius does not exceed the undrained resistance, c_u, as represented in Figure 3(a). In the present work, the tension Truncated Tresca (*TT*) criterion is used (Figura 3(b)). This criterion differs from the Classical Tresca by not allowing tension values lesser than a given value σ_{tt}. In the case of null σ_{tt}, a Tresca cut-off criterion is obtained.

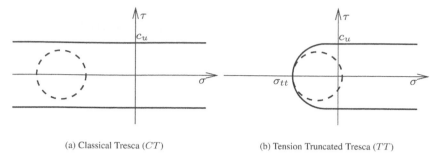

(a) Classical Tresca (CT) (b) Tension Truncated Tresca (TT)

Figure 3.   Classical Tresca (a) and Tension Truncated Tresca (b) yield criteria.

This criterion was implemented in the limit analysis numerical model previously mentioned. The main consequence of the different criterion is in the determination of the plastic dissipation, which can be written, for the Truncated Tresca criterion, as:

$$D\left(\underline{\dot{\varepsilon}}\right) = c_u \left[|\dot{\varepsilon}_I| + |\dot{\varepsilon}_{II}| + |\dot{\varepsilon}_{III}| + tr\left(\underline{\dot{\varepsilon}}\right)\right] + \sigma_{tt}\, tr\left(\underline{\dot{\varepsilon}}\right) \qquad (7)$$

where $\underline{\dot{\varepsilon}}$ is the strain rate tensor, with principal values represented by roman number subscripts and the other symbols were previously defined. Plastic admissibility implies that the equation is valid only for $tr(\underline{\dot{\varepsilon}}) \leq 0$.

2   VALIDATION OF THE NUMERICAL IMPLEMENTATION

In order to validate the numerical implementation of the Truncated Tresca criterion, the problem shown in Figure 4(a) was considered. It corresponds to a vertical cut of depth H in a homogeneous clayey soil under undrained conditions, with unit weight γ and undrained resistance c_u. The analysis of this problem is usually performed using the parameter N_{Scrit} (the critical value of N_S). For the Classical Tresca criterion, no exact solution is known. The best value of the parameter N_{Scrit} obtained by a mechanism defined by a circular slip surface (Figure 4(a)) is 3.83 (Taylor, 1948). For the case of tension cut-off Tresca criterion ($\sigma_{tt} = 0$), the value of that parameter is exactly known and equal to 2. The mechanism conducing to this value was obtained by Drucker (1953) and is also represented in Figure 4(a): it is composed by a vertical block which is detached from the soil mass and $N_{Scrit} = 2$ is obtained when the width d tends to zero.

Garnier (1995) studied the influence of the ratio $|\sigma_{tt}|/c_u$ on the parameter N_{Scrit} using both mechanisms. Results are presented in Figure 4(b). Two types of behaviour can be seen in the figure: one for low values of $|\sigma_{tt}|/c_u$ (Drucker mechanism) and another for higher values of this parameter (circular mechanism).

For the Classical Tresca criterion case, the finite element program conduces to values of N_{Scrit} lesser than 3.83 when sufficiently refined meshes are used. For the validation of the implementation of the Truncated Tresca criterion a finite element mesh allowing to obtain the value of 3.83 in the Classical Tresca case was used. Results obtained with the numerical tool are also presented in Figure 4(b).

The comparison of the two methodologies shows the practical coincidence of the results for $|\sigma_{tt}|/c_u = 2$. These results are also coincident with the 3.83 value previously referred. For most of the other values of $|\sigma_{tt}|/c_u$ the numerical results are lesser than those obtained by Garnier and, therefore, are closer to the exact solution. The lowest value of the ratio $|\sigma_{tt}|/c_u$ used was equal to 10^{-8} and the N_{Scrit} value obtained was 2.198. A lower value of this parameter could possibly be obtained using a more refined finite element mesh, in order to enable it to reproduce Drucker's mechanism as close as possible.

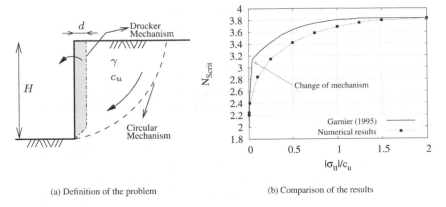

| (a) Definition of the problem | (b) Comparison of the results |

Figure 4. Validation of the numerical implementation of the Tension Truncated Tresca yield criterion through the analysis of an unsupported vertical cut.

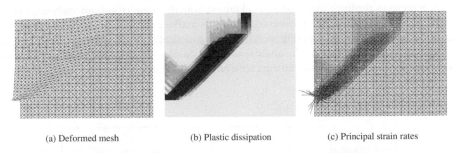

(a) Deformed mesh   (b) Plastic dissipation   (c) Principal strain rates

Figure 5. Deformed mesh, plastic dissipation and principal strain rates for $|\sigma_{tt}|/c_u = 0.75$.

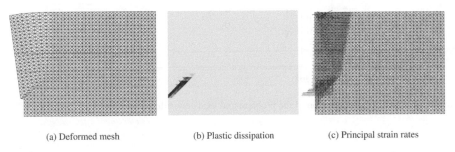

(a) Deformed mesh   (b) Plastic dissipation   (c) Principal strain rates

Figure 6. Deformed mesh, plastic dissipation and principal strain rates for $|\sigma_{tt}|/c_u = 10^{-4}$.

In Figures 5 and 6 results for $|\sigma_{tt}|/c_u = 0.75$ and $|\sigma_{tt}|/c_u = 10^{-4}$ are shown. It should be noted that for visualization purposes the finite element meshes represented are less refined than those used in the calculations presented. For each one of the two cases, the deformed mesh, the plastic dissipation and the principal strain rate values are shown.

For the case $|\sigma_{tt}|/c_u = 0.75$ the mechanism is still very similar to the one corresponding to Classical Tresca (close to circular). It is only in the upper region of the mesh that strain rates in pure tension appear, which, due to the high value of the ratio, contribute significantly to plastic dissipation (Figure 5(b)).

For $|\sigma_{tt}|/c_u = 10^{-4}$ the mechanism seems to be tending to the one proposed by Drucker, the zone in pure tension beeing extended from the surface down to the base of the mechanism. Only at the base (which dimension is as small as more refined the mesh is) shear zones arise. Due to the very small value of the tension allowed, significant plastic dissipation occurs in this zone only.

The differences between the two cases justify the important contrast in the values shown in Figure 4.

3 APPLICATION TO THE DETERMINATION OF EARTH PRESSURES IN PURELY COHESIVE MATERIALS

Figure 2 defines the geometry of the problem, as well as the loads considered for the determination of earth pressures. The problem consists of a supported vertical cut of depth H in a clayey soil with undrained shear strength c_u, tension truncated stress σ_{tt} and unit weight γ (Figure 7). The aim is the determination of the limit values of the earth pressures normal to the cut when a known tangential uniform stress is applied. The problem considered a $|\sigma_{tt}|/c_u$ ratio of 10^{-4}. From a practical point of view, this corresponds to tension cut-off. The same problem has been approached by the authors for the traditional Tresca case (Guerra et al., 2004; Cardoso et al., 2006).

Figure 8 presents the limit values of F_h/μ obtained from the numerical limit analyses. Figure 8a shows an important effect of the tension truncation on active earth pressures. On the contrary, in the case of passive earth pressures (Figure 8b) there is practically no effect for the range of χ_{am} for which a solution could be found. The absense of effect on passive earth pressures is explained by the fact that there are, for this case, no pure tension stresses on the soil mass.

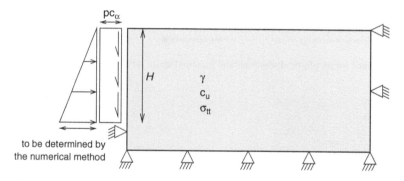

Figure 7.   Schematic representation of the problem and loads considered.

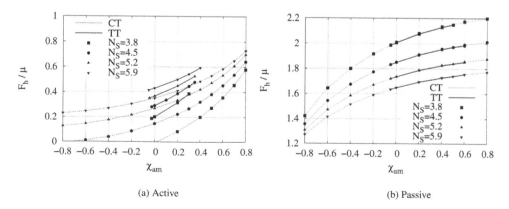

(a) Active

(b) Passive

Figure 8.   Values of F_h/μ obtained for active and passive cases.

252

As the figure shows, it was not possible to obtain results for the same range of χ_{am} for both solutions (CT and TT) presented. When attempts were made to find solutions for TT to left and right of those presented, local mechanisms were obtained, which were not representative of active or passive earth pressures and therefore are not presented in this paper.

As the active earth pressures are concerned, it can be seen that, as expected, greater values of χ_{am} lead to greater values of F_h/μ. Also, the differences in the ratio F_h/μ obtained for CT and TT are greater for greater values of χ_{am} and for lesser values of N_S. In addition, results of the ratio are slightly more variable with χ_{am} for the case TT than for CT.

These facts can be explained by the obtained mechanisms, as shown in Figure 9. In this figure, the deformed finite element meshes for the obtained active limit state are presented. In the left side of the figure results for CT are shown; in the right side results for TT are presented. The upper four graphics correspond to $\chi_{am} = 0$ and the lower to $\chi_{am} = 0.3$. In each one of the two sets of four graphics, the two upper ones are obtained for $N_S = 3.8$ and the two lower ones to $N_S = 5.9$.

The analysis of this figure shows that:

• the differences in the mechanism are clearer for $\chi_{am} = 0.3$ than for $\chi_{am} = 0$;
• for the same value of χ_{am} differences are clearer for lesser values of N_S.

The mechanism for the case CT does not depend on the value of N_S, although the ratio F_h/μ is, naturally, different. In the case TT, such dependency exists, which means that the mechanism itself contributes to a change in F_h/μ, adding to the effect of N_S. This justifies that the results of F_h/μ are more variable for TT than for CT.

It should also be noted that the results presented are mainly valid for cases where the support is flexible, as the deformed meshes show.

In the analysis of earth pressures in cohesive materials it is usual to consider the existence of tension cut-off cracks. For purely cohesive materials its depth is $z_0 = 2c_u/\gamma$. Dividing by H, the following equation is obtained:

$$\frac{z_0}{H} = \frac{2}{N_S} \tag{8}$$

For the cases of N_S presented in Figure 9, the ratio z_0/H is 0.53 for $N_S = 3.8$ and 0.34 for $N_S = 5.9$. As it can be seen, these depths can be recognized in the deformed meshes shown.

In Figure 10, deformed finite element meshes for the obtained passive limit state are presented for $\chi_{am} = 0$ and $N_S = 3.8$ for the cases CT and TT. The mechanisms are identical and the same conclusion would be drawn from the analysis of the results for other values of N_S and χ_{am}, which justifies the almost exact superposition of results seen in Figure 8.

4 APPLICATION TO THE DETERMINATION OF LIMIT VERTICAL LOADS OF FLEXIBLE RETAINING WALLS

The solution for the problem described in the previous section was motivated by the need to determine vertical limit loads of flexible retaining structures. The problem is schematically represented in Figure 2a. Vertical equilibrium requires that (Cardoso et al., 2006):

$$\frac{N_{pile} - W_w}{\mu} = \frac{F_h}{\mu} \text{tg}\beta - \frac{2\chi_{am}}{N_S} \tag{9}$$

Using in this equation the values of F_h/μ presented in the previous section for the active case, results shown in Figure 11 are obtained.

From the analysis of this figure, it can be seen that truncation leads to an increase in the minimum vertical pile resistance needed to ensure equilibrium. Such result can be justified by the loss of resistance that the truncation implies.

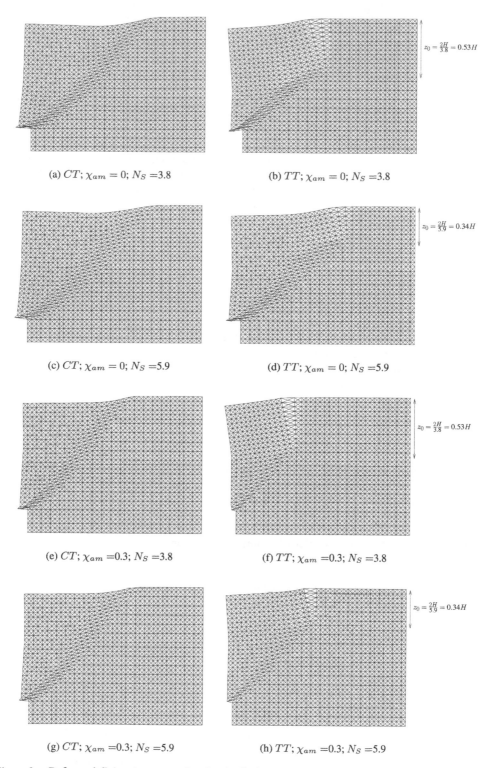

(a) CT; $\chi_{am} = 0$; $N_S = 3.8$

(b) TT; $\chi_{am} = 0$; $N_S = 3.8$

$z_0 = \frac{2H}{3.8} = 0.53H$

(c) CT; $\chi_{am} = 0$; $N_S = 5.9$

(d) TT; $\chi_{am} = 0$; $N_S = 5.9$

$z_0 = \frac{2H}{5.9} = 0.34H$

(e) CT; $\chi_{am} = 0.3$; $N_S = 3.8$

(f) TT; $\chi_{am} = 0.3$; $N_S = 3.8$

$z_0 = \frac{2H}{3.8} = 0.53H$

(g) CT; $\chi_{am} = 0.3$; $N_S = 5.9$

(h) TT; $\chi_{am} = 0.3$; $N_S = 5.9$

$z_0 = \frac{2H}{5.9} = 0.34H$

Figure 9. Deformed finite element meshes for the limit state for two values of χ_{am} (0 and 0.3) and two values of N_S (3.8 and 5.9) for the Classic Tresca (CT) and Traction Truncated Tresca (TT) with $|\sigma_{tt}|/c_u = 10^{-4}$ criteria.

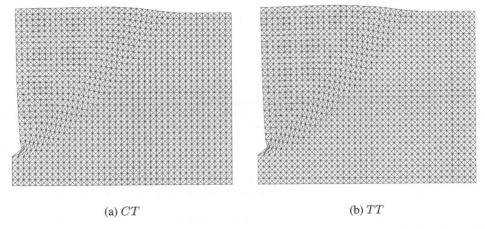

(a) CT (b) TT

Figure 10. Deformed finite element meshes of passive limite state for $\chi_{am} = 0$ and $N_S = 3.8$ for CT and TT.

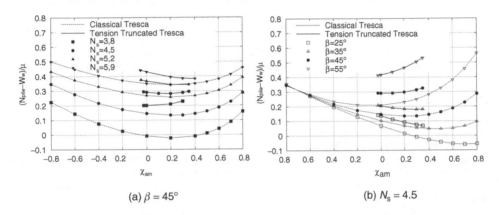

(a) $\beta = 45°$ (b) $N_s = 4.5$

Figure 11. Values of $(N_{pile} - W_w)/\mu$ obtained from the values of F_h/μ determined for the active case.

It can also be seen that the increase of the required vertical resistance due to the consideration of truncation is greater for lesser values of N_S, which can be justified by the fact that the depth of the tension zone is greater for these cases and, so, there is a greater relative resistance loss. It can finnally be seen that the value of χ_{am} to which the minimum needed resistance corresponds is lesser for the truncated case.

5 CONCLUSIONS

The implementation of the Tension Truncated Tresca criterion in a limit analysis upper bound finite element program was validated.

The program was applied to the determination of active and passive earth pressures of purely cohesive soils with no tension resistance. The results were compared with those obtained with the Classical Tresca yield criterion and it could be concluded that the range of χ_{am} for which a solution could be found is, in the TT case, much narrower than in the CT case. Values obtained for the passive case reveal almost no change, whereas in the active earth pressures the loss of tension resistance leads to an increase in the values of the earth presssures. This increase is greater for the more resistant soils.

Results of the active earth pressures were also used for the determination of limit loads of vertical piles in soldier pile retaining walls. It could be concluded that these loads are increased when no tension strength is assumed and the relative increase is greater for more resistant soils.

REFERENCES

Antão, A. N. (1997). *Analyse de la Stabilité des Ouvrages Souterrains par une Méthode Cinématique Régularisée*. PhD thesis, École Nationale des Ponts et Chaussées, Paris. In French.
Antão, A. N., Guerra, N. M. C., Matos Fernandes, M., and Cardoso, A. S. (2005). Limit analysis of concrete soldier-pile walls in clay: influence of the height of the excavation levels on vertical stability. In Owen et al., editor, *Computational Plasticity VIII, Fundamentals and Applications, Proceedings of the VIII International Conference on Computational Plasticity*, volume 1, pages 165–168, Barcelona, Espanha, 5–7 Setembro.
Broms, B. B. and Stille, H. (1976). Failure of anchored sheet pile walls. *ASCE Journal of Geotechnical Engineering Division*, 102(3):235–251.
Cardoso, A. S., Guerra, N. M. C., Antão, A. N., and Matos Fernandes, M. (2006). Limit analysis of anchored concrete soldier-pile walls in clay under vertical loading. *Canadian Geotechnical Journal*, 43(5):516–530.
Clough, G. W. and O'Rourke, T. D. (1990). Construction induced movements of in situ walls. *ASCE Geotechnical Special Publication No 25, Design and Performance of Earth Retaining Structures*, pages 439–470.
Drucker, D. C. (1953). Limit analysis of two and three-dimensional soil mechanics problems. *J. Mech. Phys. Solids*, 1:217–226.
Garnier, D. (1995). *Analyse par la Théorie du Calcul à la Rupture des Facteurs de Réduction de la Capacité Portante de Fondations Superficielles*. PhD thesis, École Nationale des Ponts et Chaussées, Paris. In French.
Gould, J. P., Tamaro, G. J., and Powers, J. P. (1992). Excavation and support systems in urban settings. *ASCE Geotechnical Special Publication No 33, Excavation and Support for the Urban Infrastructure*, pages 144–171.
Guerra, N. M. C., Cardoso, A. S., Matos Fernandes, M., and Gomes Correia, A. (2004). Vertical stability of anchored concrete soldier-pile walls in clay. *ASCE Journal of Geotechnical and Geoenvironmental Engineering*, 130(12):1259–1270.
Guerra, N. M. C., Gomes Correia, A., Matos Fernandes, M., and Cardoso, A. S. (2001). "Modelling the collapse of Berlin-type walls by loss of vertical equilibrium: a few preliminary results". In *3rd International Workshop Applications of Computational Mechanics in Geotechnical Engineering*, pages 231–238, Oporto, Portugal.
Stocker, M. F. (1991). Contribuição para a discussão na sessão n. 4b. In *Proceedings of 10th European Conference of Soil Mechanics and Foundation Engineering*, volume 4, page 1368, Firenze.
Taylor, D. W. (1948). *Fundamentals of Soil Mechanics*. John Wiley.
Ulrich, E. J. J. (1989). Tieback supported cuts in overconsolidated soils. *ASCE Journal of Geotechnical Engineering Division*, 115(4):521–545.
Winter, D. G. (1990). Pacific First Center performance of the tieback shoring wall. *ASCE Geotechnical Special Publication No 25, Design and Performance of Earth Retaining Structures*, pages 764–777.

Applications of Computational Mechanics in Geotechnical Engineering – Sousa,
Fernandes, Vargas Jr & Azevedo (eds)
© 2007 Taylor & Francis Group, London, ISBN 978-0-415-43789-9

Methods of two-dimensional modelling of soil anchors: preliminary results of the application to flexible retaining walls

Nuno M. da C. Guerra
ICIST, IST, Technical University of Lisbon, Lisbon, Portugal

Cláudia Santos Josefino
IST MSc student, Technical University of Lisbon, Lisbon, Portugal

Manuel Matos Fernandes
FEUP, University of Porto, Porto, Portugal

ABSTRACT: Soil-anchors apply to the soil mass stresses that are clearly three-dimensional. However, soil anchors are often modelled in two dimensional plane strain conditions, particularly in the finite element analysis of flexible retaining walls. In this paper, methods of 2D anchor modelling are reviewed, described and compared. Advantages and inconvenients of each method are presented and discussed. Results obtained from finite element modelling of excavations using flexible retaining structures with different soil anchor modelling techniques are presented and compared.

1 INTRODUCTION

Design of flexible retaining structures is growingly supported on results of numerical analyses, usually performed using the finite element method. In most cases the analyses are two-dimensional, assuming plane strain conditions. This is due to: (i) the significant difficulties of geometry definition and result analysis of 3D calculations; (ii) significantly greater CPU time of 3D calculations; (iii) a reasonably frequent correspondence between the excavation geometry and plane strain conditions; (iv) the conservative nature of most plane strain analyses.

Most of the urban excavations using flexible retaining structures use pre-stressed anchors, which apply a concentrated force on the wall and on the seal zone, causing a difficulty to 2D plane strain modelling. The importance of this difficulty depends on whether the seal zone can be assumed fixed for the loads and geometry changes induced by the excavation (see Figure 1(a)). When the seal is formed in a soil mass with mechanical characteristics significantly better than the supported soil (see Figure 1(a), left), displacements of the seal due to the excavation are very small; on the other hand, if the seal is disposed in a soil mass with mechanical characteristics similar to the supported soil and relatively close to the wall (see Figure 1(a), right), its displacements towards the excavation will affect the global behaviour of the retaining structure and the ground.

The problem which arises from the 3D concentrated nature of the force on the wall exists in both cases and is solved by considering beams distributing the loads. This will not be focused on this work.

The issue of the 3D nature of the seal is treated differently in fixed and movable seals.

If the seal is fixed, modelling is usually performed by applying to the wall the load corresponding to the initial pre-stress, simulating the seal through a fixed support and using a bar element to represent the free length of the anchor (see Figure 1(b), left). This way of modelling has no open issues of the knowledge of the authors and therefore will not be studied in this work.

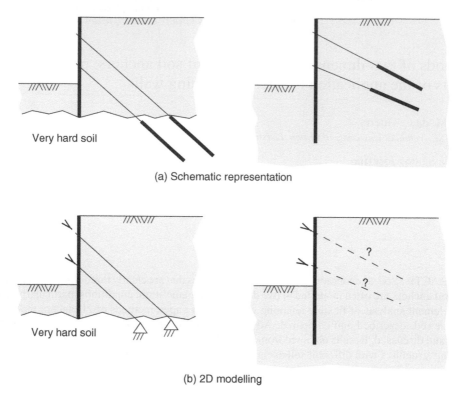

(a) Schematic representation

(b) 2D modelling

Figure 1.   Pre-stressed anchors sealed in the soil mass with fixed seal (left) and movable seal (right).

If the seal is not fixed (see Figure 1(b), right) pre-stress is also simulated by applying a con-centrated force on the wall at the location of the anchor head whereas the simulation of the other elements of the anchor can be performed by using different methods. In this paper, some of these methods are described and results of excavations modelled using different methods are compared.

2   MAIN MODELLING METHODS OF PRE-STRESSED ANCHORS SEALED IN THE SOIL MASS

Figure 2 shows the modelling methods of pre-stressed anchors sealed in the soil mass analised in this work. Methods F and FF are references for the others and do not correspond to modelling procedures frequently used.

In the first of these methods, the anchor is modelled solely by a force applied on the wall at the anchor head. Its value is equal to the initial pre-stress and therefore the anchor loads will not change through out the analysis.

In the second method the seal is not really movable: a fixed support is used to model it, and therefore movements of the soil mass in the zone of the seal are not considered. The anchor is, in this case, modelled by a force, a bar element and a fixed support.

Method FN considers the anchor modelled by a bar element (representing the free-length) from the anchor head to a node N of the finite element mesh representative of the seal zone (Guerra, 1993). In this method (as in method FF) the bar element will only be loaded with anchor load changes and consequently only these changes are applied to the soil mass through node N.

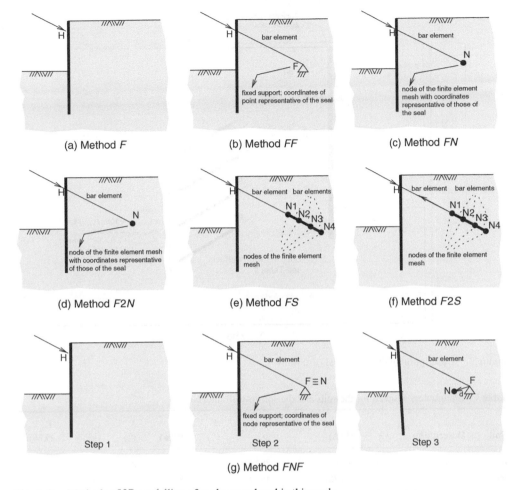

(a) Method *F* (b) Method *FF* (c) Method *FN*

(d) Method *F2N* (e) Method *FS* (f) Method *F2S*

Step 1 Step 2 Step 3

(g) Method *FNF*

Figure 2. Methods of 2D modelling of anchors analysed in this work.

Method *F2N* is a variant of method *FN* in which forces representing the initial pre-stress load are applied to both the anchor head and the soil mass, concentrated in node *N*.

Method *FS* considers the seal explicitly modelled by bar elements connecting nodes which belong also to the finite elements representing the soil mass.

In a variant *F2S* of this method (Clough and Tsui, 1974; Mineiro et al., 1981), which is considered in Plaxis finite element program, the force of the initial pre-stress is applied in both the anchor head and the seal.

Method *FNF* is the method proposed by Matos Fernandes (1983) in the version presented by Guerra (1999). In this method, pre-stress is applied to the wall through a force and a bar element is afterwards activated. The properties of this bar element are derived from anchor load tests, so that the desired curve load-displacements is obtained (incorporating both the behaviour of the free-length and the seal). The bar element connects the anchor head to a point *F* initially fixed, with coordinates that are equal to the ones of a point of the soil mass representative of the seal zone *N*. In the next stages, displacements of node *N* are applied to node *F* in the next iterations of the calculation. This way, the behaviour of the anchor considering both the free-length and displacements of the seal are taken into account.

Methods using joint elements to model the interface between the seal and the soil mass are not considered in this work. Methods considering the seal modelled explicitly with 2D finite elements

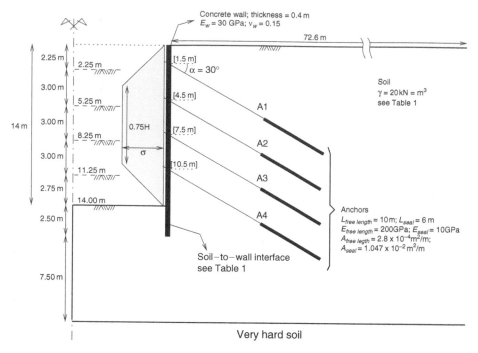

Figure 3. Description of the numerical case study.

Table 1. Parameters adopted for the soils and the soil-to-wall interfaces.

Soil	Description	K_0	E (kPa)	v	ϕ' (°)	c_u (kPa)	δ (°)	c_a (kPa)	K_s (kN/m³)
A	sand	0.6	$50000\left(\frac{\sigma'_v}{p_a}\right)^{0.5}$	0.333	30	—	20	—	22928
B	clay	0.7	24000	0.49	—	80	—	40	

(instead of the 1D elements of methods FS and $F2S$) are also not considered; the effect of the small thickness of these elements is expected to be negligible.

3 DESCRIPTION OF THE CASE STUDY

To test the methods presented in the previous section, the numerical case study represented in Figure 3 is considered: a symmetrical excavation, 14 m deep and 16 m wide supported by a concrete diaphragm wall, 0.4 m thick. The supported soil mass was assumed of two types, as represented in Table 1. In this table, K_0 is the at-rest earth pressure coefficient, E is the soil modulus, v is Poisson ratio of the soil, ϕ' is the soil friction angle, c_u is the undrained resistance, δ is the soil-to-wall interface friction angle, c_a is the soil-to-wall interface adhesion and K_s is the tangential stiffness of the interface.

Soils were modelled considering elastic-perfectly plastic behaviour. A rigid layer was assumed under the soil mass. The applied pre-stress was 122.6 kN/m for the anchors of the first level and 286.0 kN/m for the second, third and fourth levels. The total horizontal component of the anchor pre-stress is aproximately equal to the resultant of the trapezoidal diagram represented in Figure 3 with horizontal stress, σ, of $0.3\gamma H$.

260

Table 2.　Adopted construction stages.

Stage	Description
1	Excavation of 1st level
2	Activation of the seal of the 1st anchor level (*FS* e *F2S*) and application of pre-stress
3	Activation of the free length of the 1st anchor level (except *F*) and excavation of the 2nd level
4	Activation of the seal of the 2nd anchor level (*FS* e *F2S*) and application of pre-stress
5	Activation of the free length of the 2nd anchor level (except *F*) and excavation of the 3rd level
6	Activation of the seal of the 3rd anchor level (*FS* e *F2S*) and application of pre-stress
7	Activation of the free length of the 3rd anchor level (except *F*) and excavation of the 4th level
8	Activation of the seal of the 4th anchor level (*FS* e *F2S*) and application of pre-stress
9	Activation of the free length of the 4th anchor level (except *F*) and excavation of the 5th level

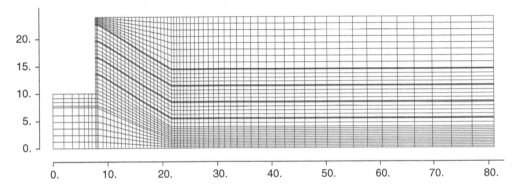

Figure 4.　Finite element mesh at the last excavation stage.

4　FINITE ELEMENT PROGRAM

The finite element program used is written in Fortran 77 and was developed for geotechnical applications in plane strain, plane stress and axissymmetry (Cardoso, 1987; Almeida e Sousa 1998; Guerra, 1999). The analyses can be performed using an elastoplastic model, with hardening and softening, with Von Mises, Drucker-Prager, Tresca or Mohr-Coulomb criteria.

Soil and wall were modelled by subparametric finite elements of 5 nodes, the contact between the soil and the wall with 4-noded interface elements and the anchors (free length and seal) with 2-noded bar elements. Analyses were carried out in drained (soil *A*) and undrained (soil *B*) conditions. Soil modulus was assumed increasing with depth (soil *A*) and constant (soil *B*). Mohr-Coulomb (soil *A*) and Tresca (soil *B*) criteria were used.

It was assumed that the excavation would be performed in 5 levels, indicated in Figure 3, through the sequence indicated in Table 2. Figure 4 shows the finite element mesh for the last excavation stage.

5　CALCULATION RESULTS AND DISCUSSION

For both soils (*A* and *B*) presented in the previous section, the methods presented in section 2 were applied. Points *F* and *N* were considered the connection between the free length and the seal. In method *FNF* the anchor stiffness was assumed equal to 90% of the theoretical stiffness (axial stiffness of the tendon with a length equal to the free length).

Figure 5 presents the results of the horizontal displacements of the wall and the settlements of the surface of the supported soil at the last excavation stage, obtained for the different methods

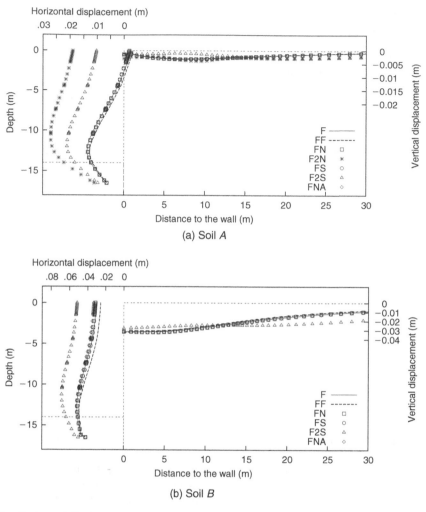

Figure 5. Horizontal displacements of the wall and settlements of the supported soil at the last excavation stage ($\sigma \simeq 0.3\gamma H$).

and for soils A and B. Results for method $F2N$ (in case of soil B) are not presented because the calculation could not end with the desired convergence level.

The analysis of the figure shows that the displacements provided by methods $F2N$ and $F2S$ are clearly different from the other methods.

As it was seen, methods $F2N$ and $F2S$ are different from their correspondent FN and FS in the loads due to pre-stress: they are applied not only to the wall, at the anchor head (as in FN and FS) but also to the seal. They are transmitted to the soil mass in one node (method $F2N$) and through the seal length (method $F2S$). Even if method $F2N$ is not considered valid (it considers a large force applied in one single node of the mesh and results could not be obtained for soil B), it is interesting to observe that the effect of this load (in method $F2S$) is very significant in the displacements of the wall and of the soil.

Results of the other methods show almost identical displacements, with only minor differences for method FF in the displacements in case of soil B. This can only be explained by the fact that anchor load changes are relatively small.

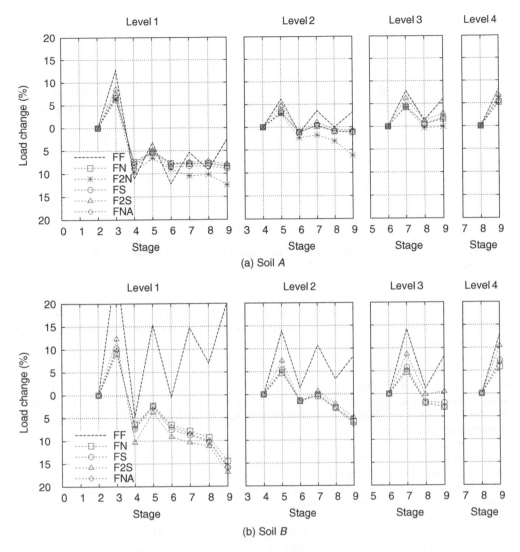

Figure 6.   Anchor load changes ($\sigma \simeq 0.3\gamma H$).

In fact, Figure 6 illustrates that the greater anchor load changes are those of method *FF*, due to the greater global stiffness of the anchor caused by the fact that seals are represented by fixed supports. But, in agreement with the displacements, this is more noticeable in case of soil *B*. It can also be observed that method *F2N* presents significant decreases in the loads whereas the other methods give quite similar results.

6   RE-ANALYSIS

The examination of the finite element results in the previous section lead to two main questions:

- Are methods *FF* and *F*, which were only used as reference methods, adequate to model anchors?
- Is the influence of the force applied to the seal (method *F2S*) realistic?

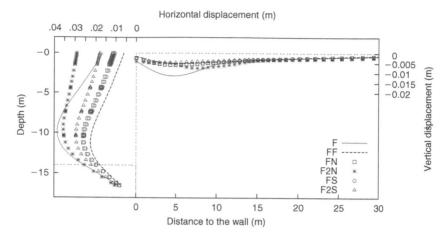

Figure 7. Horizontal displacements of the wall and settlements of the supported soil at the last excavation stage (solo B; $\sigma \simeq 0.15\gamma H$).

The answer to the first question is yes, but only when anchor load changes are small to moderate. Results presented in the previous section considered a suitable pre-stress level, and therefore anchor load changes were moderate.

Figure 7 shows the results of the displacements obtained when a pre-stress level of 50% of the one considered in the previous section is applied, for the case of soil B. It can be seen that results from methods F and FF are now very different: method F is not able to increase the anchor loads and therefore, for an insufficient pre-stress level, displacements are overestimated; method FF considers no displacements of the seal and so anchor load change is overestimated and displacements are underestimated.

It is interesting to notice that methods FN and FS continue to give very similar displacement results and that methods $F2N$ and $F2S$ give quite greater displacements than FN and FS. In the case of method $F2N$, results are probably not very realistic due, as seen before, to the fact that the force applied to the seal is not distributed along a seal length. But results of method $F2S$ continue to give important increases of the displacements relatively to those of method FS, which leads to the second question: are they realistic?

In order to try to answer the second question, the following should be considered: there is, of course, no doubt that the pre-stress mobilizes a force on the seal. But in plane strain conditions, this force is equivalent to a distributed load in the direction normal to the plane analysed, as if the seal was a plate. So, the second question could be refrased as: is the concentrated force in plane strain conditions realistic, taking into account of the three-dimensional nature of the real problem?

Two additional calculations were then performed, using Plaxis 3D finite element program for the case of soil B: in both calculations the seal was explicitly modelled using 3D finite elements (the shape of the seal was assumed square parallelipipedic, in a simplified way), the anchors applied the concentrated force at the anchor head and were assumed with a longitudinal spacing of 3 m (a 1.5 m thick slice was considered, assuming a very long excavation and taking advantage of the symmetry of the problem; see Figure 8). The finite element mesh is represented in Figure 9. In one of the calculations, no force was applied to the seal ($FS - 3D$) and in the other pre-stress was applied to it ($F2S - 3D$).

Displacement results of both calculations are presented in Figure 10a, where they are compared with the previous 2D calculations using methods FS and $F2S$. Due to the 3D nature of the calculations, results for two extreme planes of the 3D slice analysed are presented: the front plane is the plane of the anchors, where the forces are applied, and the rear plane is the plane at the back, that is, the place spaced 1.5 m from the front plane and therefore in the middle of two neighbour

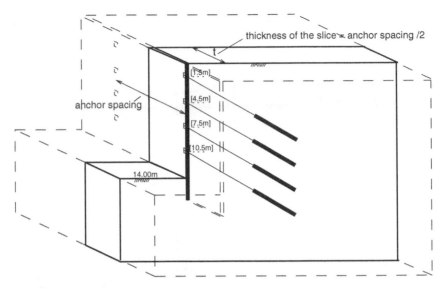

Figure 8. Geometry of the 3D problem.

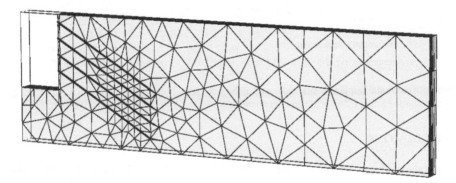

Figure 9. 3D finite element mesh.

anchor planes. The analysis of the figure shows that: (1) there is no practical difference between the results of the two planes of the 3D calculations; (2) there is no practical difference between the results of 2D and 3D calculations and therefore the differences in the displacements observed in 2D between methods *FS* and *F2S* are also observed under 3D conditions.

The fact that the results for the front and rear planes are coincident can be explained by the relatively small anchor spacing (3 m) and the relatively large stiffness of the wall. A second set of two 3D calculations was carried out, assuming anchor spacing equal to 6 m (an upper limit value, in practical excavations) and a wall stiffness ten times lower ($E_w = 3$ GPa). Results are presented in Figure 10b, compared with the 2D calculations also with a reduced wall modulus.

The analysis of this figure shows that the results of the two planes are now different and that the correspondent 2D displacement results are roughly between the results of the front and the rear planes, which seems consistent.

The most relevant conclusions of the 3D calculations are, however, that the differences between methods *FS* and *F2S* still occur in 3D conditions, as in 2D plane strain, which means that the forces applied to the seal are, in fact, realistic, and therefore should be taken into account in anchor modelling methods.

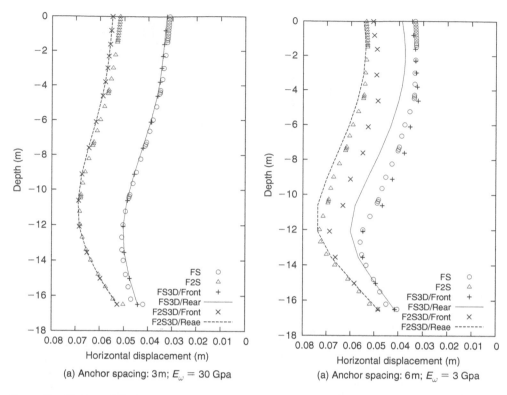

Figure 10.   Horizontal displacements of the wall at the last excavation stage (soil B; $\sigma \simeq 0.3\gamma H$): comparison between methods FS and $F2S$ for 2D and 3D conditions.

7   CONCLUSIONS

The analysis of the different methods of 2D anchor modelling allows to draw the following conclusions:

- reference methods F and FF are not suitable: the first can be valid to give a first estimate of displacements only when anchor load changes are not expected to be significant; the second does not consider the displacements of the seal zone due to the construction, and therefore can only give good results if the seal displacements are very small;
- method $F2N$ is also not suitable because the load due to anchor pre-stress causes an unrealistic response of the soil mass which leads to large displacements and (or) to the inability of obtaining any results due to lack of convergence of the finite element calculations;
- methods FN, FS and FNF give very similar results, clearly distinct from the ones of method $F2S$;
- this difference is due to the effect of anchor load forces applied to the seal on the displacements of the wall and supported soil; 3D calculations show that this influence is realistic and it should therefore be taken into account in anchor modelling.

ACKNOWLEDGMENTS

The authors would like to thank the contributions of FCT/University of Coimbra in the person of Prof. Jorge Almeida e Sousa, who made possible the use of Plaxis 3D finite element program, and of Eng. António Gonçalves Pedro, for his important help in the initial stages of the 3D calculations.

REFERENCES

Almeida e Sousa, J. (1998). *Tunnelling in soft ground. Behaviour and numerical simulation*. PhD thesis, Faculdade de Ciências e Tecnologia, University of Coimbra, Coimbra, Portugal. In Portuguese.

Cardoso, A. J. M. S. (1987). *Soil nailing applied to excavations*. PhD thesis, Faculdade de Engenharia, University of Porto, Porto, Portugal. In Portuguese.

Clough, G. W. and Tsui, Y. (1974). Performance of tied-back walls in clay. *ASCE Journal of Geotechnical Engineering Division*, 100(12):1259–1273.

Guerra, N. M. C. (1993). Berlin-type retaining walls. Analysis of three-dimensional effects. Master's thesis, Universidade Nova de Lisboa, Lisbon, Portugal. In Portuguese.

Guerra, N. M. C. (1999). *Collapse mechanism of Berlin-type retaining walls by loss of vertical equilibrium*. PhD thesis, Instituto Superior Técnico, Technical University of Lisbon, Lisbon, Portugal. In Portuguese.

Matos Fernandes, M. A. (1983). *Flexible structures for earth support. New design methods*. PhD thesis, Faculdade de Engenharia, University of Porto, Porto, Portugal. In Portuguese.

Mineiro, A. C. J., Brito, J. A. M., and Fernandes, J. S. (1981). Étude d'une paroi moulée multi-ancrée. In *Proceedings of 10th International Conference of Soil Mechanics and Foundation Engineering*, volume 2, pages 187–192, Stockholm, Sweden. A. A. Balkema.

Applications of Computational Mechanics in Geotechnical Engineering – Sousa,
Fernandes, Vargas Jr & Azevedo (eds)
© *2007 Taylor & Francis Group, London, ISBN 978-0-415-43789-9*

Consolidation and overall stability of unretained excavations in clayey soils by finite element method

J.L. Borges
Department of Civil Engineering, Faculty of Engineering, University of Porto, Porto, Portugal

ABSTRACT: The behaviour of an unretained sloped excavation in an overconsolidated clay is analysed by a numerical model based on the finite element method. The model incorporates the Biot's consolidation theory (coupled formulation of the flow and equilibrium equations) and constitutive relations simulated by the p-q-θ critical state model. Special emphasis is given to the analysis in time, during and after the construction period, of the excess pore pressures, effective stresses, displacements and stress levels. On the other hand, taking into account the influence of the consolidation on the shear strength of soil, the variation in time of the safety is assessed using a computer program of overall stability analysis. This program, based on limit equilibrium formulations, uses the numerical results obtained from the finite element applications and formulations of the critical state soil mechanics. Finally, comparisons of results are analysed by changing some parameters, namely the problem geometry and the over-consolidation ratio of soil.

1 INTRODUCTION

When an unretained sloped excavation in a saturated clayey soil is undertaken, the variation of the stress state in the ground is basically determined by the decrease of total mean stress and the increase of shear stress.

The decrease of total mean stress generates negative excess pore pressure in the clay, whereas the increase of shear stress may generate positive or negative values (Lambe and Whitman, 1969). Usually, these values are negative in medially to strongly overconsolidated clays (plastic volumetric strain occurs with volume increase) and positive in normally consolidated or lightly overconsolidated clays (plastic volumetric strain occurs with volume decrease).

Therefore, at the end of excavation, there are pore pressure gradients in the ground that determine a water velocity field. Initial conditions of a transient flow are established and load transferences from the water (pore pressure) to the soil skeleton (effective stress) take place in time. The process (consolidation) ends when a steady flow is reached.

Mainly in strongly overconsolidated clays, these load transferences may determine swelling of the soil (decrease of effective mean stress) and, therefore, decay of its long term shear strength; the safety of the problem may then decrease in time.

However, there is an opposite effect (i.e. in favour of the structural safety) that determines the increase of the soil shear strength (increase of effective mean stress), which is the lowering of the water level associated to the excavation. Thus, overall stability usually varies in time, the variation depending on the relative magnitude of the two mentioned contrary effects.

In the paper, the geotechnical behaviour of an unretained sloped excavation in an overconsolidated clay (illustrative case) is analysed by a finite element model developed by Borges (1995). Special emphasis is given to the analysis in time, during and after the construction period, of the excess pore pressures, effective stresses, displacements and stress levels.

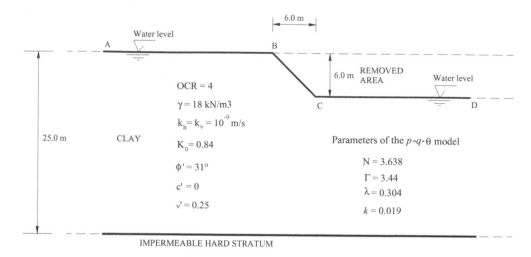

Figure 1. Geometry of the excavation and geotechnical properties of the clay.

On the other hand, taking into account the influence of the consolidation on the shear strength of soil, the variation in time of the overall safety of the problem is assessed by a computer program, developed by Borges (1995), that uses the numerical results obtained from the finite element applications and formulations of the critical state soil mechanics. The illustrative case and two more similar cases with changes in the geometry (weight of excavated soil) and over-consolidation ratio of the clay are analysed.

Basically, for the present applications, the finite element program uses the following theoretical hypotheses: a) plane strain analysis; b) coupled analysis, i.e. coupled formulation of the flow and equilibrium equations with soil constitutive relations formulated in effective stresses (Biot's consolidation theory) (Borges, 1995; Lewis and Schrefler, 1987; Britto and Gunn, 1987; Borges and Cardodo, 2000), applied to all phases of the problem, both during the excavation and in the post-construction period; c) utilisation of the p-q-θ critical state model (Borges, 1995; Lewis and Schrefler, 1987; Britto and Gunn, 1987; Borges and Cardodo, 1998), an associated plastic flow model, to simulate constitutive behaviour of the soil.

The accuracy of the finite element program and of the program of overall stability analysis has been assessed in several ground structures involving consolidation (Borges, 1995, 2004; Borges and Cardoso, 2001, 2002; Costa, 2005; Domingues, 2006). Very good agreements of numerical and field results were observed in all those studies.

2 DESCRIPTION OF THE ILLUSTRATIVE CASE

The geometry of the illustrate case (case 1), an unretained sloped excavation in an overconsolidated clay, is shown in Figure 1. Before excavation, the water level is at the ground surface. During and after excavation the water is drained and its level is assumed to follow the new ground surface (ABCD, after excavation).

The clay lies on a rigid and impermeable stratum (lower boundary) and its thickness is 25 m. The finite element mesh (with and without excavated elements) is shown in Figure 2. The six-noded triangular element for coupled analysis is used. While all six nodes of an element have displacement degrees of freedom only the three vertice nodes have excess pore pressure degrees of freedom.

No horizontal displacement is allowed on the vertical boundaries of the mesh while the bottom boundary is completely fixed in both the vertical and horizontal direction.

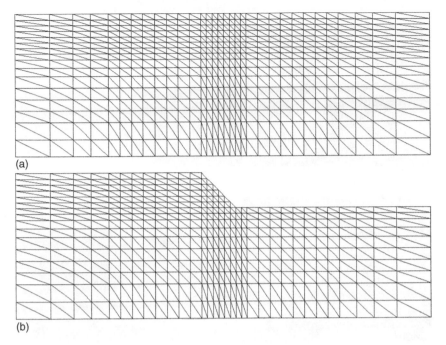

(a)

(b)

Figure 2.   Finite element mesh: (a) with all elements, (b) without excavated elements.

Excavation is simulated by removing layers of elements at a uniform rate. For a 6 m high cutting, excavation was completed in an overall time of 15 days.

Since excess pore pressure is defined as the difference between the water pressure at a particular instant and its initial hydrostatic value, on the ground surface at the end of excavation and during post-construction period, excess pore pressure is assumed as follows: (i) zero, along line AB; (ii) -60 kPa, along line CD (water unit weight adopted was 10 kN/m^3); (iii) linear variation along the slope, from zero at point B to -60 kPa at point C. The vertical and bottom boundaries of the mesh are assumed to be impermeable.

The constitutive behaviour of the clay was modelled by the p-q-θ critical state model (Borges,1995; Borges and Cardoso, 1998; Lewis and Schrefler, 1987). The parameters used are indicated in Figure 1 (λ, slope of normal consolidation line and critical state line; k, slope of swelling and recompression line; Γ, specific volume of soil on the critical state line at mean normal stress equal to 1 kPa; N, specific volume of normally consolidated soil at mean normal stress equal to 1 kPa). Figure 1 also shows other geotechnical properties: γ unit weight; ν', Poisson's ratio for drained loading; ϕ', angle of friction defined in effective terms; k_h and k_v, coefficients of permeability in horizontal and vertical directions; K_0, at rest earth pressure coefficient; OCR, over-consolidation ratio. All values of these parameters were adopted taking into account typical values reported in bibliography for this kind of soils (Lambe and Whitman, 1969; Borges, 1995).

3   RESULTS OF THE ILLUSTRATIVE CASE

Figures 3 and 4 show results of excess pore pressure at several stages, during and after excavation. As expected, excess pore pressure is negative at all points of the ground, at every stage. During excavation, the highest absolute values of excess pore pressure occur below the excavation bottom, being approximately equal to the weight of soil removed above. This occurs because the reduction of total mean stress is highest in this zone, being progressively lower below the slope and below the crest.

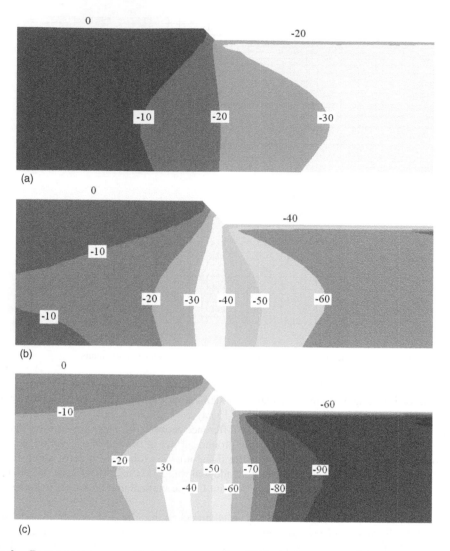

Figure 3. Excess pore pressure (Δu) during excavation (kPa): (a) 2 m excavated ($\Delta u_{min} = -39.99$ kPa), (b) 4 m excavated ($\Delta u_{min} = -73.09$ kPa), (c) 6 m excavated, end of excavation ($\Delta u_{min} = -103.04$ kPa).

After excavation, very typical shapes of isovalue curves of excess pore pressure are observed. These curves logically coincide with equipotential lines (where, for instance, -50 kPa of excess pore pressure corresponds to -5 m of total hydraulic load). Figure 4b shows the curves associated to the steady flow at the end of consolidation, imposed by the 6 m water level lowering.

Displacement vectors (amplified 20 times) are shown in Figure 5. As expected, at the end of excavation, horizontal displacements are towards the excavated area while vertical displacements are downwards below the crest and upwards below the excavation bottom. These upwards vertical displacements are clearly increased after excavation due to the swelling effect determined by the consolidation (increase of pore water pressure).

Principal effective stresses at the end of excavation and at the end of consolidation are shown in Figure 6. During excavation, very large shear stresses (deviatoric stress) occur below the slope (the angles of principal stresses with the horizontal direction vary significantly in that zone), whereas effective mean stress remains practically constant (this can be observed at any point of the ground

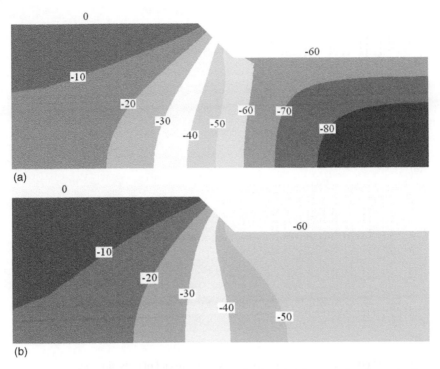

(a)

(b)

Figure 4. Excess pore pressure (Δu) after excavation (kPa): (a) 349 days after the end of excavation ($\Delta u_{min} = -89.26$ kPa), (b) end of consolidation, 5370 days after the end of excavation ($\Delta u_{min} = -60$ kPa).

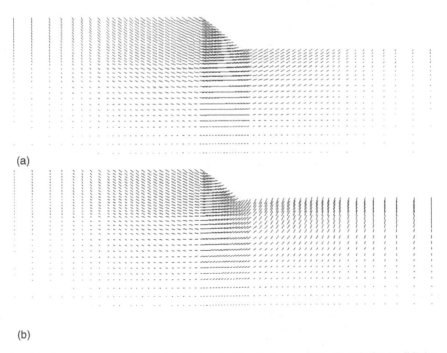

(a)

(b)

Figure 5. Displacement vectors (amplified 20 times): (a) end of excavation, (b) end of consolidation.

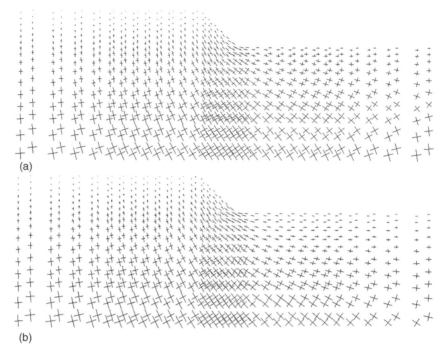

(a)

(b)

Figure 6.   Principal effective stresses: (a) end of excavation, (b) end of consolidation.

comparing its stress crosslet to the crosslet of the point, practically not affected by the excavation, near the left boundary, at the same level).

On the other hand, after excavation, the opposite occurs: shear stress does not vary significantly (expressive variations of the directions of the principal stresses do not take place), whereas there are significant reductions of effective mean stress (vertical principal stress significantly reduces, especially below the excavation bottom).

Figures 7 and 8 show distributions of stress level, SL, in the ground at several stages, during and after excavation. Stress level measures the proximity to the critical state. In normally consolidated soils, SL varies from zero to 1, the latter being the critical state level. In overconsolidated soils, because of their peak strength behaviour, stress level may be higher than 1. The results show that, during excavation, stress level significantly increases, especially near the slope (Figure 7); this is why, as reported above, deviatoric stress increases and effective mean stress does not change significantly. After excavation (Figure 8), there are also expressive increases of the stress level below the excavation bottom; during this period, this is due to the reduction of the effective mean stress (without significant changes of the deviatoric stress), as also reported.

4   OVERALL STABILITY. PARAMETRIC ANALYSIS

As said in section 1, overall stability of unretained excavations in clayey soils varies in time as consolidation takes place.

To better understand the influence of consolidation on the overall safety of the problem, three cases are studied in this section: (i) case 1, the illustrative case analysed above; (ii) cases 2 and 3, problems similar to case 1, but both considering a 4.5 m thick sandy layer overlying the clay (see Figure 9); the excavation consists of, firstly, removing the 4.5 m sandy layer (in 7.5 days) and, secondly, excavating the same slope as in case 1 (in 15 days). In case 2 all geotechnical properties

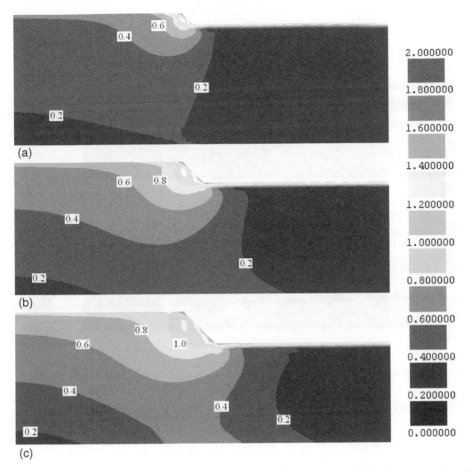

Figure 7. Stress level during excavation: (a) 2 m excavated, (b) 4 m excavated, (c) 6 m excavated, end of excavation.

of the clay are the same as in case 1 (see Figure 1), whereas in case 3 a lightly overconsolidated clay is considered, with OCR and K_0 equal to 1.2 and 0.48 respectively. All other parameters of the clay in case 3 are also equal to case 1. Obviously, in cases 2 and 3, due to the weight of overlying sandy layer, initial effective stresses of the clay (and obviously shear strength) are not the same as in case 1. Geotechnical parameters of the sand are shown in Figure 9.

The variation in time of the overall safety factor, calculated by the stability analysis program that uses the numerical results and formulations of the critical state soil mechanics (Borges, 1995), is shown in Figure 10 for cases 1 to 3. For case 1, the critical slip surface at the end of excavation is shown in Figure 11.

Figure 10 clearly shows that, while in case 1 overall safety practically remains constant during all post-excavation period, in cases 2 and 3 the stability reduces in time. In absolute terms, the safety reduction is higher in case 2, the more overconsolidated clay case. Thus, these results show that there is no typical pattern for safety variation in time for these three cases.

These differences can be explained by observing the results of excess pore pressure, at the end of excavation and at the end of consolidation, for the three cases (Figures 3, 4, 12 and 13). Absolute values of excess pore pressure at the end of excavation are higher in cases 2 and 3 than in case 1 (Figure 3c). This is why the excavated soil weight, which includes the sandy layer weight, is higher

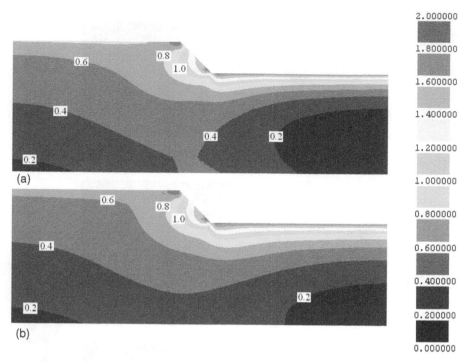

Figure 8. Stress level: (a) 349 days after the end of excavation, (b) 5370 days after end of excavation (end of consolidation).

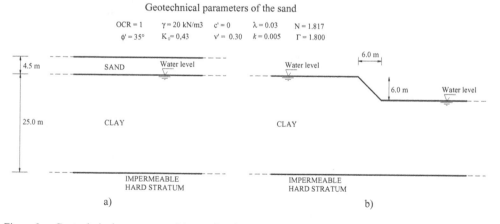

Geotechnical parameters of the sand

OCR = 1 γ = 20 kN/m3   c' = 0 λ = 0.03   N = 1.817
ϕ' = 35° K_0 = 0,43   v' = 0.30   k = 0.005 Γ = 1.800

Figure 9. Geotechnical parameters of the sand and geometry for cases 2 and 3: (a) before excavation, (b) after excavation.

in cases 2 and 3 than in case 1. However, since the long term hydraulic boundary conditions are equal for the three cases, consolidation in cases 2 and 3 occurs with higher dissipation of excess pore pressure which obviously implies higher reductions of effective mean stress and shear strength. These reductions, despite the contrary effect of water level lowering, have some expression at the points of the critical slip surface for cases 2 and 3, which explains the safety factor reductions in time plotted in Figure 10. The same does not happen in case 1. This is why the magnitude of

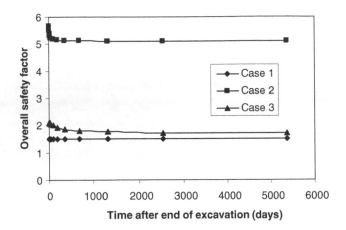

Figure 10. Evolution in time of overall safety for cases 1 to 3.

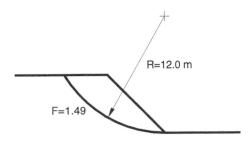

Figure 11. Slip circle for case 1, at the end of excavation (overall safety factor, F = 1.49).

negative pore pressure at the end of excavation – determined by a lower excavated soil weight – is not enough expressive to exceed the contrary effect of the water level lowering along the rupture surface.

5 CONCLUSIONS

The influence of consolidation on unretained sloped excavations in clayey soils was analysed. Two computer programs were used: (i) a finite element program that incorporates coupled analysis with constitutive relations modelled by the p-q-θ model; (ii) a stability analysis program that uses the finite element results and formulations of the critical state soil mechanics. The interpretation of the results was presented by analysing excess pore pressures, displacements, effective principal stresses, stress levels and overall safety factors. An illustrative case and two more similar cases with changes in the geometry (weight of excavated soil) and over-consolidation ratio of the clay were studied. Different patterns of the effect of consolidation on the overall stability were observed for the three cases. In case 1 the overall safety factor practically remained constant during all post-excavation period, whereas in cases 2 and 3, in which the weight of excavated soil is much higher, the stability reduced in time. In absolute terms, the reduction was more significant in case 2, the more overconsolidated clay case. These differences were explained by the relative importance on the stability of two contrary effects: the magnitude of excess pore pressure at the end of excavation and the water level lowering associated to the cut.

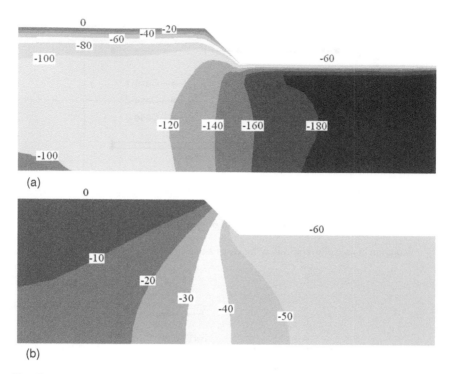

(a)

(b)

Figure 12.　Excess pore pressure (Δu) for case 2 (kPa): (a) at the end of excavation ($\Delta u_{min} = -192.7\,\text{kPa}$), (b) at the end of consolidation ($\Delta u_{min} = -60.0\,\text{kPa}$).

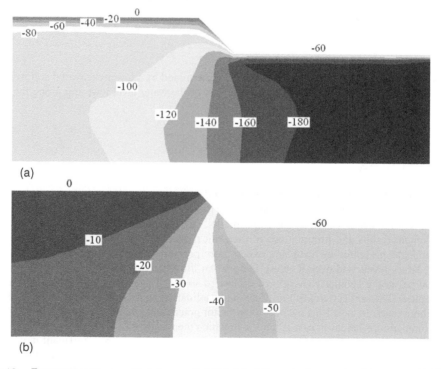

(a)

(b)

Figure 13.　Excess pore pressure (Δu) for case 3 (kPa): (a) at the end of excavation ($\Delta u_{min} = -196.8\,\text{kPa}$), (b) at the end of consolidation ($\Delta u_{min} = -60.0\,\text{kPa}$).

REFERENCES

Borges, J.L. 1995. Geosynthetic-reinforced embankments on soft soils. Analysis and design. PhD Thesis in Civil Engineering, Faculty of Engineering, University of Porto, Portugal (in Portuguese).

Borges, JL. 2004. Three-dimensional analysis of embankments on soft soils incorporating vertical drains by finite element method. Computers and Geotechnics, 31(8): 665–676, Elsevier.

Borges, J.L., Cardoso, A.S. 1998. Numerical simulation of the p-q-θ critical state model in embankments on soft soils. Geotecnia, Journal of the Portuguese Geotechnical Society, n° 84, pp. 39–63 (in Portuguese).

Borges, J.L., Cardoso, A.S. 2000. Numerical simulation of the consolidation processes in embankments on soft soils. Geotecnia, Journal of the Portuguese Geotechnical Society, n° 89, pp 57–75 (in Portuguese).

Borges, J.L, Cardoso, A.S. 2001. Structural behaviour and parametric study of reinforced embankments on soft clays. Computers and Geotechnics, 28(3): 209–233.

Borges, J.L., Cardoso, A.S. 2002. Overall stability of geosynthetic-reinforced embankments on soft soils. Geotextiles and Geomembranes, 20(6): 395–421.

Britto, A.M., Gunn, M.J. 1987. Critical soil mechanics via finite elements. Ellis Horwood Limited, England.

Costa, P.A. 2005. Braced excavations in soft clayey soils – Behavior analysis including the consolidation effects. MSc Thesis, Faculty of Engineering, University of Porto, Portugal (in Portuguese).

Domingues, T.S. 2006. Foundation reinforcement with stone columns in embankments on soft soils – Analysis and design. MSc Thesis, Faculty of Engineering, University of Porto, Portugal (in Portuguese).

Lambe, T.W., Whitman, R.V. 1969. Soil mechanics. John Wiley and Sons, Inc., New York.

Lewis, R.W., Schrefler, B.A. 1987. The finite element method in the deformation and consolidation of porous media. John Wiley and Sons, Inc., New York.

Applications of Computational Mechanics in Geotechnical Engineering – Sousa,
Fernandes, Vargas Jr & Azevedo (eds)
© 2007 Taylor & Francis Group, London, ISBN 978-0-415-43789-9

Parametric study of stone columns in embankments on soft soils by finite element method

T.S. Domingues
Department of Civil Engineering, Higher Institute of Engineering of Porto, Portugal

J.L. Borges & A.S. Cardoso
Department of Civil Engineering, Faculty of Engineering, University of Porto, Portugal

ABSTRACT: The use of stone columns in embankments on soft soils is one of the most adequate techniques when the main purpose is to increase overall stability and decrease and accelerate settlements. In the paper, a parametric study in an embankment on soft soils reinforced with stone columns is performed using a computer program based on the finite element method. The unit cell formulation is used, considering one column and its surrounding soil with confined axisymmetric behaviour. The computer program incorporates the Biot consolidation theory (coupled formulation of the flow and equilibrium equations) with constitutive relations simulated by the p-q-θ critical state model. The following parameters are analysed: the replacement area ratio and the deformability of the column material. In order to understand the influence of these parameters on the problem, settlements, horizontal displacements, excess pore pressures and improvement factor are analysed.

1 INTRODUCTION

In spite of all experience obtained over the last decades, designing embankments on soft clayey soils still raises several concerns related to the weak geotechnical characteristics of the soft soils: (i) the low shear strength significantly limits the embankment height that is possible to consider with adequate safety for short term stability and (ii) the high deformability and the low permeability determine large settlements that develop slowly as pore water flows and excess pore pressure dissipates (consolidation).

The use of stone columns in embankments on soft soils is one of the most adequate techniques when the main purpose is to increase overall stability and decrease and accelerate settlements.

In the paper, a numerical study is conducted to investigate the two major influence factors in this kind of works: the replacement area ratio (varying the column spacing with the same column diameter) and the deformability of the column material. Besides these two factors, other studies were performed by Domingues (2006), which revealed a minor influence on the problem: the soft soil thickness, the deformability and the friction angle of the embankment fill.

Basically, for the present applications, the finite element program, developed by Borges (1995), uses the following theoretical hypotheses: (a) coupled analysis, i.e. coupled formulation of the flow and equilibrium equations with soil constitutive relations formulated in effective stresses (Biot consolidation theory) (Borges, 1995; Borges and Cardoso, 2000; Lewis and Schrefler, 1987; Britto and Gunn, 1987), applied to all phases of the problem, both during the construction and in the post-construction period; (b) utilisation of the p-q-θ critical state model (Borges, 1995; Borges and Cardoso, 1998; Lewis and Schrefler, 1987), an associated plastic flow model, to simulate constitutive behaviour of the materials.

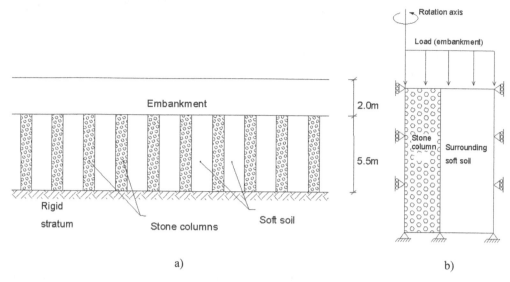

Figure 1.   (a) Embankment on a soft ground reinforced with stone columns, (b) unit cell.

In this study, the axisymmetric cylindrical unit cell is used, which consists of considering one column and its surrounding soil, with fixed horizontal displacements on the lateral boundary.

In the p-q-θ model – which is a extension of the Modified Cam-Clay model into the three-dimensional stress space using the Mohr-Coulomb failure criteria – the parameter that defines the slope of the critical state line, M, is not constant (which happens in the Modified Cam-Clay model), but depends on the angular stress invariant θ and effective friction angle, ϕ'. This is an important feature of the p-q-θ model because, as shown in triaxial and plane strain tests (Mita et al., 2004), the soil critical state depends on θ (Drucker-Prager is the failure criteria of the Modified Cam-Clay model and does not depend on θ).

2   NUMERICAL MODELLING

The problem concerns a 2 m height embankment on a soft ground reinforced with stone columns (Figure 1a). The soft ground is a 5.5 m thick normally consolidated clay lying on a rigid and impermeable stratum (lower boundary). The water level is at the ground surface. The column depth is 5.5 m, equal to the thickness of the soft ground. The diameter of the column and the diameter of the unit cell are 1.0 and 2.3 m, respectively. The unit cell diameter corresponds to a 2.2 m column spacing in a triangular grid or 2.03 m spacing in a square grid.

As said above, because a very large embankment width is considered, the axisymmetric cylindrical unit cell is used (Figure 1b). The finite element mesh for several stages is shown in Figure 2. Coupled analysis in the clay and drained analysis in the stone column and in the embankment fill are modelled. The six-noded triangular element is used. All six nodes of a drained element (at the vertices and at the middle of the edges) only have displacement degrees of freedom, whereas the three vertice nodes of a coupled element also have excess pore pressure degrees of freedom.

No horizontal displacement is allowed on the vertical boundaries of the mesh while the bottom boundary is completely fixed in both the vertical and horizontal direction. The ground surface is a drainage boundary (zero value of excess pore pressure) while the vertical and bottom boundaries of the mesh are assumed to be impermeable.

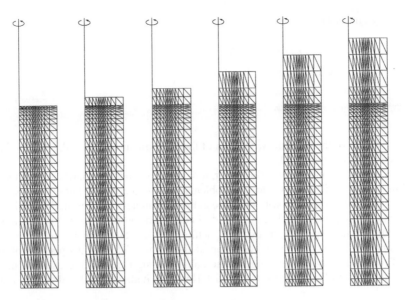

Figure 2.　Finite element mesh for several stages of the problem.

Table 1.　Mechanical and hydraulic properties of the materials.

	k	λ	Γ	ϕ' °	v'	N	γ kN/m³	$k_x = k_y$ m/s	OCR
Stone column	0.00275	0.011	1.8942	38	0.3	1.9	20	–	1
Soft soil	0.02	0.22	3.26	30	0.25	3.40	17	10^{-9}	1
Embankment fill	0.005	0.03	1.8	35	0.3	1.81733	20	–	1

Embankment construction is simulated by adding layers of elements at a uniform rate. The construction of the embankment is completed in an overall time of 28 days.

The constitutive behaviour of the materials is modelled by the p-q-θ critical state model (Borges, 1995; Borges and Cardoso, 1998; Lewis and Schrefler, 1987). The parameters used are indicated in Table 1 (λ, slope of normal consolidation line and critical state line; k, slope of swelling and recompression line; Γ, specific volume of soil on the critical state line at mean normal stress equal to 1 kPa; N, specific volume of normally consolidated soil at mean normal stress equal to 1 kPa). Table 1 also shows other geotechnical properties: γ, unit weight; v', Poisson's ratio for drained loading; ϕ', angle of friction defined in effective terms; k_h and k_v, coefficients of permeability in horizontal and vertical directions; K_0, at rest earth pressure coefficient; OCR, over-consolidation ratio. All values of these parameters were adopted taking into account typical values reported in bibliography for this kind of soils (Lambe and Whitman, 1969; Borges, 1995).

In order to consider the increase of the initial effective horizontal stresses in the soft ground determined by the installation of the stone columns, the value of 0.7 is adopted for the at-rest earth pressure coefficient after the installation of the columns, a higher value than the one estimated by the Jaky's equation ($K_0 = 1 - sen\phi'$) for normally consolidated soils. It should be noted that the choice of this coefficient is controversial among authors. Some authors consider a conservative perspective by adopting the same value before and after the installation of the columns (Besançon et al, 1984; Nayak, 1982), whereas others adopt the hydrostatic value of 1 (Priebe, 1995; Goughnour and Bayuk, 1979). Therefore, in this study, an intermediate value of these two values was adopted.

3 ANALYSIS OF RESULTS

3.1 Influence of the replacement area ratio

The replacement area ratio, r, is defined as follows:

$$r = \frac{A_c}{A} = \frac{A_c}{A_c + A_s} \tag{1}$$

where A_c is the area of the column, A_s the area of the soft soil surrounding the column in the unit cell and A the area of the unit cell.

In order to analyse the influence of this factor, four cases are considered: (i) case C0, the problem described in section 2; (ii) cases C1 to C3, problems similar to case C0, but with different values of r (varying the column spacing with the same column diameter), as indicated in Table 2.

Settlements on the ground surface at the end of consolidation for cases C0 to C3 are shown in Figure 3. These results show that, as expected, settlements increase with $1/r$, as the effectiveness of the reinforcement decays with the reduction of the number of columns per unit of area. The variation of the settlement on the centre of the column is significant, increasing from 20 cm to 32 cm (1.6 times) when $1/r$ varies from 3.3 to 10. However, it should be noted that the normalised

Table 2. Replacement area ratio and other related factors for cases C0 to C3.

Case	r	$1/r$	b (m)	T (m)	S (m)
C0	0,19	5,3	1,15	2.20	2.03
C1	0,10	10,0	1.58	3.01	2.80
C2	0,15	6.7	1.29	2.46	2.28
C3	0,30	3.3	0.91	1.73	1.61

r – replacement area ratio; b – distance between the axis of the column and the lateral boundary of the cell; T and S – column spacing in triangular and square grids respectively.

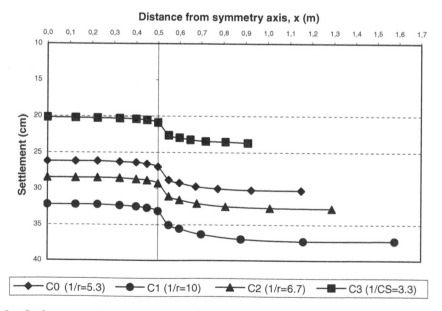

Figure 3. Settlements on the ground surface at the end of consolidation for cases C0 to C3.

differential settlement (difference between the average settlements of the soft soil and of the column, divided by the average settlement of the unit cell) practically does not change with $1/r$. Its value is 12%, approximately, for all four cases.

Complementing these results, the variation of the improvement factor, n (ratio between the average settlements of unreinforced and reinforced problems), with $1/r$ is shown in Figure 4. The results plotted in this figure clearly show that the more the area of the unit cell reduces the more the improvement factor increases.

Since the stone column works not only as reinforcement but also as a drain, the increase of the replacement area ratio (i.e. the increase of the number of columns per unit of area) also increases the acceleration of the consolidation. This is shown in Figure 5 where the variation in time of the average settlement on the ground surface is plotted for cases C0 to C3. The results show, for instance, that the average degree of consolidation at 13 weeks varies between 59% (case C1, $1/r = 10$) and 97% (case C3, $1/r = 3.3$). The results also show that the settlement at the end of construction (4 weeks) is an expressive portion of the total final settlement, varying from 24% ($1/r = 10$) to 64% ($1/r = 3.3$).

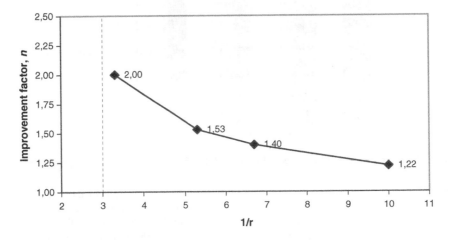

Figure 4.   Variation of the improvement factor with $1/r$.

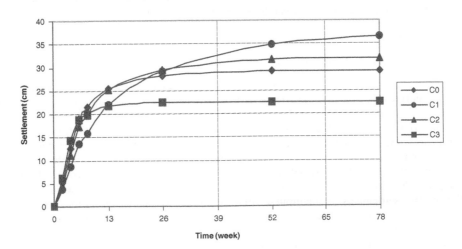

Figure 5.   Average settlement on the ground surface for cases C0 to C3.

285

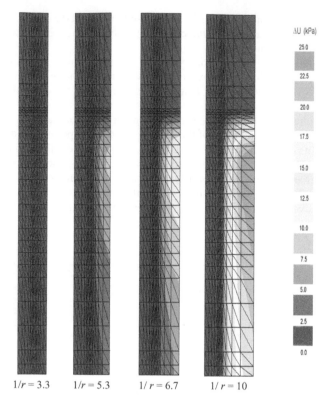

				ΔU (kPa)
$1/r = 3.3$	$1/r = 5.3$	$1/r = 6.7$	$1/r = 10$

Figure 6.   Excess pore pressure at 13 weeks for cases C0 to C3.

Figure 6 shows distributions of the excess pore pressure at 13 weeks for cases C0 to C3. The acceleration of the consolidation with the replacement area ratio is clearly observed in this figure, the highest values of excess pore pressure corresponding to the highest value of $1/r$.

Horizontal displacements across the column-soil interface at the end of consolidation for cases C0 to C3 are shown in Figure 7. These results show that: (i) the maximum horizontal displacement is similar for the four cases, occurring at about a column diameter depth; (ii) horizontal displacement on the interface tends to spread in depth as $1/r$ increases, which is due to a lower horizontal confinement.

3.2   Influence of the deformability of the column material

In the parametric study presented in this section, four cases are considered by varying the parameter (λ_{col}) that defines the slope of the normal consolidation line of the column material, as shown in Table 3 (this parameter is related to the traditional compression index, C_c, by the equation $\lambda = C_c/\ln 10$). Case D0 is the same as case C0 presented in the previous section. For the column material, the other parameters of the p-q-θ model are adjusted as follows: (i) a constant ratio between λ and k ($\lambda/k = 4$) is considered; (ii) the same values for N and ϕ' are adopted in all cases; (iii) values of the parameter Γ are obtained considering that the ellipse of the p-q-θ model contains the origin of the p and q axes. All other parameters of the problem are the same as in case D0.

The average settlements on the ground surface at the end of consolidation for cases D0 to D3 are shown in Figure 8. These results clearly show a high variation of the average settlement with m (ratio between λ_{soil} and λ_{col}), which confirms the importance in design of λ_{col} (or C_c, which is equivalent). As expected, the more m increases, the more the average settlement reduces. In Figure 8 is also plotted a logarithm curve that fits the numerical results with good agreement.

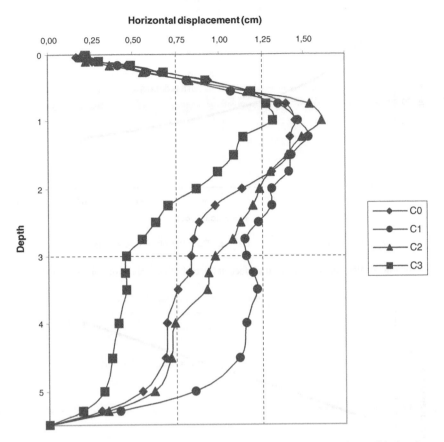

Horizontal displacement (cm)

Depth

Figure 7. Horizontal displacements across the column-soil interface at the end of consolidation for cases C0 to C3.

Table 3. Values of λ_{col} for cases D0 to D3

Case	D0	D1	D2	D3
λ_{col}	0.011	0.022	0.0055	0.0022
$m = \lambda_{soil}/\lambda_{col}$	20	10	40	100

Corroborating these results, the variation of the improvement factor (n) with m is shown in Figure 9; in this case, the results are adequately fitted by a straight line.

The radial variations of the settlement on the ground surface at the end of consolidation for cases D0 to D3 are shown in Figure 10. These results show that: (i) the more the stiffness of the column increases, the more the settlements on the ground surface reduce, not only on the column surface but also on the soft soil surface; this is due to transfer stress from the soft soil to the column, by arching effect, that develops both in the embankment fill and in the foundation, as shown in Figure 11 for case D0 (Domingues, 2006; Domingues et al., 2006); this figure shows the principal effective stresses at the end of consolidation; (ii) however, mainly because the average settlement reduces, the normalized differential settlement increases with the column stiffness (for instance, for $m = 20$ and $m = 100$, the normalized differential settlement is about 14% and 57% respectively); (iii) on the column surface, for all four cases, the settlement practically does not change with the

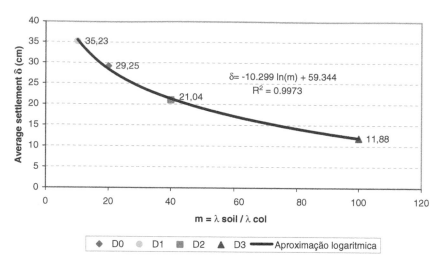

Figure 8. Average settlement on the ground surface at the end of consolidation for cases D0 to D3.

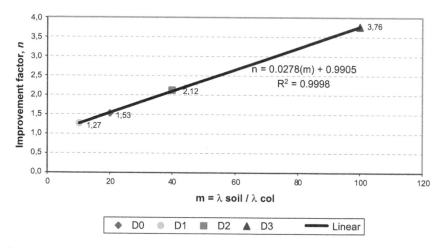

Figure 9. Variation of the improvement factor, n, with m for cases D0 to D3.

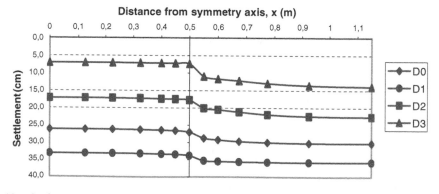

Figure 10. Settlements on the ground surface at the end of consolidation for cases D0 to D3.

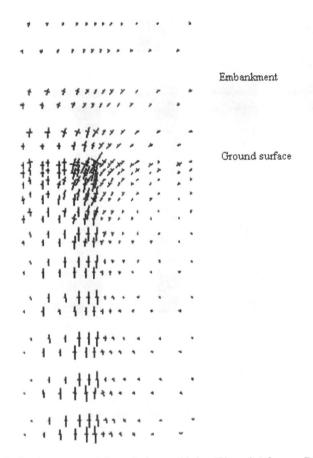

Embankment

Ground surface

Figure 11.   Principal effective stresses at the end of consolidation (78 weeks) for case D0.

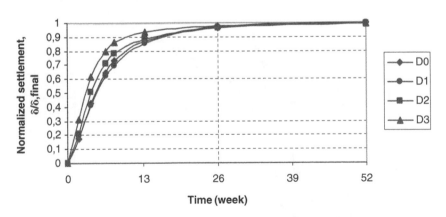

Figure 12.   Average degree of consolidation on the ground surface for cases D0 to D3.

distance from the symmetry axis, this not occurring on the surrounding soft soil surface, mainly for the cases with high values of the column stiffness.

The variation in time of the average degree of consolidation (ratio between average values of settlement, δ, and total final settlement, δ_{max}, on the ground surface) is plotted in Figure 12 for cases D0 to D3. As expected, the results show that the velocity of consolidation increases with

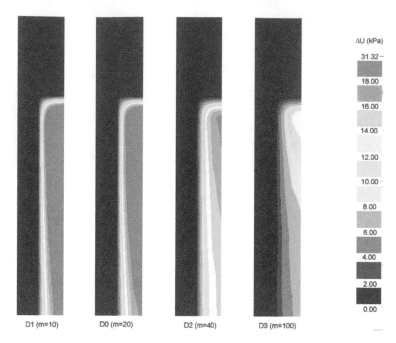

Figure 13. Excess pore pressure one week after the end of construction for cases D0 to D3.

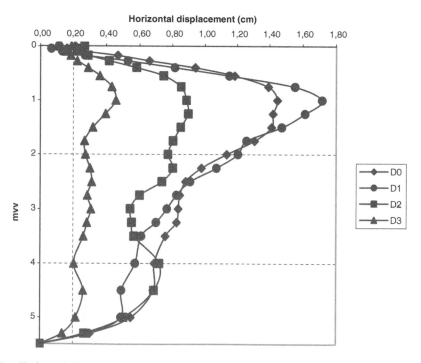

Figure 14. Horizontal displacement across the column-soil interface at the end of consolidation for cases D0 to D3.

the stiffness of the column material. For instance, 90% for δ/δ_{max} is reached in 16 and 10 weeks respectively for $m = 20$ and $m = 100$ (time reduction of 40%).

In order to better illustrate this effect, distributions of the excess pore pressure one week after the end of construction for the four cases are shown in Figure 13, where higher dissipation of excess pore pressure is clearly observed for the cases with higher values of m.

Horizontal displacements across the column-soil interface at the end of consolidation for cases D0 to D3 are shown in Figure 14. As expected, these results show that horizontal displacements (as well as settlements, as described above) significantly reduce with the column stiffness.

4 CONCLUSIONS

A numerical study was conducted to investigate the two major influence factors in embankments on soft soils reinforced with stone columns: the replacement area ratio and the deformability of the column material. The confined axisymmetric cylindrical unit cell was used. The analyses were performed by a finite element program that incorporates coupled analysis with constitutive relations simulated by the p-q-θ model. The interpretation of the results was presented by analysing settlements, horizontal displacements, excess pore pressures, improvement factor and arching effect. As expected, the following overall conclusions were observed: both the increase of the replacement area ratio and the increase of the stiffness of the column material significantly determine that: (i) settlements and horizontal displacements reduce; (ii) improvement factor increases; (iii) consolidation accelerates.

REFERENCES

Besançon, G., Iorio, J.P., Soyez, B. 1984. Analyse des paramètres de calcul intervenant dans le dimensionnement des colonnes ballastées. Renforcement en place des sols et des roches, Paris.

Borges, J.L. 1995. Geosynthetic-reinforced embankments on soft soils. Analysis and design. PhD Thesis in Civil Engineering, Faculty of Engineering, University of Porto, Portugal (in Portuguese).

Borges, J.L., Cardoso, A.S. 1998. Numerical simulation of the p-q-θ critical state model in embankments on soft soils. R. Geotecnia, n° 84, pp. 39–63 (in Portuguese).

Borges, J.L., Cardoso, A.S. 2000. Numerical simulation of the consolidation processes in embankments on soft soils. R. Geotecnia, n° 89, pp 57–75 (in Portuguese).

Britto, A.M., Gunn, M.J. 1987. Critical soil mechanics via finite elements. Ellis Horwood Limited, England.

Domingues, T.S. 2006. Foundation reinforcement with stone columns in embankments on soft soils – Analysis and design. MSc Thesis, Faculty of Engineering, University of Porto, Portugal (in Portuguese).

Domingues, T.S., Borges, J.L., Cardoso, A.S. 2006. Embankments on soft soil reinforced with stone columns – Analysis by finite element method. in 10° Congresso Nacional de Geotecnia, Lisboa, pp.1249–1258 (in Portuguese).

Goughnour, R.R., Bayuk, A.A. 1979. A field study of long-term settlement of loads supported by stone columns in soft ground. in Proc. Int. Conf. Soil Reinfor., Paris, Vol. 1, pp. 279–286.

Lambe, T.W., Whitman, R.V. 1969. Soil mechanics. John Wiley and Sons, Inc., New York.

Lewis, R.W., Schrefler, B.A. 1987. The finite element method in the deformation and consolidation of porous media. John Wiley and Sons, Inc., New York.

Mita, K. A., Dasari, G. R., Lo, K. W. 2004. Performance of a three-dimensional Hvorslev-Modified Cam Clay model for overconsolidated clay. International Journal of Geomechanics, ASCE, 4(4): 296–309.

Nayak, N.V. 1982. Recent innovations on ground improvement by stone columns. in Symposiom on recent developments in ground improvement techniques, Bangkok.

Priebe H.J. 1995. The design of vibro replacement. Ground Engineering, 28(10): 31–37.

Applications of Computational Mechanics in Geotechnical Engineering – Sousa,
Fernandes, Vargas Jr & Azevedo (eds)
© 2007 Taylor & Francis Group, London, ISBN 978-0-415-43789-9

Long term behaviour of excavations in soft clays: a numerical study on the effect of the wall embedded depth

P. Alves Costa, J.L. Borges & M. Matos Fernandes
Faculty of Engineering of University of Porto, Porto, Portugal

ABSTRACT: The paper presents a study on the long term behaviour of deep excavations in soft clays with emphasis on the effect of the wall embedded depth. A numerical study was carried out considering an excavation in soft silty clays underlain by stiff impermeable clays, supported by an impermeable braced wall whose embedded depth has distinct values, including the one corresponding to the wall tip sealed in the bedrock. By means of coupled finite element analyses, and assuming that time construction is so short that undrained conditions prevail, a comparison is made between the behaviour at the end of the construction and at the end of the consolidation. The results show that the embedded wall depth has a relevant influence on the evolution of the wall and ground displacements and on the structural stresses during the consolidation. Some conclusions are extracted.

1 INTRODUCTION

When a braced excavation is undertaken, the variation of the state of stress in the adjacent ground basically consists of a decrease of the total mean stress and an increase of the deviatoric shear stress. In saturated normally consolidated or lightly overconsolidated clays, the decrease of the total mean stress induces negative excess pore pressures whereas the increase of shear stress gives rise to positive excess pore pressures. Therefore, the excess pore pressures at the end of construction at a given point of the ground in the vicinity of the excavation may be positive or negative, depending upon the magnitude of the two above mentioned contrary effects.

However, field measurements of a number of excavations suggest that, in general, excess pore pressures are usually negative at the end of construction (Lambe and Turner, 1970; DiBiagio and Roti, 1972; Clough and Reed, 1984; Finno et al., 1989).

Thus, after construction there are pore pressure gradients in the ground that determine a consolidation process. This process is clearly dependent on both the magnitude and distribution of excess pore pressure at the end of the construction as well as the long term equilibrium conditions. These conditions may correspond to a hydrostatic pore pressure distribution or to a steady state flow, and are mainly affected by the wall embedment depth and by the permeability of the soil below the bottom of the wall (Alves Costa, 2005).

The wall embedded depth is one of the parameters that affect the excavation behaviour due to several reasons: (i) during construction, the changes on the states of stress and strain in the adjacent ground are controlled by a number of factors, among which the wall embedded length is one of the most important; (ii) the excess pore pressures generated during the construction process are a result of the stress and strain changes; (iii) at a long term, the wall embedded depth may or may not correspond to an impermeable boundary between the supported ground and the ground below the excavation.

The effect of the wall embedment on the excavation behaviour during construction is well characterized in the bibliography (Clough and Schmidt, 1984; Matos Fernandes et al., 1997; Matos Fernandes et al., 1998; Bose & Som, 1998). However, the evolution of the performance of excavations at long term is one of the few subjects related with this topic for which a clear pattern is not established yet.

In the present paper some features about the long term behaviour of an excavation are presented. The study is based on the numerical simulation of a braced excavation from the beginning of the construction period until the end of the consolidation process. Coupled analyses are performed – with a computer program based on the finite element method, developed by Borges (1995) – and the consolidation effects on the results after the construction period are presented and discussed. The study clearly illustrates the effect of the wall embedment depth on the excavation performance both during construction and consolidation.

2   DESCRIPTION OF THE PROBLEM

The present application refers to a permanent excavation performed on a soft clayey soil. Its maximum depth and width are 10.8 m and 20.0 m, respectively. It was assumed that the longitudinal development of the excavation is considerably superior to its transversal dimension, which permits to premise a plane strain analysis. During the construction stages the retaining structure consists of a concrete diaphragm wall braced at three levels. The wall embedment depth was considered distinct for the three analyses carried out, as shown in Figure 1.

Concerning to the ground, it was assumed that it is composed by: (i) a superficial rubber fill, 2 m thick; (ii) a layer of soft silty clay, extending down to a depth of about 25.0 m; (iii) a very stiff overconsolidated clay, with very low permeability. The contrast between the permeability of the soft silty clay and the one of stiff clay permits to assume the top of the latter as an impermeable boundary for the analysis of the consolidation process.

The safety factor in relation to the basal heave, calculated by the method of Terzaghi (1943) and not considering the embedded height of the wall, takes a value around 1.2.

When the maximum depth of the excavation is reached a basal slab is concreted and the struts are progressively removed in a bottom-up process of replacement of the temporary support system by the permanent internal structure, as described on the right side of Figure 1.

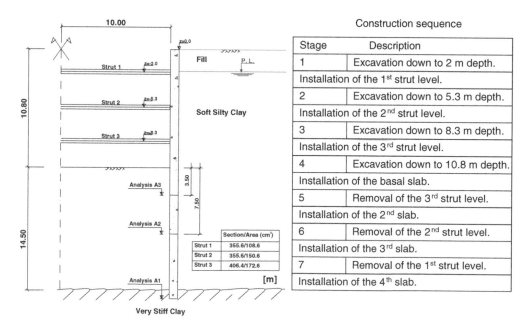

Figure 1.   Characteristics of the excavation and construction sequence.

As mentioned above, three analyses were conducted, corresponding to three distinct depths of wall embedment. In analysis A1 the wall is extended down to the layer of stiff and impermeable clay, constituting an impermeable barrier between the supported ground and the soil below the excavation. In analyses A2 and A3 this condition is not verified, since the wall tip rests on the soft silty clay, permitting the establishment of a steady state flow for the excavation at the end of the consolidation (see Figure 1).

3 NUMERICAL SIMULATION

The numerical analyses performed comprise all the stages of the construction process (see Figure 1) and were extended in time until the end of the consolidation process. During the construction period, changes of the stress state in the adjacent ground generate excess pore-pressures that will be dissipated in the time. Consolidation can obviously develop during the construction phase, but its magnitude can be considered negligible if the ground has low permeability and the construction occurs rapidly (Osaimi & Clough, 1979; Alves Costa, 2005).

Therefore, for the construction stages, the study comprised an undrained analysis, by performing a "simplified coupled analysis" where the construction time is considered approximately zero and the boundaries are defined as impermeable. After that, during the post-construction stage, the time factor was considered and consolidation was simulated. For all the analyses it was admitted that the water level remains invariable in time behind the wall, at an elevation corresponding to the upper layer of fill (see Figure 1).

With regard to the particular aspects of the finite element simulation, two types of the six-noded triangular element are considered: (i) with 12 displacement degrees of freedom (at the vertices and middle of the edges) plus 3 excess pore pressure degrees of freedom (at the vertices), for the soil elements where coupled analysis (consolidation) is considered (clay elements); (ii) with 12 displacement degrees of freedom (at the vertices and at middle of the edges), for the granular fill elements, where drained condition is assumed, and for the wall elements. The interface between the wall and the ground is simulated by mean of six-noded joint elements. The struts are modelled by three-noded bar elements.

The constitutive relations of the soils are simulated by using the p-q-θ critical state model (Lewis and Schrefler, 1987; Britto and Gunn, 1987; Borges, 1995; Borges and Cardoso, 1998), whose values are indicated in Table 1. The structural members as the braces, the wall and the slabs are simulated by using an elastic and linear constitutive relation.

Bearing in mind that the layer of stiff clay presents a great contrast of mechanical and hydraulic properties in relation to the remaining layers of the ground (more rigid, resistant and impermeable), it was simply simulated by considering its upper plane as a fixed and impermeable boundary of the finite element mesh. The lateral boundaries of the mesh as well as the wall faces were considered as impermeable. The bottom of the excavation was defined as a drainage boundary having null pore pressures; the same applies to the top plane of the soft clay outside of the excavation.

Table 1. Geotechnical properties of the soil.

	$\gamma(kN/m^3)$	K_0	OCR	$k_h = k_v$ (m/s)	$\phi'(°)$	v	λ	κ	N	Γ
Soft Clay	16	0.50	1.1	10^{-9}	26	0.25	0.18	0.025	3.16	3.05
Fill	20	0.43	1.1	∞	35	0.30	0.025	0.005	1.817	1.80

Notes: γ, unit weight; K_0, at-rest earth pressure coefficient; OCR, overconsolidation ratio; k_h and k_v, coefficients of permeability in the horizontal and vertical directions; ϕ', effective angle of shearing resistance; v', Poisson's ratio for drained loading; λ, slope of normal consolidation line and critical state line; k, slope of swelling and recompression line; Γ, specific volume of soil on the critical state line at mean normal stress equal to 1 kPa; N, specific volume of normally consolidated soil at mean normal stress equal to 1 kPa.

4 COMPARATIVE ANALYSES OF RESULTS

4.1 *Stress levels and excess pore pressures in the soil*

Figure 2 shows the distributions of the stress levels for analyses A1 to A3, at the end of construction. The stress level, *SL*, is the ratio between the acting and the critical state values of the deviatoric stress for the mobilized effective mean normal stress; therefore, in normally consolidated soils the stress level varies from zero to one, the latter being the critical state level.

As expected, stress level values increase from analysis A1 to A3. In analysis A1 the critical state is reached mainly in a region of the supported ground, while in analyses A2 and A3 the critical state comprises a larger region of the supported soil spreading up to the ground surface. In the case of analyse A3 the critical state is also reached in a significant region in front of the wall, involving its tip.

The distribution of excess pore pressures at the end of construction is depicted in Figure 3, for the three analyses. In the present paper excess pore pressures are measured in relation to the initial hydrostatic pore pressure distribution.

By comparing the results some differences can be detected. In the analyses where the wall tip does not reach the stiff layer the distribution of excess pore pressures is more irregular. This fact can be explained by two reasons: i) at the vertical plane containing the wall and below its tip, compatibility between excess pore pressure at both sides must naturally occur; ii) when the embedded depth of the wall is small, the critical state is reached in a extensive region which affects the excess pore pressure generation.

In Figure 4, the difference between the pore pressure distributions at the end of the consolidation process and before the excavation execution is showed for analyses A2 and A3. For analyses A1 this result is not presented because it would be redundant since the final distribution of pore pressures

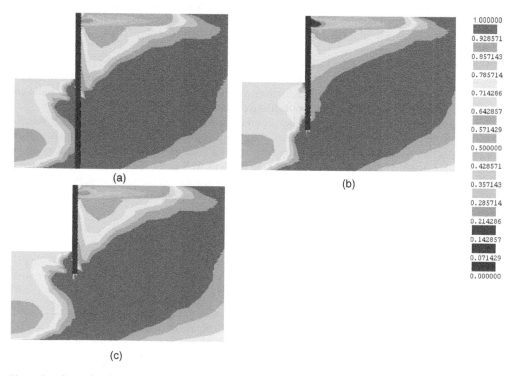

(a)

(b)

(c)

Figure 2. Stress levels mobilized at the end of construction: (a) analysis A1; (b) analysis A2; (c) analysis A3.

is hydrostatic. As expected in analyses A2 and A3 a steady flow is reached at the end of the consolidation, and pore pressures behind the wall are lower than the hydrostatic pore pressures, the opposite occurring in front of the wall.

In the analyses A2 and A3, as mentioned above, the pore pressure distribution at the end of consolidation does not correspond to the hydrostatic condition. In fact, in the supported ground near to the wall, the seepage forces are directed downwards. This fact justifies that the vertical

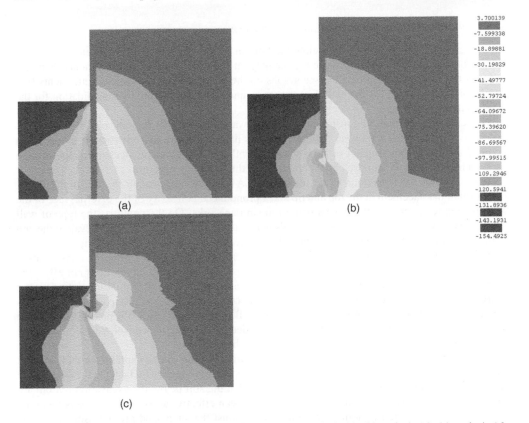

Figure 3.   Excess pore pressure at the end of construction: (a) analysis A1; (b) analysis A2; (c) analysis A3.

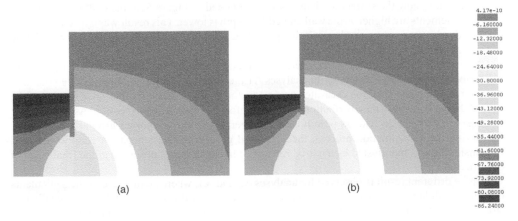

Figure 4.   Differential between initial pore pressures and pore pressures at the steady flow condition: (a) analysis A2; (b) analysis A3.

effective stresses in the ground behind the wall increase in relation to the values observed at the end of construction.

For analysis A1, the evolution of the effective stress state during consolidation is quite different from the analyses presented above. The excess pore pressures generated during the construction stages are negative, so their dissipation is accompanied by the decrease of the effective mean stress.

4.2 Lateral wall displacements and surface settlements

As expected, the depth of the wall below the bottom of the excavation plays a relevant role on the performance of the retaining structure, namely on the wall and ground movements. Figure 5 presents the lateral wall deflections (see graphs on the left side) and the surface settlements (see graphs on the right side) at the end of construction and its evolution due to consolidation for the three analyses performed.

As mentioned, during the construction stages it was assumed that the consolidation does not occur, so the differences between the computed results for the end of construction presented in Figure 5 are due to the influence of the embedded depth of the wall on the mechanical equilibrium reached by the system composed by the ground and the retaining structure.

At short term, the behaviour found for analyses A1 and A2 concerning the wall movements, with a small displacement at the top and tip and a pronounced convexity of the external face of the wall, is in agreement with the typical pattern in similar works. On the contrary, the type of wall displacement observed for analyses A3 reveals that marginal stability conditions prevail at the end of the excavation, where the maximum lateral displacement is reached by the wall tip.

With regard to the evolution of the lateral displacements of the wall during the consolidation period some comments can be done. Firstly, for analysis A1 the consolidation does not affect the lateral wall displacements. On the contrary, for analyses where the wall tip is not resting on the bedrock, the lateral wall displacements experience an evolution due to the dissipation of the excess pore pressure and the establishment of a steady state flow. For these cases it is possible to observe that the displacements below the excavation bottom decrease during the consolidation process and the deflection above this level remains practically unchanged.

This recover of the displacements below the excavation bottom can be explained taking account the excess pore pressures originated during the construction stages. In fact, at the end of the construction period the excess pore pressure installed at the soil below the excavation bottom are negative, so its dissipation implies a decrease on the mean effective stress and an expansion of the soil under the excavation bottom pushing the wall against the supported ground. Similar results were also observed in the studies developed by other authors (Martins, 1993; Ou & Lai, 1994).

Commenting now the surface settlements still contained in Figure 5, at the end of construction period settlements are higher as the wall embedded depth is lower. This result was expected because the wall displacements (see results on the left part of the same figure) increase on the inverse relation of the wall height of embedment.

By comparing the results of the long term settlements, some differences determined by the consolidation process can be observed. In analyses A1 the wall constitutes an impermeable boundary between the supported and the excavated ground, so the pore pressure distribution at the end of the consolidation is equal to the distribution prevalent before the excavation execution. As it is possible to observe in Figure 3a, excess pore pressures generated in the supported ground are negative. So, during the consolidation process, the supported ground suffers a negative volumetric strain (expansion) that is manifested by a partial recover of the settlements observed at the final stage of construction.

A quite different result is observed for analyses A2 and A3, where an increase of the settlements during consolidation can be observed. This result is induced by the increase of the vertical effective stresses on the ground due to the establishment of a steady state flow (see Figure 4). It is also possible to conclude that the increase on the magnitude of the settlements is so much intense as the reduced is the embedded depth of the wall.

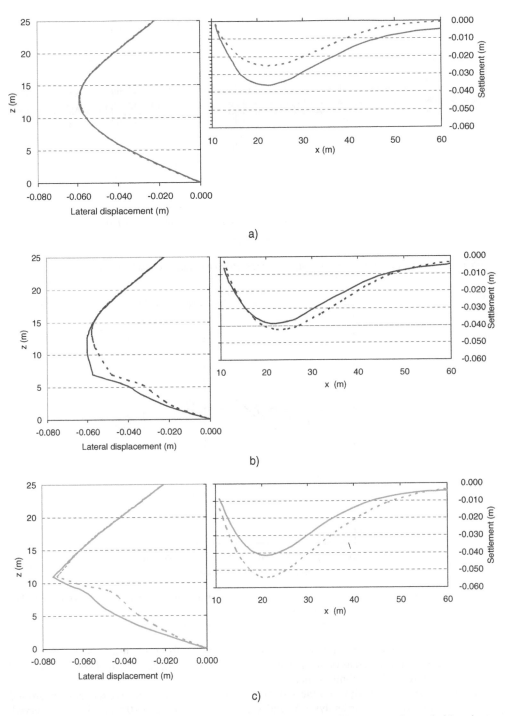

Figure 5.   Lateral wall displacements and surface settlements at the end of the construction period (continuous lines) and at the end of consolidation (dashed lines): (a) analysis A1; (b) analysis A2; (c) analysis A3.

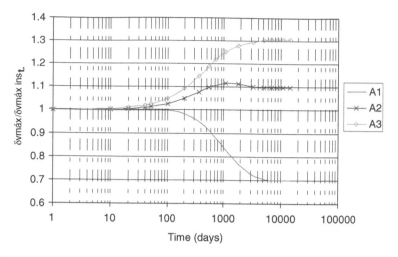

Figure 6.   Temporal evolution of the maximum surface settlement during consolidation.

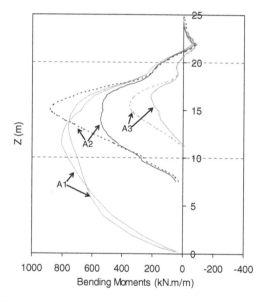

Figure 7.   Computed wall bending moments at the end of construction (continuous lines) and at the end of consolidation process (dashed lines)for the three analyses.

In Figure 6 the evolution in time of the maximum settlement from the three analyses is plotted; the settlements are presented as divided by the value observed at the last stage of construction and the time is plotted in a logarithmic scale.

As can be observed, in analysis A1 the maximum displacement experiences a reduction of around 30%, the contrary occurring in analyses A2 and A3, where an increase of 10% and 30% is observed, respectively. A curious aspect that is patent in Figure 6 is the fact that the evolution of the maximum settlements starts sooner in analyses A2 and A3 than in analysis A1. According to Henkel (1970), this result is expected in the cases where flow between the supported ground and the ground below the excavation bottom is allowed, because the hydraulic gradients between these two regions are very high, inducing an acceleration of the consolidation process.

300

4.3 Bending moments on the wall

Computed wall bending moments at the end of the construction and at the end of the consolidation are depicted in Figure 7.

As shown, in analysis A1 the magnitude of the maximum bending moment is not strongly affected by the consolidation. This result was expected taking into account the results presented in Figure 6, where the evolution of the wall displacements was shown. For analyses A2 and A3, a pronounced increase of the maximum positive bending moment is observed, mainly in analysis A2. This result is due to the increase of the convexity of the wall generated by the volumetric expansion of the soil below the bottom of the excavation.

5 CONCLUSIONS

The long-term behaviour of braced excavations in soft clayey soils was analysed with the help of a finite element program which incorporates the Biot's consolidation theory and soil constitutive relations simulated by the p-q-θ critical state model.

With the purpose of analysing the influence of the retaining wall embedded length on the long-term behaviour of the soil and the structure, the analyses performed comprised all the stages of the construction as well as the consolidation process.

Based on the results of this study several conclusions can be pointed out:

- the performance of excavations on soft clayey soils is highly time dependent;
- during the construction stages, negative excess pore pressure are generated, both in the supported ground and in the soil below the bottom of the excavation (with more extent for the latter);
- in the cases with the retaining wall constituting an impermeable boundary between the supported ground and the ground underneath the excavation, during consolidation the soil experiences a negative volumetric strain (expansion), due to the dissipation of negative excess pore pressure, accompanied by a decrease of the mean effective stress; this expansion manifests in the supported ground by a partial recover of the surface settlements that had occurred during the construction period;
- on the contrary, when a steady-state flow beneath the wall tip is allowed at the end of the consolidation, the surface settlements will increase in relation to the ones at the end of the construction, since the seepage forces in the supported ground are directed downwards, inducing an increase of the vertical effective stresses;
- on the other hand, the stresses in the structural elements of the retaining structure can be significantly affected by the consolidation process; in the cases where the steady-state flow is allowed, the maximum positive wall bending moments may experience a considerable rise in time, due to the increase of the convexity of the wall face generated by the volumetric expansion of the soil below the bottom of the excavation.

Taking into account the global results provided by this study, one can conclude that the embedment height of the retaining wall revealed to be a parameter of major importance on the performance of excavations in soft clayey soils, not just at short-term – which is a well-known fact – but also at long-term, influencing the pattern of behaviour of the excavation during the process of consolidation.

REFERENCES

Alves Costa, P. 2005. *Braced Excavations in Soft Clay. Behavior Analysis Including the Consolidation Effects.* MSc Thesis. Faculty of Engineering of University of Porto., Portugal (in Portuguese).

Borges, J. L. 1995. *Geosynthetic reinforced embankments on soft soils – Analysis and design.* PhD Thesis in Civil Engineering. Faculty of Engineering of University of Porto., Portugal, (in Portuguese).

Borges, J. L. and Cardoso A. S. 1998. Numerical simulation of the p-q-θ critical state model in embankments on soft soils. *R. Geotecnia, n° 84*: 39–63 (in Portuguese).

Bose, S. K. and Som, N. N. 1998. Parametric study of a braced cut by finite element method. Elsevier Science Publishers (eds.).*Computers and Geotechnics vol. 22 no2*: 91–107.

Britto, A. M. and Gunn, M. J. 1987. *Critical State Soil Mechanics via Finite Elements*. England: Ellis Horwood Limited.

Clough, G. W. and Reed, M. W. 1984. Measured behaviour of braced wall in very soft clay. *Journal Geotech. Eng. Div. Vol. 110, N° 1*: 1–19.

Clough, G. W. and Schmidt, B. 1981. Design and performance of excavations and tunnels in soft clay. Brand and Brenner (eds) *Soft Clay Engineering, Chapter 8*. Amsterdam: Elsevier Scientific Publishing Company.

DiBiagio, E. and Roti, J. A. 1972. Earth pressure measurements on a braced slurry-trench wall in soft clay. *Proc. 5th Europ. Conf. Soil Mech. Found. Eng. Vol 1*: 473–483.

Finno, R. J., Atmatzidis, D. K. and Perkins, S. B. 1989. Observed performance of a deep excavation in clay. *Journal Geotech. Eng. Div. Vol. 115:* 1045–1060.

Henkel, D. J. 1970. Geotechnical considerations of lateral stresses. *Proc. Specialty Conf. On Lateral Stresses in Ground and Design of Earth Retaining Structures*: 1–49, Cornell Univ., Ithaca, New York. *Journal Geothecnical Division, GT4*: 481–498.

Lambe, T. W. and Turner, E. K. 1970. Braced excavations. *Proc. Spec. Conf. On Lateral Stresses in Ground and Design of Earth Retaining Structures*: 149–218. Cornell Univ., Ithaca, New York.

Lewis, R. W. and Screfler, B. A. 1987. The finite element method in the deformation and consolidation of porous media. Chichester: John Wiley and Sons.

Martins, F. F. (1993). *Elasto-plastic consolidation – Program and applications*. PhD Thesis. Univerty of Minho. Potugal. (in Portuguese).

Matos Fernandes, M., Cardoso, A. and Fortunato, E. 1998. The wall conditions below the base of the cut and the movements induced by an excavation in clay. *Proc. of the World Tunnel Congress '98 on Tunnels and Metropolises: 399–404*. São Paulo. Brazil. Rotterdam: Balkema.

Matos Fernandes, M., Cardoso, A. S. and Fortunato, E. 1997. A reappraisal of arching around braced excavations in soft ground. Azevedo et al. (eds.) *Proc. Applications of Computational Mechanics in Geotechnical Engineering*: 333–350. Rotterdam: Balkema.

Osaimi, A. E. and Clough, G. W. 1979. Pore-pressure dissipation during excavation.

Ou, C. Y. and Lai, C. H. 1994. Finite-element analysis of deep excavation in layered sandy and clayey soil Deposits. *Canadian Geotechnical Journal Vol. 31*: 204–214.

Terzaghi, K. 1943. *Theoretical Soil Mechanics*. New York: John Wiley and Sons.

Foundations

Applications of Computational Mechanics in Geotechnical Engineering – Sousa,
Fernandes, Vargas Jr & Azevedo (eds)
© 2007 Taylor & Francis Group, London, ISBN 978-0-415-43789-9

Geotechnical and structural problems associated with bridge approach slabs

C. Sagaseta
Universidad de Cantabria. Santander, Spain

J.A. Pacheco
SENER. Madrid, Spain

L. Albajar
Universidad Politécnica de Madrid, Spain

ABSTRACT: The work deals with reinforced concrete approach slabs, used to absorb differential settlements between bridge abutments and the embankments. For excessive settlements the slab adopts a slope incompatible with road use, and sometimes, accompanied by breaking or degradation of the concrete slab. An analysis of the most common causes for embankment settlement is presented. Then, the conditions of support of the slab on the embankment are examined, with the aid of a numerical analysis. Then, the dynamic effects of the traffic loads are evaluated. As a result, a laboratory test work is carried out, at 1/2 scale, of the dynamic and fatigue failure of the slabs, for several support conditions and loads. A structural model of the slab taking into account the different support conditions, stiffness changes in the slab due to cracking, dynamic effects of moving traffic loads and fatigue damage due to high load frequency.

1 INTRODUCTION

Improvement in road quality has led to optimized geometry and continuity, increasing the number of skew and curved bridges.

Service conditions such as deflections and vibrations are carefully controlled in bridges, increasing safety and comfort of passengers. Nevertheless two points of discontinuity remain: the deck joints and the transition between bridge deck and the embankment (commonly solved with an approach slab).

The work presented herein deals with these reinforced concrete approach slabs, linked to the abutment at one side and resting on the embankment (Figure 1). Differential movements between the bridge abutment and the embankment cause a vertical and/or horizontal discontinuity of the road surface. The slab bridges this discontinuity adopting a controlled slope compatible with the road normal service.

This solution works well in most cases, but problems have been detected in many countries that produce significant maintenance costs and troubles to users. Excessive embankment settlement induces slab slope incompatible with safe and comfortable road use. Sometimes, this is accompanied by breaking or degradation of the concrete slab.

Most of the past research work deals with geotechnical questions related to these settlements. However, the effects on the slab, leading eventually to its breaking or degradation, have received less attention. This is the aim of this paper. This structural problem is closely linked to geotechnical aspects that govern the support of the slab on the embankment after settlement. An acceptable

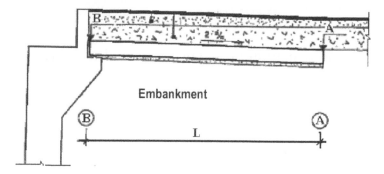

Figure 1. Approach slab. Typical conditions.

safety margin to slab failure is necessary to allow for operation repairs in the case of excessive settlements.

Approaching bridge slabs are seldom calculated. They are designed according to the guidelines for design and construction edited in many countries. Updating of these guidelines has been very useful and will be necessary in the future.

The document issued by the Virginia D.O.T. (Briaud et al., 1999) gives a realistic approach to the subject. The compilation and analysis of actual cases points out the functional and economical impact of the maintenance of the approach slabs, and indicates values of settlements and relative slope discontinuity, among other important remarks.

The description and discussion of the inquiry made in 48 States of USA provides some trends in dimensions and depth positions of approaching slabs and also in construction practice. Big differences in design criteria are detected.

In Spain, two documents have been published by the Ministerio de Fomento. The first one (1992) gives the design criteria for approach slabs in both situations of flexible asphalt or rigid concrete pavements in normal bridges, and the Guide (2000) deals with the situation of integral bridges. Details of the reinforcement are given for most frequent situations.

The paper by Muzás (2000) includes a complete presentation of structural aspects of the problem. The slab length is optimized as a function of the elastic length from the theory of beam on Winkler support.

A more complex but also more realistic method has been developed by Khodair (2002). Cracking of the reinforced concrete slab is analyzed using the non linear finite element program ABAQUS, introducing realistic constitutive behaviour equations for steel, concrete and soil and interaction mechanism like steel tension stiffening in cracked areas. Impact action from the traffic is also included.

The aim of the present work has been to elaborate a structural model of the approach slab that takes into account the different support conditions due to the differential settlement, stiffness changes in the slab due to cracking, dynamic effects of moving traffic loads and fatigue damage due to high frequency of heavy axes in long distance main highways. The model is fitted with the results of actual observations and laboratory large scale test.

2 TYPICAL CONDITIONS

In usual bridge design, the abutment is usually fixed, either on a shallow or deep foundation, so that its settlements are negligible. On the other hand, the access embankment is directly founded on the natural ground. So, immediate and differed settlements of a few centimeters are considered as acceptable.

In order to reduce the discontinuity created by this settlement, the slab is hinged to the abutment and directly supported on the embankment. However, in some instances, a simple footing (sleeper)

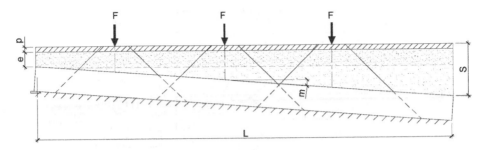

Figure 2. Geometry of the slab: shallow slabs ($e \approx 0$, $m = 0$), deep slabs ($e = 0$–1 m, $m = 0$–10%).

has been proposed to provide a uniform foundation area, avoiding a concentrated support when the slab rotates due to the settlement.

Usual dimensions for the slab are (Min. Fomento, 1992) 5 m in length and 300 mm thick, reinforced in the lower face with N20@0.20 m longitudinally and N10@0.20 m transversally, and with N10@0.20 m in both directions in the upper face. Concrete characteristic strength is usually 20 MPa.

There are two preferred situations for the slab: shallow or deep. In the first case the slab is horizontal, placed directly under the pavement. Deep slabs are inclined, typically with a slope of 10%, and at a depth of the order of 1 m (Figure 2). The main difference between these situations is related to the acting loads. Overburden pressures are obviously higher for deep slabs, but traffic loads are more concentrated in the case of shallow slabs.

3 SOURCES OF DIFFERENTIAL MOVEMENTS

Bridge abutments are usually well founded, and under normal conditions they hardly settle. Hence, differential displacements between the embankment and the abutment are mainly due to movements of the embankment after construction of the pavement. The main sources of movements are depicted in Figure 3, and they can be classified as follows.

- Related to the foundation ground:
  - Consolidation of the embankment foundation. This is the main cause of embankment movements. In clayey soils, primary and secondary consolidation settlements can be large and require a long time to develop. Typically, the final settlement of a 10 m high embankment on a 10 m thick deposit of soft clay can be of the order of 1 meter. Time for reaching 90% of the final settlement ranges from weeks to months or years, depending on soil permeability and presence of interbedded draining layers. Measures against this include the use of prefabricated vertical drains or stone columns to accelerate consolidation, and in the case of columns, to reduce the final settlement. Other techniques of soil improvement are also available (soil replacement, deep mixing, etc.). A complete description of the available solutions falls out of the scope of the present work.
  - Lateral movements of the foundation ground. This includes lateral squeeze due to the surcharge of the embankment (in very soft soils), or lateral movements associated to creep in embankment on sloping ground.
- Deformation of the embankment. The embankments deform by compression under self-weight. However, most of this deformation is very rapid, and it is fully completed before the pavement construction. As a general rule, delayed deformation of an embankment can be avoided by a proper selection of materials (particularly avoiding use of expansive, collapsible or dispersive soils) and a careful construction. This is specially true in the zone of the embankment close to the bridge. The same consideration applies for horizontal movements, which can be controlled

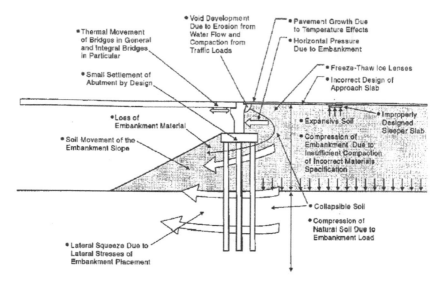

Figure 3. Typical sources of movements at the bridge-embankment transition (Briaud et al., 1997).

also with a proper selection of the slope angles. Hence, this source of delayed movements must be seen as representative of exceptional conditions, such as lack of suitable materials, or action of extremely adverse weathering (dessication-wetting or thawing-freezing cycles).

From the above considerations, well designed and constructed embankments can have delayed settlements of a few centimeters. However, the settlements can be larger in unfavourable cases, particularly in cases of foundation on soft soils.

4 MECHANICAL ANALYSIS OF THE SLAB

A numerical analysis of the stresses in the slab was performed by FEM. Several details of the analysis are commented below.

4.1 Embankment settlement

Based on the considerations given in the preceding section, a typical settlement of 40 mm is considered as representative of good average conditions.

4.2 End support

The first point is the conditions of the support of the slab on the embankment. In the initial situation, for the undeformed embankment, the entire slab is resting on the ground. However, the settlement of the embankment makes the slab being supported only at its end. This produces infinite stresses below the contact point, and some finite contact area develops, due to soil yielding and to slab flexural deformation.

The extension of the contact area has been evaluated first by a limit analysis of the soil. The slab end is taken as a strip footing. For the existing traffic loads, and for a friction angle of 33° for the soil, a width of about 0.5 m is required.

This is compared with a more refined FEM analysis. The program ANSYS was used, in two dimensions. The slab deformation is modeled by geometry updating, and with contact elements to

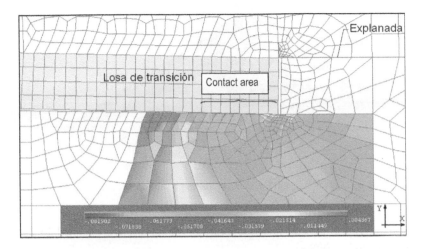

Figure 4. Stresses near the end of the slab. FEM analysis.

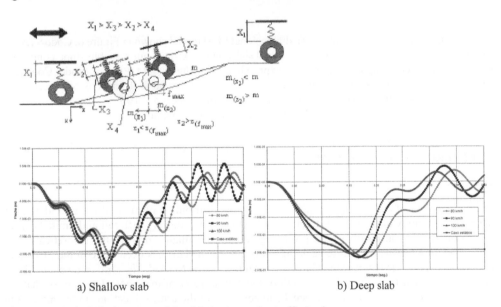

a) Shallow slab   b) Deep slab

Figure 5. Vertical displacement of the slab due to moving traffic loads.

follow the evolution of the soil-slab contact. Figure 4 shows a representative result. As it can be seen, the contact area extends to about 1.5 times the slab thickness, i.e., to about 0.50 m.

In a first step, this area is taken constant in transversal direction. However, additional calculations are done to include the possibility of uneven settlement of the embankment, giving rise to partial loss of support, either in the central zone of the embankment (in cases with settlement at the axis being greater than at the sides), or at the sides (in cases of settling shoulders).

4.3 Dynamic traffic loads

The effect of moving loads has also been analyzed, with the method commonly used for railway bridges. A dinamic amplification occurs due to the abrupt change in slope, and also to the road

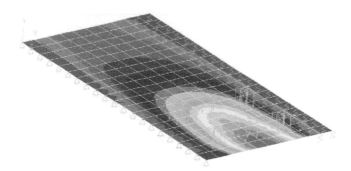

Figure 6.   Bending moments in the slab for a typical loading condition.

roughness. An increase of 15% the force for normal operating speeds is obtained for shallow slabs, and 9% for deep ones (Figure 5).

4.4   *Analysis*

All these data were introduced into a two dimensional FEM model shown in Figure 6. Code STAAD Pro 2003 was used. The fatigue life of standard slabs under extreme but possible slope conditions has been evaluated, using the S-N fatigue expressions from the actual CEB FIP Model Code. For shallow slabs the model predicts fatigue fracture of the steel for up to 4 millions of cycles while for deep slabs the fatigue problem appears in the concrete for $N > 5 \times 106$.

The number of life cycles increases more than one order of magnitude.

5   EXPERIMENTAL PROGRAM

5.1   *Test setup*

Fatigue behavior of concrete is complex, affecting the distribution of internal forces and making the predictive study of reinforced structures submitted to fatigue very difficult and limited (7). To clarify the theoretical results of the preceding section and finally improve the theoretical tools, a series of five fatigue tests of ½ scale of bridge approach slabs has been done in the Structural Laboratory of the ETSICCP of the Universidad Politécnica de Madrid. The reinforcement has been designed to fulfill the same relative depth of the neutral fiber of the cracked section so that the same stress oscillation is obtained in steel and concrete for the critical locations in the model and in the reality. Dimensions of tested slabs are $2.5 \times 3.5 \times 0.15$ m.

A frequency of 4 Hertz is used in the tests to avoid resonance problems. The dynamic amplification is introduced by increasing the cyclic loads according to the foregoing theoretical considerations.

Another amplification coefficient is included to reproduce the effect of non-uniform transversal settlement of the embankment on the lack of uniformity of the band support of the slab on the embankment in the transversal direction. This last question has been theoretically evaluated.

The loads are applied through a three-leg steel device coupled to the dynamic jack, so that the individual leg force is known. The above described FEM model was used to fix the dimensions of the device and magnitude of applied load in order to fit the similarity laws between model and reality. In particular, the bending moments are properly scaled in a significant area of the slab (including real dimensions of lorry axes and wheels).

Data instrumentation is implemented in the test slabs. Accurate reference points for mechanical extensometers are positioned in the upper face of the slab, whilst LVDTs are used in the lower part, due to access limitations. Resistance strain-gauges are glued to some reinforcement bars and

Figure 7. Experimental setup. Three-leg dynamic loading device on the 1/2 scale slab.

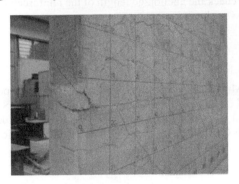 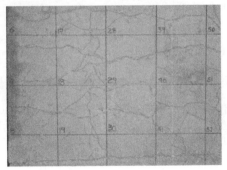

Figure 8. Cracking pattern after a test.

deflections are measured at six points. These data, mainly the curvatures, are correlated to the predictions of the FEM model.

Moments are also checked with the help of well established moment–curvature relationship. Tests number 1 and 4 simulate deep slabs, and test 2 and 3 superficial slabs, while test 5 tries to reproduce transversal failure due to bad lateral support conditions.

5.2 *Results*

Shallow slabs break by steel fatigue under a number of load cycles in agreement to the numerical model, with very low previous deflections. Figure 8 shows the crack pattern of the slab.

Deep slabs show a more complicated behavior. Numerical predictions indicate concrete fatigue failure. In the tests the concrete does not break by fatigue but the slab suffers important permanent deflections (0.01 m in test number 4) due to fatigue concrete detriment.

Final failure comes from fatigue of reinforcing steel under a number of load cycles higher than the predicted for concrete fatigue but lower than the necessary for steel fatigue fracture. Figure 9 shows these brittle steel failures.

Figure 9.   Brittle fracture in the reinforcement steel bars.

6   CONCLUSIONS

Regarding the practical implications of the presented results, some consequences are obvious referring to recommended actions to improve fatigue life under unfavorable support conditions. The most evident ones are: (i) increase concrete strength; (ii) avoid long time working with excessive slab slope; (iii) avoid welding the reinforcement; (iv) check the anchorage length of the reinforcement; and also sufficient proportion of secondary transversal reinforcement.

ACKNOWLEDGEMENTS

This research has been sponsored by the Spanish Ministerio de Fomento in the framework of support given to civil works and maintenance of cultural and historical buildings. (National Plan for Scientific Research, Development and Technological Innovation 2000-2003, Orden/FOM/1540/2002 de 6 de Junio), including a three years Research Grant for the second author at the Universidad Politécnica de Madrid.

REFERENCES

Briaud, J.l., R.W. James & S.B. Hoffman 1997. *Settlement of Bridge Approaches (the bumb and the end of the bridge)*. NCHRP Synthesis of Highway Practice 234. Transportation Research Board, Washington, D.C.

Hoppe, E.J. 1999. *Guidelines for the use, design and Construction of bridge approach slabs*. Virginia Transportation Research Council.

Khodair,Y.A., 2002. *Finite Element analysis of bridge approach slabs*. Doctoral Thesis. University of Rutgers NJ.

Min. Fomento 1992. *Nota de Servicio sobre losas de transición en obras de paso*. Dirección General de Carreteras. MFOM. Spain.

Min. Fomento, 2000. *Guía para la concepción de puentes integrales en carreteras y autopistas*. Dirección General de Carreteras. MFOM. Spain.

Muzás, F. 2000. Comportamiento y diseño de losas de transición. Revista de Obras Públicas, 3397,51–56.

Pacheco, J.A. 2006. *Estudio de la problemática estructural de las losas de transición. Interacción terreno-losa.* Doctoral Thesis. Univ. Polit. Madrid.

*Applications of Computational Mechanics in Geotechnical Engineering – Sousa,
Fernandes, Vargas Jr & Azevedo (eds)
© 2007 Taylor & Francis Group, London, ISBN 978-0-415-43789-9*

Three-dimensional finite-difference modeling of a piled embankment on soft ground

O. Jenck
Polytech'Clermont-Ferrand, France. Previously: INSA Lyon, France

D. Dias, R. Kastner, R. Vert & J. Benhamou
INSA Lyon, Villeurbanne, France

ABSTRACT: Piled embankment is an alternative solution to construct on soft ground. The soil reinforcement consists of a grid of rigid piles driven trough soft soil down to a substratum, on which an embankment is built. Arching occurs in the granular soil constituting the embankment, due to the differential settlements at the embankment base, leading to partial load transfer onto the piles and surface settlement reduction and homogenization. A three-dimensional finite-difference numerical modeling is performed to study the load transfer and the settlement reduction mechanisms. The embankment material behavior is simulated using a non-linear elastic perfectly plastic model with Mohr-Coulomb failure criterion and the soft soil behavior is simulated using the modified Cam Clay model. A parametric study is performed on the embankment parameters and on the soft soil compressibility to assess the impact of the geotechnical parameters on the mechanisms occurring in the embankment.

1 INTRODUCTION

One of the ground improvement techniques to construct on soft soil is the use of vertical rigid piles in combination with a load transfer granular platform or an embankment (Fig. 1), in which arching

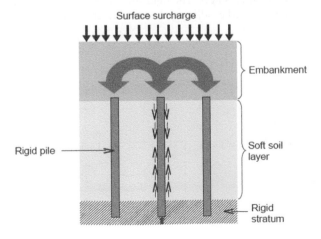

Figure 1. Sketch profile and improvement principle.

occurs due to the differential settlement at its base between the soft soil and the rigid piles. The skin friction along the piles also contributes to the system behavior: the upper part of the piles is submitted to negative skin friction (Combarieu 1988), whereas the lower part is submitted to positive skin friction. In some cases, the embankment base or load transfer platform is reinforced with one or several geosynthetic layers. The main application areas for this technique are road and railway embankments and industrial areas, especially when time is a major constraint.

Several design methods exist to determine the load transfer onto the piles by arching in the embankment (Hewlett & Randolph 1988, Russell & Pierpoint 1997, Svano et al. 2000, etc.), but they usually lead to dissimilar results (Briançon et al. 2004, Jenck et al. 2006). Moreover, there is no method available to assess the surface settlements.

The aim of this study is thus to examine more precisely the mechanisms occurring in the embankment fill and the influence of the geotechnical parameters. A three-dimensional numerical approach is proposed, which simulates explicitly the piles, the soft ground and the embankment in a numerical model representing an elementary part of a squared grid of piles.

This study is part of a French research project, the ASIRI project, acronym for "Amélioration des Sols par Inclusions RIgides" (ground improvement using rigid piles), which has brought together academic, public and industrial partners, to edit guidelines for the design and the construction of platforms and embankments over soft soil improved by vertical rigid piles.

2 NUMERICAL MODEL

2.1 Geometry

A fictitious case but with realistic dimensions and materials is simulated. The grid of piles considered is squared with a pile spacing of 2 m and the pile diameter is equal to 350 mm (Fig. 2). The area ratio, which is the proportion of the total surface treated by piles, is then equal to 2.4%. We consider a regular grid and a part situated in a central zone of the embankment. Only a quarter of an elementary part of the grid needs therefore to be simulated thanks to the symmetry conditions.

The soft soil layer is 5 m- thick, over a rigid stratum. The soft ground layer is made up of a superficial dry crust and a saturated soft soil deposit. The embankment height is 5 m and 20 kPa uniform surface surcharge is applied.

Figure 3 presents the numerical model which consists of a quarter of a pile, the 5 m-thick soft ground layer and an embankment erected in ten successive 0.5 m-thick layers. Static equilibrium under self weight is reached at each loading stage. The surface surcharge is then applied on the top of the embankment by two 10 kPa-stages. The rigid stratum is simulated by fixing the nodes. The calculation is performed in drained conditions: the time effect of the consolidation in the soft soil deposit is not taken into account.

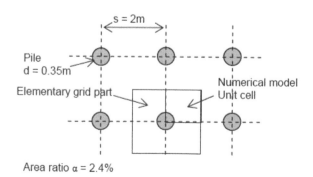

Figure 2.   Top view of a regular squared grid of piles.

2.2 The embankment material

The coarse embankment material called "Lake Valley", described and tested by Fragaszy et al. (1992), is simulated in our study. The authors performed triaxial tests on compacted samples at confinement pressures equal to 75 and 150 kPa. The material parameters can be identified using these experimental results.

The embankment material behavior is simulated with an elastic perfectly plastic model with a Mohr-Coulomb type failure criterion, as in most of the numerical models of embankments over soft ground improved using piles found in the literature (Russell & Pierpoint 1997, Kempton et al. 1998, Rogbeck et al. 1998, Laurent et al. 2003). However, some authors used the Duncan & Chang (1970) model to take account for the stress dependent behavior of the embankment material (Jones et al. 1990, Han & Gabr 2002).

To take account for the non-linear behaviour of the granular material, the Janbu (1963) formula (Equ. 1) can be used (Varadarajan et al. 1998). It links the Young's modulus E to the minor principal stress σ_3.

$$\frac{E}{P_a} = k \cdot \left(\frac{\sigma_3}{P_a}\right)^m \tag{1}$$

where E = Young's modulus; P_a = atmospheric pressure (100 kPa); σ_3 = confinement pressure; k and m = Janbu's parameters.

In our calculation, when using the "Mohr-Coulomb" model for the embankment, the Young's modulus is recalculated at each embankment loading stage using Equation 1 according to the actual stress in every soil zone. A hardening elastoplastic model with non linear elasticity can also be

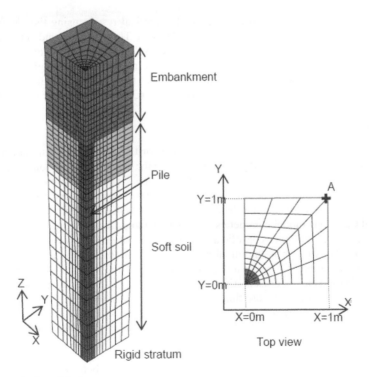

Figure 3.   Numerical model view.

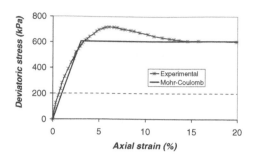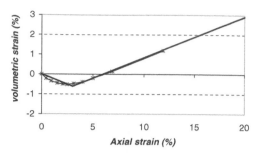

Figure 4. Triaxial test results on Lake Valley coarse soil and numerical results using "Mohr-Coulomb" model, at confinement pressure 150 kPa.

Table 1. Modified Cam Clay model parameters.

	λ	κ	M	e_λ	v
Dry crust	0.12	0.017	1.2	1.47	0.35
Soft soil	0.53	0.048	1.2	4.11	0.35

λ: slope of the normal consolidation line, κ: slope of the swelling line, M: frictional constant, e_λ: void ratio at normal consolidation for p = 1 kPa, v: Poisson's ratio

implemented (Jenck 2005) to avoid this modulus modification (however, equivalent results were found).

Figure 4 presents the experimental results and numerical results using the "Mohr-Coulomb" model of the triaxial test at a confinement equal to 150 kPa. The authors found out that the material was cohesionless (using the 75 kPa-confinement test results). The experimental results permit us to determine a Young's modulus E equal to 20 MPa, an internal friction angle φ equal to 42°, a Poisson ratio v equal to 0.4 and a dilation angle ψ equal to 5° . A usual value for the m Janbu's parameter is 0.5, the k parameter is thus taken equal to 163 (E = 16.3 MPa for $\sigma_3 = 100$ kPa).

2.3 The soft soil layer

The experimental site of Cubzac-les-Ponts is chosen for the soft ground simulation. Magnan & Belkeziz (1982) describe the site and give parameters for the behavior simulation using the Modified Cam Clay model (Roscoe & Burland 1968). The site presents an 8-m-thick soft gray clay layer, which rests on a substratum. A 1-m-thick over-consolidated silty clay dry crust lies at the top of the layer. We use the given values for the dry crust and the values at depth 4–6 m for the soft soil. The modified Cam Clay model parameters are given in Table 1.

The site presents a light over-consolidation: the preconsolidation pressure is then initially set equal to p + 10 kPa, where p is the initial mean effective pressure. The compressibility is indicated by the λ parameter. Compressibility is very high in the soft soil (compressibility index $C_c = 0.35 - 0.48$) and the dry crust is stiffer ($C_c = 0.13$). The earth pressure at rest is $K_0 = 0.5$. The unit weight is $\gamma = 18$ kN/m^3. We use this value in the dry crust and $\gamma' = \gamma - \gamma_w = 8$ kN/m^3 under the ground water level, in the soft soil layer, which corresponds to a drained calculation.

2.4 The rigid piles

The piles are reinforced concrete piles and they are assumed to behave elastically (E = 10 GPa, $v = 0.2$).

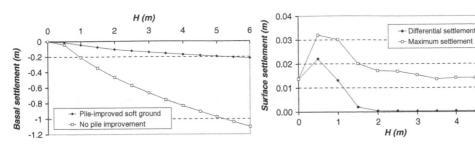

Figure 5. Maximum embankment base settlement according to the equivalent embankment height.

Figure 6. Maximum and differential surface settlement due to the next embankment layer.

3 NUMERICAL RESULTS OF THE REFERENCE CASE

3.1 Settlement reduction

The settlements obtained in the numerical model of the embankment over pile-improved soft ground are compared to the settlements obtained on a simple numerical model without piles. As the embankment is erected in successive layers, only the embankment base settlement can be known from the embankment installation beginning. The maximum settlement at embankment base is reached mid-span between piles (point A from Fig. 3). Figure 5 depicts this settlement evolution according to the equivalent embankment height H. H is the real embankment height from 0 to 5 m, and H = 6 m corresponds to 5 m embankment and 20 kPa surface surcharge. The settlements are reduced compared to the non-reinforced case due to arching that occurs in the embankment, which reduces the load applied on the soft soil layer. For H = 6 m, this settlement is reduced by about 80%.

At each new embankment layer installation, the surface settlements are computed. Figure 6 depicts the maximum and the differential surface settlements due to the next 0.5 m-thick fill layer according to the current embankment height H. The maximum settlement is reached above point A from Figure 3 and the differential settlement corresponds to the difference of surface settlement between the point above the pile and point A. Figure 6 shows that the surface differential settlement can be neglected from an embankment height of 2 m. The maximum surface settlement varies between 14 mm and 32 mm for H = 0.5 m, whereas it reaches 160 mm for the case without improvement by piles.

3.2 Load transfer

The load transfer onto the piles can be estimated by the term efficacy, which is defined as the proportion of the total load (embankment + surface surcharge) transmitted to the piles. When no arching occurs, the efficacy is equal to the capping ratio (here 2.4%). Figure 7 depicts the efficacy according to the equivalent embankment height H. This figure highlights the embankment height influence on the arching in the fill: the efficacy increases with H. The efficacy value for H = 6 m is 0.72, which means that 72% of the total load is supported by the piles due to arching.

4 PARAMETRIC STUDY

A parametric study on the numerical model presented was performed by Vert & Benhamou (2006) by varying the five "Mohr-Coulomb" constitutive model parameters of the embankment material and the soft soil compressibility (by varying the Cam Clay parameters λ and κ).

Table 2 gives the range of value of each of the parameters for the "Mohr-Coulomb" model for the embankment, as well as the values used in the reference calculation. By varying the k-parameter between 100 and 1000, the Young's modulus at confinement 100 kPa varies between 10 and 100 MPa.

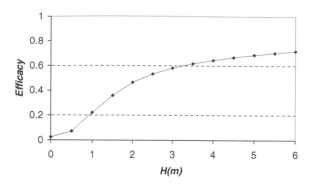

Figure 7. Efficacy according to the equivalent embankment height.

Table 2. "Mohr-Coulomb" model parameters for the embankment material.

Setting		reference	case parametric study
E	k	163	100–1000
	m	0.5	–
v		0.4	0.2–0.4
φ		42°	20–45°
c		0 kPa	0–40 kPa
ψ		5°	0–30°

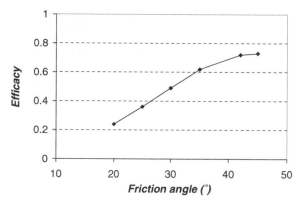

Figure 8. Efficacy for H = 5 m + 20 kPa surcharge according to the embankment soil friction angle.

The soft soil compressibility influence is analyzed by varying the λ parameter of the modified Cam Clay model between 0.1 and 1, as it is directly related to the soil compressibility (compressibility index $C_c = 2.3\,\lambda$). The κ parameter was also varied according to the relation $\kappa = 0.1\,\lambda$. The dry crust properties are maintained constant.

4.1 Influence of the embankment material friction angle

Figure 8 depicts the efficacy values obtained for the 5 m-height embankment and 20 kPa surface surcharge according to the embankment friction angle. The efficacy – thus the soil arching in the embankment fill – considerably increases with the friction angle value: the maximum efficacy is equal to 0.24 for $\varphi = 20°$, and it reaches 0.73 for $\varphi = 45°$, or an increase by 300%.

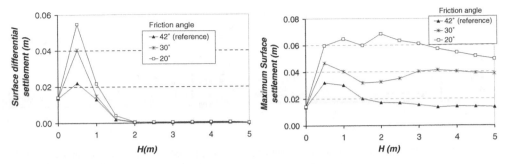

Figure 9.   Differential surface settlement due to the
next embankment layer for different values of φ.

Figure 10.   Maximum surface settlement due to the
next embankment layer for different values of φ.

Table 3.   Influence of the cohesion of the embankment material.

Friction angle	cohesion	Efficacy for H = 6 m	Max. surf. diff. settlement	Max. surf. settlement
42°	0 kPa	0.72	23 mm	32 mm
	10 kPa	0.70	13 mm	25 mm
	40 kPa	0.69	13 mm	15 mm
20°	0 kPa	0.25	55 mm	69 mm
	10 kPa	0.31	13 mm	65 mm
	40 kPa	0.47	13 mm	56 mm

Figure 9 and 10 respectively present the differential and the maximum surface settlement due to the next layer application. The more the friction angle, the less are the differential and maximum surface settlements. The surface differential settlement due to next layer application (Fig. 9) reaches 0.054 m for $\varphi = 20°$, and only 0.022 m for $\varphi = 42°$, or a reduction by 60%. However, the embankment height required to annihilate the differential settlement is not affected by the friction angle. The total surface settlement (Fig. 10) reaches 0.069 m for $\varphi = 20°$, and only 0.032 m for $\varphi = 42°$, or a reduction by 54%.

4.2   Influence of cohesion in the embankment material

The traditional embankment materials generally do not have cohesion. However, treated material (with lime and/or cement) can be used for the embankment, which is thus a cohesive material (Dano et al. 2004). The parametric study on the cohesion influence is performed on two reference cases:

– with a friction angle equal to the reference value (42°),
– with a lower value of the friction angle (20°), which corresponds to the case of a material which would be treated.

Table 3 gives the values of the efficacy for H = 6 m and the maximum values obtained for both the differential and maximum surface settlement due to next layer application, according to the friction angle and the cohesion of the embankment material. When the friction angle is high (42°), there is no influence of the cohesion on the efficacy, whereas the influence of the cohesion is more important when the friction angle is low (20°). For both values of friction angle, the more the cohesion, the less are the differential and maximum surface settlements. Figures 11 and 12 give the values of the surface settlements according to the embankment height for $\varphi = 20$ or 42° and for c = 0 or 40 kPa.

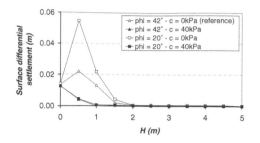

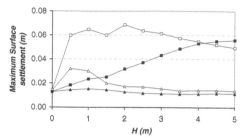

Figure 11. Differential surface settlement due to the next embankment layer for different couples (c, φ).

Figure 12. Maximum surface settlement due to the next embankment layer for different couples (c, φ).

Table 4. Influence of the embankment material dilatancy angle.

Dilatancy angle	Efficacy for H = 6 m	Maximum surf. diff. settlement
0°	0.72	42 mm
5°	0.70	23 mm
30°	0.63	12 mm

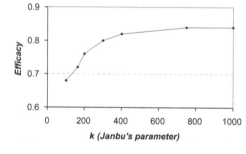

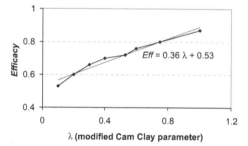

Figure 13. Efficacy for H = 6 m according to the k-parameter (proportional to the Young's modulus).

Figure 14. Efficacy for 5 m embankment and 20 kPa surface surcharge according to the λ parameter.

4.3 Influence of the embankment material dilatancy angle

Table 4 gives the values of the efficacy and the differential surface settlement according to different values for the embankment dilatancy material. There is almost no influence on the efficacy (a small reduction by 12% is observed for an increase of the dilatancy angle from 0 to 30°), whereas the settlement are greatly reduced. In fact, arching effect mobilizes shearing mechanisms, and the more the dilatancy angle, the more the volumetric strains during shearing.

4.4 Influence of the embankment material elastic parameters

The Young's modulus varies when the Janbu's k-parameter varies (Equ. 1). Figure 13 gives the maximum efficacy value according to the parameter k. The efficacy obtained for H = 6 m ranges between 0.68 for $k = 100$ and 0.84 for k = 1000, but there is almost no efficacy increase from $k = 400$. The surface settlements are lightly reduced when the value of k increases, but there is no noticeable influence on the differential surface settlements.

The embankment material Poisson's ratio has no influence on the efficacy but a small impact on the displacement field in the embankment. The settlements are increased by about 35% for a Poisson ratio of 0.2 instead of 0.4, for a constant k value.

4.5 Influence of the soft soil compressibility

Figure 14 indicates that the efficacy greatly increases for increasing compressibility.

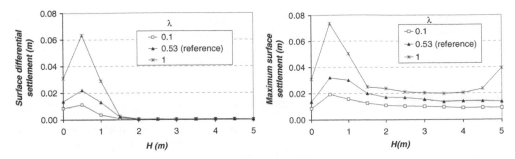

Figure 15. Differential surface settlement due to the next embankment layer for different λ values.

Figure 16. Maximum surface settlement due to the next embankment layer for λ values.

For the range of compressibility investigated, the relation between the efficacy and the λ parameter is almost linear. However, the current design methods to assess the soil arching in the embankment fill do not consider the soft soil compressibility (they assume that the foundation soil provides no support). This study shows that not considering the soft soil support can lead to an overestimation of the arching effect. Stewart & Filz (2005) also showed by a numerical analysis that the compressibility of the ground had a large impact on the load repartition in the embankment and they stated that this parameter should be a factor of design.

Figures 15 and 16 show the foundation soil compressibility influence on the embankment surface settlements. For a limited height embankment (less than 1.5 m), the more the soft soil compressibility, the more the differential and total surface settlements.

5 CONCLUSIONS AND OUTLOOK

The proposed three-dimensional numerical study of a piled embankment highlights the mechanisms occurring in the granular embankment fill: arching occurs, leading to partial load transfer onto the piles – evaluated using the term 'efficacy' – and settlement reduction. The efficacy greatly increases with the embankment height, simultaneously the differential surface settlements are reduced and can disappear by setting up a sufficient embankment height (here H = 1.5 m). The surface total settlements are greatly reduced compared to the non reinforced case. The studied case is fictitious, but with realistic geometry, materials and appropriate constitutive models (Cam Clay for the soft soil and a non-linear elastic perfectly plastic model with a Mohr-Coulomb failure criterion for the embankment). However, such a model should be compared to field testing results in order to be validated.

A parametric study on the embankment and foundation soil geotechnical parameters was then performed. This study shows that the embankment friction angle has a great impact on both the load transfer and surface settlement reduction. The increase of cohesion (up to 40 kPa) has an impact on the efficacy only for a low friction angle value (20°), but leads to a settlement reduction for both values of friction angle studied (20 and 42°). The increase of the embankment material dilatancy angle (up to 30°) leads to a surface settlement reduction, as expected as arching effect mobilizes shearing mechanisms in the embankment fill. However, the increase of dilatancy do not lead to a larger efficacy (a decrease of efficacy is rather observed). The influence of the embankment Young's modulus was studied by varying the Janbu parameter k. An increase of k leads to a larger efficacy value, but a threshold value is reached from $k = 400$. No influence is recorded on the differential settlement, whereas the total surface settlements are lightly reduced when k increases. Lastly, the influence of the soft soil compressibility was investigated by varying the parameter λ of the modified Cam Clay model (directly related to the compression index C_c). The efficacy greatly increases with the soft soil compressibility, as well as the differential and total surface settlements.

In conclusion, the geotechnical parameters which must be taken into account in a design method to determine the load transfer onto the piles are the embankment material shearing parameters (friction and cohesion), the embankment material rigidity and also the soft soil compressibility, whereas most of the current design methods only consider the embankment friction angle. No design method yet exists to assess the surface settlement in such a system, except numerical analysis. All the geotechnical parameters considered in this study have an impact on the settlements.

REFERENCES

Briançon, L., Kastner, R., Simon, B. & Dias, D. 2004. Etat des connaissances – Amélioration des sols par inclusions rigides. In Dhouib et al. (eds). *Proc. Symp. Int. sur l'Amélioration des Sols en Place (ASEP-GI), Paris, 9–10 Sept. 2004*: 15–44. Paris : Presses de l'ENPC.

Combarieu, O. 1988. Amélioration des sols par inclusions rigides verticales. Application à l'édification de remblais sur sols médiocres. *Revue Française de Géotechnique* 44: 57–79.

Dano, C., Hicher, P.-Y. & Taillez, S. 2004. Engineering properties of grouted sands. *Journal of Geotechnical and Geoenvironmental Engineering* 130(3): 328–338.

Ducan, J. M. & Chang, C. Y. 1970. Non linear analysis of stress and strain in soil. *ASCE, Journal of Soil Mechanics and Foundations* 96: 1629–1653.

Fragaszy, R. J., Su, J., Siddigi, H. & Ho, C. J. 1992. Modelling strength of sandy gravel. *Journal of Geotechnical and Geoenvironmental Engineering* 118(6): 920–935.

Han, J. & Gabr, M. A. 2002. Numerical analysis of geosynthetic-reinforced and pile-supported earth platforms over soft soil. *Journal of Geotechnical and Geoenvironmental Engineering* 128: 44–53.

Hewlett, W. J.& Randolph, M. F. 1988. Analysis of piled embankment. *Ground Engineering* 21(3): 12–18.

Janbu, N. 1963. Soil compressibility as determined by oedometer and triaxial tests. In *Proc. of the European Conf. on Soil Mechanics and Foundations Engineering, Wiesbaden.* Essen: Deutsche Gesellschaft für Erdund Grundbau: 19–25.

Jenck, 0. 2005. *Le renforcement des sols compressibles par inclusions rigides verticales. Modélisations physique et numérique.* Thèse de doctorat, INSA de Lyon, Villeurbanne.

Jenck, 0., Dias, D. & Kastner, R. 2006. Numerical modeling of an embankment on soft ground improved by vertical rigid piles. In *Proc. of the 4th Intern. Conf. on Soft Soil Engineering, Vancouver, 4–6 October 2006.*

Jones, C. J. F. P., Lawson, C. R. & Ayres, D. J. 1990. Geotextile reinforced piled embankments. In Den Hoedt (ed). *Proc. of the 4th Int. Conf. on Geotextiles Geomembranes and related Products, Den Haag, 28 May–1 June 1990*: 155–160 (vol. 1).

Kempton, G., Russell, D., Pierpoint, N. D. & Jones, C. J. F. P. 1998. Two- and three-dimensional numerical analysis of the performance of piled embankments. In Rowe (ed). *Proc. of the 6th Int. Conf. on Geosynthetics, Atlanta, 25–29 March 1998:* 767–772.

Laurent, Y., Dias, D., Simon, B. & Kastner, R. 2003. A 3D finite difference analysis of embankments over pile-reinforced soft soil. In Vermeer et al. (eds). *Proc. of the Int. Workshop on Geotechnics of Soft Soils – Theory and Practice, Noordwijkerhout, 17–19 Sept. 2003:* 271–276. Essen: Verlag Glückauf.

Magnan, J.-P. & Belkeziz, A. 1982. Consolidation d'un sol élastoplastique. *Revue Française de Géotechnique* 19: 39–49.

Rogbeck, Y., Gustavsson, S., Soedergren, I. & Lindquist, D. 1998. Reinforced piled embankments in Sweden – Design aspects. In Rowe (ed). *Proc. of the 6th Int. Conf. on Geosynthetics, Atlanta, 25–29 March 1998*: 755–762.

Roscoe, K. H & Burland, J. B. 1968. On the generalised stress-strain behaviour of 'wet' clay. *Engineering Plasticity*: 535–609. Cambridge University Press.

Russell D., Pierpoint N. 1997. An assessment of design methods for piled embankments. *Ground Engineering* (November 1997): 39–44.

Stewart, M. E. & Filz, G. M. 2005. Influence of clay compressibility on geosynthetic loads in bridging layers for column-supported embankments. In Anderson et al. (eds). *Proc. of Geo-Frontiers, Austin, 24–26 January 2005.*

Svano, G., Ilstad, T., Eiksund, G. & Want, A. 2000. Alternative calculation principle for design of piles embankments with base reinforcement. In *Proc. of the 4th Int. Conf. of Ground Improvement Geosystem, Helsinki, 7–9 juin 2000.*

Varadarajan, A., Sharma, G. & Aly, M. A. A. 1999. Finite element analysis of reinforced embankment foundation. *International Journal for Numerical and Analytical Methods in Geomechanics* 23(2) : 103–114.

Vert, R. & Benhamou, J. 2006. *Le renforcement des sols par inclusions rigides. Etude paramétrique sur les paramètres géotechniques.* Mémoire de fin d'études. INSA de Lyon, Villeurbanne.

Applications of Computational Mechanics in Geotechnical Engineering – Sousa,
Fernandes, Vargas Jr & Azevedo (eds)
© 2007 Taylor & Francis Group, London, ISBN 978-0-415-43789-9

Finite element analysis of a footing load test

A.A.L. Burnier
Federal Universitiy of Viçosa, MG, Brazil

R.R.V. Oliveira
Federal Universitiy of Ouro Preto, MG, Brazil

R.F. Azevedo & I.D. Azevedo
Federal Universitiy of Viçosa, MG, Brazil

C.L. Nogueira
Federal Universitiy of Ouro Preto, MG, Brazil

ABSTRACT: This work presents the finite element analysis of a shallow foundation load test. The soil stress-strain behavior was modeled by non-linear-elastic (hyperbolic), elastic-perfectly-plastic (Mohr-Coulomb) and Lade-Kim elasto-plastic work hardening models. The constitutive model parameters were determined from triaxial tests. Numerical results were compared to the load-settlement curves obtained with the field tests leading to the conclusion, that the analysis performed using the hyperbolic model gave rise to the best results.

1 INTRODUCTION

Soil constitutive models are formulations that try to represent important aspects of soil stress-strain-strength behavior (Desai and Siriwardane, 1984).

The degree of complexity of a constitutive model and its ability to adequately model a given problem are the main factors that influence its use in geotechnical projects. For instance, the linear-elastic model, in spite of its simplicity, is normally suitable to represent the behavior of geotechnical problems where over consolidated soils predominate and are loaded up to stress levels far from failure. However, geotechnical structures involving normally consolidated soils loaded up to rupture require complex and difficult to be used elasto-plastic models.

It is, therefore, interesting that geotechnical problems are instrumented and monitored, so that numerical analyses using different constitutive models provide comparisons between field and numerical results indicating the constitutive model that best represent the problem.

Following this reasoning, this work presents the numerical analysis of an instrumented load tests performed on a shallow foundation (Duarte, 2006).

The analysis was accomplished using the finite element method, considering three constitutive models: the non linear elastic hyperbolic (Duncan and Chang, 1970, Duncan, 1980), the Mohr-Coulomb elastic perfectly plastic (Desai and Siriwardane, 1984) and the elasto-plastic Lade-Kim (Lade and Kim, 1995).

The main objective of the work is to indicate what constitutive model, calibrated by laboratory tests, best represents the load-settlement curves observed in the field load test.

2 MATERIALS AND METHODS

The load test was executed with a rigid, square footing made of reinforced concrete and sizing 80 cm by 80 cm. The footing was placed on a residual soil at a depth of 0,50 m. Soil physical indexes are presented in Table 1.

A series of conventional triaxial, isotropically consolidated, drained tests was performed in samples with the characteristics presented in Table 2. Undisturbed soil samples were used in the tests. All the samples were non-saturated and tested with the natural water content. Test results will be shown in the next section, together with the constitutive model reconstitutions.

The load – displacement curve given by the load test is shown in Figure 1. As it can be seen, the rupture load is not clearly defined, because the footing presented a failure of the punching shear type.

2.1 Numerical analysis methodology

The numerical analysis was accomplished in three steps. The first consisted of calculating the initial stress state admitting, for each soil layer, different values of over consolidation ratio. The second step considered an overload of 8.2 kPa on the soil surface to simulate the footing embedment. The

Table 1. Soil characteristics (Duarte, 2006).

Depth (m)	Clay (%)	Silt (%)	Sand (%)	LL	LP	IP	w_{nat} (%)	γ_s (kN/m^3)	γ_{nat} (kN/m^3)
0,40–2,20	51	19	30	60,80	35,06	25,74	31,21	27,72	16,44

Table 2. Triaxial test parameters.

Depth (m)	W_{nat} (%)	γ_{nat} (kN/m^3)
0,50	35,53	16,44

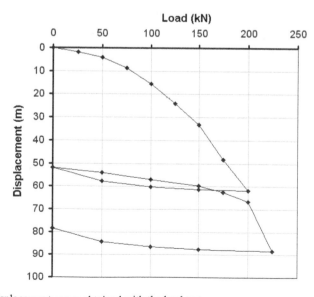

Figure 1. Load-displacement curve obtained with the load test.

third and final step consisted of simulating the footing loading up to a vertical footing displacement equal to 90mm (vertical stress equal to 360 kPa).

Due to the lack of experimental data for the soils located at depths deeper than, approximately, 3.0 m, the linear-elastic model was used to simulate the behavior of these soils in all analyses.

3 CONSTITUTIVE MODEL PARAMETERS

Tables 3, 4 and 5 present the parameters for Mohr-Coulomb, hyperbolic and Lade-Kim models, respectively. These parameters were obtained using the triaxial test results and, for Lade-Kim model, results of a hydrostatic compression test using well established methodologies (Duarte, 2006; Duncan, 1980; Lade and Kim, 1995; Desai and Siriwardane, 1984). As it is recommended by Terzaghi (1943), since the footing failed by punching, the Mohr-Coulomb strength parameters were reduced by a factor of 1/3 when the Mohr-Coulomb elasto-perfectly plastic model was used.

After determining the parameters, the models were used to reconstitute the laboratory stress-strain curves (Figures 2 and 3).

The hyperbolic model was able to represent reasonably well the relation between the deviatoric stress and the axial strain for all confining pressure values. However, the volumetric behavior was not satisfactorily represented, except at the very beginning of the test (Figure 2).

In Figure 3, it can be observed that Lade-Kim model was able to represent accurately the relation between the deviatoric stress and the axial strain for all values of confining pressure. However, the volumetric behavior was only satisfactorily represented when the confining pressure was equal to 50 kPa. For other confining pressure values, the model well represented the laboratory results only for axial strains up to 5%.

It has to be emphasized that the laboratory results do not seem to be consistent since the test with confining pressure equal to 250 kPa gave rise to larger volumetric strains than the one with confining pressure equal to 500 kPa.

Table 3. Mohr-Coulomb model parameters.

Sample	v	E (kPa)	ϕ (°)	c (kPa)
D-01	0.2	12.500	14	22,8

Table 4. Hyperbolic model parameters.

Sample	c (kPa)	ϕ (°)	R_f	K	n	K_b	m
D-01	34,18	20,57	0,75	121,60	0,3615	32,87	0,417

Table 5. Lade-Kim model parameters.

Sample	Elastic		Hardening		Rupture		Yielding			Plastic potential	
	M	λ	c	ρ	η_1	m	ψ_1	h	α	ψ_2	μ
D-01	206,1	0,179	0,001	1,33	30,6	0,43	0,004	0,50	1	−3,02	2,31

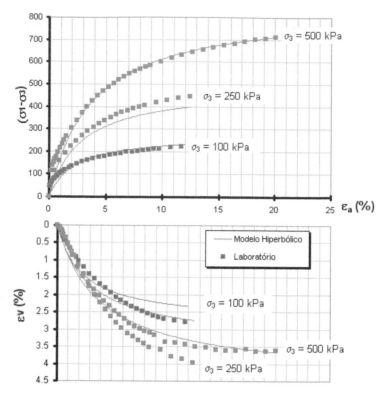

Figure 2.   Comparisons between laboratory and model results. Hyperbolic model.

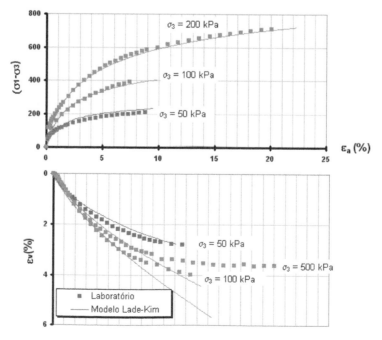

Figure 3.   Comparisons between laboratory and model results. Lade and Kim model.

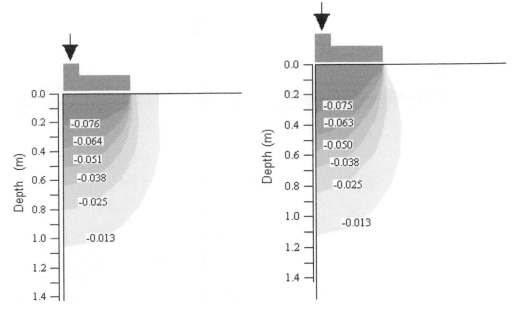

Figure 4. Vertical displacement isocurves (m). Hyperbolic model.

Figure 5. Vertical displacement isocurves (m). Mohr-Coulomb.

4 NUMERICAL ANALYSES RESULTS

Figures 4 to 6 show isocurves of vertical displacements obtained with the numerical analyses for the different constitutive models when the footing displacement was equal to 90 mm.

As it can be seen, the results are very similar, although the displacements obtained when Lade and Kim model was used are slightly greater than those obtained with the other models.

Figures 7, 8 and 9 show isocurves of vertical stresses obtained with the numerical analyses when the footing displacement was equal 90 mm.

Analyzing Figures 7 to 9 it is possible to see that the results are very similar. Besides, it can be observed that the isocurve that passes at depth 2B, where B is the footing width, corresponds to a vertical stress equal to approximately 25% of the footing load. This value is significantly greater than the one given by Boussinesq solution in which, at this depth, the vertical stress is around 10% of the applied load. Therefore the non-linear solutions provided deeper stress bulbs than the one given by Boussinesq classical elastic solution.

In Figure 10 it is presented a comparison between the field test load-settlement curve and the numerical analyses. It can be observed that the numerical results for the different constitutive models are not far from the load test result. However, the hyperbolic model provided the best comparisons.

The numerical analyses confirmed that the footing failed by punching, as it can be seen from the displacement field presented in Figure 11, where, clearly, there is no rotational motion evidence. In this figure, the displacement field corresponds to the Mohr-Coulomb model, however, field displacements obtained with the other two models were similar to this.

Comparison between numerical and load test relative vertical displacements on the vertical line passing through the footing central axis are presented in Figure 12. It is observed that, although the curve shapes are similar, quantitatively the comparisons are not good. Near soil surface the discrepancies are larger than in deeper depths. For instance, 50% of the relative settlement occurs in the field at a depth of 0.4 m whereas in the numerical solutions the depths were around 0.6 m. Moreover, at a depth of 1.6 m, the field relative settlement was, approximately, 98%, whereas in

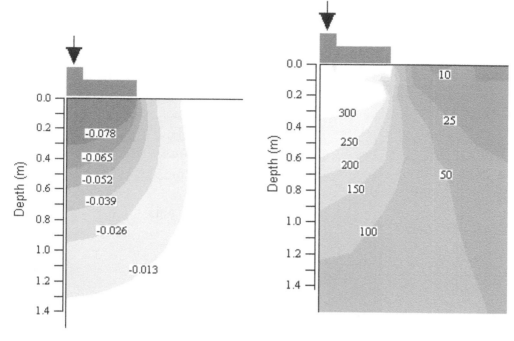

Figure 6. Vertical displacement isocurves (m). Lade-Kim.

Figure 7. Vertical stress isocurves (m). Hyperbolic model.

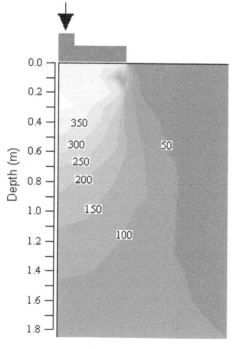

Figure 8. Vertical stress isocurves (m). Mohr-Coulomb.

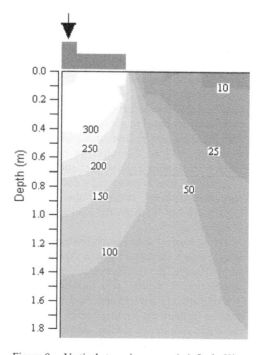

Figure 9. Vertical stress isocurves (m). Lade-Kim.

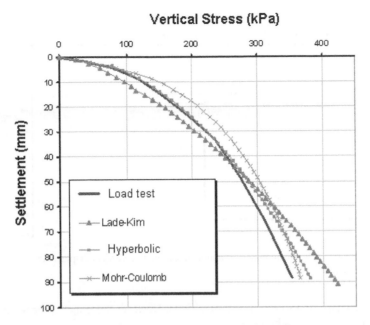

Figure 10. Comparisons between the load test and numerical analysis results.

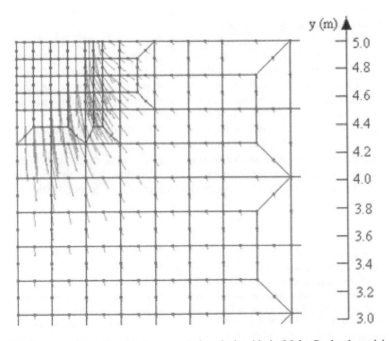

Figure 11. Displacement field given by the numerical analysis with the Mohr-Coulomb model.

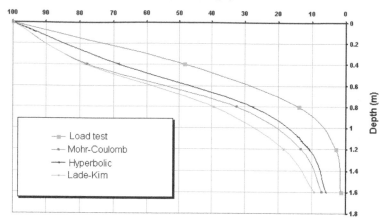

Figure 12. Comparisons between numerical and load test relative vertical displacements at vertical passing by the footing center.

the numerical analyses the values were around 93%. However, once again, the solution given by the hyperbolic model was the best.

5 CONCLUSIONS

In this work, the finite element method was used with different constitutive models to analyze a footing load test executed by Duarte (2006).

It was verified that the best comparison between the numerical analyses and the field test was achieved when the hyperbolic model was used to represent the soil behavior, although the results obtained with the other models were not so different from the one obtained with this model.

The vertical stress bulbs numerically obtained were deeper than the ones used in practice using Boussinesq solution, based on linear elasticity. Considering the soil behavior non-linear, the load affects the ground in deeper depths than twice the width of the footing.

The numerical analyses confirmed that the footing failed by punching as it was observed in the load-test.

Comparisons between numerical and load test relative vertical displacements at the vertical line passing by the footing center are not quantitatively good, although the shapes of the curves are similar. Near to the soil surface the discrepancies are larger than in deeper depths.

REFERENCES

Desai, C. S. E Siriwardane, H. J., 1984. *Constitutive Laws for Engineering Materials with Emphasis on Geologic Materials*. Prentice-Hall, Inc., Englewood Cliffs, New Jersey

Duarte, L. N. 2006. Instrumented Load Test in a Footing Placed in a Residual Soil – MSc. Thesis, UFV/Viçosa.

Duncan, M. J. 1980. Hyperbolic stress-strain relationships. *Proceedings: Workshop on Limit Equilibrium, Plasticity and Generalized Stress-Strain in Geotechnical Engineering*. 443–460, New York.

Duncan, M. J. E Chang, C. Y. (1970). Nonlinear analysis of stress and strain in soils. *Journal of Soil Mechanics and Foundation Division – ASCE*, SM5. 1629–1653.

Lade, P. V. e Kim, M. K. 1995. Single Hardening Constitutive Model for Soil, Rock and Concrete – *International Journal Solids and Structures*. 32(14). 1963–1978.

Nogueira, C. L. 1998 – Non Linear Analysis of Excavations and Embankments – Ph.D.Thesis, PUC-Rio de Janeiro (in Portuguese).

Terzaghi, K.L. 1943 – Evaluation of Coefficient of Subgrade Reaction. *Geotechnique,* 5(4). 297–326.

Applications of Computational Mechanics in Geotechnical Engineering – Sousa,
Fernandes, Vargas Jr & Azevedo (eds)
© 2007 Taylor & Francis Group, London, ISBN 978-0-415-43789-9

On the pullout of footings backfilled with cemented sand

N.C. Consoli
Federal University of Rio Grande do Sul, Porto Alegre, Brazil

B.M. Lehane, D. L'Amante & M. Helinski
The University of Western Australia, Perth, Australia

D.J. Richards & M.J. Rattley
University of Southampton, Southampton, UK

ABSTRACT: The kinematics of failure and the uplift response of shallow footings embedded in loose sand, as well as on layers of weakly to moderately cemented loose sands, when subjected to vertical pullout tests carried out on a drum centrifuge are discussed in this paper. The utilization of a cemented top layer increased uplift capacity and changed soil behavior to a noticeable brittle behavior. An attempt has been made to delineate the rupture surfaces of the shallow footings embedded in artificially cemented layers according to level of cementation. The mode of failure was also drastically affected according to the degree of cementation, varying from a vertical punching pattern for the uncemented loose sand to the development of fissures followed by small blocks for the weakly cemented sand (1% cement – q_u = 25 kPa), formation of a rounded fissure with the incidence of encircle radial fissures for 3% cement (q_u = 87 kPa) and finally formation of a few radial fissures starting at the center of loading giving rise to the formation of large blocks extending the failure to almost all the cement improved area for the moderately cemented (5% cement – q_u = 365 kPa) sand.

1 INTRODUCTION

During the last three decades, important contributions (e.g. Dupas & Pecker 1979, Clough et al. 1981, Coop & Atkinson 1993, Cuccovillo & Coop 1993, Huang & Airey 1993, 1998, Consoli et al. 2000, 2001, 2006, 2007-a, Schnaid et al. 2001, Rotta et al. 2003) have helped improved understanding of the mechanisms involved and the parameters affecting the behaviour of artificially cemented soils, in the laboratory.

Field studies of spread footings have been also carried out evaluating the use of compacted layers made up of soil mixed with cementing agents such as Portland cement (e.g. Stefanoff et al. 1983, Sales 1998, Tessari 1998, Consoli et al. 2003) and lime (e.g. Thomé 1999, Thomé et al. 2005, Consoli et al. 2007-b). Such studies have shown a noteworthy increase in the bearing capacity of foundations when placed on improved layers built over a weak soil stratum, and were restricted to layered systems under compressive forces.

Several studies are available in the literature regarding uplift capacity of foundations in sand (e.g. Balla 1961, Meyerhof & Adams 1968, Clemence & Veesaert 1977, Rowe & Davis 1982, Ilamparuthi & Muthukrishnaiah 1999, Merifield et al. 2006). Ovesen (1981) started investigations utilizing the centrifugal testing technique to study the uplift capacity of anchor slabs in sand. Lehane et al. (2006) have also used the centrifuge to study the rate effects on the vertical uplift capacity of footings founded in clay. Uplift theories are generally based on an assumed failure surface. This requires verification of the shape of the true rupture surface for varying soil parameters. Recent work by Ilamparuthi & Muthukrishnaiah (1999) for shallow foundations in sand have defined that for sands the rupture surface is a gentle curve, convex upwards, which can be closely approximated to a plane surface. The plane surface makes an angle of $\emptyset/2 \pm 2°$ (in which $\emptyset$ is the friction angle of the sand) with the vertical irrespective of density and shape of foundation.

Present work is aimed at analysing an alternative view to increasing the uplift capacity of shallow footings on soils through backfilling the excavation made for embedding the foundation with sand improved with cement addition. This investigation includes delineating the rupture surface (and kinematics of failure) for shallow footings embedded in artificially cemented layers (when compared to an uncemented sand layer), considering degree of cementing varying from weakly to moderately cemented sand, when subjected to vertical pullout.

2 EXPERIMENTAL PROGRAM

The experimental program includes the classification of the material used in the research, laboratory testing considering unconfined compression tests, direct shear tests and bender element tests, as well as centrifuge testing with CPT on cover layers and pullout tests of shallow footings embedded on backfilled layers of sand and cemented sand.

2.1 *Materials*

The soil used in this study was a non-plastic uniform fine sand. D_{50} is 0.19 mm. The minimum and maximum void ratios are 0.52 and 0.81, respectively. Mineralogical analysis showed that sand particles are predominantly quartz. The specific gravity of the solids is 2.65.

Portland cement of high initial strength was used as the cementing agent. Its fast gain of strength allowed the adoption of 20 hours as the curing time. The specific gravity of the cement grains is 3.15.

Tap water was used throughout this research, for laboratory and centrifuge tests.

All laboratory and centrifuge tests were carried out with the sand molded at loose conditions (voids ratio of about 0.74 for the uncemented sand and in the range of 0.73 to 0.68 for cemented sand mixtures, respectively for 1% to 5% in weight of dry sand).

2.2 *Laboratory testing*

In order to characterize the strength and stiffness of the cemented (molded in 3 different cement percentage: 1%, 3% and 5% regarding the weight of dry sand) and uncemented sand, unconfined compression and direct shear tests, as well as Bender elements tests were carried out in this research.

The compacted soil specimens used in the unconfined compression, direct shear tests and the isotropic compression tests with Bender element measurements of shear wave velocity were prepared by hand-mixing dry soil and water. The undercompaction process (Ladd 1978) was used to produce homogeneous specimens that could be used for a parametric study in the laboratory-testing program. The specimens molded for the unconfined compression and bender elements tests were statically compacted in three layers into a 80 mm diameter by 160 mm high split mould, at a moisture content of 12% and a voids ratio ranging from 0.74 to 0.68, respectively for sand and sand with 5% cement. The specimens molded for the direct shear tests were statically compacted in a unique layer into a 71 mm diameter by 35 mm high shear box, at the same moisture content, voids ratio range and cement contents as the previous tests. The sand was initially mixed dry with rapid-hardening Portland cement (for cemented samples), which, with an initial setting time of 3.25 hours, allows homogeneous cementing in the specimens in a very short period. Water was then added to the mixture and further mixing was performed until it was homogeneous in appearance. For the isotropic compression tests with bender element the specimens were installed in the triaxial chamber immediately after preparation (total time dispended after adding water of about 1 hour), and were left to cure for 20 hours (under an isotropic effective stress of 20 kN/m^2), which is the time necessary to gain 60% of the total strength. The specimens of the shear box tests also where installed in the direct shear box after preparation (total time dispended after adding water of about

1 hour), and were left to cure for 20 hours under the following effective normal stresses: 50 kN/m^2, 150 kN/m^2 and 300 kN/m^2.

Unconfined compression tests have been used in most of the experimental programs reported in the literature in order to verify the effectiveness of the stabilization with cement or to access the importance of influencing factors on the strength of cemented soils. An automatic loading machine, with maximum capacity of 50 kN and proving rings with capacities of 10 kN and resolution of 0.005 kN, were used for the unconfined compression tests. After curing for 16 hours, the specimens were submerged in a water tank for 4 hours for saturation and to minimize suction, totalizing 20 hours as the curing time period. The water temperature was controlled and maintained at 23°C ± 3°C. Immediately before the test, the specimens were taken out the tank and dried superficially with an absorbent cloth. Then, the unconfined compression test was carried out and the maximum load reached by the specimen recorded. As acceptance criteria, it was stipulated that the individual strengths of three specimens, molded with the same characteristics, should not deviate by more than 10% from the mean strength.

The direct shear tests were carried out under controlled displacement rate of 0.5 mm/minute. The execution of the shear box tests followed the general procedures described by BS 1377 (1990). The specimens were left immerse in order to guarantee saturation.

The Bender Element tests were introduced by Shirley & Hampton (1977), bender elements are now a standard technique for deriving the elastic shear modulus G_0 of a soil. The velocity of a shear wave propagating across the specimen V_s is measured from which G_0 may be determined:

$$G_0 = \rho V_s^2 = \rho \left(\frac{L^2}{t^2} \right) \qquad (1)$$

where ρ is the total mass density of the soil, L is the tip to tip length between the elements and t is the travel time of the shear wave trough the sample. The bender element tests were carried out in a stress path apparatus. The test procedures and methods of interpretation followed those of Jovicic et al. (1996), using a single shot sine wave with measurement of the wave velocity at the first arrival, taking care to use sufficiently high frequencies to avoid near field effects. For each reading a range of frequencies was tried, ensuring that the measured arrival time was not frequency dependent.

The unconfined compression tests carried out on cemented specimens containing 1%, 3% and 5% cement have given average unconfined compressive strengths (q_u) of 25 kPa, 87 kPa and 365 kPa, respectively. This degree of cementing characterizes a weakly to moderately cemented soil, depending on the classification considered (e.g. Beckwith & Hansen 1982; Rad & Clough 1985; Hardingham 1994). The shear strength envelope parameters obtained from direct shear tests on loose sand and cemented loose sand were the peak friction angle of 34.7° for uncemented sand, increasing to 35.3°, 39.8° and 41.5°, respectively, for sand containing 1%, 3% and 5% cement. The cohesion intercept changed from zero for the uncemented sand to 17.7 kPa, 28.2 kPa and 57.4 kPa, respectively for sand with 1%, 3% and 5% cement.

The elastic shear modulus G_0 obtained after bender element tests on sand and sand-cement (1%, 3% and 5% cement) specimens, where cemented specimens were cured under a 20 kPa stress, were 49.6 MPa for the sand and 249.4 MPa, 565.7 MPa and 972.9 MPa for sand with 1%, 3% and 5% cement respectively. So, based on the initial stiffness of the sand, the sand-cemented specimens were about 5, 11 and 20 times stiffer respectively for 1%, 3% and 5% cement addition.

2.3 Centrifuge testing

The testing program was performed in the drum centrifuge at University of Western Australia (UWA) as this centrifuge offered the possibility of conducting multiple foundation pullout tests and CPT tests within each sample. A complete technical description of this facility is presented by Stewart et al. (1998) and just a brief outline follows. The drum centrifuge features a 0.30 m wide (measured vertically) × 0.20 m deep (measured radially) × 1.20 m diameter channel used to contain a soil sample. The channel rotates about a vertical axis so that radial centrifugal acceleration field down into the soil sample. A key feature of the drum centrifuge is the availability of a central tool table that may be controlled independently to the sample containment channel, and is achieved via two concentric drive shafts. This independent control allows it to be stopped without interrupting the rotation of the main sample channel or to allow the tool table to move or remain stationary at one point relative to the soil sample. This control allows aspects of the modeling procedure to be modified without disrupting the main soil sample, and to carry out multiple experiments on one soil sample. A range of actuators can be attached to the central tool table that enables movement in three directions relative to a fixed point on the sample: vertical, radial and circumferential.

The experiments were undertaken at a scale of 1:50, so that linear dimensions were reduced by a factor of 50 and a centrifugal acceleration of 50 times the earth's gravity was applied to ensure similarity of the soil's stress/strain behaviour in the model to that in a full size equivalent prototype.

The special characteristics of the centrifuge tests carried out in the present research were that the cemented sand placed in the centrifuge was cured under stress. The total time dispended after adding water to the sand-cement mixtures, placing them in the centrifuge and start spinning was about 1 hour, in order to guarantee that the cure of the cement would occur under stress. The sand was initially mixed dry with rapid-hardening Portland cement (for cemented samples) for a period of about 15 minutes. Water was then added to the mixture and further mixing was performed until it was homogeneous in appearance. The cemented (1%, 3% and 5% cement related to the dry weight of the sand) and uncemented sand were then placed in the centrifuge in previously opened excavations made on the drum channel filled with kaolin. Four trunk-conical excavations of about 250 mm diameter in the top, 150 mm diameter in the bottom and 45 mm depth (for the pullout tests) were open (and backfilled with uncemented and cemented sand – in very loose condition – voids ratio varying from 0.74 to 0.68, respectively for sand and sand containing 5% cement – after positioning the square footings) on the kaolin – below such depth the kaolin was replaced by sand (the bottom of the channel allows drainage) in order not to have any suction in the base of the footings.

A cone penetrometer was used in-flight to measure the characteristics of the sand and cemented sand soil samples. The cone penetrometer adopted in the drum centrifuge has a diameter of 6 mm, which gives and end area of about 28.3 mm^2. A load cell is attached inside the penetrometer rod, behind the tip of the cone, measuring the tip resistance. The penetrometer is mounted on the drum actuator after the sample is prepared, and can be pushed in-flight through the soil at any position. The penetration rate used in present tests was 1.0 mm/s. Cone maximum values of 0.9 MPa, 1.5 MPa, 4.4 MPa and 6.3 MPa were recorded for the studied range of cementation (0%, 1%, 3% and 5%). As could be expected, the cone resistance and the initial stiffness increase with increasing cement content. The maximum cone resistances were mobilized at penetrations equivalent to about 3–4 cone diameters for the cemented backfills. However, for the uncemented sand backfilling, larger displacements were required to mobilize the maximum cone resistance. Similar results have been observed by Joer et al. (1999) for CPT tests on cemented calcareous sediments. At small depth (about 5 mm), in the range that the cone resistance was increasing almost linearly with penetration, the cone values were 0.08 MPa, 0.31 MPa, 0.80 MPa and 2.0 MPa respectively for sand and sand-cement, containing 1, 3 and 5% cement content. So, the cone resistance increases about 4, 10 and 25 times respectively for 1%, 3% and 5% cement addition when related to the cone resistance on sand. It seems that a similar rate of increase in elastic shear modulus and cone resistance at small penetration occur due to incorporation of rising amounts of cement in the sand.

The pullout testing program involved 4 tests, in which the embedment depth of the footings was 45 mm. The model foundations were fabricated from aluminum with 5 mm thick square bases and side widths of 30 mm. The tests were loaded to failure at a constant displacement rate of 0.1 mm/s. The utilization of cemented top layers increased uplift peak stress capacity of the footings subjected to pullout from 55 kPa for the uncemented sand, to 100 kPa for 1% cement content, 164 kPa for 3% cement content and 252 kPa for 5% cement content. The increase in uplift capacity due to insertion of cement, having the uncemented sand as a comparison, was of 82% for 1% cement, 198% for 3% cement and 358% for 5% cement. The increase in the uplift capacity was not linear with the amount of cement added. This is possibly due to variation of rupture surface angles with the vertical and different kinematics of failure after the formation of different fissuring that conducted to small and

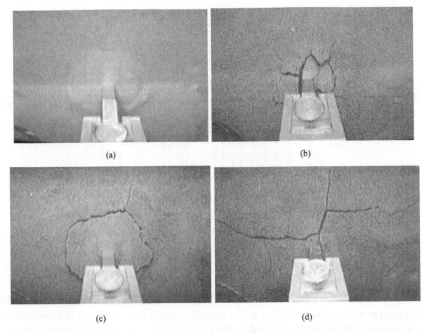

Figure 1. Failure fissures observed on the surface during shallow footing pullout for (a) sand, (b) sand with 1% cement, (c) sand with 3% cement and (d) sand with 5% cement.

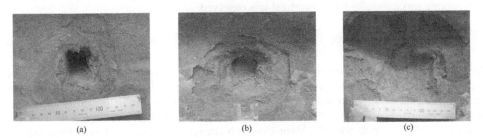

Figure 2. Failure surfaces observed for (a) sand with 1% cement, (b) sand with 3% cement and (c) sand with 5% cement.

big blocks during failure, due to diverse amount of cement addiction. Other changes were observed on the pullout load-displacement behaviour of shallow footings embedded in artificially cemented layers. Due to cement introduction the displacement at failure was changed, and a noticeable brittle behavior was observed.

Figures 1 (a), 1 (b), 1 (c) and 1 (d) present the fissures observed on the soil surface at some stage in shallow footing pullout during failure respectively for sand, sand with 1% cement, sand with 3% cement and sand with 5% cement. The mode of failure was also drastically affected according to the degree of cementation, varying from a vertical punching pattern for the uncemented loose sand to the development of fissures followed by small blocks for the weakly cemented sand (1% cement – $q_u = 25$ kPa), formation of a rounded fissure with the incidence of encircle radial fissures for 3% cement ($q_u = 87$ kPa) and finally formation of a few radial fissures starting at the center of loading giving rise to the formation of large blocks extending the failure to all the cement improved area for the moderately cemented (5% cement – $q_u = 365$ kPa) sand.

Figures 2 (a), 2 (b) and 2 (c) show pictures that illustrate an overview of the failure planes after shallow footing pullout respectively for sand with 1% cement, sand with 3% cement and sand with

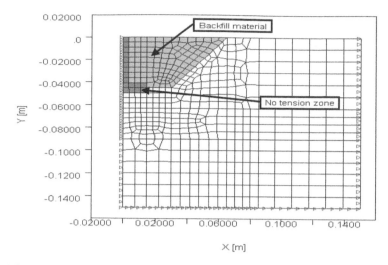

Figure 3. Finite element mesh.

5% cement. For the sand with 1% cement backfilling, the rupture surface angle of about 35° with the vertical considering a plane starting at the top perimeter of the footing (at 40 mm depth) and finishing at the external perimeter affected by the pullout at the top sand-cement bed. Such rupture surface angle increased to about 45° with the vertical for sand containing 3% cement and to 60° with the vertical for sand with 5% cement. For the uncemented sand backfilling, a punching failure with rupture surface angles near to vertical was observed.

3 NUMERICAL ANALYSES

The centrifuge uplift tests provide a clear indication of the benefits of the addition of cement to sand backfill for the anchor type under consideration. To facilitate generalization of these findings, Finite Element analyses of the centrifuge tests were performed to investigate if the observed response could be replicated. All soils in the analyses were assumed to behave as isotropic linear elastic-perfectly plastic materials with the Mohr-Coulomb strength parameters inferred from the direct shear box tests. For each anchor test, one analysis was performed using the peak strength parameters (cohesion intercept and friction angle) to predict the peak uplift capacity.

The Finite Element analyses were performed using the SAFE Finite Element program (OASYS 2002). The analyses adopted an axisymmetric mode of deformation and therefore the square footings were represented by equivalent circular footings with the same area. Fully rough interfaces were assumed between the footing and surrounding soil and tension was not permitted at the footing base interface. The Young's modulus, E', for each type of backfill was varied as a set multiple of the very small strain shear modulus (G_0); a best-fit to the predictions discussed below was found by setting $E' = G_0/30$. A nominal E' value of 20 MPa was specified for the clay outside the excavated area (this value had no effect on the predicted anchor response) and E' values for the aluminum stem and steel base of 70 GPa and 200 GPa were employed.

The Finite Element mesh is shown in Figure 3. The unit weights of soil and pore water input into the numerical model were factored by $n = 50$ in keeping with the centrifugal acceleration applied in the centrifuge tests. In this way, the numerical analyses directly modeled the centrifuge tests rather than their equivalent prototypes. Each effective stress analysis assumed fully drained conditions and displacements at the top of the anchor stem were increased incrementally until failure occurred.

The uplift peak stress capacity of the footings predicted by the FE analyses are 48 kPa for the uncemented sand, 115 kPa for 1% cement content, 165 kPa for 3% cement content and 259 kPa for

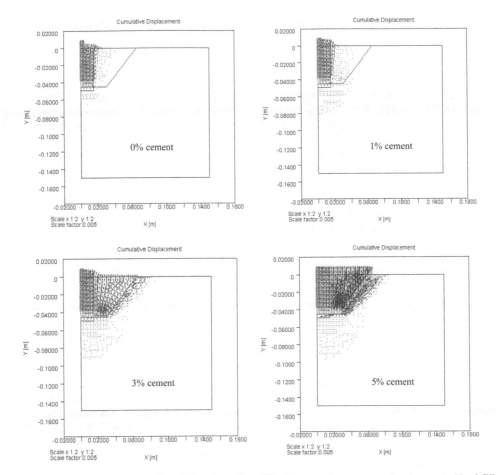

Figure 4. Displacement vectors from FE analyses for uplift of footings in uncemented and cemented backfill.

5% cement content. It is apparent that the predicted peak capacities are within 15% of the observed peak values for all cases and almost perfectly match the capacities measured with backfill cement contents of 3% and 5%.

The computed displacement vectors at peak uplift load for the un-cemented and cemented backfill are presented in Figure 4. It is evident that the addition of cement to the backfill sand leads to a progressive outward shift of the failure mechanism and it is this shift that provides the additional anchor stiffness and strength. For the case when the backfill cement content is 5%, it is apparent from the numerical analysis that the mechanism simply involves lifting of the entire block within the excavation. The finite element analysis carried out in the present work, as any other numerical analysis based on continuum mechanics, was not able to capture fissuring and block formation at failure. Even thought, the uplift peak capacity was well predicted through the numerical simulation.

4 CONCLUDING REMARKS

The following observations and conclusions are made regarding pullout tests of shallow footings embedded in loose sand, as well as on layers of weakly to moderately cemented sands:

– The uplift capacity increased due to insertion of cement, besides displacement at failure was reduced, and a noticeable brittle behavior was observed. When compared to the uncemented

337

embedment, uplift capacity increasing was 82% for 1% cement, 198% for 3% cement and 358% for 5% cement;
- The mode of failure was drastically affected according to the degree of cementation, varying from a vertical punching pattern for the uncemented loose sand to the development of fissures followed by development of small blocks for the weakly cemented sand (1% cement – $q_u = 25$ kPa), formation of a rounded fissure with the incidence of encircle radial fissures for 3% cement ($q_u = 87$ kPa) and finally formation of a few radial fissures starting at the center of loading giving rise to the formation of large blocks extending the failure to almost all the cement improved area for the moderately cemented (5% cement – $q_u = 365$ kPa) sand;
- The computed displacement vectors at peak uplift load for the un-cemented and cemented backfill sand leads to the idea of a progressive outward shift of the failure mechanism and it is this shift that provides the additional anchor stiffness and strength;
- Finally, once the excavation made for embedding the foundation needs inevitably to be backfilled, the idea of making a sand-cement backfilling is shown to be a good alternative for improving the load-displacement behaviour of shallow footings submitted to tensile loads.

ACKNOWLEDGEMENTS

The first author had a scholarship from CNPq (Conselho Nacional de Desenvolvimento Cientifico e Tecnologico – Brasil).

REFERENCES

Balla, A. 1961. The resistance to breaking out of mushroom foundations for pylons. *Proceedings of the 5th International Conference on Soil Mechanics and Foundation Engineering*, Paris: Vol. 1, 569–576.
Beckwith, G.H. & Hansen, L.A. 1982. Calcareous soils of the southwestern United States. *Proceedings Geotechnical Properties, Behaviour and Performance of Calcareous Soils*, Philadelphia: ASTM, Vol. 1, 16–35.
British Standard Methods of Test. 1990. Soil for civil engineering purposes. *BS 1377*.
Clemence, S.P. & Veesaert, C.J. 1977. Dynamic pullout resistance of anchors in sand. *Proceedings of the International Conference on Soil-Structure Interaction*, Roorkee, India, 389–397.
Clough, G.W.; Sitar, N.; Bachus, R.C. & Rad, N.S. 1981. Cemented sands under static loading. *Journal of Geotechnical Engineering Division*, New York: ASCE, 107(6), 799–817.
Consoli, N.C.; Rotta, G.V. & Prietto, P.D.M. 2000. The influence of curing under stress on the triaxial response of cemented soils. *Géotechnique*, 50(1), 99–105.
Consoli, N.C.; Prietto, P.D.M.; Carraro, J.A.H. & Heineck K.S. 2001. Behavior of compacted soil-fly ash-carbide lime-fly ash mixtures. *Journal of Geotechnical and Geoenvironmental Engineering*, ASCE, 127(9), 774–782.
Consoli, N.C.; Vendruscolo, M.A. & Prietto, P.D.M. 2003. Behavior of plate load tests on soil layers improved with cement and fiber. *Journal of Geotechnical and Geoenvironmental Engineering*, ASCE, 129(1), 96–101.
Consoli, N.C.; Rotta, G.V. & Prietto, P.D.M. 2006. Yielding-compressibility-strength relationship for an artificially cemented soil cured under stress. *Géotechnique*, London, 56(1), 69–72.
Consoli, N.C.; Foppa, D.; Festugato, L. and Heineck, K.S. 2007-a. Key parameters for the strength control of artificially cemented soils. *Journal of Geotechnical and Geoenvironmental Engineering*, ASCE, 133 (accepted for publication).
Consoli, N.C.; Thomé, A.; Donato, M.; & Graham, J. 2007-b. "Loading tests on compacted soil-ash-Carbide lime layers". *Proceedings of ICE: Geotechnical Engineering*, London, (submitted to publication).
Coop, M.R. & Atkinson, J.H. 1993. The mechanics of cemented carbonate sands. *Géotechnique*, London, 43(1), 53–67.
Cuccovillo, T. and Coop, M.R. 1993. The influence of bond strength on the mechanics of carbonate soft rocks. *Proceedings of the International Symposium on Geotechnical Engineering of Hard Soils – Soft Rocks*, Athens. Proceedings... Rotterdam: A. A. Balkema, Vol. 1, 447–455.
Dupas, J. & Pecker, A. 1979. "Static and dynamic properties of sand-cement". *Journal of Geotechnical Engineering Division*, New York, 105(3), 419–436.

Hardingham, A.D. 1994. Development of an engineering description of cemented soils and calcrete duricrust. *Proceedings of the 1ST International Symposium on Engineering Characteristics of Arid Soils*, Rotterdam, pp. 87–90.

Huang, J.T. & Airey, D.W. 1993. Effects of cement and density on an artificially cemented sand. *Proceedings of the International Symposium on Geotechnical Engineering of Hard Soils – Soft Rocks*, Athens. Proceedings... Rotterdam: A. A. Balkema, Vol. 1, 553–560.

Huang, J.T. & Airey, D.W. 1998. Properties of artificially cemented carbonate sand. *Journal of Geotechnical and Geoenvironmental Engineering*, New York, v. 124, n.6, p.492–499.

Ilamparuthi, K. & Muthukrishnaiah, K. 1999. "Anchors in sand bed: delineation of rupture surface". *Ocean Engineering*, Elsevier Science, 26, 1249–1273.

Joer, H.A.; Jewell, R.J. & Randolph, M.F. 1999. Cone penetrometer testing in calcareous sediments. *Proceedings of the International Conference Engineering for Calcareous Sediments*, Al-Shafei (ed.), Balkema, Rotterdam, Vol. 1, 243–252.

Jovicic, V.; Coop, M.R. & Simic, M. 1996. Objective criteria for determining G_{max} from bender element tests. *Géotechnique*, 46 (2), 357–362.

Ladd, R.S. (1978). Preparing test specimens using undercompaction. *Geotechnical Testing Journal*, ASTM, 1 (1), 16–23.

Lehane, B.M.; Gaudin, C.; Richards, D.J. & Rattley, M.J. 2006. Rate effects on the vertical uplift capacity of footings founded in clay. *Géotechnique*, London (submitted to publication).

Merifield, R.S.; Lyamin, A.V. & Sloan, S.W. 2006. Three-dimensional lower-bound solutions for the stability of plate anchors in sand. *Géotechnique*, 56(2), 123–132.

Meyerhof, G.G. & Adams, J.I. 1968. "The ultimate uplift capacity of foundation". *Canadian Geotechnical Journal*, 5, 225–244.

Oasys. 2002. SAFE *Users manual*, Ove Arup & Partners, London, UK

Ovesen, N.K. 1981. Centrifuge tests of the uplift capacity of anchors. *Proc., 10th Int. Conf. on Soil Mech.and Found. Engng.*, International Society of Soil Mechanics and Foundation Engineering, Stockholm, 717–722.

Rad, N.S. & Clough, G.W. 1985. Static behavior of variably cemented beach sands. Proceedings of the Symposium on Strength Testing of Marine Soils: Laboratory and In-situ Measurement, Philadelphia: ASTM, Vol. 1, 306–317.

Rotta, G.V.; Consoli, N.C.; Prietto, P.D.M.; Coop, M.R. & Graham, J. 2003. Isotropic yielding in an artificially cemented soil cured under stress. *Géotechnique*, London, 53(5), 493–501.

Rowe, R.K.; and Davis, E.H. 1982. "The behaviour of anchor plates in sand". *Géotechnique*, 32(1), 25–41.

Sales, L.F.P. 1998. *The behavior of shallow foundations placed on improved soil layers*. M.Sc. thesis, Federal University of Rio Grande do Sul, Porto Alegre, Brazil. (in Portuguese).

Schnaid, F., Prietto, P.D.M. & Consoli, N.C. 2001. Prediction of cemented sand behavior in triaxial compression. *Journal of Geotechnical and Geoenvironmental Engineering*, New York: ASCE, 127(10), 857–868.

Shirley, D.J. & Hampton, L.D. 1977. Shear-wave measurements in laboratory sediments. *Journal of Acoustics Society of America*, 63 (2), 607–613.

Stefanoff, G; Jellev, J.; Tsankova, N.; Karachorov, P. & Slavov, P. 1983. Stress and strain state of a cement-loess cushion. *Proceedings 8th European Conference of Soil Mechanics and Foundation Engineering*, Helsinki, Rotterdam: A. A. Balkema, 811–816.

Stewart, D.P.; Boyle, R.S. & Randolph, M.F. 1998. Experience with a new drum centrifuge. *Proceedings of the International Conference Centrifuge'98*, Tokyo, Vol. 1, 35–40.

Tessari, M.A. 1998. *The use of coal bottom ash improved with cement as a base for shallow foundations*. M.Sc. thesis, Federal University of Rio Grande do Sul, Porto Alegre, Brazil. (in Portuguese).

Thomé, A. 1999. *Behavior of spread footings bearing on lime stabilized layers*. Ph.D. thesis, Federal University of Rio Grande do Sul, Porto Alegre, Brazil (in Portuguese).

Thomé, A.; Donato, M.; Consoli, N.C. & Graham J. 2005. Circular footings on a cemented layer above weak foundation soil. *Canadian Geotechnical Journal*, 42, 1569–1584.

Applications of Computational Mechanics in Geotechnical Engineering – Sousa,
Fernandes, Vargas Jr & Azevedo (eds)
© 2007 Taylor & Francis Group, London, ISBN 978-0-415-43789-9

Modeling seismic failure scenarios of concrete dam foundations

J.V. Lemos & J.P. Gomes
LNEC – Laboratório Nacional de Engenharia Civil, Lisbon, Portugal

ABSTRACT: Safety evaluation of concrete dams under intense earthquake loading requires the analysis of possible failure scenarios involving the concrete-rock interface and, in some cases, the foundation rock mass. Discontinuum numerical models provide the means to take into account all structural and rock mass joints. The application of discrete element models in this type of dynamic analysis is discussed, namely the numerical representation of the concrete structure, discontinuities, water pressure effects, and dynamic boundary conditions. Experimental results of a scale model of a gravity dam tested on a shaking table are presented, and compared with a discrete element model. An example of a seismic analysis of a concrete arch dam considering the non-elastic behavior of rock mass joints is also presented.

1 INTRODUCTION

Concrete dams rely on sound rock foundations to support important loads, whether the massive weight of gravity dams, or the concentrated thrust at the abutment of arch dams. Failure through the foundation is one of scenarios that must be considered, as the few, but serious, past accidents have shown (e.g. Londe 1987). The best-known case of an arch dam subject to strong motion is the 111 m high Pacoima dam, which withstood two large seismic events, the 1971 San Fernando and the 1994 Northridge earthquakes, when peak accelerations measured at the abutments exceeded 1 g. A major sliding movement was observed on an inclined discontinuity in the left abutment rock mass, which caused several millimeters of separation in the joint between the dam the thrust block, but the dam itself suffered only minor damage (Sharma et al. 1997).

Discrete element models are nowadays a common tool in rock engineering analysis, given their ability to represent deformation and failure modes involving discontinuities. The study of dam foundations is an area of application with specific needs. Firstly, flow from the reservoir induces significant fluid pressures in joints. In addition to the rock mass discontinuities, the concrete-rock interface is also likely to be involved in failure mechanisms. Cracking at the upstream heel, with the possibility of extending along the interface, and consequent installation of reservoir water pressures, are phenomena that need to be taken into account.

Earthquakes are a decisive design action in many regions around the world. In recent years, as more seismological measurements have been available, particularly close to the source, dam engineers have been faced with the need to consider larger peak accelerations for their designs. Safety criteria impose that for the Maximum Design Earthquake (MDE) the structure must not collapse or release the reservoir water, even if seriously damaged (e.g. ICOLD 1986). Nonlinear models are needed to account for the effects of separation and slip on joints, or cracking, namely at the dam-rock contact or the horizontal lift joints in the dam body. In models of arch dams, built as separate cantilevers, the vertical contraction joints are likely to open and close under strong seismic actions. The hydrodynamic interaction of the reservoir water and the concrete upstream face has also to be taken into account, as it may substantially increase the load on the structure, especially for thin arch dams.

The present paper addresses the main issues involved in the seismic analysis of concrete dams, taking into account the non-elastic response of the discontinuities, in the rock mass or the concrete structure. The options available in discrete element models are discussed. Experimental evidence is becoming available to test these numerical models. Results from a dynamic test of a gravity monolith on a jointed foundation, performed at LNEC's shaking table (Gomes 2006), provide a means to assess the performance of the discrete element representations. Finally, a model employed in the safety assessment of an existing arch dam under earthquake action is presented, which provides an estimation of the effects of non-elastic behavior of the rock mass joints during a dynamic event.

2 HYDROMECHANICAL AND CONSTITUTIVE MODELING OF DISCONTINUITIES

Numerical modeling of the hydromechanical behavior of rock joints is essentially based on the assumption of a flow law and the effective stress hypothesis (e.g. Cook 1992). The flow law is generally based on the analogy of parallel-plate flow, and is expressed by a relation of flow rate and pressure gradient, strongly dependant on hydraulic aperture. The effective stress hypothesis allows the decomposition of the total normal stress into the effective stress, carried by the contact between rock walls, and the water pressure. In the elastic range, the normal joint stiffness (k_n) establishes a linear relation between normal effective stress and joint normal displacement. The elastic stiffness of the water in a saturated joint can be calculated as K_w/a, where K_w is the bulk modulus of water, and a is the joint aperture. The decomposition of the total stress corresponds to the association in parallel of 2 "springs", with stiffness k_n and K_w/a.

For steady-state conditions, invoked in most static analyses, the water stiffness plays no role. The pressure distribution depends on the conductivity contrasts between the various discontinuities. Models assuming impervious blocks, thus concentrating all flow in the joints, as in the code UDEC, have been applied in studies of seepage in dam foundations under normal operating conditions by various authors (e.g. Gimenes & Fernandez 2006). However, in transient conditions, the water stiffness has a major effect. During an earthquake, with intense motion lasting only a few seconds, no significant flow is expected to take place in joints with typical apertures. But the cyclic opening and closing of the joint will generate substantial variations in water pressure, and thus effective normal stress, particularly for small apertures, for which the K_w/a ratio can become quite large.

Laboratory experiments by Javanmardi et al. (2005) allowed the measurement of transient water pressures inside a crack created in a concrete specimen subject to a cyclic dynamic movement. These authors were able to quantify the changes in water pressure at various points inside the crack as it opens and closes, fed by a water reservoir. As expected, as the crack opens, the water pressure is reduced, while in the closing mode, the pressure increases (Fig. 1). If we consider a crack starting at the upstream face of a gravity dam, it is the opening mode that is critical for sliding, since it corresponds to a dam movement towards downstream. The authors conclude that, in the opening mode, the water pressure in the cracked length is somewhere between zero uplift and the static uplift pressure, therefore regulations that prescribe full reservoir pressure in the cracked length are overly conservative.

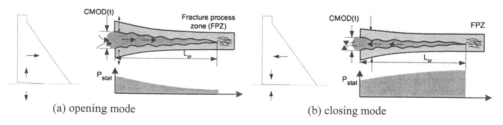

(a) opening mode   (b) closing mode

Figure 1.   Water pressure in the cracked length of the concrete-rock interface, during opening and closing modes of cyclic movement (adapted from Javanmardi et al. 2005).

In a study of the seismic behavior of a gravity dam on a rock mass with 2 joint sets (Lemos 1999), a comparison was made between two assumptions regarding the water pressure in the rock mass joints during the dynamic analysis: (i) keeping the water pressure constant and equal to the steady-state flow case (ii) allowing the transient variation of water pressures, governed by the K_w/a term of joint stiffness. The joint dynamic slip displacement were larger in case (i), which is compatible with the Javanmardi et al. (2005) experimental results, since joint slip during a seismic event tends to occur in a few discrete episodes when the normal effective stress is low. Accounting for the water stiffness in the joint opening mode reduces water pressure and increases effective stress, thus preventing the continuation of slip. Furthermore, the scant information on joint apertures makes difficult the quantification of the joint stiffness term, which also recommends the simpler assumption of constant water pressures during the dynamic run.

Although various elaborate constitutive models have been proposed for rock joints, most of these are still not available in commercial codes for geomechanical analysis. The complexity of dynamic response also advises the use of simpler constitutive models, as long as they are compatible with the known field data. The Mohr-Coulomb model remains the most straightforward option in many practical cases. Regarding the rock-concrete interface, there is limited experimental data. The Mohr-Coulomb model, taking into account cohesion and tensile strength is typically adopted. Some regulatory codes prescribe a null cohesive strength for failure scenarios, which obviously simplifies the analysis. In fact, when a crack with nonzero tensile strength fails, the drop from peak to residual (usually zero) strength may cause problems in the numerical solution. It is advisable to use a constitutive model with a softening branch in accordance with a physically meaningful fracture energy (e.g. Resende et al. 2004).

3 MODELING THE DAM-FOUNDATION-RESERVOIR SYSTEM

3.1 Modeling the rock mass

The seismic failure scenarios of concrete gravity dams include sliding on concrete lift joints, the concrete-rock interface or rock mass discontinuities. The foundation joint, or a few lift joints, can be easily accommodated by continuum codes with joint elements or interfaces (e.g. Bureau 2005). A blocky foundation is more often approached with discrete elements (e.g. Lemos 1999). In order to have an accurate simulation of the wave propagation in the rock medium, it is better to use deformable blocks with an internal finite element mesh. In codes resorting to explicit solution algorithms, such UDEC/3DEC (Itasca 2003), the time step in a dynamic analysis is governed by the stiffer elements in the system, therefore thin elements or high joint stiffnesses tend to make the analysis more expensive or impractical. Particularly for 3D models of arch dam foundations, considerable simplifications may be unavoidable, as discussed below. However, the maximum size of the elements is always constrained by wave propagation requirements, which usually translate into a minimum of 8–10 elements per wavelength of the highest frequency of interest.

3.2 Selection of discontinuities for explicit representation

Discontinuum models represent explicitly the joints, interfaces and other discontinuity surfaces, in contrast to the equivalent continuum approach. The selection of the discontinuities to be included in the model involves a good understanding of the geological and structural system under study, and must be directed by the aims of the analysis. It is not just the computational run-times that need to be maintained within reasonable limits, but also the effort to generate and verify the model, and to interpret the results. It is often more efficient to build separate models to check different failure scenarios, than to include all features in a single, more complex representation. Dynamic analysis always implies more intricate types of behavior and larger run-times, so the effort to simplify the idealizations is even more critical.

For gravity dam models, cracking and sliding of the concrete-rock interface is usually of primary concern, and a simple numerical model can be devised for such purpose. Failure mechanisms involving rock mass joints lead to more elaborate models, but as long as a 2D approximation is viable, several discontinuities of 2 sets can be easily incorporated. For arch dams, the 3D failure modes in the abutments increase substantially the computational effort in dynamic analyses, as the number of blocks created by, say, 3 sets of joints increases dramatically with the model volume. Therefore, it is usually advisable to limit the blocky system to the region that may be participate in the failure mechanism, and represent the surrounding rock mass by an equivalent continuum model, for example, a group or joined deformable blocks with suitable elastic properties in 3DEC. The case study illustrated in section 5 follows this methodology.

3.3 Concrete structure

The representation of the deformability of the concrete structure is important in dynamic models. Sometimes, a rigid block model is used to assess the global sliding safety of a gravity monolith on a rigid foundation, but a deformable block discretized into an interior finite element mesh is generally a preferable option. Note that to analyze the possible cracking of the upper part of gravity or buttress dams a fine mesh in that region is needed. If only the rock-concrete interface is of concern, a relatively coarse mesh may be used for the dam.

Modeling of an arch dam requires a good representation of bending behavior, which implies the use of higher order finite elements (e.g. 20 node bricks) or several low order elements across the thickness which is somewhat impractical. Intense seismic action causes the opening of the vertical contract joints in the arch, which have to be included in the model.

3.4 Dam-reservoir dynamic interaction

If the primary aim of the analysis is the foundation behavior, the dynamic dam-reservoir interaction may be represented with the simplest model, the Westergaard (1933) added-mass assumption. In general, this technique produces higher dam stresses than more accurate representations that take into account water compressibility (e.g. Câmara 2000).

3.5 Boundary conditions for seismic analysis

When linear behavior is assumed for the rock foundation, it may be represented by massless finite elements, which provide essentially an elastic support for the structure. If the analysis includes the failure of rock joints or the foundation interface, then a more complicated model is required (Lemos 1999). The inertial behavior of the rock needs to be considered, and the boundaries at the top and sides of the model, located at a somewhat arbitrary distance, must be able to simulate the energy radiation into the far-field. The seismic action may be represented as an upward propagating wave, a S-wave for horizontal component of particle motion, and a P-wave for vertical component. The bottom boundary is made a non-reflecting boundary, e.g. with a viscous boundary formulation, and the seismic input is applied as a stress record. At the sides of the model, free-field boundaries are provided to apply the dynamic stresses of the propagating seismic waves (Fig. 2).

4 SHAKING TABLE TESTS OF A GRAVITY MONOLITH ON A JOINTED FOUNDATION

Given the shortage of data on the response of actual dams under strong ground motion, scale models provide the best means for validation of the numerical representations for failure scenarios. Gomes (2006) presents the results of a series of tests on a scale model of a gravity monolith performed at LNEC's shaking table, which were also compared with a 3DEC model.

The physical model, shown Figure 3, is composed of a typical gravity dam section, 1.5 m high, standing on a foundation composed of 4 blocks, defined by 3 horizontal and 2 inclined joints. All 5 blocks are made of cement mortar, and are supported by a concrete base block attached

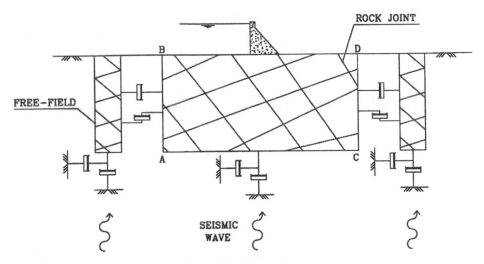

Figure 2. Boundary conditions for seismic analysis of a dam foundation.

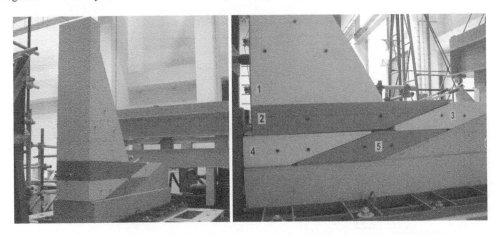

Figure 3. (left) Scale model of gravity monolith and blocky foundation on the shaking table. (right) Position after a dynamic test with top joint locked, showing sliding/rotational failure (after Gomes 2006).

to the shaking table. Loads corresponding to the hydrostatic pressure were applied to the dam upstream face, while the dynamic input was composed of sinusoidal waves, of different amplitude and frequency, applied at the base in the upstream-downstream direction.

In a first series of tests, all joints had the same properties, thus sliding took place primarily along the top joint, representing the concrete-rock interface. In a second group of tests, this joint was locked, allowing failure mechanisms involving the other discontinuities. The final position of one of the latter tests (Fig. 3, right) displays a sliding and rotational movement of the dam and top rock block, in the order of 20 mm, along a path defined by a horizontal and an inclined joint path. The horizontal input was a 10 Hz sine wave, with an acceleration amplitude of 10 m/s^2. The 3DEC model was formed by elastic blocks and purely frictional discontinuities, with friction angles of $\varphi = 32°$ for the joints between mortar blocks and $\varphi = 38°$ for the joints between the mortar blocks and the concrete base (with the dam foundation joint locked).

Experimental and numerical displacements of a point in the middle of the dam block are compared in Figure 4, where the applied motion is also plotted, showing a good agreement. Accelerations at the top of the dam (point 1) and at a point just above the foundation joint (point 4), are also

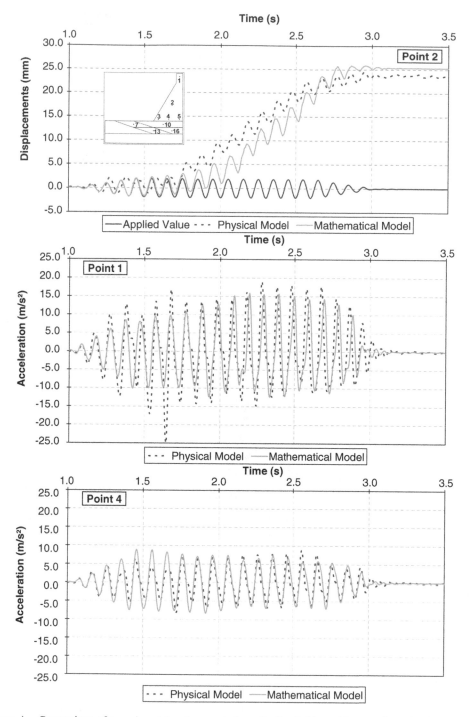

Figure 4. Comparison of experimental and numerical results for shaking table test of gravity dam model. (top) Horizontal displacement at point 2, at mid-height of the dam; (middle) acceleration at point 1, at the top of the dam; (bottom) acceleration at point 4, just above the foundation joint (after Gomes 2006).

presented. At point 4, the agreement of the numerical model with the measured values is very good, while at the top the measured peaks are slightly higher, possibly due to the inability of the relatively coarse mesh to reproduce higher vibration modes of the dam upper section. A discussion of the various experiments, including analyses of the numerical results sensitivity to joint friction, is presented by Gomes (2006).

5   ARCH DAM SEISMIC ANALYSIS

Arch dams apply important thrust forces to the abutments requiring locations where sound rock support may be relied on, which justifies the assumption of elastic rock mass implicit in most seismic analyses currently performed. There are, however, particular geological conditions that recommend a verification of failure scenarios involving the foundation, especially if intense ground motion has to be considered. One such case is El Frayle dam, in Peru, for which a study of seismic safety was performed at LNEC, on request of INADE/Autodema (LNEC 2004, Pina et al. 2006).

El Frayle dam is a double curvature arch with a maximum height of 74 m and a crest development of 51.5 m, at an elevation of 4012 m. The dam is composed of 5 blocks; the thickness of the main cantilever varies from 1.5 m at the crest to 6.2 m at the base. The dam is built in a very narrow valley with steep slopes, in an andesitic rock mass with predominantly vertical jointing. A significant feature is the presence of 3 buttresses to sustain the left slope, built following a rock fall that occurred after the first filling of the reservoir. The main aim of the 3DEC model (Fig. 5) was to examine the possible sliding on the vertical discontinuities of the left bank, under an earthquake with a peak acceleration of 0.6 g. As discussed in section 3.2, the conception of the model was directed towards this objective, implying a drastic simplification of the rock mass representation. While the model of the arch, divided in 5 cantilever blocks, and buttresses presents no computational difficulties, the consideration of 3 sets of joints in the rock mass would lead to a large number of blocks and excessive run-times for a nonlinear dynamic analysis. Therefore, only 2 joints were selected from the main set of vertical joints, and 1 from the nearly orthogonal vertical set, with traces near the left abutment, as can be seen on the top view of the model (Fig. 5). Two horizontal

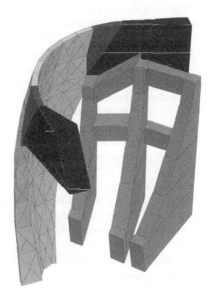

Figure 5.   Numerical model of El Frayle arch dam. (left) Top view, showing the joints represented in the left abutment; (right) Dam and buttresses viewed from downstream (after Pina et al. 2006).

347

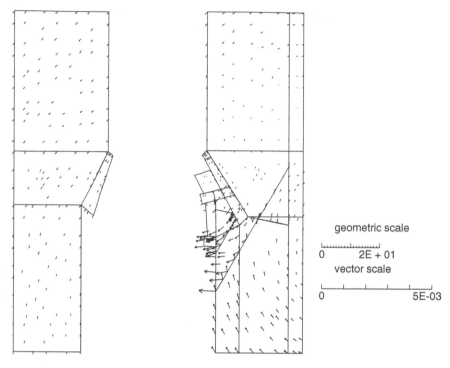

geometric scale

0   2E + 01

vector scale

0   5E-03

Figure 6. Permanent displacements after seismic action. Horizontal cross-section, h = 62 m. (Max. displacement = 0.9 mm).

joints were also included in order to create blocks susceptible of displacing into the valley, shown in darker color. The other lines on the figure are construction lines to define the model geometry, not joints. All blocks were deformable, with an internal finite element mesh.

Even with the simplified representation of the foundation, the model shown has about 6000 grid-points in the deformable blocks, and 4000 contact points at the joints. The time step, required for numerical stability of the explicit algorithm was 1.5×10^{-5} s, implying about 700000 steps for a 10 s earthquake record. As several models always have to be analyzed in this type of study, often with multiple runs performed to evaluate sensitivity to less-known properties, the computational effort easily becomes untenable for very complex models.

The seismic input was applied as a vertically propagating shear wave at the bottom absorbing boundary, as described in section 3. The reservoir effect on the dam was represented by Westergaard added-masses. Water pressures in the rock joints were set in accordance to field monitoring, and assumed constant during the dynamic event. The analysis indicated no significant movement in the rock discontinuities, with only incipient slip. Figure 6 shows the permanent displacements in the rock mass induced by the seismic action, plotted on a horizontal cross-section 12 m below the dam crest. Incipient slip can be observed on the left abutment joints, with a small magnitude (under 1 mm). The presence of the buttresses effectively contributes to the stability of the abutment region. Compressive stresses in the dam and buttresses were well below concrete strength; the main arch vertical contraction joints underwent transient episodes of opening of a few millimeters (Pina et al. 2006).

6 CONCLUSIONS

The regulatory requirements for dam design in seismic areas have pressed the development of analysis tools capable of handling the foundation failure scenarios, generally associated with nonlinear dynamic response of rock discontinuities. The application of discrete element models involves extending some of the features commonly used in rock engineering. The concrete dam and reservoir interaction have to be properly represented, and dynamic boundary conditions considered. The dam and rock mass need to be idealized with deformable blocks, to allow the correct wave propagation and the dam vibration modes. Water pressures in the rock joints, either calculated by a fluid flow analysis or set in accordance with monitored values, play an important role.

The installation of seismic monitoring systems in many dams is starting to provide vital information on the dynamic response under actual field conditions, mainly for low and moderate intensity motions, which is essential for model validation and calibration. For failure scenarios, experiments with scale models remain an important source of information. Gravity model tests have been reported, and arch dam models are now focusing the attention of experimental research. Laboratory testing of rock joints under dynamic loads is also important to verify and improve constitutive assumptions.

REFERENCES

Bureau, G.J. 2006. Use of FLAC for seismic analysis of concrete gravity dams. In Hart & Varona (eds.) *Proc. 4th Int. FLAC Symp.*, Madrid. Minneapolis:Itasca.

Câmara, R.C. 2000. A method for coupled arch-dam-foundation-reservoir seismic behaviour analysis. *Earthquake Engineering and Structural Dynamic*, 29(4):441–460.

Cook, N.G.W. 1992. Natural joints in rock: mechanical, hydraulic and seismic behaviour and properties under normal stress. *Int. J. Rock Mech. Min. Sci.* 29(3):198–223.

Gimenes, E. & Fernandez, G. 2006. Hydromechanical analysis of flow behavior in concrete gravity dam foundations. *Can. Geotech. J.* 43:244–259.

Gomes, J.M.N.P. 2006. *Experimental analysis of failure scenarios of concrete dam foundations – Static and dynamic tests*, Ph.D. Thesis, Universidade Federal Rio de Janeiro (in Portuguese).

ICOLD 1986. *Earthquake analysis for dams*. Bulletin no. 52. Paris.

Itasca 2003. *3DEC – Three Dimensional Distinct Element Code*, Version 3.0. Minneapolis:Itasca.

Javanmardi, F., Léger, P., Tinawi, R. 2005. Seismic water pressure in cracked concrete gravity dams: experimental study and theoretical modeling. *J. Struct. Eng. (ASCE)* 131(1):139–150.

Lemos, J.V. 1999. Discrete element analysis of dam foundations. In V.M. Sharma, K.R. Saxena & R.D. Woods (eds.) *Distinct Element Modelling in Geomechanics*, 89–115. Rotterdam:Balkema.

LNEC 2004. Análisis dinámico considerando un modelo no lineal de la presa de El Frayle y de sus contrafuertes. Report to INADE/Autodema, Peru. Lisbon:LNEC.

Londe, P. 1987. The Malpasset Dam failure, in *Proc. Int. Workshop on Dam Failures*, Purdue, Engineering Geology, 24:295–329.

Pina, C., Lemos, J.V., Leitão, N.S., Câmara, R., Centeno, F.V., Cevallos, F.G., Rodríguez, L.C. & Tejada, J.O. 2006. Seismic analysis of El Frayle Dam considering the non-linear behaviour of the rock majoints, In *Proc. 20th ICOLD Congress*, Barcelona, Q. 84, R. 22, 345–355.

Resende, R., Lemos, J.V. & Dinis, P.B. 2004. Application of a discontinuity model with softening to the analysis of dam foundations using the discrete element method. In H. Konietzky (ed.), *Numerical Modelling of Discrete Materials in Geotechnical Engineering, Civil Engineering and Earth Sciences*, 249–255, Rotterdam:Balkema.

Sharma, R.P., Jackson, H.E. & Kumar, S. 1997. Effects of the January 17, 1994 Northridge earthquake on Pacoima arch dam and interim remedial repairs. In *Proc. 19th ICOLD Congress*, Florence, vol. 4, 127–151.

Westergaard, H.M. 1933. Water pressures on dams during earthquakes. *Trans. ASCE* 98:418–433.

Design of spread foundations according to the EC7 and the reliability evaluation

F.F. Martins
University of Minho, Guimarães, Portugal

ABSTRACT: This paper presents the philosophy and the concepts of Eurocode 7 (EC7) applied to spread foundations. The five ultimate limit states presented in the EC7 are defined. More emphasis is devoted to the ultimate limit state GEO. It is presented an overview of the reliability methods and the three levels corresponding to the probabilistic methods are defined. It is analysed a spread foundation submitted to vertical loading using all the design approaches of the EC7, the traditional Portuguese approach based on global safety factors and the Spanish codes. The results obtained are compared. Using the Excel's solver it is calculated the Hasofer-Lind second moment reliability index β for three foundation dimensions obtained in the EC7 design and with the other approaches. Several coefficients of variation for soil parameters are used to evaluate the reliability index β.

1 INTRODUCTION

The Eurocode 7 is applied to geotechnical design of buildings and civil engineering works. It is composed by two main parts. The first part (EN 1997-1 2004) is related to general rules of geotechnical design and describes the general principles and requirements that ensure the safety, the serviceability and the durability of the supported structures. The second part includes the geotechnical investigations and field and laboratory testing. The Eurocode 7 must be used in combination with the Eurocode 0 related to bases of structural design, with Eurocode 1 related with actions on structures and with other eurocodes of design of materials. The Eurocode 8 devoted to structural and geotechnical design in seismic regions includes in its part 5 the foundations, the retaining structures and other geotechnical aspects.

According to the system of eurocodes, the design fulfils the requirements of the ultimate limit states when the design value of the actions or the effect of the actions E_d is lower than or equal to the design value of the resistance of the ground and/or the structure, R_d:

$$E_d \leq R_d \tag{1}$$

The Eurocode 7 allows the evaluation of R_d and E_d using three different design approaches. This design approaches are related to different partial factors applied to the representative values of the actions, the characteristic values of the material properties and the resistances. However the choice of the design approach and the partial factors must be defined by the countries through the National Annex.

The EC7 (EN 1997-1 2004) requires that, where relevant, five ultimate limit states shall be considered:

Loss of equilibrium of the structure or the ground, considered as rigid body, in which the strengths of structural materials and the ground are insignificant in providing resistance (EQU);

Internal failure or excessive deformation of the structure or structural elements, including e.g. footings, piles or basement walls, in which the strength of structural materials is significant in providing resistance (STR);

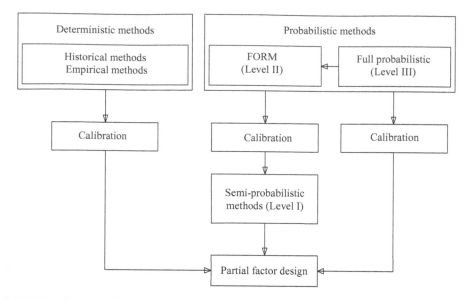

Figure 1. Overview of reliability methods (EN 1990 2002).

Failure or excessive deformation of the ground, in which the strength of soil or rock is significant in providing resistance (GEO);

Loss of equilibrium of the structure or the ground due to uplift by water pressure (buoyancy) or other vertical actions (UPL);

Hydraulic heave, internal erosion and piping in the ground caused by hydraulic gradients (HYD).

The safety in relation to ultimate limits states is applied mainly to persistent and transient situations, and the factors given in Annex A of Eurocode 7 are only valid for these situations.

In this paper it is only considered the ultimate limit state GEO.

The check against the serviceability limit states can be performed in two ways. One way is to require that the design values of the effect of the actions E_d, such as the deformations and settlements, are lower than or equal to the limit values, C_d. Another way is to use a simplified method based on the comparable experience.

2 RELIABILITY METHODS

According to the EN 1990 (2002) the partial factors and the combination factors ψ used in Equation 1 can be evaluated either on the basis of calibration related to a long experience tradition or on the basis of statistical evaluation of experimental data and field observations. These two ways can be used separately or combined. However, most of the factors proposed in the eurocodes are based on the calibration based on long experience tradition.

The available methods for calibration of factors and the relation between them are presented schematically in Figure 1 (EN 1990 2002). The abbreviate designation FORM means first order reliability methods.

It can be seen in Figure 1 that the probabilistic methods are divided in three levels. In relation to the level I, semi-probabilistic methods are used to define the partial factors. In both levels II and III the measure of reliability should be identified with the survival probability $P_s = 1 - P_f$, where P_f is the failure probability for the considered failure mode and within an appropriate reference period. The structure should be considered unsafe when P_f is larger than a pre-set target value P_0. It must be underlined that the probability of failure is only a reference value that does not necessarily represent the actual failure. It is used for code calibration purposes and comparison of reliability levels of structures.

The level III uses exact probabilistic methods and the evaluation of the probability of failure or reliability of the structure is based on the statistical distributions of all basic variables.

In the level II approximated probabilistic methods are used and the reliability is measured using the reliability index β instead of the probability of failure. The reliability index β is related to P_f by:

$$P_f = \Phi(-\beta) \tag{2}$$

where Φ is the cumulative distribution function of the standardised normal distribution. The increasing of the β values corresponds to the decreasing of the probability of failure P_f.

The probability of failure can be expressed by:

$$P_f = \text{Pr ob}(g \leq 0) \tag{3}$$

where g is a performance function given by:

$$g = R - E \tag{4}$$

where R is the resistance and E the effect of actions. R, E and g are random variables.

The structure is considered to survive if $g > 0$ and to fail if $g \leq 0$.

According to Favre (2004), Cornel, in 1969, was the first that proposed a measure of safety under the form of the index β:

$$\beta = \frac{\mu_g}{\sigma_g} \tag{5}$$

where μ_g is the mean value of g and σ_g is the standard deviation of g.

This index β is not invariable in relation to the formulation of the performance function.

To overcome this situation, five years later, Hasofer and Lind proposed a new definition for β (Favre 2004):

"The shortest Euclidian distance, in the reduced Gaussian space, from the origin to the performance equation $\Sigma(y) = 0$"

where $\Sigma(y)$ is the transformed performance function in the reduced Gaussian space.

The theory related with the evaluation of β can be found in the specialized bibliography as, for example, in Favre (2004).

More recently Low & Tang (1997) proposed an efficient method using spreadsheet software for calculating the Hasofer-Lind second moment reliability index. The method is based on the perspective of an ellipse that is tangent to the failure surface in the original space of variables. Iterative searching and numerical partial differentiation are performed automatically by a spreadsheet's optimization tool.

The matrix formulation is the following:

$$\beta_{HL} = \min \sqrt{\left(X - \mu_X^N\right)^T C^{-1} \left(X - \mu_X^N\right)} \tag{6}$$

Restrained to g(X)=0

where X is a vector representing the set of random basic variables which include the effect of actions E and resistances R; μ_X^N is the vector of the mean values of the basic variables X with the upper index N meaning normal or equivalent normal distribution; and C is the covariance matrix.

In the case of variables of non-normal distribution it is necessary to establish relationships between non-normal distribution and its equivalent normal distribution. This can be obtained by equating the cumulative probability and the probability density ordinate of the equivalent normal distribution with those of the corresponding non-normal distribution at the design point X*. This leads to the following equations:

$$\sigma_{X_i}^N \left(x_i^* \right) = \frac{\phi \left\{ \Phi^{-1} \left[F_{X_i} \left(x_i^* \right) \right] \right\}}{f_{X_i} \left(x_i^* \right)} \tag{7}$$

$$\mu_{X_i}^N \left(x_i^* \right) = x_i^* - \sigma_{X_i}^N \left(x_i^* \right) \Phi^{-1} \left[F_{X_i} \left(x_i^* \right) \right] \tag{8}$$

where $\Phi^{-1}[.]$ is the inverse of the cumulative probability of a standard normal distribution; F_{X_i} (x_i^*) is the original cumulative probability evaluated at x_i^*; $\phi\{.\}$ is the probability density function of the standard normal distribution; and $f(x_i^*)$ is the original probability density ordinates at x_i^*.

This method was implemented in this paper using the Excel's solver which is invoked to minimize β, by changing the values of the X vector, subject to $g(X) = 0$.

3 DESIGN METHODS OF SPREAD FOUNDATIONS

The methods used to verify the design of foundations are the direct method, the indirect method and the prescriptive method.

In relation to the direct method it is necessary to perform two separate verifications. One for ultimate limit states and other for serviceability limit states. For both limit states it is necessary to use a calculation model that may be numerical, analytical or semi-empirical. The last model is based on in situ test results.

The indirect method is based on comparable experience and uses the results of field or laboratory tests or other observations. This method covers both the ultimate limit states and the serviceability limit states and in the calculations may be used analytical and semi-empirical models.

In the prescriptive method the design is evaluated on the basis of comparable experience. The calculation model may include charts or tables.

4 BEARING RESISTENCE USING THE DIRECT METHOD

In relation to the bearing resistance of a spread foundation the following inequality shall be satisfied:

$$V_d \leq R_d \tag{9}$$

where V_d is the design value of vertical load or component of the total action acting normal to the foundation base and R_d is the design value of the resistance.

The design value of any component F_d of V_d shall be derived from representative values using the following equation:

$$F_d = \gamma_F \times F_{rep} \tag{10}$$

with

$$F_{rep} = \psi \times F_k \tag{11}$$

Values of ψ are given by EN 1990 (2002) and values of partial factor γ_F are given in Table 1.

R_d can be calculated through the analytical expressions presented in the sample given in Annex D of EC7 (Part 1). The design values of the strength parameters of the ground used in these expressions are obtained by dividing its characteristic values by the partial factors presented in Table 2. The resistance is also divided by a partial factor given in Table 3.

Table 1. Partial factors on actions (γ_F) or the effects of actions (γ_E).

Action		Symbol	Set	
			A1	A2
Permanent	Unfavourable	γ_G	1.35	1.0
	Favourable		1.0	1.0
Variable	Unfavourable	γ_Q	1.5	1.3
	Favourable		0.0	0.0

Table 2. Partial factors for soil parameters (γ_M).

Soil parameter	Symbol	Set	
		M1	M2
Angle of shearing resistance*	$\gamma_{\varphi'}$	1.0	1.25
Effective cohesion	$\gamma_{c'}$	1.0	1.25
Undrained shear strength	γ_{cu}	1.0	1.4
Unconfined strength	γ_{qu}	1.0	1.4
Weight density	γ_γ	1.0	1.0

*This factor is applied to $\tan \varphi'$.

Table 3. Partial resistance factors (γ_M) for spread foundations.

Resistance	Symbol	Set		
		R1	R2	R3
Bearing	$\gamma_{R,v}$	1.0	1.4	1.0
Sliding	$\gamma_{R,h}$	1.0	1.1	1.0

Table 4. Combinations for the different design approaches.

Design approach	Combination
1	$A1' + M1' + R1$
	$A2' + M2' + R1$
2	$A1' + M1' + R2$
3	A1 or $A2' + M2' + R3$

The manner in which the partial factors are applied shall be determined using one of three design approaches given in Table 4.

In relation to Design Approach 3 the partial factor A1 is applied on structural actions and A2 is applied on geotechnical actions.

As was mentioned before the Annex D of EC7 presents a sample analytical method for bearing resistance calculation of a spread foundation both on drained and undrained conditions. As the example presented in this paper is related to drained conditions it is presented below the analytical equation used in these conditions:

$$R/A' = c'N_c b_c s_c i_c + q'N_q b_q s_q i_q + 0,5\,\gamma'B'N_\gamma b_\gamma s_\gamma i_\gamma \tag{12}$$

Table 5. Characteristics of the analysed foundation and settlements.

B	L	D	p	V	q_c	Es	Settlements (cm)	
m	m	m	kPa	kN	kPa	kPa	Computed	Measured
2.6	22.8	2.0	179	10611.12	3924	14842.9	3.68	3.89

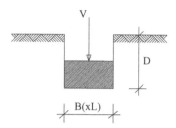

Figure 2. Geometrical characteristics and applied load on foundation.

where c′ is the effective cohesion, q′ is the overburden pressure at the level of the foundation base, γ′ is the effective weight density of the soil below the foundation level, B′ is the effective foundation width, N are the bearing capacity factors, b are the factors for the inclination of the base and i are the factors for the inclination of the load. The subscripts c, q, γ used with b, s and i are related to cohesion c, overburden pressure q and weight density γ.

In the example presented here are performed analyses based on EC7, traditional Portuguese analyses using the global safety factors and the Hansen and Vesić methods (Bowles 1996) and analyses based on Spanish geotechnical codes (Recomendaciones Geotécnicas para el Proyecto de Obras Marítimas y Portuarias, ROM 05-94, Documento Básico DB-4 "Cimentaciones" do Código Técnico de La Edificación e Guia de Cimentaciones de Obras de Carretera (Perucho & Estaire 2005)).

5 EXAMPLE

The example presented here is related to a case studied by Schmertmann and presented by Bowles (1996). It was already presented by the author in the part related to the EC7 (Martins 2006) which is presented here. However the part related to the reliability evaluation is only introduced in this paper. It is a shallow foundation of a bridge pier settled on silty sand. Table 5 presents the geometric characteristics of the foundation (Fig. 2), the value of contact pressure, p, the corresponding vertical action, V, the tip resistance from CPT-test, q_c, the Young's modulus, E_s, the computed settlements by a method presented by Bowles (1996) as well as the measured values presented by the same author.

For the EC7 calculations the values presented in Table 5 are considered as characteristics.

Bowles (1996) says neither the percentages of V corresponding to the permanent and the variable actions nor the angle of shearing resistance. In relation to the actions it is considered here 60% for the permanent actions and 40% for the variable actions. The value of the angle of shearing resistance was established on the base of the Table 6 presented in a provisory version of EC7-Part 3 (ENV 1997-3 1995) which related the tip resistance from CPT-test, q_c, with the angle of shearing resistance, φ′, and with the Young's modulus, E_s. Whereas the soil is silty it was considered a value slightly less than those obtained in Table 6. Therefore, it was considered φ′ = 32°.

Table 6. Angle of shearing resistance φ' and Young's modulus E_s for sands from cone resistance q_c.

Relative density	q_c MPa	φ' °	E_s MPa
Very low	0.0–2.5	29–32	<10
Low	2.5–5.0	32–35	10–20
Medium	5.0–10	35–37	20–30
High	10.0–20.0	37–40	30–60
Very high	>20.0	40–42	60–90

Table 7. Information related to random basic variables.

	G_v kN	Q_v kN	$\tan \varphi'$	γ' kN/m³
X_k	6366.67	4244.45	0.6249	10
$V_x\%$	1	10	5–10–15	5–7.5–10

To evaluate the reliability it is used the following vector X representing the set of basic random variables for this situation:

$$X^T = \{G_v, Q_v, \tan \varphi', \gamma'\} \tag{13}$$

where G_v is the vertical permanent action; Q_v is the vertical variable action; $\tan\varphi'$ is the tangent of the angle of shearing resistance of the soil and γ' is the effective weight density of the soil. All the other variables were considered as constant.

According to Schneider (1997) cited by Orr and Farrel (1999) the coefficient of variation of $\tan\varphi'$ ranges from 0.05 and 0.15 and of γ ranges from 0.01 to 0.10. Based on these limits and the values presented by Serra & Caldeira (2005), the values presented in Table 7 were considered for the random basic variables.

According to the EN 1990 (2002) normal distributions have usually been used for self-weight and extreme values are more appropriated for variable actions. However, lognormal and Weibull distributions have usually been used for material and structural resistance parameters and model uncertainties. Based on these considerations it was assumed a normal distribution for G_v and γ' and, for sake of simplicity, a lognormal distribution for $\tan\varphi'$ and Q_v.

The equations used to evaluate the mean of these variables are the following (Serra & Caldeira 2005):

$$\mu_{G_v} = \frac{G_{vk}}{1 \pm 1.645 V_{Gv}} \tag{14}$$

$$\mu_{Q_v} = e^{\ln(Q_{vk}) \mp 1.645 V_{Qv} + 0.5\ln(V_{Qv}^2 + 1)} \tag{15}$$

$$\mu_{\tan\varphi} = e^{\ln(\tan\varphi_k) + 0.67 V_{\tan\varphi} + 0.5\ln(V_{\tan\varphi}^2 + 1)} \tag{16}$$

$$\mu_\gamma = \frac{\gamma_k}{1 - 0.67 V_\gamma} \tag{17}$$

The sign $\pm$ in Equation 14 and $\mp$ in Equation 15 allows considering the vertical action as favourable (upper sign) and unfavourable (lower sign) for the foundation safety.

357

Table 8. Synopsis of the results for the bearing resistance.

Approach	V kN	R kN	R/V
DA-1-Comb2	11884.45	24307.20	2.05
DA-1-Comb1	14961.68	49766.63	3.33
DA2	14961.68	35547.59	2.38
DA3	14961.68	24307.20	1.62
Hansen	10611.12	44427.02	4.19
Vesić	10611.12	51658.93	4.87
Código Técnico	10611.12	45886.23	4.32
Guia Cimentaciones	10611.12	50152.45	4.73
ROM 0.5-94	10611.12	44812.84	4.22

The equivalent normal parameters can be obtained by:

$$\mu^N_{X_i} = X_i\left(1 - \ln(X_i) + \lambda_{X_i}\right) \tag{18}$$

$$\sigma^N_{X_i} = X_i\zeta_{X_i} \tag{19}$$

where λ_{X_i} and ζ_{X_i} represent the mean value and the standard deviation of the normal variable $Y_i = \ln(X_i)$.

5.1 Design according to the EC7 (Level I)

Table 8 presents the results obtained for the three design approaches of EC7, the traditional method (Hansen and Vesić) used in Portugal and the Spanish codes. For the EC7 approaches both the actions V and R are design values whereas for the other situations those values are not affected by any safety factor.

To obtain the foundation allowable load either in the traditional Portuguese calculations or using the Spanish codes it is used a safety factor equal to 3 to lower the resistance R. In the Portuguese case this factor ensures implicitly that the maximum allowable settlement is not surpassed. That's why the settlement was not computed. In the Spanish case, according to Perucho & Estaire (2005), it is also necessary to verify the settlements.

In all the computations performed by the Hansen method, Vesić method and Spanish codes it was obtained a safety factor greater than 3, surpassing in all of them 4. In these cases the higher safety is obtained using the Vesić method (4.87) and the lower safety is obtained through the code ROM 0.5-94 (4.22). In the case of EC7 is the design approach 3 that presents lower safety (R/V = 1.62) and, therefore, it determines the design in this study.

In relation to the serviceability limit states, as it can be seen, the measured settlement is lower than the value considered allowable for bridge piers, which is 5 cm (Seco e Pinto 1997).

Next it will be maintained the ratio L/B and the "optimal" width of the foundation will be calculated. This width, in the case of EC7, is that that lead to the equating between the design vertical action, V_d, and the design bearing resistance, R_d. In the Hansen method, Vesić method and Spanish codes that width will correspond to a ratio R/V = 3. The obtained results are presented in Table 9.

As it can be seen, in the EC7 case, is the Design Approach 3 that determines the dimensions of the foundation (B = 2.11 m) and in the other approaches are the Hansen method and the code ROM 05-94 that lead to larger width (B = 2.25 m), nevertheless the other analysis lead to values very close to this.

In relation to the serviceability limit states the values obtained for the settlements using the procedure of Bowles (1996) are also presented in Table 9 and, as it can be seen, are all lower than 5 cm.

Table 9. "Optimal" foundation width maintaining the ratio L/B and corresponding settlements.

Approach	B m	Settlement cm
DA-1-Comb2	1.91	–
DA-1-Comb1	1.56	–
DA2	1.81	–
DA3	2.11	4.45
Hansen	2.25	4.20
Vesić	2.13	4.41
Código Técnico	2.22	4.26
Guia Cimentaciones	2.15	4.37
ROM 0.5-94	2.25	4.20

Table 10. Values of β_{HL} for favourable actions.

B	$V_{\gamma'} \setminus V_{\tan\varphi'}$	5%	10%	15%
1.91	5%	6.81	5.46	5.09
	7.5%	6.61	5.46	5.09
	10%	6.33	5.44	5.10
2.11	5%	8.36	6.69	6.24
	7.5%	8.06	6.68	6.25
	10%	7.67	6.65	6.24
2.25	5%	9.11	7.00	6.40
	7.5%	8.68	6.97	6.41
	10%	8.13	6.93	6.40

Therefore, in the analysed foundation, the design according to EC7 is determined by the ultimate limit states and in the traditional Portuguese approach the use of a global safety factor equal to 3 covers the serviceability limit states. The computed settlements for the Spanish codes are also below the settlements considered allowable.

5.2 Reliability evaluation (Level II)

Two sets of computations were performed. In the first set the actions were considered as favourable and in the second set the actions were considered as unfavourable. For each set were considered the values corresponding to the "optimal" values obtained in Design Approach 1 (1.91 m), Design Approach 3 (2.11 m) and Hansen and ROM Approaches (2.25 m).

To analyse the sensibility of β to the variation of the geotechnical parameters $\tan\varphi'$ and γ', nine combinations of the coefficient of variation of $\tan\varphi'$ and γ' were performed for each B value. Due to lack of information, it wasn't established in this paper any correlation between the soil parameters.

The obtained values are presented in Tables 10 and 11.

It can be seen in Tables 10 and 11 that the reliability index decreases more pronouncedly with the increase of $V_{\tan\varphi'}$ than with the increase of $V_{\gamma'}$. For higher values of the coefficient of variation of $\tan\varphi'$ the reliability index almost doesn't change with the change of the coefficient of variation of γ' for the same foundation width.

Generally the β values are lower for unfavourable actions and increase with the increase of the foundation area. This is in accordance with the expected because lower β values lead to greater probability of failure.

359

Table 11. Values of β_{HL} for unfavourable actions.

B	$V_{\gamma'} \backslash V_{\tan \varphi'}$	5%	10%	15%
1.91	5%	5.39	3.54	2.93
	7.5%	5.14	3.54	2.95
	10%	4.78	3.52	2.96
2.11	5%	7.26	5.26	3.50
	7.5%	6.93	5.27	3.51
	10%	6.51	5.27	3.51
2.25	5%	8.12	5.64	3.77
	7.5%	7.64	5.63	3.78
	10%	7.04	5.61	3.78

The design using EN 1990 (2002) with partial factors given in Tables 1 to 3 is considered generally to lead to a structure with a β value greater than 3.8 for a 50 year reference period. In the case of favourable actions all the β values are greater than this minimum recommended value. Nevertheless, in the case of unfavourable actions there are cases where the β values are greater than the minimum recommended value. This is the case corresponding to B equal to 1.91 m and the coefficient of variation of $\tan\varphi'$ is equal to 10% and 15% and the case corresponding to B equal to 2.11 m and 2.25 m and the coefficient of variation of $\tan\varphi'$ equal to 15%.

Nevertheless, the higher values of $V_{\tan \varphi'}$ lead to a very broad variation of φ' (Serra & Caldeira 2005). This can be explained by an incorrect evaluation of the soil parameters or an important heterogeneity of the soil. In the latter case the formulation of the Annex D of the EN 1997-1 is meaningless because it is only applicable to homogeneous soil.

6 CONCLUSIONS

As it can be seen, for vertical loading, in the EC7 case, is the Design Approach 3 that led to larger foundation dimensions. However, the traditional approaches (Hansen and Vesić) and the Spanish codes led to larger values of the foundation dimensions. The computed settlements obtained in all the approaches are very close. However the higher settlement is obtained with the dimensions obtained with the EC7 calculations.

The mean of resistances obtained through the Hansen and Vesić methods is about 66% of the value of the bearing resistance obtained for EC7. In relation to the Spanish codes this value is about 65%. The design bearing resistance R_d obtained through EC7 is close to the medium value of the resistance obtained with the Spanish codes dividing the resistance value by 2. The difference is of 3.4% and is similar to that obtained by Perucho & Estaire (2005) that is around 3%. Considering the mean of the resistance values obtained through the Hansen and Vesić formulae the difference is of 1.2%.

The verification in relation to the serviceability limit state doesn't govern the design for any of the design approaches whereas the total settlement is lower than the allowable settlement of 5 cm.

In the traditional Portuguese calculations it is current practice to adopt a global factor of safety of 3 for drained conditions considering that the use of this factor ensures that the allowable settlement is not exceeded. In the analysed case this allowable settlement is not exceeded.

In relation to the reliability, as the values of β were obtained based on initial simplifications and assumptions, its values less than 3.8 don't necessary mean that the minimum safety is not satisfied. It must be stressed that the results are influenced by several factors such as the assumptions related to the actions, the statistical distributions of the basic variables and the correlation between them. Therefore, the conclusions presented here shouldn't be generalized to all the situations.

ACKNOWLEDGEMENTS

This work was financed by Foundation for Science and Technology. Project POCI/ECM/57495/ 2004 – Geotechnical risk for tunnels in high-speed trains.

REFERENCES

Bowles, J. E. 1996. *Foundation Analysis and Design*, 5th ed., New York: McGraw Hill.

EN 1990 2002. *Eurocode – Basis of Structural Design*. CEN.

EN 1997-1 2004. *Eurocode 7 Geotechnical Design – Part 1: General rules*. CEN.

ENV 1997-3. 1995. *Eurocode 7 – Part 3: Geotechnical design assisted by field tests* (Draft). CEN.

Favre, J. L. 2004. *Sécurité des ouvrages. Risques. Modélisation de l'incertain, fiabilité, analyse des risques*. Paris: Ellipses Édition Marketing S. A.

Low, B. K. & Tang, W. H. 1997. Efficient reliability evaluation using spreadsheet. *Journal of Engineering Mechanics* 123 (7): 749–752.

Martins, F. F. 2006. Dimensionamento de fundações superficiais recorrendo ao Eurocódigo 7. *XIII Congresso Brasileiro de Mecânica dos Solos e Engenharia Geotécnica, Curitiba, Brasil, 27–31 August 2006*, Vol. 2, 997–1002.

Perucho, A. & Estaire, J. 2005. Comparación del dimensionamiento de cimentaciones aplicando el eurocódigo EC7 y las normativas españolas. In Gomes Correia et al. (eds.), *Segundas Jornadas Luso-Espanholas de Geotecnia, Lisbon, 29–30 September 2005*. Lisbon: SPG, SEMSIG, LNEC, 359–368.

Seco e Pinto, P. S. 1997. Fundações em estacas. Dimensionamento de Estacas em Compressão Segundo o Eurocódigo 7. *Sessão Comemorativa dos 25 da Sociedade Portuguesa de Geotecnia (SPG) – EC7 Projecto Geotécnico. Lisbon: 22–24 October 1997*. Lisbon: SPG, LNEC, II-1-II-67.

Serra, J. B. & Caldeira, L. 2005. Capacidade de carga de fundações superficiais. Dimensionamento de acordo com as abordagens de cálculo do EC7 e avaliação da fiabilidade. In Gomes Correia et al. (eds.), *Segundas Jornadas Luso-Espanholas de Geotecnia, Lisbon, 29–30 September 2005*. Lisbon: SPG, SEMSIG, LNEC, 377–386.

Orr, T. L. L. & Farrel, E. R. 1999. *Geotechnical Design to Eurocode 7*, London: Springer-Verlag.

Applications of Computational Mechanics in Geotechnical Engineering – Sousa,
Fernandes, Vargas Jr & Azevedo (eds)
© 2007 Taylor & Francis Group, London, ISBN 978-0-415-43789-9

Geometrically non-linear analysis – application to a shallow foundation

P.J. Venda Oliveira & L.J.L. Lemos
University of Coimbra, Portugal

ABSTRACT: The paper presents a brief description of a finite elements formulation capable of performing non-linear geometrical analysis. The formulation is applied to the study of a shallow foundation in a drained condition. The influence of the load level, stiffness and the width of the foundation is analysed in terms of vertical and horizontal displacements and evolution at depth of the vertical and horizontal stresses.

1 INTRODUCTION

The technological development noted in the past few decades has helped to improve our understanding of a wide diversity of phenomena, some of which intersect with various spheres of scientific knowledge. Knowledge of geotechnics has also benefited from technological progress, in terms of forecasting the behaviour of works, their execution with the implementation of advanced technological processes, and the observation of their behaviour.

At the moment, the behaviour of a great many situations can be predicted, no matter how complex they are, by means of automatic calculation software based on the finite elements method. This method has also developed considerably, to the point where it is currently capable of simulating most geotechnical works, in all their variants and circumstances, including those with complex boundary conditions, heterogeneities, highly complicated constitutive models, time-dependent phenomena, etc.

But the analysis of most of these problems supposes that the strains are infinitesimal, i.e. that the geometry of the elements remains unchanged during the calculation process. In the case of structures built on very deformable ground, like highly compressible clays, this theory is unworkable, since this type of soil exhibits strains of some significance that will influence subsequent calculation phases. When the formulation depends on the process of calculating the strains determined in the preceding phases, the influence is held to be geometrically non-linear.

The formulation and the analysis described in this paper set out to show the importance of geometrical non-linearity in the calculation of a surface foundation on deformable ground. The load level, stiffness and width of the foundation are analysed in terms of the development of vertical and lateral displacements and additional vertical and horizontal stresses.

2 GEOMETRICALLY NON-LINEAR ANALYSIS FORMULATION

No limits on the magnitude of the displacements (u, v) and their gradients, which induce the initial change in the geometry of the domain, are imposed on the finite elements method formulation presented (Zienkienkiewicz & Taylor, 1991). The part relating to the product of the positional derivatives of the displacement is therefore not neglected, with the strain vector being given by:

$$
\{\varepsilon\}=\left\{\begin{array}{c}\varepsilon_x \\ \varepsilon_y \\ \gamma_{xy}\end{array}\right\}=\left[\begin{array}{c}\frac{\partial u}{\partial x}+\frac{1}{2}\left[\left(\frac{\partial u}{\partial x}\right)^2+\left(\frac{\partial v}{\partial x}\right)^2\right] \\ \frac{\partial v}{\partial y}+\frac{1}{2}\left[\left(\frac{\partial u}{\partial y}\right)^2+\left(\frac{\partial v}{\partial y}\right)^2\right] \\ \frac{\partial u}{\partial y}+\frac{\partial v}{\partial x}+\left[\frac{\partial u}{\partial x}\cdot\frac{\partial u}{\partial y}+\frac{\partial v}{\partial x}\cdot\frac{\partial v}{\partial y}\right]\end{array}\right]=\underbrace{\left[\begin{array}{c}\frac{\partial u}{\partial x} \\ \frac{\partial v}{\partial y} \\ \frac{\partial u}{\partial y}+\frac{\partial v}{\partial x}\end{array}\right]}_{\{\varepsilon_L\}}+\underbrace{\left[\begin{array}{c}\frac{1}{2}\left[\left(\frac{\partial u}{\partial x}\right)^2+\left(\frac{\partial v}{\partial x}\right)^2\right] \\ \frac{1}{2}\left[\left(\frac{\partial u}{\partial y}\right)^2+\left(\frac{\partial v}{\partial y}\right)^2\right] \\ \frac{\partial u}{\partial x}\cdot\frac{\partial u}{\partial y}+\frac{\partial v}{\partial x}\cdot\frac{\partial v}{\partial y}\end{array}\right]}_{\{\varepsilon_{NL}\}}
\tag{1}
$$

where $\{\varepsilon_L\}$ represents the vector of the infinitesimal strains and $\{\varepsilon_{NL}\}$ the non-linear strain vectors, which are defined by the following expressions:

$$
\{\varepsilon_L\}=[B_L]\{U_e^n\}
\tag{2}
$$

$$
\{\varepsilon_{NL}\}=\frac{1}{2}[A]\{\theta_e\}
\tag{3}
$$

where $[B_L]$ is the matrix of the infinitesimal strains defined in terms of functions of form N_i. $[A]$ and $\{\theta_e\}$ are described by the expressions:

$$
[A]=\left[\begin{array}{cccc}\frac{\partial u}{\partial x} & \frac{\partial v}{\partial x} & 0 & 0 \\ 0 & 0 & \frac{\partial u}{\partial y} & \frac{\partial v}{\partial y} \\ \frac{\partial u}{\partial y} & \frac{\partial v}{\partial y} & \frac{\partial u}{\partial x} & \frac{\partial v}{\partial x}\end{array}\right]
\tag{4}
$$

$$
\{\theta_e\}=\left\{\begin{array}{c}\frac{\partial u}{\partial x} \\ \frac{\partial v}{\partial x} \\ \frac{\partial u}{\partial y} \\ \frac{\partial v}{\partial y}\end{array}\right\}=\left[\begin{array}{cc}\frac{\partial}{\partial x} & 0 \\ 0 & \frac{\partial}{\partial x} \\ \frac{\partial}{\partial y} & 0 \\ 0 & \frac{\partial}{\partial y}\end{array}\right][N]\{U_e^n\}=\underbrace{\left[\begin{array}{cc}\frac{\partial N_i}{\partial x} & 0 \\ 0 & \frac{\partial N_i}{\partial x} \\ \frac{\partial N_i}{\partial y} & 0 \\ 0 & \frac{\partial N_i}{\partial y}\end{array}\right]}_{[G]}\{U_e^n\}
\tag{5}
$$

Bearing the above expressions in mind, the relation (1) takes the following form:

$$
\{\varepsilon\}=[B_L]\{U_e^n\}+\frac{1}{2}[A][G]\{U_e^n\}
\tag{6}
$$

the differentiation of which leads to (Almeida e Sousa, 1998):

$$
\{\delta\varepsilon\}=[B_L]\{\delta U_e^n\}+[A][G]\{\delta U_e^n\}=([B_L]+[B_{NL}])\{\delta U_e^n\}=[\bar{B}]\{\delta U_e^n\}
\tag{7}
$$

with the matrix $[\bar{B}]$ in each nodal point, i, being expressed as:

$$
[\bar{B}_i]=[B_{Li}]+[A_i][G_i]
\tag{8}
$$

Considering this formulation, the expression of the change in the internal strain energy is given by:

$$
\delta W_d = \int_{V_e}\{\sigma\}\cdot\{\delta\varepsilon\}dV_e = \int_{V_e}[\bar{B}]^T\cdot\{\delta U_e^n\}\{\sigma\}dV_e = \{\delta U_e^n\}\int_{V_e}[\bar{B}]^T\cdot\{\sigma\}dV_e
\tag{9}
$$

with the change of energy resulting from exterior loads expressed by:

$$\delta W_e = \int_{V_e} \{V_e\}_i \cdot \{\delta U_e^n\} dV_e + \int_{S_e} \{T_e\} \{\delta U_e^n\} dS_e$$

$$= \{\delta U_e^n\} \left[\int_{V_e} \{V_e\} dV_e + \int_{S_e} \{T_e\} dS_e \right] = \{\delta U_e^n\} \{F_e\} \tag{10}$$

The equilibrium condition, which corresponds to equality between the energy of interior strain and the energy of the exterior loads, leads to:

$$\delta W_e = \delta W_i \quad \Rightarrow \quad \{F_e\} = \int_{V_e} [\overline{B}]^T \cdot \{\sigma\} dV_e \tag{11}$$

The stiffness matrix of an element is found by differentiating expression (11):

$$\{\delta F_e\} = \int_{V_e} [\overline{B}]^T \cdot \{\delta\sigma\} + [\delta\overline{B}]^T \cdot \{\sigma\} dV_e \tag{12}$$

where the "stress-strain" relation is described by:

$$\{\delta\sigma\} = [D]\{\delta\varepsilon\} = [D][\overline{B}] \{\delta U_e^n\} \tag{13}$$

and

$$[\delta\overline{B}]^T \{\sigma\} = [G]^T \cdot [M] [\delta\theta_e] = [G]^T \cdot [M][G] \{\delta U_e^n\} \tag{14}$$

with:

$$[M] = \begin{bmatrix} \sigma_x & 0 & \tau_{xy} & 0 \\ 0 & \sigma_x & 0 & \tau_{xy} \\ \tau_{xy} & 0 & \sigma_y & 0 \\ 0 & \tau_{xy} & 0 & \sigma_y \end{bmatrix}$$

Combining expressions (12), (13) and (14), we obtain:

$$\{\delta F_e\} = [R_e] \{\delta U_e^n\} = \{\delta U_e^n\} \underbrace{\int_{V_e} \left([\overline{B}]^T \cdot [D][\overline{B}] \right) dV_e}_{[R_{eI}]} + \{\delta U_e^n\} \underbrace{\int_{V_e} \left([G]^T \cdot [M][G] \right) dV_e}_{[R_{eII}]} \tag{15}$$

Where $[R_e]$ is the stiffness matrix of the element that is decomposed in:

$$[R_e] = [R_{eI}] + [R_{eII}] \tag{16}$$

with the component $[R_{eI}]$ being related to the rheological characteristics of the material and the geometry of the element, and the part $[R_{eII}]$ to the stress state of the element.

3 CASE STUDY

The influence of the consideration of geometrical non-linearity is studied in a problem of a continuous foundation sustained on very deformable soil. The correlations between the load level, stiffness and width of the foundation are analysed in terms of the strain and stress field. In all the analysis

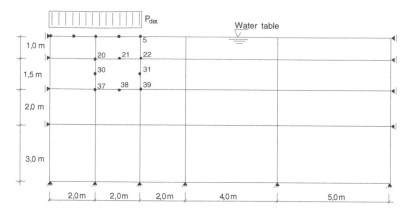

Figure 1.   Finite elements mesh used in the drained analysis of a continuous foundation.

a comparison was drawn between the infinitesimal analysis, that is, linear (LA), and the analysis with finite strains, to which the inclusion of geometrical non-linearity (NLA) corresponds.

In this problem it is taken that the water table coincides with the ground surface and that the material of the foundation soil exhibits isotropic elastic behaviour, with parameters: $E' = 1500 \, \text{kPa}$, $v' = 0,3$, $Ko = 0,5$ e $\gamma_{sat} = 20,0 \, \text{KN/m}^3$.

The finite elements mesh (Figure 1) is composed of 20 isoparametric elements with 8 nodes, with 79 nodal points. The load applied to the foundation is simulated: (i) for the case of a foundation with null stiffness ($E_f = 0$) for nodal forces in the vertical direction, regardless of the strain level, (ii) and in the other cases by elements with a unit weight equal to the intended load.

In the calculation the load is subdivided into 50 equal increments and the value 0.0001% is adopted for the tolerance of the convergence criterion.

4   INFLUENCE OF FOUNDATION STIFFNESS

The influence of the foundation stiffness is studied for three cases: (i) null stiffness ($E_f = 0$); (ii) $E_f = 100 \times E_s$; (iii) $E_f = 10000 \times E_s$. The effect of changing load level for two situations, 200 kPa and 600 kPa, is studied at the same time.

4.1   *Settlements*

Figures 2a and 3a show the evolution of the settlements over the ground surface, obtained respectively by linear and non-linear analysis. In both analysis (LA and NLA), and in qualitative terms, can be seen that the increase in the foundation stiffness induce the uniformity of the settlements and a reduction in their maximum value. It can be seen in Table 1 that with the increment in foundation stiffness ($E_f = 10000 \times E_s$) there is a reduction of over 19% in the settlement value, and this difference increases for non-linear analysis and load level.

Figure 4a gives a direct comparison of the LA and NLA results. For the 200 kPa load it can be seen that the results of the two analysis almost coincide, regardless of the foundation stiffness, although there was a slight tendency for the NLA to generate smaller settlements. This effect is slightly more relevant with increasing foundation stiffness.

A greater influence of the foundation stiffness on the results is observed with the load level increment (P = 600 kPa). For $E_f = 0$, therefore, the NLA caused a slight increase in maximum strains, with this tendency being reversed for $E_f = 10000 \times Es$, i.e., with the foundation stiffness increment the NLA leads to progressively smaller settlements relative to the LA.

These results are in agreement with those of Carter et al. (1977) and Asaoka et al. (1995, 1997), showing that the inclusion of geometric non-linearity tends to reduce the settlements in relation to

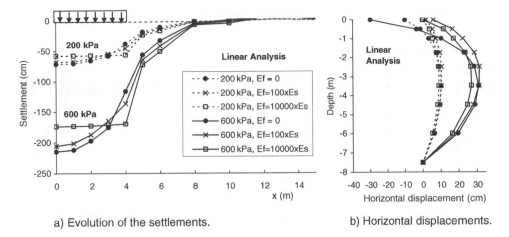

a) Evolution of the settlements. b) Horizontal displacements.

Figure 2.   Linear analysis. Influence of foundation stiffness.

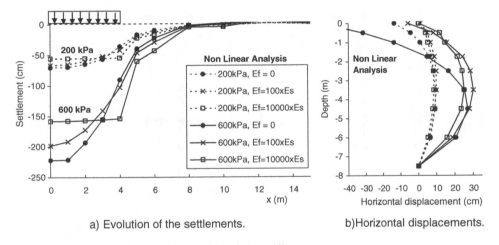

a) Evolution of the settlements. b)Horizontal displacements.

Figure 3.   Finite strain analysis. Influence of foundation stiffness.

Table 1.   Reduction of maximum settlement between $E_f = 10000 \times E_s$ and $E_f = 0$.

Load (kPa)	Linear analysis	Non-linear analysis
200	19,0 %	20,9 %
600	19,0 %	28,8 %

infinitesimal analysis, even though the last-mentioned researchers did not find such a clear trend in the analysis of embankments.

4.2   Horizontal displacements

Figures 2b and 3b show the evolution of the horizontal displacements on the vertical boundary of the foundation, obtained respectively by infinitesimal and finite strain analysis. On the whole, the same qualitative behaviour is seen in both analysis with the differences being more pronounced for

367

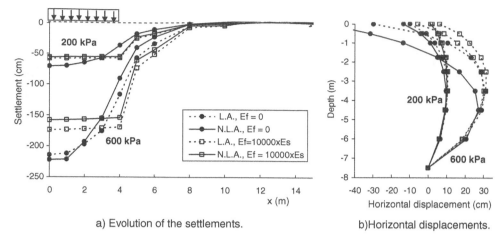

a) Evolution of the settlements.

b)Horizontal displacements.

Figure 4.   Comparison of infinitesimal analysis (LA) and finite strain analysis (NLA). Influence of foundation stiffness.

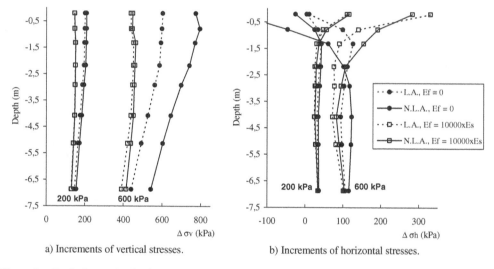

a) Increments of vertical stresses.

b) Increments of horizontal stresses.

Figure 5.   Evolution at depth of $\Delta\sigma_v$ on the axis, of $\Delta\sigma_h$ and τ on the vertical under the foundation side. Influence of foundation stiffness.

the increment in load level. It can also be seen that the consideration of null foundation stiffness ($E_f = 0$) tends to "drag" the surface points towards its axis. In the two cases analysed it can be seen that the displacement horizontal to the surface decreases with the stiffness increment, which is reflected in a slight fall in the value of the maximum horizontal displacements.

Figure 4b gives a direct comparison of both analysis (LA and NLA). It can be seen that the NLA leads to the appearance of smaller lateral displacements towards the exterior of the foundation, and these differences increase with additional load and lower foundation stiffness.

4.3   *Stress analysis*

Figure 5 shows the evolutions at depth of the increased vertical stress ($\Delta\sigma_v$) under the foundation axis, and of the increased horizontal stress ($\Delta\sigma_h$) under the foundation side, for $E_f = 0$ e de $E_f = 10000 \times E_s$.

368

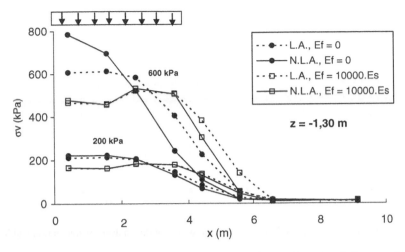

Figure 6. Horizontal evolution of vertical stresses (σ_v) at a depth of 1.30 metres. Influence of foundation stiffness.

For $E_f = 0$, it is found that the NLAs generate vertical stress growth diagrams characterised by increases of $\Delta\sigma_v$ greater than the applied stress (for $P = 600$ kPa the difference is 197 kPa, which is 33.2%). This effect was also noted by Asaoka et al. (1997), and termed "*load concentration effect*". From Figure 6 it can be seen that this effect occurs immediately around the foundation axis (<B/2). With lateral separation the compensation for this effect can be seen, expressed in the reduction of $\Delta\sigma_v$.

The "*load concentration effect*" is directly related to the magnitude of the differential settlements beneath the foundation, which cause the horizontal displacement of the surface nodal points in the direction of its axis. The width over which the nodal loads are applied decreases as the strains evolve, leading to a greater load concentration beneath the foundation. This effect is obviously invisible in infinitesimal analysis, since in these the loads are applied on the non-deformed structure, which does not happen in the NLAs.

In the $\Delta\sigma_v$ diagram, it can be seen that with the increase in foundation stiffness there is a decrease in the vertical stresses. This fact produce smaller settlements and the attenuation of differences between the two analysis (LA and NLA). This behaviour is explained by the increase in foundation stiffness reducing the differential settlements, thereby producing a substantial reduction in the "*load concentration effect*". Behaviour of this kind can also be observed in Figure 6, where for $E_f = 10000 \times E_s$ there is an increase in vertical stress very close to the foundation boundary, and greater vertical stresses are generated with lateral separation. This offsets the fact that $\Delta\sigma_v$ lower than the applied stress are obtained on the foundation axis.

Examination of the horizontal stress evolution diagrams for $E_f = 0$ (Figure 5b) confirms that the consideration of geometric non-linearity gives rise to a slight increase in $\Delta\sigma_h$ in the lower and middle zones of the stratum, and a significant decrease at the surface, leading to negative values (tensions). The above effect is more relevant for the increased applied load. The evolution of the horizontal stress diagram is directly related to the lateral displacement diagram, and it can be seen that the increase in tension stresses at the surface is linked to the surface horizontal displacements towards the foundation axis.

It can also be seen that the consideration of foundation stiffness completely reverses the direction of horizontal stress evolution in the points nearer the surface, with a substantial increase in $\Delta\sigma_h$ at the surface for $E_f = 10000 \times E_s$ (giving rise to important compressive stresses), contrary to what occurs for $E_f = 0$. This behaviour is more marked with the increased load level and the consideration of geometric non-linearity, and is explained by the fact that the foundation stiffness hampers the "dragging" of the surface points towards its axis.

369

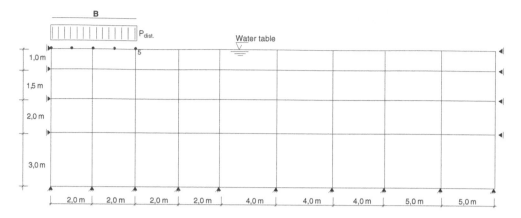

Figure 7.   Finite elements mesh used in the parametric study, for the analysis of foundation width.

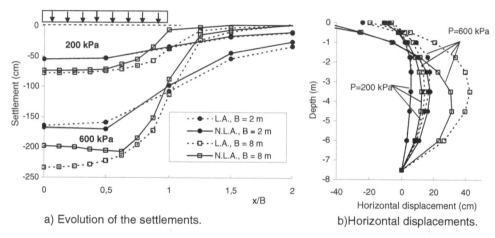

a) Evolution of the settlements.   b)Horizontal displacements.

Figure 8.   Comparison of infinitesimal analysis (LA) and finite strain analysis (NLA). Influence of foundation width.

5   INFLUENCE OF FOUNDATION WIDTH

The influence of the foundation width (B) on the behaviour of the underlying ground, with and without the consideration of geometric non-linearity, is analysed. In order to study foundations more than 4.0 metres width, the parametric analysis described in this section is carried out with a bigger mesh, 30 m wide (Figure 7). This mesh is composed of 36 isoparametric elements with 8 nodes, totalling 135 nodal points.

Foundations 2, 4, 8 and 12 metres of width have been studied and all the calculations assumed loads applied directly on the ground, i.e. a null stiffness foundation.

5.1   *Settlements*

Figures 8a and 9 compare the results of the linear and non-linear analysis as a function of foundation width (B). As expected, it was found that with increasing B there was greater uniformity of the vertical strains beneath the foundation. There was also a tendency for the settlements to decrease with the geometrical non-linearity, and this effect was more relevant with increasing B and load level. This behaviour agrees with that found by Asaoka et al. (1997) in their study of embankments,

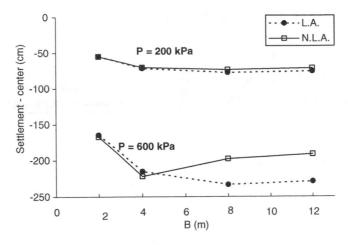

Figure 9. Relation between foundation width and settlement at the axis.

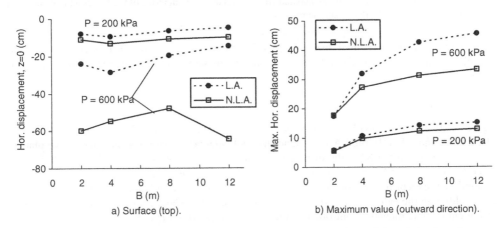

a) Surface (top).   b) Maximum value (outward direction).

Figure 10. Relation between foundation width and lateral displacements beneath foundation boundary.

making it possible to conclude that the embankment width has a considerable influence on the non-linear behaviour (in geometric terms), especially for high load levels.

5.2 Horizontal displacements

Figure 8b gives the comparison of the two analysis in terms of horizontal displacements, for Bs of 2 and 8 metres. It can be seen that with the increased width there is greater discrepancy between the LA and the NLA, and this effect is enhanced with load level. These aspects are also explained in Figure 10, which gives the evolution of the lateral strains at the top of the foundation soil (Fig. 10a) and the maximum lateral displacements (Fig. 10b) as a function of foundation width. Analysis of the figures shows that the behaviour is not linearly dependent on B, and that in the LAs for B of less than 4.0 metres there is increased surface lateral displacement inwards. But for B of more than 4 metres, the opposite occurs. In the NLAs, for high load levels, a different behaviour was observed, characterised by reduced surface horizontal displacement as B changed from 2 to 8 metres, with this trend being reversed for greater widths.

The analysis of Figure 10b permits the conclusion that the maximum horizontal displacements (towards the exterior of the foundation) tend to grow with foundation width, and shows that the

371

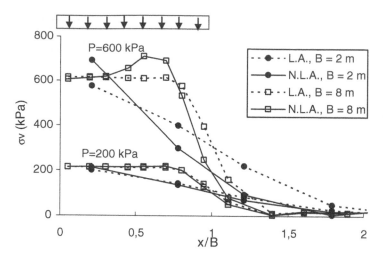

Figure 11. Comparison of the infinitesimal and finite strain analysis, in terms of the horizontal evolution of vertical stresses (σ_v) at a depth of 1.30 metres.

LA generates greater horizontal displacements than the NLA. For values of B over 8 metres, and regardless of the analysis method, there is a trend towards stabilisation of the maximum horizontal strain.

5.3 *Stress analysis*

Figure 11 illustrates the evolution of the vertical stress at a depth of 1.3 metres as a function of x/B, for LA and NLA. In the LA it is found that as B increases there is a more extensive plateau of stresses, while in the NLA the displacing of the *"load concentration effect"* towards the side of the foundation can be seen, which is the zone where there is greater differential settlement. It is found that, except for the zone of *"load concentration effect"*, both analysis give similar stress levels, although the values in the NLA are slightly lower. In the NLA, for lower Bs, the maximum settlement is fundamentally controlled by the *"load concentration effect"*.

6 CONCLUSIONS

The analysis performed allow the following conclusions to be drawn:

(i) in the analysis using geometric non-linearity, a *"load concentration effect"* is observed which depends on the magnitude of differential settlements, and increases with them; so this effect is amplified with increasing load level and decreasing foundation stiffness;
(ii) the non-linear analysis induces the horizontal "dragging" of the surface points towards the foundation axis, an effect that increased with load level;
(iii) with greater foundation stiffness, the NLA tends progressively to give rise to smaller settlements than the infinitesimal analysis does, and the approximation of the curves for the lateral displacements;
(iv) foundation stiffness influences the development of horizontal stresses, especially close to the foundation boundary;
(v) the width of the foundation affects behaviour in qualitative terms, from the linear to the non-linear analysis; so in the NLA, the broader foundations generate smaller maximum settlements than the LA, with this qualitative relation being reversed for smaller foundation width;
(vi) horizontal displacements are more marked for greater foundation width, as are the differences between the maximum displacements obtained in both analysis (LA and NLA).

REFERENCES

Almeida e Sousa, J.N. 1998. Túneis em maciços terrosos : Comportamento e modelação numérica. *PhD thesis.* Faculdade de Ciências e Tecnologia da Universidade de Coimbra.

Asaoka, A.; Noda, T.; Fernando, G.S.K 1995. Effects of changes in geometry on deformation behaviour under embankment loading. *Numerical Models in Geomechanics – NUMOG V, Pande e Pietruszczak, pp. 545–550.* Rotterdam: Balkema.

Asaoka, A.; Noda, T.; Fernando, G.S.K. 1997. "Effects of changes in geometry on the linear elastic consolidation deformation". *Soils and Foundations, Vol. 37, N° 1, pp 29–39.*

Carter, J.P.; Small, J.C; Booker, J.R. 1977. A theory of finite elastic consolidation. *International Journal Solids Structures, Vol. 13, pp. 467–478.*

Venda Oliveira, P.J. 2000. Aterros sobre solos moles – Modelação numérica. *PhD thesis.* Faculdade de Ciências e Tecnologia da Universidade de Coimbra.

Zienkiewicz, O.C.; Taylor, 1991. The finite element method. *Fourth edition, Vol. 2.*

Ground reinforcement

Applications of Computational Mechanics in Geotechnical Engineering – Sousa,
Fernandes, Vargas Jr & Azevedo (eds)
© 2007 Taylor & Francis Group, London, ISBN 978-0-415-43789-9

Influence of vertical rigid piles as ground improvement technique over a roadway embankment. 2D & 3D numerical modelling

M.A. Nunez, D. Dias & R. Kastner
URGC – INSA de Lyon, FRANCE

C Poilpré
GTS, St Priest, FRANCE

ABSTRACT: Settlements remain the major concern during the construction and along the life of a roadway embankment built over soft soils. A foundation ground reinforced by a system of Rigid Inclusions is an effective solution to reduce the relative settlements and increase the bearing capacity of the ground. The principle of the technique consists in associate a network of vertical rigid inclusions and a platform made up of granular materials, to transfer the loads downwards to a more resistant layer.

Modern constructions reinforced by this system began to be developed in the middle of the seventies, mainly in the Scandinavian countries. In France, it has become a common use technique in the last fifteen years.

Confrontations of the various designing methods used in France highlighted differences in results. This resulted in implementing a French National Research Project A.S.I.R.I. (Ground Improvement by Rigid Inclusions) gathering construction companies, engineering and design departments, universities and research centers. This project aims to propose guidelines for the design and construction of reinforced soils by rigid inclusions.

This article presents the case study of the reinforcement of a compressible soil intended to support the roadway embankment of the Senette's street in the community of Carrières –sous- Poissy, in France. Two numerical methods are confronted and discussed; an axisymmetrical configuration (FLAC2D) versus a full three dimensional approach (FLAC3D). A complementary study permits to know the influence of the substratum rigidity.

1 INTRODUCTION

Columns-supported embankments have been used to allow fast embankment construction over soft compressible soils. It combines three components: (1) embankment material, (2) a Load Transfer Platform (LTP), (3) vertical elements extending from the LTP to the firm substratum.

The surface and embankment loads are partially transferred onto the piles by arching which occurs in the granular material (LTP) constituting the embankment. This causes homogenization and reduction of surface settlements. Soil arching is a natural phenomenon encountered in geotechnical engineering (Terzaghi, 1943). Friction along the piles is also involved in the improvement mechanism, leading to a complex soil/structure interaction phenomenon (Figure 1).

As a part of the construction of a residential building at Carrières-sous-Poissy, France, soil treatment was necessary to ensure stability of embankments, pipes and roads. Rigids inclusions technique was selected over stone columns and precharging with drains, because of poor soil properties, the presence of pipelines and mainly due to time restrain of this project.

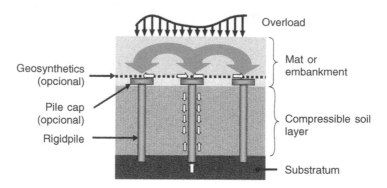

Figure 1.   Rigid inclusions supported embankment.

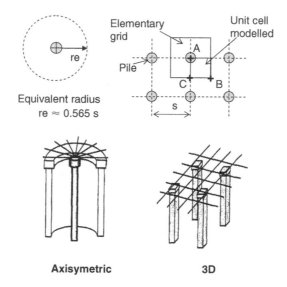

Equivalent radius
re ≈ 0.565 s

Axisymetric   **3D**

Figure 2.   Different symmetric approaches.

Guidelines do not yet exist in France for this technique of soil improvement. A French national research project "ASIRI" (Amelioration de sols par Inclusions Rigides) is starting to solve this problem.

Usually, designing of this technique is made with numerical Finite Elements Methods tools in 2D axymetrical configurations while it's a fully three dimensional problem (Kempton *et al.*, 1998).

In this paper, we compare and discuss results obtained with two and three dimensions numerical simulations made with Difference Finite Methods computer software FLAC2D and FLAC3D (Figure 2).

2   CASE STUDY

2.1   *General project information*

The site is situated next to the Seine River in an urban site so many challenges were to overcome. Geological information shows the presence of alluvional deposits over plastic clays, Meudon's marls and Campanienne's Silts.

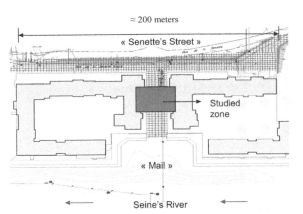

Em (MPa)   pl (MPa)

Figure 3. Residential Building Project at Carrières sous Poissy. Rigid inclusions treated zones.

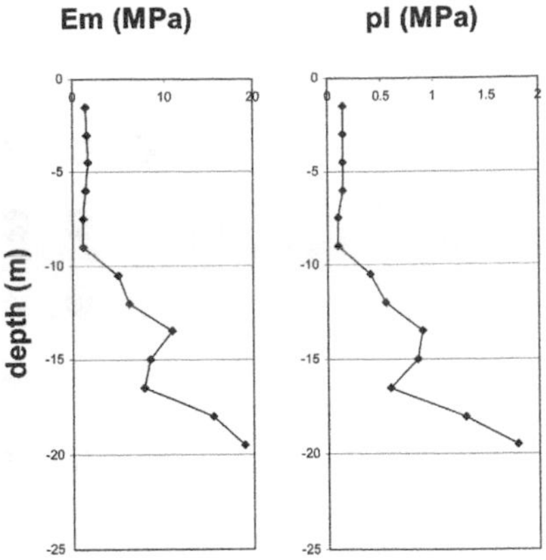

Figure 4. Pressuremeter tests. (Em = Menard's pressuremeter Modulus, pl = Menard's limit presssure)

The project consists in the construction of two residential buildings, a new road "Senette's street" and a pedestrian street "Mail" (Figure 3). The streets were built on a fill of 3.5 meters average height. The inclusions (35 centimeters of diameter) are made up with a Driving Back Auger.

Originally the embankments were proposed to be supported by 1.9 meters square spaced inclusions. After reviewing design parameters it was decided to delimit two zones, "Senette's Street" with 1.9 meters square spaced inclusions, and "Mail" with 2 × 2.2 meters rectangular spaced inclusions.

2.2 Ground conditions

Many soil campaigns were made, mainly using Menard's pressuremeter in-situ test (Figure 4). Oedometer tests were also made to define compressive soils. Water table is followed with piezometers.

The soils met near the future project and taken into account for the study are as follows:

1. Fills: constituted by silts, sands and gravels. Average thicknesses 2 m.
2. Modern alluvional deposits of the Seine: constituted by muddy silts, with finely sandy layers. Thicknesses 4 to 8 meters.
3. Old alluvional deposits: constituted by sands and beige, yellowish gravels. Thicknesses 2 to 4 meters.

The thickness of the modern alluvional deposits is variable. It evolves while approaching the river. Six meters at 140 m of the border, to twelve meters at 60 m.

The watertable level was located two meters under the ground surface, which corresponds to the level of the alluvial watertable, in connection with the Seine River.

The oedometric tests made on undisturbed samples show that the first two meters of silty clay above the watertablecloth are overconsolidated (OCR = 2). It is considered that the grounds under the two meters of depth are normally consolidated.

3 EXPERIMENTAL SURVEY

Survey instrumentation was installed after rigid inclusion's construction (Figure 5); it will permit the analysis of the load transfer phenomenon on pile caps and verification of our settlements prediction.

This survey is composed by: Geodetec horizontal strip, pressure cells and settlement plates.

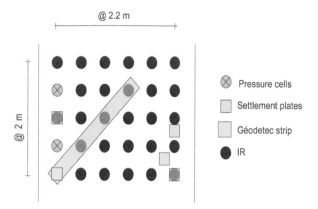

Figure 5.   Situation of the surveillance instrumentation.

A follow-up program was implemented to guarantee the correct operation of the system, i.e. in situ tests routines as single pile axial loading to check the quality of the rigid inclusions embedment concrete's volume was measured for each column to ensure continuity.

4   NUMERICAL MODELLING

4.1   *Introduction*

Numerical analyses were performed to predict settlements and stresses on the system. Also it was important to evaluate the performance of two different materials proposed to form the Load Transfert Plateform (LTP). First, a dense well graded gravelly sand and then a silt-cement treated mixture. Three dimensional and two dimensional axisymmetrical numerical analyses were performed using FLAC3D and FLAC computers programs (ITASCA). Results were compared and discuss.

4.2   *Behaviour modelling*

The embankment fill and the substratum were modelled as an elastic linear perfectly plastic constitutive material, with Mohr-Coulomb failure criteria. Soft soils were modelled with the incremental hardening/softening elastoplastic model, Modified Cam Clay.

The study was driven for a specific zone previously chosen to be instrumented to monitor the behaviour of the embankment; its profile is shown in Figure 6.

Parameters for the Mohr-Coulomb materials are base on the literature and empirical correlations with the pressuremeter tests results (Table 1). Cam Clay parameters where obtain by fitting laboratory oedometric tests to numerical simulations (Table 2).

The numerical analysis included the substratum soil so strains under the inclusions are permitted. However some simplifications were made in our model:

- Setting up the piles by Driving Back Auger is not simulated. We do not take into account pile expansion due to this method.
- Geosynthetics were omitted. French experience has been based mostly on piled structural fills systems without geosynthetic reinforcements (Briançon, 2002).

The construction of the embankment was simulated by layers of 0.5 meters high each, at the end of its construction an overburden of 5 KPa is added to its surface. The capping ratio α, defined as the proportion of the surface covered by the pile caps, is of 2.7% for this case.

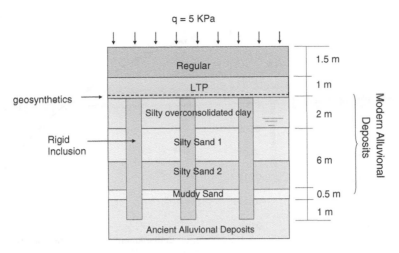

q = 5 KPa

| Regular | 1.5 m |
| LTP | 1 m |

geosynthetics →

Silty overconsolidated clay	2 m
Silty Sand 1	
Silty Sand 2	6 m
Muddy Sand	0.5 m
	1 m

Rigid Inclusion →

Ancient Alluvional Deposits

Modern Alluvional Deposits

Figure 6. Studied case.

Table 1. Fill's Mohr-Coulomb parameters.

	C′(KPa)	ϕ'	ψ'	E (MPa)	γ(KN/m³)
Granular LTP	0	40	10	50	20
Cohesive LTP	20	30	0	50	20
Regular fill	5	30	0	30	20

Table 2. Parameters fitting example and results.

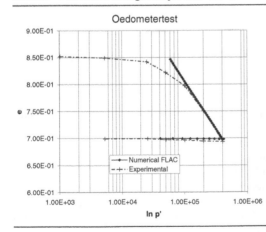

Oedometertest

	Dry Crust	Silty Sand 1	Silty Sand 2	Muddy Sand
σ'p en kPa	55	80	55	85
σ'0 (en kPa)	25,5	48	52	82
OCR	2.16	1.67	1.06	1.04
v	1.85	1.97	1.97	4.31
λ	0,08	0,13	0,08	0,60
vλ	2.7	3.4	2.8	11.2
φ'	25	25	30	20
M	0.98	0.98	1.2	0.77
K (10⁻³)	1.4	3.1	2.6	3
γ (KN/m³)	18.5	17	17	12

Only the 1.9 meters square spaced inclusions configuration is presented with a 2.5 meters high embankment. The rigid inclusions were driven down to 9.5 meters in order to fully cross 8.5 meters of compressible soils and be embedded in the stiff gravelly sand layer (Ancient Alluvional Deposits E = 50 MPa v = 0.3 C = 0 ϕ = 35° ψ = 0°). The concrete modulus was set to 7000 MPa, i.e. a low cement dosed formula.

381

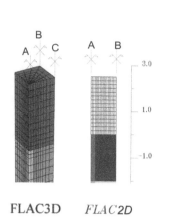

FLAC3D   *FLAC2D*

Figure 7. Studied points and axis for 3D and 2D FLAC models.

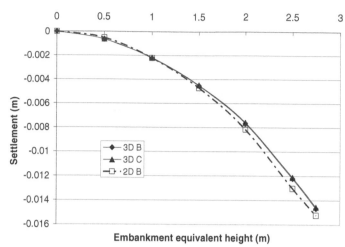

Figure 8. Settlement previsions 3D vs 2D axysimetrical. (a) Rigids Inclusions with a granular LTP.

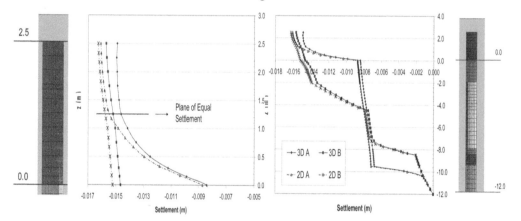

Figure 9. Final vertical displacements all along the rigid inclusion and maximum settlement's axis.

4.3 *Modelling results*

First, a simulation without soil reinforcement was made; the calculated settlement at the interface between compressible soil and LTP was around 16 centimetres by both calculations. This result justified a soil treatment to avoid structures break up (pipelines, walls, etc.).

Next step was to simulate the soil behaviour with rigid inclusions. Table 1 summarizes the characteristics of the two types of LTP proposed for the project.

We studied the points were maximum settlement was expected to happen; in the middle of the square formed by four inclusions for the 3D case (3D B) and at the exterior limit for the axisymetrical model (2D B). Settlements over main axis were also followed, i.e. rigid inclusion and maximums settlements axis. Figure 7 shows the observed points and axis.

Figure 8 shows settlement's evolution at surface due to granular LTP construction for both 2D and 3D models. It can be seen that main maximum settlements are divided by ten. For both cohesive and granular LTP two-dimension results are higher than the three-dimension configuration; 3.7% in the granular case and 5.9 % for the cohesive one.

Figure 9 compares final verticals displacement along rigid inclusion and maximum settlement's axis for the 2D and 3D configuration. The upper plane of equal settlement defined as the plane

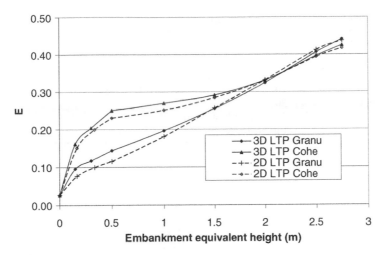

Figure 10. Efficacy evolution for a granular and a cement treated fill. 3D and 2D results.

where differential settlements at the embankment surface are neglected is situated for the granular LTP case 1.25 meters above the ground surface. For Coherent LTPs this plane has the tendency to be at a lower level. The other plane of equal settlement, "the neutral point", defined as the plane where inclusions and compressible soil's settlement are similar is situated 6 meters under the ground surface. Until this limit negative friction is to be expected, under it positive friction is developed. At this plane the inclusion is submitted to a maximum normal stress which has to be considered for the concrete's resistance design.

Efficacy is used to assess the degree of arching in the fill. The efficacy E of the pile support was defined by Hewlett & Randolph (1988) as the proportion of the mat weight carried by the piles (eq. 1).

$$E = \frac{F_p}{W} = \frac{F_p}{A\gamma Hr} \qquad (1)$$

where F_p is the load applied on a pile and W the weight of the embankment's surface (A) covered by a single inclusion. This parameter has a value equal to the capping ratio α when there is no arching effect.

Figure 10 depicts efficacy evolution during embankments construction for both, granular or coherent fill. The graph highlights the significance of the coherent fill's efficacy in the first meter, about 85% higher than for a granular fill at the first 50 centimetres and around 40% at 1 meter. After the first meter, efficacies become very similar. This confirms the important influence of cohesion (C') demonstrated by parametric works done by Jenck, 2005.

4.4 Influence of the substratum rigidity

Several cases of embedment were made to show the influence of the substratum by modifying the elastic properties of the Ancient Alluvional Deposits (AAD). The parametric study deals with the previous studied case; the numerical tests are presented in Table 3. The constitutive model taken into account for the ADD was elastic linear perfectly plastic with a failure criteria of Mohr Coulomb type.

Results were compared for each studied case (Figure 11). Obtained settlements are smaller for rigid bedrock than those resulting of a so called "floating inclusion" (E_{10} E_{50} & $E\infty$). However, results shows a limit value when an increase of the bedrock stiffness do not affect the efficacy value. From this limit, and for this particular study, the embedding substratum seems not to be

Table 3. Studied cases and parameters.

Substratum properties

Numerical test	E (Young Modulus MPa)	ν	ϕ	C (KPa)	ψ
E10	10	0.3	35°	0	0
E30	30				
E50	50 (previously studied)				
E∞	Infinite (nodes fixed)				

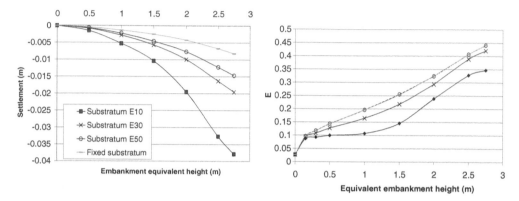

Figure 11. Influence of the bedrock stiffness on rigid inclusions performance.

a very important parameter for the Efficacy. If we take the previous case E_{50} as a reference, the differences with the E_{10} and E∞ cases are about 30% and 1% respectively at the final state. The plane of equal settlement was also observed; significant variation was remarked between these cases, softer the substratum is, quicker the plane of equal settlements is obtained.

5 CONCLUSIONS AND PERSPECTIVES

Rigid inclusions soil reinforcement technique was selected to treat alluvional soil intended to support a two to three meters high road embankment. Simulations and experience show that this technique will reduce settlements so they will not be dangerous for structures stability and performance (Simon, 2006).

For this case two-dimensional axisymetrical model seems to be sufficient to make an accurate prediction of settlements and performance of the rigid inclusions soil reinforcement system.

Cement treated fill was selected to be used as the Load Transfer Platform. Economics reasons were the principal argument for this choice. Concrete mechanical's characteristics were tested in laboratory. A low cement dosed concrete was used in this case.

Even if the final performance of both LTP materials was similar for this case, uncertainties exist on their Young modulus evolution due to loading, not represented here by a Mohr Coulomb criteria; Jenck (2005), demonstrated that non linear behaviour has significant importance in soil arching development.

Experience has shown that the Driven Back Auger rigid inclusions system has a positive influence in the performance of the column reinforced embankments, in this paper their effect was not consider but it began to be studied for further cases. Survey instrumentation was installed after rigid inclusion's construction; we will obtain experimental data soon that will be compared to numerical simulations.

ACKNOWLEDGEMENTS

The authors wish to extend special thanks to Professor C. Plumelle and to Bouyges Immobilier for the cooperation given to realize this study.

REFERENCES

Briançon L. (2002): Renforcement des sols par inclusions rigides – Etat de l'art en France et à l'étranger, *IREX*, 185p.

Camp W. M., Siegel T. (2006): "Failure of a column-supported embankment over soft ground" In: Soft Soil Engineering – Chan & Law (eds).

Hewlett W. J., Randolph M. F. "Analysis of pile embankment" Ground Engineering, 1988, vol. 21, n° 3, pp 12–18.

Itasca (2002). FLAC3D Fast Lagrangian Analysis of Continua, User's guide. ITASCA Consulting Group, Minneapolis, USA.

Itasca (1993). FLAC-Fast Lagrangian Analysis of Continua, User's guide. ITASCA Consulting Group, Minneapolis, USA. User's guide.

Jenck O., Dias D., Kastner R. (2005) Two Dimensional Physical and Numerical Modelling of a Pile-Supported Earth Platform over Soft Soil. Journal of geotechnical and geoenvironmental engineering. ASCE/March 2007.

Kempton, G.T., Russell, D., Pierpoint, N. & Jones, C.J.P.F. (1998). "Two and three dimensional numerical analysis of the performance of geosynthetics carrying embankment loads over piles." *Proc. of the 6th Int. Conf. on Geosynthetics*, Atlanta, Georgia.

Mendoza M. J. (2006) "On the soil arching and bearing mechanisms in a structural fill over piled foundations" in: *SYMPOSIUM Rigid Inclusions in Difficult Subsoil Conditions.* Mexico City.

Simon B., Schlosser F. (2006). State of the Art "Soil reinforcement by vertical stiff inclusions in France" in: *SYMPOSIUM Rigid Inclusions in Difficult Subsoil Conditions.* Mexico City.

Applications of Computational Mechanics in Geotechnical Engineering – Sousa,
Fernandes, Vargas Jr & Azevedo (eds)
© 2007 Taylor & Francis Group, London, ISBN 978-0-415-43789-9

FE prediction of bearing capacity over reinforced soil

C.L. Nogueira
Department of Mines Engineering, School of Mine, UFOP, Ouro Preto/MG, Brazil

R.R.V. Oliveira & L.G. Araújo
Department of Civil Engineering, School of Mines, UFOP, Ouro Preto/MG, Brazil

P.O. Faria
Department of Civil Engineering – CTTMar – Univali, Itajaí/SC, Brazil

J.G. Zornberg
Department of Civil Engineering, University of Texas at Austin, Austin/TX, USA

ABSTRACT: This paper presents the numerical simulation using an elastoplastic analysis of the bearing capacity of shallow foundations. The problem involves axisymmetric conditions on reinforced soil using finite element method (FEM). The foundation soil is modeled as a non-associative elastoplastic Mohr-Coulomb material. The reinforcement is modeled as a linear elastic material. The ultimate bearing capacity obtained in this study is compared to solutions obtained using limit equilibrium and limit analysis. A parametric study was conducted for different configurations of reinforcement for a special case of frictionless foundation soil. The numerical results show good agreement with analytical results indicating the suitability of the numerical model used in this study and implemented into the code ANLOG – Non-Linear Analysis of Geotechnical Problems.

1 INTRODUCTION

An application of the finite element method (FEM) for non-linear elastoplastic analysis of reinforced soil structures under axisymmetric condition is presented in this paper.

The Mohr-Coulomb criterion suggested by Sloan & Booker (1986) and Abbo & Sloan (1995), which includes treatment of the singularities of the original Morh-Coulomb criterion, is used for modeling the foundation soil. A general formulation that considers associative and non-associative elastoplastic models for soil was adopted. Hence, the influence of the dilatancy angle on the bearing capacity of reinforced soil could be investigated. The reinforcement is considered as linear elastic and the soil-reinforcement interface was considered rigid; thus interface elements were not considered in these analyses.

The numerical simulation was conducted using the code ANLOG – *Non Linear Analysis of Geotechnical Problems* (Zornberg, 1989; Nogueira, 1998; Pereira, 2003; Oliveira, 2006).

2 FINITE ELEMENT REPRESENTATION OF REINFORCED SOIL

A discrete representation for reinforced soil structures is adopted in this study. Each component of reinforced soil structure – the soil, the reinforcement and the soil-reinforcement interface – can be represented using a specific finite element with its own kinematic and constitutive equations. In the specific case of a bearing capacity problem of shallow foundations, the soil-reinforcement interface was considered rigid and therefore is not discussed in this paper.

In considering an incremental formulation by FEM, the kinematic equation that describes the relationship between the increment of strain ($\Delta\varepsilon$) and the increment of nodal displacement ($\Delta\hat{u}$) in each finite element can be written as:

$$\Delta\varepsilon = -\mathbf{B}\Delta\hat{u} \tag{1}$$

where

$$\mathbf{B} = \nabla\mathbf{N} \tag{2}$$

∇ is a differential operator and $\mathbf{N}$ is the matrix that contains the interpolation functions N_i. Both the operator and matrix depend on the type of element adopted. The negative sign in Equation 1 is a conventional indicator of positive compression.

The increment of stress ($\Delta\sigma$) can be obtained using the incremental constitutive equation:

$$\Delta\sigma = \mathbf{D}_t\Delta\varepsilon \tag{3}$$

where $\mathbf{D}_t$ is the constitutive matrix defined in terms of the elastoplasticity formulation as:

$$\mathbf{D}_t = \mathbf{D}_e - \mathbf{D}_p \tag{4}$$

where $\mathbf{D}_e$ is the elastic matrix and $\mathbf{D}_p$ is the plastic parcel of the constitutive matrix defined as:

$$\mathbf{D}_p = \frac{\mathbf{D}_e\mathbf{b}(\mathbf{D}_e\mathbf{a})^T}{\mathbf{a}^T\mathbf{D}_e\mathbf{b}+H} \tag{5}$$

$\mathbf{a}$ is the gradient of the yield function ($F(\sigma,h)$), $\mathbf{b}$ is the gradient of the potential plastic function ($G(\sigma,h)$), h is the hardening parameter and H is the hardening modulus. In the case of perfect plasticity, since hardening is not considered, H equals zero.

Starting from an equilibrium configuration where the displacement field, the strain state, and the stress state are all known, a new equilibrium configuration, in terms of displacements, can be obtained using the modified Newton Raphson procedure with automatic load increment (Nogueira, 1998). In this paper, only the elastic parcel of the constitutive matrix was considered in the iterative procedure used to obtain the global stiffness matrix.

At each increment the iterative scheme satisfies, for a selected tolerance, the global equilibrium, compatibility conditions, boundary conditions and constitutive relationships. Yet attention must be given to the stress integration scheme adopted to obtain the stress increments (Equation 3), in order to guarantee the Kuhn-Tucker conditions and the consistency condition.

2.1 Soil representation

The soil is represented by the quadratic quadrilateral isoparametric element (Q8). This element has two degree of freedom, u and v, in the directions r and y (radial and axial), respectively. The stress and strain vectors are defined as:

$$\sigma^T = \begin{bmatrix} \sigma_r & \sigma_y & \sigma_\theta & \tau_{ry} \end{bmatrix} \tag{6}$$

$$\varepsilon^T = \begin{bmatrix} \varepsilon_r & \varepsilon_y & \varepsilon_\theta & \gamma_{ry} \end{bmatrix} \tag{7}$$

In which $\varepsilon_\theta = u/r$. The kinematic matrix $\mathbf{B}$ can be written as:

$$\mathbf{B} = \begin{bmatrix} \partial N_1/\partial r & 0 & & \partial N_8/\partial r & 0 \\ 0 & \partial N_1/\partial y & & 0 & \partial N_8/\partial y \\ N_1/r & 0 & \cdots & N_8/r & 0 \\ \partial N_1/\partial y & \partial N_1/\partial r & & \partial N_8/\partial y & \partial N_8/\partial r \end{bmatrix} \tag{8}$$

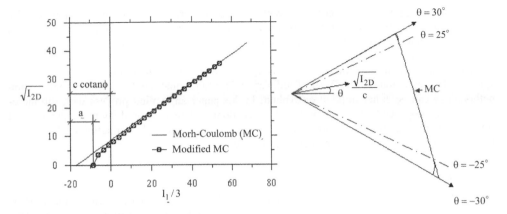

Figure 1. Mohr-Coulomb yield function (Abbo & Sloan, 1995).

where N_i is the i node shape function by the finite element Q8 (Nogueira, 1998). The stiffness matrix for axisymmetric condition for this element is given by:

$$\mathbf{K} = \int_{-1}^{+1}\int_{-1}^{+1}\left[\mathbf{B}^T\mathbf{D}_t\mathbf{B}(2\pi(\mathbf{N}\hat{r})\det\mathbf{J}\right]d\xi d\eta \qquad (9)$$

where $\hat{r}$ is the nodal global coordinate vector; (ξ, η) is the natural coordinate system and $\mathbf{J}$ is the Jacobian operator.

To describe the stress-strain relationship a perfectly elastoplastic model with non-associative plasticity was adopted. The plastic parcel of the constitutive matrix is obtained using the modified Mohr-Coulomb criterion proposed by Sloan & Booker (1986) and Abbo & Sloan (1995) (Figure 1). The modified version of the Mohr-Coulomb model involves removal of the singularities at the edges ($\theta = \pm\pi/6$) and the apex of the original model. Its yield function is written as:

$$F = \sqrt{I_{2D}(K(\theta))^2 + (a\sin\phi)^2} - (I_1/3)\sin\phi - c\cos\phi \qquad (10)$$

where

$$\theta = (1/3)\sin^{-1}\left((-1.5\sqrt{3})I_{3D}(I_{2D})^{-3/2}\right) \quad \theta \in [-\pi/6;\pi/6] \qquad (11)$$

θ is the Lode angle, I_1 is the first invariant of the stress tensor; I_{2D} is the second invariant of the desviator stress tensor, I_{3D} is the third invariant of the desviator stress tensor, c and ϕ are the material cohesion and internal friction angle, respectively. A transition angle (θ_T) was introduced to define the $K(\theta)$ function on the Equation 10. Sloan & Booker (1986) suggest θ_T value range from 25° to 29°. For the case in which $|\theta| > \theta_T$,

$$K(\theta) = A + B\sin 3\theta \qquad (12)$$

where

$$A = (1/3)\cos\theta_T\left(3 + \tan\theta_T\tan 3\theta_T + (1/\sqrt{3})\text{signal}(\theta)(3\tan\theta_T - \tan 3\theta_T)\sin\phi\right) \qquad (13)$$

$$B = (1/(3\cos 3\theta_T))\left(\text{signal}(\theta)\sin\theta_T - (1/\sqrt{3})\cos\theta_T\sin\phi\right) \qquad (14)$$

Or, for the case in which $|\theta| \leq \theta_T$

$$K(\theta) = \cos\theta + \frac{1}{\sqrt{3}}\sin\theta\sin\phi \qquad (15)$$

389

The parcel a sin ϕ was introduced to prevent the singularity related to the surface apex. For the parameter "a" Abbo & Sloan (1995) recommend 5% of (c cotanϕ). The potential plastic function (G) can be written the same way as the yield function (F) but using the dilatancy angle (ψ) instead of the friction angle (ϕ).

An important step in a non linear analysis using MEF relates to the integration of the constitutive equation. This equation defines a set of ordinary differential equations for which the integration methodology can be either implicit or explicit. In this paper an explicit process with sub incrementation, as proposed by Sloan et al. (2001), was adopted. This methodology uses the modified Euler scheme that determines the size of the sub increment automatically evaluating the local error induced during integration of the parcel stress plastic.

2.2 Reinforcement representation

The reinforcement is represented by quadratic one-dimensional isoparametric elements (R3) (Oliveira, 2006). The reinforcement thickness is considered in the constitutive equation. This element has one degree of freedom, u', on its own longitudinal direction r'. The longitudinal direction is related to the radial direction on the local coordinate system according to the following transformation:

$$r' = \mathbf{NT}\hat{r} \tag{16}$$

In which $\mathbf{N}$ is the matrix that contains the shape functions (N_i) for this element (Oliveira, 2006), $\hat{r}^T = [r_1\ y_1 \cdots r_3\ y_3]$ is the nodal global coordinate vector, and

$$\mathbf{T} = \begin{bmatrix} \cos\beta & \operatorname{sen}\beta & \cdots & \cdots & 0 & 0 \\ \vdots & \vdots & \vdots & \vdots & \vdots & \vdots \\ 0 & 0 & \cdots & \cdots & \cos\beta & \operatorname{sen}\beta \end{bmatrix} \tag{17}$$

where $\cos\beta = (dr/d\xi)/\det \mathbf{J}$; $\operatorname{sen}\beta = (dy/d\xi)/\det \mathbf{J}$; $\det \mathbf{J} = \sqrt{(dr/d\xi)^2 + (dy/d\xi)^2}$; $\dfrac{dr}{d\xi} = \sum_{i=1}^{3} \left(\dfrac{dN_i}{d\xi} r_i\right)$ and $\dfrac{dy}{d\xi} = \sum_{i=1}^{3} \left(\dfrac{dN_i}{d\xi} y_i\right)$.

The R3 element has two components of strain and stress: longitudinal ($\varepsilon_{r'}$ and $\sigma_{r'}$) and circumferential ($\varepsilon_{\theta'}$ and $\sigma_{\theta'}$). The kinematic condition is given by the relation:

$$\varepsilon' = -\mathbf{B}\hat{u}' = -\begin{bmatrix} \dfrac{1}{\det \mathbf{J}} \dfrac{\partial N_1}{\partial \xi} & \cdots & \dfrac{1}{\det \mathbf{J}} \dfrac{\partial N_3}{\partial \xi} \\ \dfrac{N_1}{r'} & \cdots & \dfrac{N_3}{r'} \end{bmatrix} \mathbf{T}\hat{u} \tag{18}$$

where $\hat{u}$ is the vector of the nodal global displacement components (u,v).

The constitutive matrix for the reinforcement element is given by:

$$\mathbf{D}_t = \frac{(J/t)}{(1-v^2)} \begin{bmatrix} 1 & v \\ v & 1 \end{bmatrix} \tag{19}$$

where J is the reinforcement stiffness (kN/m), t is the reinforcement thickness and v is the Poisson ratio.

The reinforcement stiffness matrix under axisymmetric condition is given by:

$$\mathbf{K} = \int_{-1}^{+1} \left[(\mathbf{BT})^T \mathbf{D}_t (\mathbf{BT})(2\pi t)(\mathbf{N}\hat{r}) \det \mathbf{J} \right] d\xi \tag{20}$$

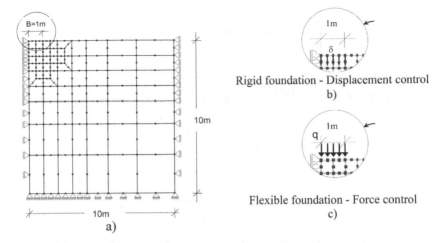

Figure 2.   Finite element mesh: (a) full mesh; (b) detail of the displacement imposed boundary condition; (c) detail of stress imposed boundary condition.

3   BEARING CAPACITY ON UNREINFORCED SOIL

The analyses presented in this study involve smooth circular foundation subjected to vertical loading acting on the ground surface. The problem is analyzed under axisymmetric condition and is modeled as both flexible and rigid foundation using load and displacement controls respectively (Figure 2). The foundation soil is considered weightless. As mentioned the soil is considered as an elastic perfectly plastic material described by a non-associative modified Mohr-Coulomb model. Both the friction and dilatancy angle were varied to assess their influence on the bearing capacity of the shallow foundation. According to Houlsby (1991) the dilatancy is a key factor in geotechnical problems involving kinematic movement restrictions, such as the bearing capacity of shallow and deep foundations. Results of this study are compared with results of studies that utilize equilibrium limit and limit analysis theories.

An incremental-iterative modified Newton Raphson scheme with automatic loading increments is used considering a tolerance of 10^{-4} for the force criterion of convergence. For the stress integration algorithm the following tolerances are used: $FTOL = 10^{-9}$ and $STOL = 10^{-8}$. The FTOL tolerance is related to the transition condition from elastic to plastic state which is affected by the finite precision arithmetic. The STOL tolerance is related to the local error in the stresses in the Euler modified schemes.

Numerical results are presented in terms of the κ factor which is a normalized stress defined as:

$$\kappa = Q/(Ac) = q/c \qquad (21)$$

in which A is the footing area and Q is the reaction force at the foundation, defined as:

$$Q = \sum_{e=1}^{n}\left(\int_{V_e} \mathbf{B}^T \sigma_e dV_e \right) \qquad (22)$$

The reaction force is evaluated as the sum of the internal force's vertical components equivalent to the elements's stress state right beneath the foundation. The cohesion is adopted to normalize the results but in the case of cohesionless soil the atmospheric pressure can be adopted instead.

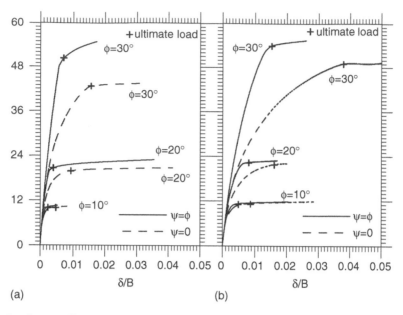

Figure 3. Load versus displacement curves: a) Flexible foundation; b) Rigid foundation.

Table 1. Normalized ultimate bearing capacity ($\kappa_{ult} = q_{ult}/c$).

$\phi(°)$	$\psi(°)$	Flexible foundation	Rigid foundation
10	0	10.1	11.4
	10	10.4	11.4
20	0	20.0	22.0
	20	20.4	22.4
30	0	43.4	49.5
	30	49.5	54.0

Figure 3 presents the κ factor versus normalized settlement (δ/B) curves obtained by ANLOG for flexible and rigid foundations and for different values of friction and dilatancy angles. It can be observed that associative analysis ($\psi = \phi$) provides the lowest displacement at failure.

Table 1 presents normalized ultimate bearing capacity (κ_{ult}), as shown in Figure 3. As expected, the κ_{ult} value obtained for rigid foundation is higher than that obtained for a flexible foundation. The difference in κ_{ult} values was approximately 9.5%, but the highest difference was observed in non-associative plasticity (approximately 11.2%).

Analyses conducted in this study show that when the friction angle was decreased to 10° and 20°, the ultimate bearing capacity factor (κ_{ult}) was no longer affected by the dilatancy angle. For friction angle of 30° the associate plasticity analysis ($\psi = \phi$) provided the highest ultimate bearing capacity factor and the lowest displacement at failure. Zienkiewicz et al. (1975) observed a similar response for friction angles of 40°. Monahan & Dasgupta (1993) reported such behavior for friction angles higher than 25°.

Table 2 presents a comparison between results obtained using ANLOG and those from a classical solution from equilibrium limit by Terzaghi (1943), limit analyses solution by Chen (1975), and a recent numerical solution based on limit analyses using FEM by Ribeiro (2005). Good agreement can be observed among these results.

392

Table 2. Ultimate bearing capacity values for flexible circular footing and associative plasticity.

$\phi(°)$	This study κ_{ult}	Terzaghi (1943) $(1.3N_c)$	Chen (1975)	Ribeiro (2005)
10	10.4	10.86	9.98	11.91
20	20.4	19.29	20.1	24.87
30	49.5	39.18	49.3	52.76

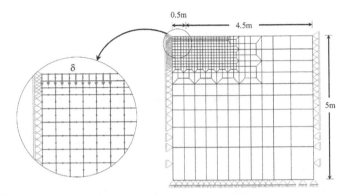

Figure 4. Finite element mesh – Rigid rough circular without embedment shallow foundation.

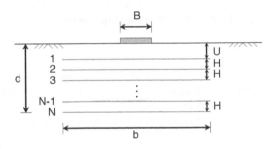

Figure 5. Layout of the rigid foundation on reinforced soil.

4 BEARING CAPACITY ON REINFORCED SOIL

A rigid rough circular shallow foundation subjected to vertical loading is analyzed using different reinforcement configurations. The soil is considered frictionless, weightless and elastic perfectly plastic with the following properties: E = 10 MPa; $\nu = 0.49$, c = 30 kPa, $\phi = 0°$, a = 0, $\theta_T = 28°$. The dilatancy angle (ψ) was varied during the study. A reinforcement of 4 m in diameter is considered linear elastic with: t = 2.5 mm, J = 2500 kN/m and $\nu = 0$. The interface soil-reinforcement was considered rigid and therefore interface elements are not considered in these analyses. The finite element mesh and the boundary conditions are presented in Figure 4.

Figure 5 illustrates the reinforcement layout considered in this study. B is the circular footing diameter, U is the depth to the first reinforcement layer, H is the space between each reinforcement layer, N is the number of reinforcement layers; b is the diameter of reinforced zone and d is the depth of the last reinforcement layer.

393

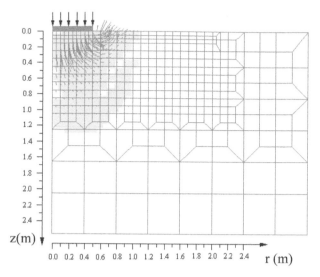

Figure 6.   Failure mechanism – unreinforced soil.

4.1   Unreinforced foundation

For an unreinforced foundation, the κ_{ult} value obtained by ANLOG was 5.42. Potts and Zdravković (2001) have obtained 5.39. The difference, approximately 0.5%, is considered negligible. At this level the settlement obtained by ANLOG was 0.025 m.

Figure 6 illustrates the failure mechanism with displacement vectors. The failure mechanism is consistent with that proposed by Prandtl (1920) for strip footing.

4.2   Reinforced foundation

Prediction of the bearing capacity was initially conducted considering a single layer of reinforcement under axisymmetric condition. The diameter of the reinforced zone is constant (b = 4B) while the reinforcement depth varies from 0.05 B to 0.9 B.

The bearing capacity improvement is evaluated by quantifying the bearing capacity ratio (BCR) defined as:

$$BCR = \frac{\kappa}{\kappa_{ult}^{0}} \tag{23}$$

in terms of the κ factor for the reinforced soil foundation and the ultimate bearing capacity for the unreinforced soil foundation (κ_{ult}^{0}). For consistency, the BCR must be evaluated at a particular settlement level. For instance, $BCR_{0.1}$ means the bearing capacity improvement is being evaluated with the κ factor at a normalized settlement (δ/B) of 0.1.

The settlement reduction improvement is evaluated by the settlement reduction ratio (SRR) which is defined as:

$$SRR(\%) = \left[\frac{\delta^0 - \delta^r}{\delta^0}\right]100 \tag{24}$$

where δ^0 is the settlement at the ultimate load of unreinforced foundation soil and δ^r is the settlement of reinforced foundation soil at the ultimate load of unreinforced soil foundation. Figure 7 illustrates these indexes.

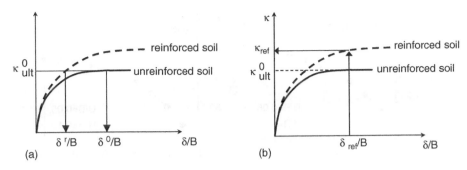

Figure 7.   References parameters: (a) δ^r/B definition; (b) κ_{ref} definition.

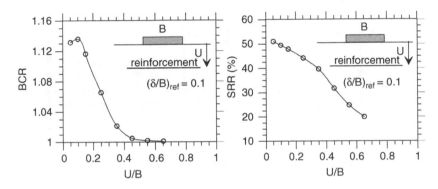

Figure 8.   U/B influence on the BCR and the SRR.

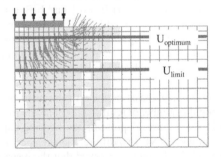

Figure 9.   Optimum and limit reinforcement position.

As expected, the settlement decreases because of the reinforcement of the foundation soil. A region can be defined through where the reinforcement location maximizes the SRR. In this case the higher value of SRR was around 40% from 0.05B to 0.35B (Figure 8). In terms of bearing capacity improvement, the results provided in Figure 8 indicate that there is little improvement for a single layer of reinforcement (maximum BCR was 14%). It should also be noted that there is an optimum depth as well as a limit depth, beyond which no improvement is verified.

Figure 9 presents the displacement field at the failure for unreinforced soil foundation and the optimum and limit reinforcement positions. Note that the limit depth (U_{limit}) coincides with the lowest point of the failure wedge and the optimum depth ($U_{optimum}$) coincides with a high level of mobilized shear stress for the unreinforced foundation soil.

The numerical results suggest that the reinforcement starts to work after the soil deforms plastically. To investigate the influence of the number of reinforcement layers, the foundation settlement

395

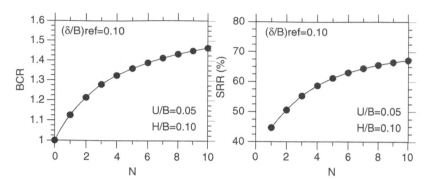

Figure 10.   Influence of the number of reinforcement layers.

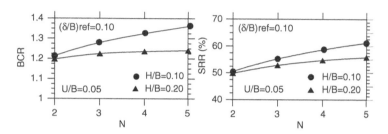

Figure 11.   Influence of the space between each reinforcement layer.

reference of 0.05 m is adopted. Although this value is high in terms of allowed settlement for a shallow foundation it was adopted to ensure that the load in the reinforcement is mobilized. The baseline geometry was: B = 1 m, U/B = 0.05, H/B = 0.1 and b/B = 4. The parametric evaluation involved varying the number of the reinforcement layers (N). Results are shown in Figure 10.

In this case, which involved circular footing and frictionless soil foundation, the bearing capacity improvement (BCR) was around 10% and the settlement reduction ratio (SRR) was around 6% as the number of reinforcements was increased from 5 to 10. Accordingly, the number of reinforcement layers should not exceed 4 to 7.

A second parametric evaluation involved assessment of the influence of the space between each reinforcement layer (H) varying the number of reinforcement layers from 2 to 5. The geometry was: B = 1 m, U/B = 0.05 and b/B = 4. Figure 11 presents the bearing capacity ratio (BCR) in terms of the number of reinforcement layers (N). Two different spaces between the reinforcement layers are considered: H/B = 0.10 and 0.20. It can be observed that the bearing capacity increases as the spacing decreases.

Figure 12 shows the horizontal displacement field of the unreinforced soil and of two configurations of reinforced soil (H/B = 0.10 and H/B = 0.20). This displacement field is at the settlement level corresponding to an ultimate level of unreinforced soil (δ/B = 0.05). The number of reinforcement layers (N = 5) and the position of the first reinforcement layer (U/B = 0.05) are constant. The lowest horizontal displacement was observed when the H/B is 0.10. As expected, the results confirmed that high confinement improves bearing capacity.

Adopting the foundation settlement reference (δ_{ref}) of 0.05 m and maintaining B = 1 m, H/B = 0.1, N = 5 and b/B = 4 as constant, the influence of the depth of the first reinforcement layer (U) was analyzed. In order to explain this influence the curve BCR versus U/B (Figure 13) was divided into 3 zones in terms of the bounded values (U/B)$_{optimum}$ and (U/B)$_{limit}$. Zone 1 defines the suitable values for the position of the first layer. Zone 2 is characterized by a significant decrease in the bearing capacity ratio. In Zone 3 shows no improvement in bearing capacity. In this case the bounded values, (U/B)$_{optimum}$ and (U/B)$_{limit}$, was respectively around 0.05 and 0.25.

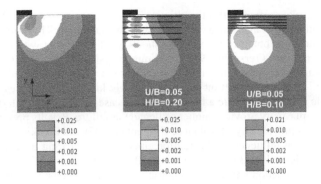

Figure 12. Horizontal displacement (m) – rigid foundation ($\delta_{ref}/B = 0.1$).

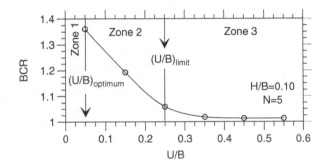

Figure 13. Influence of the first reinforcement layer depth – ($\delta_{ref}/B = 0.1$).

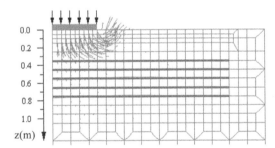

Figure 14. Influence of the first reinforcement layer depth – ($\delta_{ref}/B = 0.1$).

Figure 14 shows the failure mechanisms for the case in which the depth of the top reinforcement layer exceeds $(U/B)_{limit}$. Note that the reinforced layer of soil works as a rigid and rough base. In this region both vertical and horizontal displacements are approximately zero.

5 CONCLUSIONS

This paper presented a numerical simulation using FEM to analyze the bearing capacity of shallow foundations on reinforced soil under axisymmetric conditions. The modified Mohr-Coulomb constitutive model was implemented into ANLOG. The implementation of the explicit integration stress algorithm proposed by Sloan et al (2001) was needed in order to obtain good performance of the Newton Raphson algorithm at the global level.

The numerical results confirmed that the ultimate bearing capacity of a rigid shallow foundation on unreinforced soil is higher than that on a flexible shallow foundation. The ultimate bearing capacity of flexible foundations obtained numerically shows good agreement with the results obtained by equilibrium limit theory (Terzaghi, 1943) and limit analysis (Chen, 1975; Ribeiro, 2005).

The ultimate bearing capacity of unreinforced soil was not affected by the dilatancy angle when the friction angle is low but is relevant for comparatively high friction angles. Therefore, for a high friction angle the κ_{ult} values are a little high in the case of associative plasticity. In general, the non-associative plasticity provides higher settlement at failure. Results presented in this paper agree with the results provided by Monahan & Dasgupta (1995) and Zienkiewics et al (1975).

In order to show the influence of the reinforcement on the bearing capacity and settlement reduction, a parametric study was conducted using different reinforcement configurations. A rigid, rough, and shallow foundation under axisymmetric condition was considered in the analysis. The soil foundation was considered weightless and purely cohesive ($\phi = 0°$) and the interface soil-reinforcement was considered rigid. Based on the results, it may be concluded that:

The bearing capacity increases and the settlement reduction increases as the number of reinforcement layers increase. A cost-benefit analysis should be conducted to define the optimum number of reinforcement layers to be used.

The bearing capacity ratio, which indicates the improvement on the bearing capacity, was approximately 14% for just one reinforcement layer; it may be considered modest. In this case the optimum depth for placing it is 0.1B and the limit depth is 0.5B. The reinforcement influence on the settlement, however, is significant (around 40% to 50%). The reinforcement starts to work after the soil deforms plastically, which often occur at a high level of settlement.

The results show the existence of three regions related to the depth of the first reinforcement layer to consider. The first, Zone 1, defines the suitable value for this depth. Zone 2 is characterized by a significant decrease in the bearing capacity ratio. Zone 3 corresponds to a lack of no influence on the bearing capacity. Effort should be made to identify these zones in order to define the best position for the reinforcement layer.

ACKNOWLEDGEMENTS

The authors are grateful for the financial supports received by the first author from CAPES (Coordinating Agency for Advanced Training of High-Level Personnel – Brazil) and by the second author from Maccaferri do Brasil LTDA. They also acknowledge the Prof. John Whites for the English revision of this text.

REFERENCES

Abbo, A.J. & Sloan, S.W. 1995. A smooth hyperbolic approximation to the Mohr-Coulomb yield criterion. *Computers & Structures*, 54(3): 427–441.

Chen, W.F. 1975. *Limit Analysis and Soil Plasticity*. Elsevier Science Publishers, BV, Amsterdam, The Netherlands.

Houlsby, G.T. 1991. How the dilatancy of soils affects their behaviour, *Proc. X Euro. Conf. Soil Mech. Found. Engng.*, Firenze 1991, (4):1189–1202.

Manohan, N. & Dasgupta, S.P. 1995. Bearing capacity of surface footings by finite element. *Computers & Structures*, (4): 563–586.

Nogueira, C.L. 1998. Non-linear analysis of excavation and fill. *DS. Thesis*. PUC/Rio, RJ, 265p (in Portuguese).

Oliveira, R.R.V. 2006. Elastoplastic analysis of reinforced soil structures by FEM. *MS. Thesis*. PROPEC/UFOP (in Portuguese).

Pereira, A.R. 2003. Physical non-linear analysis of reinforced soil structures. *MS. Thesis*. PROPEC/UFOP (in Portuguese).

Potts. D.M. & Zdravković, L. 2001. *Finite Element analysis in geotechnical engineering: application*. Thomas Telford Ltda.

Ribeiro, W.N. 2005. Applications of the numerical limit analysis for axisymmetric stability problems in Geotechnical Engeneering. *MS. Thesis*, UFOP (in Portuguese).

Sloan, S.W. & Booker, J.R. 1986. Removal of singularities in Tresca and Mohr-Coulomb yield criteria, *Comunications in Applied Numerical Methods*, (2): 173–179.

Sloan, S.W.; Abbo, A.J. & Sheng, D. 2001. Refined explicit integration of elastoplastic models with automatic error control, *Engineering Computations*, 18(1): 121–154.

Terzaghi, K. 1943. *Theoretical Soil Mechanics*. Wiley.

Zienkiewicz, O.C., Humpheson, C. & Lewis, R.W. 1975. Associated and non-associated visco-plasticity and plasticity in soil mechanics. *Geotechnique*, 25(4): 671–689.

Zornberg, J.G. 1989. Finite element analysis of excavations using an elasto-plastic model. *MS. Thesis*. PUC-Rio, Rio de Janeiro (in Portuguese).

Environmental geotechnics

Applications of Computational Mechanics in Geotechnical Engineering – Sousa,
Fernandes, Vargas Jr & Azevedo (eds)
© 2007 Taylor & Francis Group, London, ISBN 978-0-415-43789-9

Numerical stochastic analysis of sediment storage in large water dam reservoirs

J.P. Laquini & R.F. Azevedo
Federal University of Viçosa, Viçosa, Minas Gerais, Brazil

ABSTRACT: The paper deals with a model to predict the deposition of sediment in water reservoirs as a function of time. It shows a stochastic model for the total deposition of sediment in reservoirs developed by Soares (1975). The model is analyzed to obtain the mean and the variance of the cumulative deposition as a function of discrete time. Besides the geometry of the water course bed, the model considers the inflow, outflow, the concentration of sediments in the inflow and in the outflow, and features of the sediments such as solids specific weight and grain-size distribution of the deposited sediments. Finally, the model is used to determine the sediment deposition in a reservoirs for which the bed elevations were surveyed. Comparisons between model results and actual measurements are found satisfactory.

1 INTRODUCTION

Erosion, transport and deposition of sediment are natural processes which have occurred throughout geologic times. The amount of sediment which is moved from a watershed varies greatly from one area to another. This variation depends upon several factors, such as geologic and climatic conditions, vegetation and physical characteristics of the watershed. When a dam is constructed in a natural water course to store water, it changes the hydraulic characteristics of flow immediately upstream of the dam and, consequently, the sediment-transport capacity. Sedimentation of the reservoir created by the dam is then inevitable, reducing the storage capacity of the reservoir. The sedimentation of reservoir is complex and depends upon several interrelated factors (Soares et al. 1982). Excellent reviews of the overall aspects of this process have been presented by Brown (1950) and Gottschalk (1964).

The amount of sediment trapped in a reservoir during a certain period of time is the difference between the sediment carried into the reservoir by the inflow and the sediment released with the outflow. Several deterministic methods have been developed to predict the amount of sediment carried into a reservoir as a function of watershed characteristics (Ackermann & Corinth 1962, Paulet 1971, Gupta 1974). In any real situation the inflow will vary from period to period and, consequently, the amount of sediment carried into the reservoir will also change. The same is true concerning the amount of sediment released, even under the assumption of constant outflow, because the level in a reservoir varies, changing the concentration of sediment of the outflow. Therefore, the long-term storage of sediment in a reservoir is stochastic in character and a model to predict the amount of sediment storage in reservoirs must involve the theory of probability (Soares et al. 1982).

Under the assumptions of Moran's (1954) dam storage model in discrete time, a stochastic model for the storage of sediment in reservoirs was developed by Soares (1975). The inputs to this model are the events representing the amount of sediment trapped in the reservoir for different combinations of inflows and the levels in the reservoir. Like output, the expectation and variance of the total sediment deposit is obtained as a function of discrete time.

2 STOCHASTIC SEDIMENT STORAGE MODEL

The model developed by Soares (1975) is presented as an additive stochastic process defined on a finite Markov chain. The model is based on Moran's (1954) dam storage model in discrete time, modified to deal with uniform release (Moran 1956). The main theoretical aspects are presented to determine the expected value and the variance of the amount of sediment deposit.

2.1 Moran's dam storage model in discrete time with uniform release

The storage function of the Moran's model, under same assumptions concerning inflow, outflow and release, with uniform release satisfies:

$$Z_{n+1} = \text{Min}[\{Z_n + X_n - \text{Min}[R; Z_n + X_n]\}; H] \tag{1}$$

where n = discrete time; Z_n = content of the dam; H = finite capacity of the dam; X_n = amount of inflow during period $(n, n+1)$; and R = continuous release.

Under the assumption that $X_0, X_1, X_2 \ldots$ are random variables that are independent and identically distributed, the sequence $\{Z_n; n = 0, 1, 2, \ldots\}$ defined by Equation 1 forms a time-homogeneous Markov chain with reflecting barriers at $Z = 0$ and $Z = H$.

It is supposed that the quantity of water is measured in terms of some unit amount, say a, that H is an integral multiple, h, of this unit, and that the inflow X has a discrete distribution with:

$$\text{Prob}\{X_n = R + ia\} = p_i \qquad (i = -m, -m+1, \ldots, 0, \ldots, N-m-1) \tag{2}$$

where N = number of distinct quantities used to represent the inflow. As the inflow cannot be negative, then $ma \leq R$.

2.2 The sediment storage model

The sediment trapped in a reservoir is the difference between the sediment carried into the reservoir by the inflow and the amount of sediment released with the outflow.

In the Soares's (1975) model, the following assumptions are made: (1) the inflow concentration is related only to the specified inflow; and (2) the outflow concentration is a function of the reservoir levels at the beginning and end of the interval $(n, n+1)$.

Let $W_n = S_{il}$ be the sediment that is deposited in a reservoir during period $(n, n+1)$ during which the inflow was $X_n = R + ia$ and the reservoir level, at time n, was $Z_n = la$. If CI and CO are, respectively, the inflow and outflow concentration of sediment, the values S_{il} are expressed as:

$$S_{il} = [R + ia]\,\text{CI}(i) - [R + (l + i)\,a]\,\text{CO}(l, 0); \qquad l + i < 0 \tag{3a}$$

$$S_{il} = [R + ia]\,\text{CI}(i) - [R]\,\text{CO}(l, l + i); \qquad 0 \leq l + i \leq h \tag{3b}$$

$$S_{il} = [R + ia]\,\text{CI}(i) - [R + (l + i - h)\,a]\,\text{CO}(l, h); \qquad l + i > h \tag{3c}$$

The values S_{il} are bounded, i.e. $S_{il} \leq s$ with probability one for some $s > 0$ and, under assumptions (1) and (2), W_n is a function only of Z_n and X_n. Recalling that Z_n depends only on Z_{n-1} and X_{n-1} it follows that the sequence $\{W_n; n = 0, 1, \ldots\}$ forms a Markov chain. Hence,

$$\text{Prob}\{W_{n+1} = S_{il} \,|\, W_n = S_{i'l'}\} = \text{Prob}\{Z_{n+1} = la; X_{n+1} = R + ia \,|\, Z_n = l'a; X_n = R + i'a\} \tag{4}$$

404

As X_{n+1} is an independent random event and does not depend on Z_n or Z_{n+1},

$$\text{Prob}\{W_{n+1} = S_{il} \mid W_n = S_{i'l'}\} = q_{i'i} \ \text{Prob}\{Z_{n+1} = la \mid Z_n = l'a; X_n = R + i'a\} \tag{5}$$

where,

$$q_{i'i} = \text{Prob}\{X_{n+1} = R + ia \mid X_n = R + i'a\} = \text{Prob}\{X_n = R + ia\} = p_i \tag{6}$$

Furthermore,

$$\text{Prob}\{Z_{n+1} = la \mid Z_n = l'a; X_n = R + i'a\} = \delta_{l\omega} \tag{7}$$

where

$$\delta_{l\omega} = \begin{cases} 0, & \text{se } l \neq \omega \\ 1, & \text{se } l = \omega \end{cases} \qquad \text{and} \qquad \omega = \begin{cases} 0, & \text{se} \quad i' + l' \leq 0 \\ i' + l', & \text{se} \quad 0 < i' + l' < h \\ h, & \text{se} \quad i' + l' \geq h \end{cases} \tag{8}$$

Thus,

$$\text{Prob}\{W_{n+1} = S_{il} \mid W_n = S_{i'l'}\} = p_i \, \delta_{l\omega} \tag{9}$$

In order to simplify the notation, the variable $W_n = S_{il}$ is redefined as $W_n = S_j$, with:

$$j = lN + i + m + 1; \qquad \begin{cases} i = -m, -m+1, ..., 0, ..., N - m - 1 \\ l = 0, 1, ..., h \\ j = 1, 2, ..., N(h+1) \end{cases} \tag{10}$$

The chain has a finite number of states, $N_s = N(h + 1)$, and its matrix of transition probabilities is written as:

$$P = [p_{jk}] = \text{Prob}\{W_{n+1} = S_k \mid W_n = S_j\} \tag{11}$$

Soares (1975) shows that the chain defined by Equation 11 is homogeneous, irreducible with all states being aperiodic, and ergodic. Thus, there is a unique row vector $u = \{u_k\}$ of stationary probability distribution, such that:

$$u \, P = u \tag{12}$$

2.3 The model for total sediment

The total sediment at time n is the sediment trapped up to time $(n - 1)$ plus the sedimentation that occurred during $(n - 1, n)$, i.e. W_{n-1}. Denoting the total sediment at time n by T_n, we have:

$$T_n = T_{n-1} + W_{n-1} \tag{13}$$

This is equivalent to:

$$T_n = \sum_{i=0}^{n-1} W_i = \sum_{k=1}^{N_s} S_k Y_k^{(n)} \tag{14}$$

where $Y_k^{(n)}$ is the number of times the event S_k occurred among the n periods $(0,1), ..., (n - 1, n)$.

Although the numerical approach developed below allows the determination of higher-order moments of T_n, the analysis will deal with the calculation of the expected value and variance of T_n. From Equation 14 the expectation and variance of T_n are:

$$E\{T_n\} = \sum_{k=1}^{N_s} S_k E\{Y_k^{(n)}\} \tag{15}$$

and

$$\text{Var}\{T_n\} = \text{Var}\left\{\sum_{k=1}^{N_s} S_k Y_k^{(n)}\right\} \tag{16a}$$

that can be rewritten as

$$\text{Var}\{T_n\} = \sum_k S_k^2 \text{Var}\{Y_k^{(n)}\} + \sum_{j<k}\sum_k 2 S_j S_k \sigma_{jk}^{(n)} \tag{16b}$$

where

$$\text{Var}\{Y_k^{(n)}\} = E\left\{[Y_k^{(n)}]^2\right\} - \left[E\{Y_k^{(n)}\}\right]^2 \tag{17}$$

and

$$\sigma_{jk}^{(n)} = E\left\{[Y_j^{(n)} - E\{Y_j^{(n)}\}][Y_k^{(n)} - E\{Y_k^{(n)}\}]\right\} \tag{18}$$

is the covariance of $Y_j^{(n)}$ and $Y_k^{(n)}$. Using the fact that:

$$E\left\{[Y_j^{(n)} + Y_k^{(n)}]^2\right\} = 2E\{Y_j^{(n)} Y_k^{(n)}\} + E\{[Y_j^{(n)}]^2\} + E\{[Y_k^{(n)}]^2\} \tag{19}$$

we have that

$$2\sigma_{jk}^{(n)} = E\left\{[Y_j^{(n)} + Y_k^{(n)}]^2\right\} - E\{[Y_j^{(n)}]^2\} - E\{[Y_k^{(n)}]^2\} - 2E\{Y_j^{(n)}\}E\{Y_k^{(n)}\} \tag{20}$$

From Equations 15–20 it is seen that in order to compute $E\{T_n\}$ and $\text{Var}\{T_n\}$, the quantities $E\{Y_k^{(n)}\}$, $E\{[Y_k^{(n)}]^2\}$ and $E\{[Y_j^{(n)} + Y_k^{(n)}]^2\}$ must be computed for all j and k.

The model as specified by Equation 13 is seen to be an additive process defined on a finite ergodic Markov chain (Volkov 1958) with transition matrix $P = [p_{ij}]$, which has been shown to be primitive and irreducible.

Let the state of the chain corresponding to time n be denoted by s_n. Now let $s_1, s_2, \ldots, s_n$ be a realization of the Markov chain with transition matrix $P = [p_{ij}]$ given that $s_0 = i$. If attention is focused say, to state k ($k = 1, \ldots, N_s$) of the chain and a score one is noted each time the state k is occupied and zero otherwise, then the cumulative score is the number of times state k is occupied among times $1, 2, \ldots, n$, i.e. $Y_k^{(n)}$ conditional on the initial state $s_0 = i$.

The moment-generating function of the density function of the cumulative score associated with an n-fold transition from state i to state j is:

$$\phi_{ij}^{(n)}(k,\omega) = E\{e^{-\omega Y_k^{(n)}} | s_0 = i, s_n = j\} \tag{21}$$

Defining the matrices:

$$P_k^{(1)}(\omega) = [\, p_{ij}^{(1)} \phi_{ij}^{(1)}(k,\omega)\,] \tag{22}$$

406

and

$$P_k^{(n)}(\omega) = [\, p_{ij}^{(n)} \phi_{ij}^{(n)}(k, \omega)] \tag{23}$$

It is proved (Cox & Miller 1965) that $P_k^{(n)}(\omega) = \{P_k^{(1)}(\omega)\}^n$. Volkov (1958) shown that the sum of the elements in any given row of $P_k^{(n)}(\omega)$ is the moment-generating function of $Y_k^{(n)}$ conditional on the initial state corresponding to that row, i.e:

$$E\{e^{-\omega Y_k^{(n)}} | s_0 = i\} = \sum_j \{p_{ij}^{(n)} \phi_{ij}^{(n)}(k, \omega)\} = \sum_j \{P_k^{(1)}(\omega)\}^n \tag{24}$$

Since:

$$\phi_{ij}^{(1)}(k, \omega) = E\{e^{-\omega Y_k^{(1)}} | s_0 = i, s_1 = j\} = \begin{cases} 1, & se \ j \neq k \\ e^{-\omega}, & se \ j = k \end{cases} \tag{25}$$

then $P_k^{(1)}(\omega)$ is equal to matrix $P = [p_{ij}]$ with its kth column multiplied by $(e^{-\omega})$.

For a given state k, the matrix $P_k^{(1)}(\omega)$ is computed and for a given value of n, the power $\{P_k^{(1)}(\omega)\}^n$ is also calculated. Then the Equation 24 is computed to determining the following relations:

$$E\{Y_k^{(n)} | s_0 = i\} = -\frac{\partial}{\partial \omega} \Big[E\{e^{-\omega Y_k^{(n)}} | s_0 = i\} \Big]_{\omega = 0} \tag{26}$$

and

$$E\{[Y_k^{(n)}]^2 | s_0 = i\} = \frac{\partial^2}{\partial^2 \omega} \Big[E\{e^{-\omega Y_k^{(n)}} | s_0 = i\} \Big]_{\omega = 0} \tag{27}$$

Now focusing the attention on two different states, say l and k, a score one is noted each time one of the two states l or k is occupied and zero otherwise. As the two states are mutually exclusive, the cumulative score is the number of times states l and k are occupied among times 1, 2, ..., n, i.e. $Y_l^{(n)} + Y_k^{(n)}$ conditional on the initial state $s_0 = i$. The joint moment-generating function of the joint density function of the scores $Y_l^{(n)}$ and $Y_k^{(n)}$ associated with an n-fold transition from state i to state j is:

$$\phi_{ij}^{(n)}(l, k, \omega) = E\{e^{-\omega[Y_l^{(n)} + Y_k^{(n)}]} | s_0 = i, s_n = j\} \tag{28}$$

The sum of the elements in any given row of the matrix:

$$P_{lk}^{(n)}(\omega) = [\, p_{ij}^{(n)} \phi_{ij}^{(n)}(l, k, \omega)] \tag{29}$$

is the joint moment-generating function of $Y_l^{(n)}$ and $Y_k^{(n)}$ conditional on the initial state corresponding to that row, i.e:

$$E\{e^{-\omega[Y_l^{(n)} + Y_k^{(n)}]} | s_0 = i\} = \sum_j \{p_{ij}^{(n)} \phi_{ij}^{(n)}(l, k, \omega)\} = \sum_j \{P_{lk}^{(1)}(\omega)\}^n \tag{30}$$

Since:

$$\phi_{ij}^{(1)}(l, k, \omega) = E\{e^{-\omega[Y_l^{(n)} + Y_k^{(n)}]} | s_0 = i, s_1 = j\} = \begin{cases} 1, & se \ j \neq l, k \\ e^{-\omega}, & se \ j = l, k \end{cases} \tag{31}$$

then $P_{lk}^{(1)}(\omega)$ is equal to matrix $P = [p_{ij}]$ with columns l and k multiplied by $(e^{-\omega})$.

For all combinations of states $l \neq k$ the matrix $\boldsymbol{P}_{lk}^{(1)}(\omega)$ is computed and for a given value of n, the power $\{\boldsymbol{P}_{lk}^{(1)}(\omega)\}^n$ is also calculated. Then the Equation 30 is computed to determining the following relation:

$$E\{[Y_l^{(n)} + Y_k^{(n)}]^2 \mid s_0 = i\} = \frac{\partial^2}{\partial^2 \omega}\left[E\{e^{-\omega[Y_l^{(n)} + Y_k^{(n)}]} \mid s_0 = i\}\right]_{\omega=0} \tag{32}$$

In order to computing $E\{T_n\}$ and $Var\{T_n\}$, the corresponding unconditional expected values are:

$$E\{Y_k^{(n)}\} = \sum_i E\{Y_k^{(n)} \mid s_0 = i\} \, Prob\{s_0 = i\} \tag{33}$$

$$E\{[Y_k^{(n)}]^2\} = \sum_i E\{[Y_k^{(n)}]^2 \mid s_0 = i\} \, Prob\{s_0 = i\} \tag{34}$$

$$E\{[Y_l^{(n)} + Y_k^{(n)}]^2\} = \sum_i E\{[Y_l^{(n)} + Y_k^{(n)}]^2 \mid s_0 = i\} \, Prob\{s_0 = i\} \tag{35}$$

where $Prob\{s_0 = i\} =$ initial distribution of states. The values of $Prob\{s_0 = i\}$ are unknown that can be obtained under two assumptions:

If it is considered that the chain has already reached its stationary distribution at the initial time considered, then $Prob\{s_0 = i\}$ is given by the stationary distribution u, solution of Equation 12.

If the level in the reservoir at a time previous to the initial time considered is known, say l, then:

$$Prob\{s_0 = i\} = \sum_{j=-m}^{N-m-1} p_j \, Prob\{s_n = j \mid s_{n-1} = lN + j + m + 1\} \tag{36}$$

2.4 The model with serially correlated inflow

If the assumption of independent inflow is dropped and it is assumed that the inflows are dependent with correlation pattern assumed to be Markovian, then the model is affected only in the representation of the matrix of transition probabilities of W_n (Soares 1975), that becomes:

$$Prob\{W_{n+1} = S_{il} \mid W_n = S_{i'l'}\} = q_{i'i} \, \delta_{l\omega} \tag{37}$$

with $\delta_{l\omega}$ defined as previously and with the transition probabilities of the inflow $[q_{i'i}]$ being compatible with the stable distribution, $Prob\{X_n = R + ia\} = p_i$. Hence,

$$\sum_{i'=-m}^{N-m-1} q_{i'i} p_{i'} = p_i \qquad i = -m, -$$

$$m + 1, ..., 0, ..., N - m - 2 \tag{38}$$

The equation for $i = N - m - 1$ is implied by the other $(N - 1)$ equations since $\sum_i p_i$ is already equal to one.

The elements in each row of matrix $[q_{i'i}]$ must add to one, i.e.:

$$\sum_{i'=-m}^{N-m-1} q_{i'i} = 1 \qquad i = -m, -m+1, ..., 0, ..., N-m-1 \tag{39}$$

There are N^2 values of $q_{i'i}$ restricted by $(N - 1)$ conditions given by Equation 38, and N conditions given by Equation 39. Therefore $(N - 1)^2$ additional parameters may be used to describe the inflow according to the correlation pattern assumed or determined from data.

With the new matrix of the transition probability defined by Equation 37, the procedure outlined in Section 2.3 can be used to determine $E\{T_n\}$ and $Var\{T_n\}$ for the case in which the inflows are Markovian-correlated. However, in computing the unconditional expected values, Equations 33–35, only the first assumption for the initial probability Prob$\{s_0 = i\}$ is valid because the inflows are no longer independent.

2.5 The model with a three ordinate input

In any real situation the amount of water flowing into a reservoir in an interval $(n, n+1)$ is a continuous random variable, capable of taking any value in some finite range. For the purpose of the stochastic model, shown herein, the continuous input must be replaced by a discrete variable. The larger the set of discrete values used the more accurately will this approximation fit the continuous input. The number of parameters involved in the model will also be larger. The estimates of these parameters will then become less exact; besides the number of states of the Markov chain in which the model is defined increases with N (the number of discrete values used to represent the continuous inflow), increasing considerably the computer time for solution. Hence, some compromise must be reached, which will depend on the extent of data and on the facilities available (Soares et al. 1982).

To illustrate the model a three-valued discrete representation of the inflow is adopted for both cases of inflow considered (independent and serially correlated, respectively) like Soares (1975) had done. The three-ordinate representation of the continuous inflow is rather a cruder approximation but the methods outlined for this simple input representation would apply equally to the more complicated case.

The continuous inflow, say, Y, is replaced by an approximating discrete variable X, assumed to be capable of taking only three possible values, namely, $(R - a)$, R and $(R + a)$, where R is the constant release and a is a value to be determined from historical data. This inflow representation means that in the model $N = 3$ and $m = 1$.

With the value a determined, the reservoir is divided in levels, such that the volume between two consecutive levels is equal to a. This can be done through the use of the capacity curve for the reservoir under consideration. This procedure also gives the value h, the number of discrete intervals in which the reservoir is divided.

To compute the values S_{il} defined by Equations 3a, b, c, an estimate of the inflow and out-flow concentrations of sediment is necessary. It could be mentioned here that for some streams suspended-sediment rating curves have been established by measurement, giving the sediment load either in volume or weight against discharge (Miller 1951). From these curves, the respective concentration of sediment for the discrete inflow representation can be obtained. In any case a determination of these concentrations has to be made. Thus, if no measurements are available then these quantities can be estimated by use of one of the various sediment-load equations which express the sediment load as a function of sediment and flow characteristics.

Finally, to compute the quantities $\mathrm{E}\{T_n\}$ and $\mathrm{Var}\{T_n\}$ it is required to have the matrix of transition probabilities $\boldsymbol{P} = [p_{jk}]$ as defined by Equation 11. This will be done separately for the two cases of inflow under consideration.

2.5.1 Independent inflows

Let the probabilities of the three-ordinate inflow representation be:

$$\mathrm{Prob}\{X = R - a\} = p_{-1} \tag{40a}$$

$$\mathrm{Prob}\{X = R\} = p_0 \tag{40b}$$

$$\mathrm{Prob}\{X = R + a\} = p_1 \tag{40c}$$

with $p_{-1} + p_0 + p_1 = 1$.

The transition probability matrix $\boldsymbol{P} = [p_{jk}]$ is presented in Table 1; its stationary distribution is obtained by solving the system of Equations 12 yielding:

$$\boldsymbol{u} = u_1\{1, s, r, r, sr, r^2, r^2, \ldots, r^h, r^h, sr^h, r^{h+1}\} \tag{41}$$

Table 1. Transition probability matrix $P = [p_{jk}]$ for independent inflows.

| | | l | 0 | 0 | 0 | 1 | 1 | 1...$h-1$ | $h-1$ | $h-1$ | h | h | h |
| | | i | -1 | 0 | 1 | -1 | 0 | 1...-1 | 0 | 1 | -1 | 0 | 1 |
l'	i'	j,k	1	2	3	4	5	6...$3h-2$	$3h-1$	$3h$	$3h+1$	$3h+2$	$3h+3$
0	-1	1	p_{-1}	p_0	p_1	0	0	0					
0	0	2	p_{-1}	p_0	p_1	0	0	0					
0	1	3	0	0	0	p_{-1}	p_0	p_1					
1	-1	4	p_{-1}	p_0	p_1	0	0	0					
1	0	5	0	0	0	p_{-1}	p_0	p_1					
1	1	6	0	0	0	0	0	0					
⋮	⋮	⋮						⋱					
$h-1$	1	$3h$						0	0	0	p_{-1}	p_0	p_1
h	-1	$3h+1$						p_{-1}	p_0	p_1	0	0	0
h	0	$3h+2$						0	0	0	p_{-1}	p_0	p_1
h	1	$3h+3$						0	0	0	p_{-1}	p_0	p_1

where

$$s = p_0 / p_{-1}, \qquad r = p_1 / p_{-1} \qquad \text{and} \qquad u_1 = p_{-1} / \sum_{k=0}^{h} r^h \tag{42}$$

The initial distributions under assumption (2), mentioned at the end of Section 2.3, are as follows:
Level 0 at time previous to the initial time:

$$u = \{p_{-1}(p_{-1}+p_{-1}); \; p_0(p_{-1}+p_0); \; p_1(p_{-1}+p_0); \; p_{-1}p_1; \; p_0p_1; \; p_1^2; \; 0; \; 0; \; ...; \; 0\} \tag{43}$$

Level h at time previous to the initial time:

$$u = \{0; \; ...; \; 0; \; p_{-1}^2; \; p_{-1}p_0; \; p_{-1}p_1; \; p_{-1}(p_0+p_1); \; p_0(p_0+p_1); \; p_1(p_0+p_1)\} \tag{44}$$

Level $0 < l < h$ at time previous to the initial time:

$$u_{3l-2} = p_{-1}^2; \quad u_{3l-1} = p_{-1}p_0; \quad u_{3l} = p_{-1}p_1; \quad u_{3l+1} = p_0p_{-1}; \quad u_{3l+2} = p_0^2;$$

$$u_{3l+3} = p_0p_1; \quad u_{3l+4} = p_1p_{-1}; \quad u_{3l+5} = p_1p_0; \quad u_{3l+6} = p_1^2; \tag{45}$$

and zero otherwise.

2.5.2 Correlated inflows

With the correlation structure assumed to be Markovian, the general transition probability matrix for the inflow simplified according Soares (1975) is:

$$\begin{bmatrix} 1-2\alpha & \alpha & \alpha \\ \beta & 1-2\beta & \beta \\ \gamma & \gamma & 1-2\gamma \end{bmatrix} \tag{46}$$

where is shown that:

$$\alpha = K / p_{-1}; \qquad \beta = K / p_0; \qquad \gamma = K / p_1; \tag{47}$$

and $K =$ the only parameter to be determined.

410

Table 2. Transition probability matrix $P = [p_{jk}]$ for correlated inflows.

			l: 0	0	0	1	1	1...h−1	h−1	h−1	h	h	h
			i: −1	0	1	−1	0	1...−1	0	1	−1	0	1
l'	i'	j,k:	1	2	3	4	5	6...3h−2	3h−1	3h	3h+1	3h+2	3h+3
0	−1	1	$1-2\alpha$	α	α	0	0	0					
0	0	2	β	$1-2\beta$	β	0	0	0					
0	1	3	0	0	0	γ	γ	$1-2\gamma$					
1	−1	4	$1-2\alpha$	α	α	0	0	0					
1	0	5	0	0	0	β	$1-2\beta$	β					
1	1	6	0	0	0	0	0	0					
⋮	⋮	⋮						⋱					
h−1	1	3h						0	0	0	γ	γ	$1-2\gamma$
h	−1	3h+1						$1-2\alpha$	α	α	0	0	0
h	0	3h+2						0	0	0	β	$1-2\beta$	β
h	1	3h+3						0	0	0	γ	γ	$1-2\gamma$

For this case the matrix of transition probability is presented in Table 2, and its stationary distribution is found to be:

$$u = u_0\{(2 - 3\alpha)/3\alpha,\ \tfrac{1}{3}\beta^{-1},\ 1,\ 1,\ s/\beta,\ r,\ r,\ ...,\ r^{h-2},\ r^{h-2}s/\beta,$$
$$r^{h-1},\ r^{h-1},\ r^{h-1}/3\beta,\ r^{h-1}[(2-3\gamma)/3\gamma]\} \qquad (48)$$

where

$$r = (2 - 3\gamma)/(2 - 3\alpha), \qquad s = (\gamma + \alpha - 3\alpha\gamma)/(2 - 3\alpha) \qquad (49)$$

and $u_0 =$ a normalizing constant, such that

$$\sum_{k=1}^{N_s} u_k = 1 \qquad (50)$$

3 APPLICATION OF THE MODEL

The sediment storage in the John Martin Reservoir is analyzed by the model. The data are collected from Soares (1975). It is located on the Arkansas River in Bent Country in Colorado and projected for flood control and irrigation. It is 25 km long and two major tributaries, the Arkansas and Purgatore rivers at las Animas, feed it as well as a small ephemeral tributary, Rule Creek, which enters the sides of the narrow reservoir.

The data of inflow, outflow, sediment inflow and sediment outflow for the water-years 1943–1966 are reported in Soares (1975). Figure 1 shows the histogram of the log-transformed data of inflow. The plot of the logarithm of the annual inflow against the accumulated frequency fits approximately a straight line. The skewness coefficient of the log-transformed data is found to be 0.046. A two-parameter log-normal distribution is assumed for the inflow that gives $\mu_n = 19.283$ and $\sigma_n^2 = 0.356$.

The three-ordinate representation is chosen such that it is log-normal distributed with mean μ_n and variance σ_n^2, i.e.:

$$\sum p_i \ln(R + ia) = \mu_n \qquad (51)$$

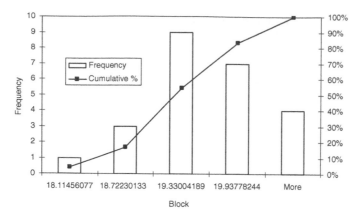

Figure 1. Histogram of the logarithm of the annual inflow.

$$\sum p_i \left[\ln(R + ia) - \mu_n\right]^2 = \sigma_n^2 \qquad\qquad i = -1, 0, 1 \tag{52}$$

with

$$\sum p_i = 1 \tag{53}$$

With R made equal to the average annual inflow, i.e. $R = 283{,}027{,}649\,\mathrm{m}^3$ and adopting $a = 191{,}207{,}475\,\mathrm{m}^3$, the p_i's are calculated from Equations 51–53. Thus $p_{-1} = 0.259$, $p_0 = 0.520$ and $p_1 = 0.221$.

With the value adopted for a, the reservoir was divided into three levels by using its capacity curve; hence $h = 2$ and Markovian chain defined by Equation 11 has nine states.

The parameter K (Eq. 47) for correlated inflows representation, is estimated following Lloyd (1963) by making the serial correlation at lag 1 of the three-ordinate inflow representation equal to the estimated correlation coefficient for the measured data. From Soares et al. (1982), we have

$$K = 1/6 \left[p_1 + p_{-1} - (p_1 - p_{-1})^2\right] / (1 - r_1) \tag{54}$$

where $r_1 =$ the lag-1 correlation coefficient of the three-ordinate inflow.

The estimated correlation coefficient is found to be equal to -0.032. Using this value as the estimator of r_1, it is obtained that $K = 0.082$ and from Equation 47, $\alpha = 0.318$, $\beta = 0.158$ and $\gamma = 0.373$.

The concentration of sediment $CI(i)$ of the inflow $X_i = R + ia$; $i = -1, 0, 1$, is estimated as the average measured concentration of the inflow that is represented by the discretized quantity X_i, i.e. $CI(-1)$ is equal to the average concentration of the inflows $0 < X < \eta_1$, $CI(0)$ is equal to the average concentration of the inflows $\eta_1 < X < \eta_2$ and $CI(1)$ is equal to the average concentration of the inflows $X > \eta_2$, with

$$\eta_1 = \sqrt{R(R - a)} \qquad \text{and} \qquad \eta_2 = \sqrt{R(R + a)} \tag{55}$$

that returns $\eta_1 = 130{,}744$ acre-ft and $\eta_2 = 297{,}131$ acre-ft. The measured data thus give $CI(-1) = 4.7 \cdot 10^{-3}$, $CI(0) = 5.0 \cdot 10^{-3}$ and $CI(1) = 7.7 \cdot 10^{-3}$.

The concentration of sediment in outflow is estimated by use of the following relation:

$$C = \chi \frac{C_b n_c U}{\sqrt{S_b}\, D^{2/3}} \frac{1}{15\beta/\alpha} \left[(1 + 2.5\alpha)[1 - e^{-15\beta/\alpha}] + 2.5\alpha \int_0^1 \ln\eta\ e^{-15(\beta/\alpha)\eta}d\eta \right] \tag{56}$$

Table 3. Size distribution of the bed at dam site.

Size range (mm)	d_{50} (mm)	Class	Percentage
<0.002	0.001	clay	32
0.002–0.006	0.003	fine silt	40
0.006–0.020	0.010	medium silt	21
0.020–0.060	0.030	coarse silt	6
0.060–0.200	0.100	fine sand	1

Table 4. Concentration of sediment in outflow and values of the events S_k.

l	i	$l+i$	k	$CO(l, l+i)$ $(\times 10^{-4})$	$S_k (m^3)$
0	−1	−1	1	3.31	881,836
0	0	0	2	10.2	2,446,408
0	1	1	3	1.53	7,816,662
1	−1	0	4	1.53	853,769
1	0	1	5	0.14	3,058,943
1	1	2	6	0.11	7,958,484
2	−1	1	7	0.11	936,566
2	0	2	8	0.07	3,063,250
2	1	3	9	0.12	7,952,049

where C_b = concentration at the bed; n_c = Manning's coefficient; U = velocity of the outflow; S_b = slope of the channel bed at the dam site; D = depth of flow; χ = correction factor to allow application of this expression to fine sediment particles; and;

$$\alpha = n_c g^{1/2} / D^{1/6} = (f/8)^{1/2} \tag{57}$$

$$\beta = \omega / U \tag{58}$$

$$\eta = z / D \tag{59}$$

with g = gravitational acceleration; f = the Darcy-Weisbach friction coefficient; ω = fall velocity of sediment; and z = depth in the flow.

The concentration at the bed, C_b, was computed based upon an average density of 1213 kg/m³ as reported in Anonymous (1966), thus giving a value $C_b = 0.46$. The Manning's coefficient, n_c, was set equal to 0.015 and the slope of the channel bed at the dam site, S_b, was take as 0.001. The grain-size distribution of the bed material at the dam site is shown in Table 3. The velocity U is obtained through dividing the constant outflow by the respective cross-sectional area for each elevation. The outflow was assumed to take place during one-fourth of the year, giving a discharge of 35.91 m³ s⁻¹ equivalent to 283,027,649 m³.

Soares (1982) shows how to obtain the concentration of sediment in outflow. The computed values are summarized in Table 4.

The values S_k; $k = 1, 2, \ldots, 9$, are obtained from Equations 3a, b, c with the concentration values estimated above. As the estimated concentrations are of solid volume, i.e. the voids are not included, these values S_k must be divided by the concentration of sediment at the bed level, $C_b = 0.46$, in order to obtain the volume occupied by the deposited sediment. The values S_k are presented in Table 4.

Table 5. Results for independent inflows and initial distribution given in Equation 41.

n (yr.)	$E\{T_n\}$ (m^3)	$Var\{T_n\}^{1/2}$ (m^3)
2	6,888,327.5	3,626,963.5
4	13,776,655.0	5,271,700.8
8	27,553,310.0	7,694,626.6
16	55,106,620.0	11,171,975.7

Table 6. Results for independent inflows and initial distribution given in Equation 43.

n (yr.)	$E\{T_n\}$ (m^3)	$Var\{T_n\}^{1/2}$ (m^3)
2	6,714,298.5	3,661,234.4
4	13,504,791.4	5,329,856.1
8	27,192,862.8	7,761,051.7
16	54,706,969.3	11,228,276.0

Table 7. Results for independent inflows and initial distribution given in Equation 44.

n (yr.)	$E\{T_n\}$ (m^3)	$Var\{T_n\}^{1/2}$ (m^3)
2	7,096,446.2	3,570,747.5
4	14,107,898.1	5,174,604.9
8	27,996,854.8	7,581,936.0
16	55,599,902.0	11,076,829.9

Table 8. Results for independent inflows and initial distribution given in Equation 45 (for $l = 1$).

n (yr.)	$E\{T_n\}$ (m^3)	$Var\{T_n\}^{1/2}$ (m^3)
2	6,914,696.2	3,624,121.5
4	13,812,620.6	5,268,037.9
8	27,597,265.8	7,690,968.2
16	55,154,080.3	11,168,897.7

3.1 *Independent inflows*

The matrix $P = [p_{jk}]$, Table 1, is obtained with the compute values p_{-1}, p_0 and p_1 and $h = 2$. This is used to compute $E\{Y_k^{(n)}|s_0 = i\}$, $E\{[Y_k^{(n)}]^2|s_0 = i\}$ and $E\{[Y_l^{(n)} + Y_k^{(n)}]^2|s_0 = i\}$ for $n = 2, 4, 8$ and 16. The unconditional expectations $E\{Y_k^{(n)}\}$, $E\{[Y_k^{(n)}]^2\}$ and $E\{[Y_l^{(n)} + Y_k^{(n)}]^2\}$ and the covariance matrices $\sigma_{jk}^{(n)}$ are then computed for each n and for the two assumptions concerning the initial distribution Prob$\{s_0 = i\}$.

Substituting these values into Equations 15–20, the following solutions, depending on the initial distribution considered, are obtained. These are given in Tables 5–8.

The effects of the initial distribution are attenuated quickly, since the reservoir was divided into only three levels. This can be seen by comparing the results of the two extreme levels, Table 6 and 7. The percentage difference in the expectation between the results for 2, 4, 8, 16 yr are, respectively, 5.4%, 4.3%, 2.9% and 1.6%.

414

Table 9. Results for independent inflows and initial distribution given in Equation 48.

n (yr.)	$E\{T_n\}$ (m³)	$Var\{T_n\}^{1/2}$ (m³)
2	6,907,274.9	3,559,144.1
4	13,814,549.7	5,135,539.1
8	27,629,099.5	7,480,669.4
16	55,258,198.9	10,882,001.4

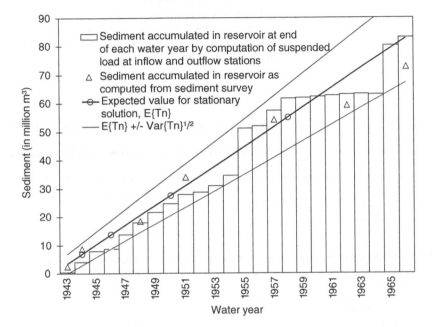

Figure 2. Accumulated sediment in the reservoir.

3.2 Correlated inflows

The matrix $P = [p_{jk}]$, Table 2, is obtained with the compute values of α, β and γ and $h = 2$. This is used to compute $E\{Y_k^{(n)}\}$, $E\{[Y_k^{(n)}]^2\}$, $E\{[Y_l^{(n)} + Y_k^{(n)}]^2\}$ and $\sigma_{jk}^{(n)}$ for $n = 2, 4, 8$ and 16 for the stationary distribution.

The computed value -0.032 for the serial correlation coefficient at lag 1 for inflows shows that the correlation between two consecutive years is very weak. Therefore, the results in Table 9 for correlated inflows are almost the same that for the case of independent inflows, especially because the assumption that a Markovian correlation structure for the inflows exists is not confirmed by the data.

3.3 Comparison between results and measurements

Figure 2 shows the sediment accumulated in the reservoir at the end of each water-year as computed by the measured suspended load at inflow and outflow stations from the years 1943–1966, and as computed from sediment surveys. In this figure the calculated expected values of the stationary process for independent inflows (Tab. 5), as well as the values of $E\{T_n\} \pm Var\{T_n\}^{1/2}$, are plotted for the purpose of comparison.

It is evident from Figure 2 that the stochastic model is capable of predicting the cumulation of sediment deposition in the reservoir. The measured data values are found to be derived from a population that has the expected value and variance specified by the model. It should be mentioned hear that the model does not consider consolidation effects of sediment. This effect can, however, be incorporated into the model by applying to the computed results a consolidation factor (Soares et al. 1982). The sediment surveys give results that include such a consolidation.

4 CONCLUSIONS

The stochastic sediment storage model developed by Soares (1975) predicted, in spite of the simple inflow representation adopted, satisfactorily the total volume of sediment deposition as a function of time for the reservoir analyzed, as shown in Figure 2. In order to apply the model to predict sedimentation in new reservoirs, the parameters characterizing the inflow and the concentration of sediment in the inflow are required. These parameters can be easily evaluated if necessary measurements are carried out before the construction of the dam.

REFERENCES

Ackermann, C.W. & Corinth, R.L. 1962. An empirical equation for reservoir sedimentation. *Int. Assoc. Sci. Hydrol.* 59: 359–366.
Anonymous 1966. *Sedimentation in John Martin Reservoir, Arkansas River Basin*. Albuquerque: U.S. Army Corps Eng.
Brown, C.B. 1950. Sedimentation in reservoirs. In H. Rouse (ed.), *Engineering Hydraulics*. New York: Wiley
Cox, D.R. & Miller, H.D. 1965. *The theory of stochastic processes*. London: Methuen.
Gottschalk, L.C. 1964. Reservoir sedimentation. In V.T. Chow (ed.), *Handbook of Applied Hydrology*. New York: McGraw-Hill.
Gupta, S.K. 1974. *A distributed digital model for estimation of flows and sediment load from large ungaged watershed. Ph.D. Thesis*. Waterloo: University of Waterloo.
Lloyd, E.H. 1963. A probability theory of reservoirs with serially correlated inputs. *J. Hidrol.* 1: 98–128.
Miller, C.R. 1951. *Analysis of flow-duration sediment rating curve method of computing sediment Yield*. Denver: U.S. Bur. Reclam.
Moran, P.A.P. 1954. A probability theory of dams and storage systems. *Aust. J. Appl. Sci.* 5: 116–124.
Moran, P.A.P. 1956. A probability theory of a dam with a continuous release. *Q.J. Math.* 7: 130–137.
Paulet, M. 1971. *An interpretation of reservoir sedimentation as a function of watershed characteristics. Ph.D. Thesis*. Lafayette: Purdue University.
Soares, E.F. 1975. *A deterministic-stochastic model for sediment storage in large reservoirs. Ph.D. Thesis*. Waterloo: University of Waterloo.
Soares, E.F., Unny, T.E. & Lennox, W.C. 1982. Conjunctive use of deterministic and stochastic models for predicting sediment storage in large reservoirs, 1. A stochastic sediment storage model. *J. Hydrol.* 59: 49–82.
Volkov, I.S. 1958. On the distribution of sums of random variables defined on a homogeneous Markov chain with a finite number of states. *Theory Probab. Its Appl.* 3(4): 384–399.

Applications of Computational Mechanics in Geotechnical Engineering – Sousa,
Fernandes, Vargas Jr & Azevedo (eds)
© 2007 Taylor & Francis Group, London, ISBN 978-0-415-43789-9

Numerical schemes for the solution of advection problems

A.L.B. Cavalcante & M.M. Farias
University of Brasília

ABSTRACT: Hyperbolic equations are related to advection problems, in which dissipative phenomena are minimal or may be disregarded. There are many Geotechnical problems involving this type of equation. The solution of such equations using some classical methods, such as Finite Differences and Lax-Wendroff, produce dissipation and other spurious effects that are purely numerical. In order to solve this problem the Cubic Interpolated Pseudo-particle (CIP) method is introduced in this paper. Applications of these methods to a problem with simple initial conditions and known analytical solution show the superiority of the CIP approach.

1 INTRODUCTION

Many problems in Geotechnical Engineering involve advection phenomenon. Perhaps the best known of them is the transport of solutes by fluids percolating in a porous media (Freeze and Cherry, 1979). This phenomenon is governed by two mechanisms: diffuse molecular movements (diffusion) and macroscopic fluid flow (advection).

Advection is the process in which the solute is carried by the fluid, generally water, with or without interaction with some porous media. In advective transport without interaction with a porous media, the contamination front is abrupt and moves at a velocity that is equal to the average linear velocity (v) of the percolating fluid without any alteration of peak concentration. This is the case, for instance, of the transport of an oil spill dragged by the maritime currents and the wind. Little bubbles of oil are also transported by dispersion in the sub superficial currents.

When there is interaction between the fluid flow and some porous media, such as soil, the effective percolation velocity of the fluid may computed from the apparent velocity (obtained from Darcy's law) divided by the porosity of the media. In this case, the hydraulic conductivity of the soil is a fundamental parameter for the solute transport, since it gives a measure of the media resistance to the percolation of the fluid and consequently of the solute dissolved in it. In this case, advection may be regarded as a chemical transport caused by a hydraulic gradient.

Other kind of problems using the fluid motion in a continuous media can be formulated in parallel with a constitutive model. Problems that describe large deformation analysis of geomaterial may use these kinds of equations if the soil is regarded as a Bingham fluid. Sawada *et al.* (2004) presented a study of the simulation of slope failure due to earthquake and due to heavy rain. Moriguchi (2005) studied the gravitational flow of a soil column using the Bingham model as a basic constitutive model taking into the consideration the shear strength of geomaterials governed by cohesion c and friction angle ϕ. Moriguchi (2005) also simulated the penetration of rigid body in a soil, the flow of a real ground, soil-slump tests, the bearing capacity of cohesive ground, and flow of sands. All these phenomena were governed, at least partially by advective equations.

Uzuoka (2000) and Hadush *et al.* (2000 and 2001) also studied the behavior of geomaterials subjected to flow. They treated liquefied soil undergoing ground flow as a Bingham fluid and developed numerical methods based on fluid dynamics.

Hyperbolic equations govern advective problems (Xiao, 1996), in which dissipative phenomena are not present or may be conveniently disregarded. The solution of such problems requires initial conditions as well as boundary conditions. The general mathematical formulation of an advective problem, in one-dimensional space, is described by the following hyperbolic equation:

$$u_t + v \cdot u_x = 0 \tag{1}$$

in which $v > 0$ is the advection velocity and $v.u_x$ is the so called advective term. The independent variable u is function both of space x and time t; u_x and u_t denote its derivatives.

Equation (1) represent the transport of u along axis x towards the right-hand side when $v > 0$. Since this equation does not contain a dissipative term, u_{xx}, then the value of u should just be transported along x, without undergoing any alteration, between time t_0 and $t_0 + \Delta t$. The absence of dissipative phenomena implies that any discontinuity in the initial conditions should propagate to the solution at any time $t > 0$. This implies that hyperbolic equation admits discontinuous solutions, and the numerical method adopted to solve these equations should be able to deal with such discontinuities efficiently.

The initial value problem, or Cauchy problem, for the advection equation consists in finding a function $u(x, t)$ in the semi-space $D = \{(x, t)/t \geq 0, -\infty < x < \infty\}$, that satisfies both Equation (1) and a particular initial condition. In general the solution is continuous and sufficiently differentiable, but this is not always the case.

This paper presents the Cubic Interpolated Pseudo-particle (CIP) method as a possible tool to deal with such kind of problems. The method is compared to the analytical solution and to approximate solutions obtained by other traditional numerical schemes such as Finite Differences and Lax-Wendroff. The CIP method is not new nor an invention of the authors. It was proposed by Takewaki et al. (1985) and Takewaki & Yabe (1987) in Japan and it has since evolved and has been applied to many problems in that country. However it is still not very publicized in other countries and this paper is an opportunity to promote the method and its advantages. An application of the method to the solution of bed load transport of heterogeneous sediments in tailings dams is presented in another paper by the authors in this workshop (Cavalcante et al., 2007).

2 ADVECTION EQUATION

The analytical solution of the advection equation (1) in the semi-space D, subjected to an initial condition $u(x, 0) = f(x)$, in which $f(x)$ is any function describing the initial solution for $\forall x$ and $t = 0$, is simply given by $u(x, t) = f(x - vt)$ for any value x and for $t > 0$. In fact, in the absence of any other term and for constant advection velocity v, it is not necessary to use any numerical method to solve such equation. However this problem was chosen to illustrate the occurrence of unrealistic dissipative phenomena that appear in the numerical solution given by some classical methods, and to compare the analytical solution wit the approximation given by CIP method.

As an example, consider the phenomenon of a sine wave propagating to the right hand side as shown in Figure 1(a). The initial boundary condition may be given by:

$$u(x,0) = \begin{cases} 0, & 0 \leq x < 1 \\ 12 sen[(x-1)\pi], & 1 \leq x \leq 2 \\ 0, & x > 2 \end{cases} \tag{2}$$

Equation (1) have the same characteristics as the solution of the differential equation $dx/dt = v$, which is given by a family of straight lines $x = vt + c_1$, where c_1 is a constant. For the line passing through the point $(x_0, 0)$, then $c_1 = x_0$ the solution is illustrated in Figure 1(b) and the expression for the line is given by:

$$t = \frac{x - x_0}{v} \tag{3}$$

418

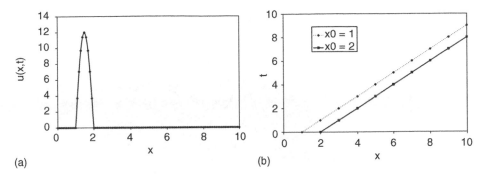

(a)   (b)

Figure 1. (a) Initial condition and (b) characteristic curves for the advection equation in space $x - t$.

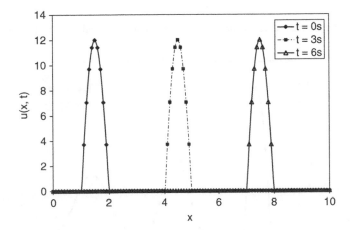

Figure 2. Analytical solution for the propagation of the advection equation.

As the initial sine wave propagates, it should not suffer any dissipation, as illustrated in Figure (2). The overall solution $(x, t, u(x, t))$ shown in the three-dimensional plot of Figure 3.

3 TRADITIONAL NUMERICAL METHODS FOR ADVECTION PROBLEMS

As emphasized previously, hyperbolic equations propagate the initial information at a constant velocity, without suffering dissipation mechanisms. Besides, the initial condition may present discontinuities which should also propagate with dissipation. The properties may result in wrong approximations if the problem is not adequately treated during numerical solutions.

The most intuitive numerical treatment for such problems is to adopt a forward scheme in time and a central scheme in the space domain. Despite its simplicity, the stability of this method can not be guaranteed with a simple criterion. A more stable solution using Finite Differences may be obtained adopting a forward scheme in time and backwards difference in space, leading to the following fully explicit algorithm:

$$u_k^{n+1} = C \cdot u_{k-1}^n + (1 - C) \cdot u_k^n \tag{4}$$

in which $C = v\Delta t / \Delta x$ is known as the Courant number and according to von Neumann it is possible to establish the stability criterion as $C \leq 1$.

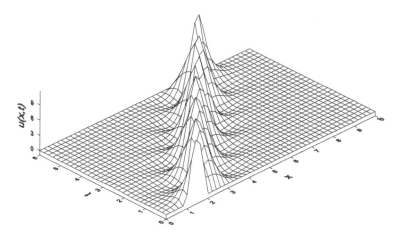

Figure 3. Three-dimensional representation of the exact solution of the advection equation.

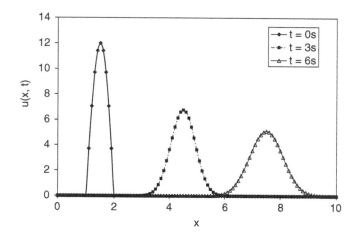

Figure 4. Numerical solution of the advection equation using the Finite Differences.

Only in the case when $C = 1$ it is possible to obtain the exact solution, $u_k^{n+1} = u_{k-1}^n$, and the solution at the upstream $k - 1$ of a given of a point k at time n, u_{k-1}^n, is transported to the point k at time $n+1$, u_k^{n+1}, without any dissipation or loss of information. In any other case with $C < 1$, there is sufficient information to compute u_k^{n+1} without numerical instabilities, however, the exact solution is not reproduced. The profile $u(x, t)$ is similar to that observed during dissipative phenomenon, but this dissipation is purely numerical and undesirable. This is illustrated in Figure 4, for the sine wave initial conditions ($t = 0$) of Eq. (2), using a Courant number $C = 0.5$, and the propagation of the initial condition to times $t = 3$ and $t = 6$ s. The complete solution at the time-space domain, $(x, t, u(x, t))$, is shown in the three-dimensional plot in Figure 5. Notice the numerical dumping of the initial condition, with decreasing amplitudes of the sine wave as the numerical solution propagates in time.

The dissipation phenomenon observed in Figures 4 and 5 is purely a numerical error, which should be avoided since the real problem does not involve any dissipation mechanism. The exact solution should be as illustrated in Figure 3. A novice in numerical analysis may be tempted to "improve" the numerical solution by using smaller time intervals Δt. This would in fact decrease the Courant number C and deteriorate even more the numerical approximation. This is illustrated for the same problem at a given time ($t = 4$ s) using different values of C (Figure 6). For smaller time intervals (and C) the more pronounced the undesirable numerical dissipation will be.

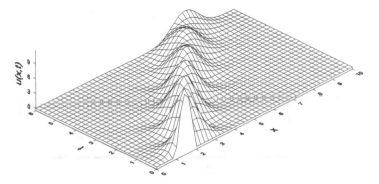

Figure 5. Complete numerical solution of the advection equation using the Finite Differences.

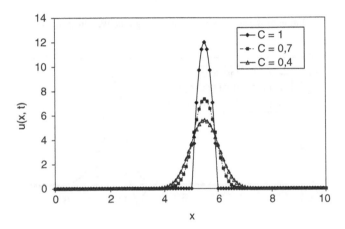

Figure 6. Numerical solution of the advection equation using the Finite Differences method and different values of Courant number.

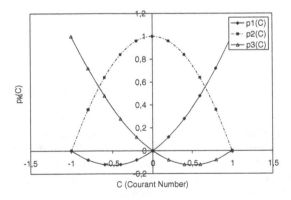

Figure 7. Polynomials $p_k(C)$ of Lax-Wendroff Method.

Another classical method is that proposed by Lax & Wendroff (1960). In this case the numerical approximation for the advection equation may be expressed as:

$$u_k^{n+1} = p_1(C) \cdot u_{k-1}^n + p_2(C) \cdot u_k^n + p_3(C) \cdot u_{k+1}^n \qquad (5a)$$

421

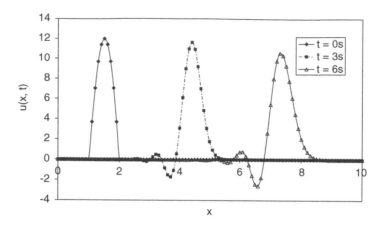

Figure 8. Numerical solution of the advection equation using Lax-Wendroff method.

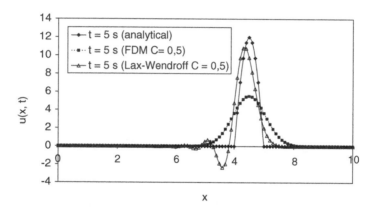

Figure 9. Comparison between analytical, Lax-Wendroff and Finite Differences solution for the advection equation.

$$p_1(C) = \frac{1}{2}C(C+1); \; p_2(C) = (1+C)(1-C); \; p_3(C) = \frac{1}{2}C(C-1) \qquad (5b)$$

in which C is the Courant number and $p_k(C)$ are the polynomials given in Eq. 5(b) and illustrated in Figure 7.

For a Courant number $C = 1$, the Lax-Wendroff polynomials assume values $p_1 = 1, p_2 = p_3 = 0$ and the exact analytical solution, $u_k^{n+1} = u_{k-1}^n$, is also reproduced as in the case of Finite Differences. However, numerical dissipation is also observed for the Lax-Wendroff method if values of $C < 1$ are assumed. This is illustrated in Figure 8, which shows the solution for the sine wave initial conditions of Eq. (2), at different times, using the scheme of Eq. (5) with $C = 0.5$.

Nevertheless the dissipation effect in the Lax-Wendroff method is less pronounced than that computed when using the Finite Differences method. On the other hand, a different time of numerical perturbation is observed in the Lax-Wendroff solution with the occurrence of oscillations to the left of the peak value. As the solution propagates in time, the peaks decrease and the oscillations in the left increase.

A comparison between analytical and the numerical solution with the methods of Lax-Wendroff and Finite Differences is presented in Figure 9 for $C = 0.5$ and time $t = 5$ s. The peak value of Lax-Wendroff method is closer to the exact analytical solution but slightly retarded in time.

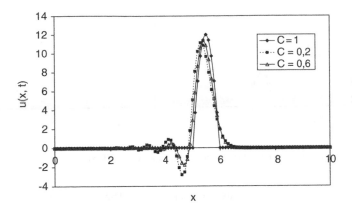

Figure 10. Numerical solution of the advection equation using the Lax-Wendroff method and different values of Courant number.

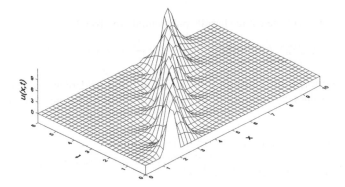

Figure 11. Complete numerical solution of the advection equation using Lax-Wendroff method.

Figure 10 shows the effect of the Courant number in the Lax-Wendroff numerical solution. The solution is shown for time $t = 4$ s and different values of C. The dissipation still occurs but is much less pronounced than in the Finite Differences method. The decrease in peak values with decreasing values of Courant number in the Lax-Wendroff method seems to be "compensated" with larger oscillations to the left of the left. The solution also seems to lag behind as time evolves and this effect increases for smaller values of C.

Figure 11 shows the complete solution, $(x, t, u(x, t))$, of the advection problem in the time-space domain using Lax-Wendroff method.

4 CIP METHOD

The cubic interpolated pseudo-particle method, CIP, as proposed by Yabe & Takei (1988), is used here to find an approximate solution $u(x, t)$ to the advection problem. For a given constant advection velocity $v > 0$, the solution should propagate the information, displacing the curve by an amount equal to $\Delta x = v \Delta t$ to the right during the time interval Δt. For this to happen, the following hypothesis is assumed as illustrated in Figure 12 (a):

$$u(x,t) \cong u(x - \Delta x, t - \Delta t) \tag{6}$$

In the CIP method (Yabe & Aoki, 1991), the discrete solution u_k^n at time n for a mesh of points x_k in the space domain x is smoothed by approximating a Hermite cubic polynomial $U(x)$ in each space

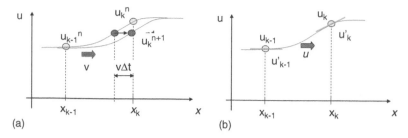

Figure 12.   Hypothesis of CIP Method (modified from Moriguchi, 2005).

interval of length Δx between successive points $[x_{k-1}, x_k]$. The general form of the polynomial is given by:

$$U(x) = a^n_{k-1}\left(x - x^n_{k-1}\right)^3 + b^n_{k-1}\left(x - x^n_{k-1}\right)^2 + c^n_{k-1}\left(x - x^n_{k-1}\right) + d^n_{k-1} \tag{7}$$

The space derivatives of the cubic Hermite polynomial are given by:

$$U'(x) = 3a^n_{k-1}\left(x - x^n_{k-1}\right)^2 + 2b^n_{k-1}\left(x - x^n_{k-1}\right) + c^n_{k-1} \tag{8}$$

The CIP method forces the polynomial approximation and its derivatives, $U(x)$ and $U'(x)$, to match the discrete values, $u(x, t)$ and $u'(x, t)$, at the extremes of each space interval $[x_{k-1}, x_k]$ (see Figure 12b):

$$u(x_{k-1}, t_k) = U\left(x^n_{k-1}\right) = u^n_{k-1} \text{ and } u'(x_{k-1}, t_k) = U'\left(x^n_{k-1}\right) = u'^n_{k-1} \tag{9a}$$

$$u(x_k, t_k) = U\left(x^n_k\right) = u^n_k \text{ and } u'(x_k, t_k) = U'\left(x^n_k\right) = u'^n_k \tag{9b}$$

From the hypothesis presented in Eq. (9a) and using Eqs. (7) and (8), it is possible to determine the coefficients c_{k-1} and d_{k-1}:

$$c^n_{k-1} = u'^n_{k-1} \text{ and } d^n_{k-1} = u^n_{k-1} \tag{10}$$

Noticing that $\Delta x = x^n_k - x^n_{k-1} = v\Delta t$ and applying the hypothesis presented in Eq. (9b) to the extreme values in Eqs. (7) and (8), it is possible to determine the other coefficients a^n_{k-1} and b^n_{k-1} as:

$$a^n_{k-1} = \frac{\left(u'^n_{k-1} + u'^n_k\right)}{\Delta x^2} + \frac{2\left(u^n_{k-1} - u^n_k\right)}{\Delta x^3} \tag{11}$$

$$b_{k-1} = \frac{3\left(u^n_k - u^n_{k-1}\right)}{\Delta x^2} - \frac{\left(2u'^n_{k-1} + u'^n_k\right)}{\Delta x} \tag{12}$$

Now that all constants are determined, the discrete values of $u(x, t)$ and $u'(x, t)$ may be propagated to the next time step $n + 1$ as follows:

$$u^{n+1}_k = a^n_{k-1}(v\Delta t)^3 + b^n_{k-1}(v\Delta t)^2 + u'^n_{k-1}(v\Delta t) + u^n_{k-1} \tag{13}$$

$$u'^{n+1}_k = 3a^n_{k-1}(v\Delta t)^2 + 2b^n_{k-1}(v\Delta t) + u'^n_{k-1} \tag{14}$$

The operations previously explained should be performed for all points of the mesh, for each interval in the space domain. This allows to estimate values of u and u' explicitly at the next time

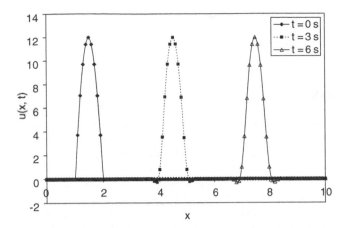

Figure 13.   Numerical solution of the advection equation using the CIP method and Courant number ($C = 0.5$).

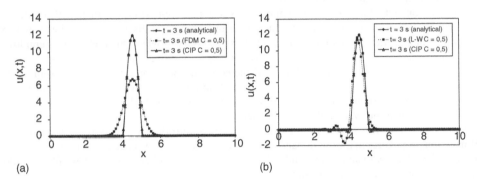

(a)   (b)

Figure 14.   (a) Comparison between CIP and Finite Differences. (b) Comparison between CIP and Lax-Wendroff.

step $n + 1$, given these values at time step n. Therefore the initial solution at $t = 0$ is propagated in time during as many time steps as necessary. Notice, however, that the CIP scheme requires not only the values of the function at all space points as an initial condition u_k^0, but also the values of the derivatives, $u_k'^0$. If these derivatives are not explicitly given as a continuous function, $u'(x, 0) = f'(x)$, it is still possible to estimate them using the central finite differences of the initial discrete function values (Moriguchi, 2005):

$$u_k'^n = \frac{u_{k+1}^n - u_{k-1}^n}{2\Delta x} \tag{15}$$

The CIP method can eliminated the dissipation and oscillation problems observed when solving the advection equation with the Finite Differences and Lax-Wendroff methods with Courant numbers less than unit. This is fundamental for an independent and efficient discretization of the problem in both time and space domain.

The solution for the advection problem with the initial conditions of Eq. (2), was obtained using the CIP method and a Courant number $C = 0.5$. This solution is shown in Figure 13 for different times. Notice that the initial solution propagated in time with any loss of information, dissipation or oscillations.

Figure 14 shows a comparison of the solutions of the advection problem using the methods present in this paper. Figure 14(a) shows a comparison between CIP and Finite Differences for

$C = 0.5$. Figure 14(b) shows the comparison between CIP and Lax-Wendroff method also for $C = 0.5$. In both cases the analytical solution is also indicated in the figures. Notice the perfect match obtained when the CIP method is adopted.

5 CONCLUSIONS

This paper presented three different numerical methods that may be used to solve the hyperbolic differential equation that represents advection problems. Namely the solution schemes investigated were: (a) the explicit Finite Differences Method with forward derivatives in time and backwards derivatives in the space domain; (b) the Lax-Wendroff method; and (c) the Cubic Interpolated Pseudo-particle (CIP) method. The numerical solutions were compared to the exact analytical solution for an initial condition given by a sine wave.

The solution using Finite Differences is conditionally stable but present spurious numerical dissipation for Courant number ($C = v \Delta t / \Delta x$) less than unit. This undesirable dissipation effect increasing, dumping the solution, if smaller Courant numbers are adopted. Therefore the solution gets worse if smaller time intervals are used.

The solution with Lax-Wendroff method gives a better approximation than with Finite Differences. However this method also produces spurious numerical dissipation. Furthermore,some oscillation and a little retardation of the solution were observed when this method was employed. These undesirable effects are more pronounced for smaller Courant numbers.

Both Finite Differences and Lax-Wendroff method can only reproduce the exact solution if a value of Courant number $C = 1$ is adopted. This will in fact make the time and space discretization interdependent. Such an ideal discretization can only be achieved in simple one-dimensional problem with equal intervals in the space domain.

The CIP method can overcome the numerical errors introduced by the other schemes. The information propagates in time and space without any undue dissipation or oscillation. The solution matched the analytical values independently of the Courant number adopted. This allows a more efficient and independent discretization in time and space.

The CIP method has a great potential for application in Geotechnical Engineering. Many problems have a least part of the system of equations that is governed by advective phenomena. These equations can be adequately solved using CIP without introducing significant numerical perturbations.

ACKNOWLEDGEMENTS

The authors acknowledge the support of the following institutions: Brazilian National Research Council (CNPq), University of Brasilia, UPIS University and FINATEC. The authors are also indebted to Dr. Shuji Moriguchi and Professor Takayuki Aoki from Tokyo Institute of Technology, Japan, for their kindness in teaching us the details of the CIP method.

REFERENCES

Cavalcante, A.L.B., Farias, M.M. & Assis, A.P. (2007). Heterogeneous Sediment Transport Model Solved by CIP Method. *Proc. of the 5th Int. Workshop on Applications of Computational Mechanics in geotechnical Engineering*, Guimarães, Portugal, 9 p.
Freeze, R.A. & Cherry, J.A. (1979). *Groundwater*. Prentice Hall, Inc., U.S. 604 p.
Hadush, S., Yashima, A. & Uzuoka, R. (2000). Importance of Viscous Fluid Characteristics in Liquefaction Induced Lateral Spreading Analysis. *Computers and Geotechnics*. Vol. 27, pp. 199–224.
Hadush, S., Yashima, A., Uzuoka, R., Moriguchi, S. & Sawada, K. (2001). Liquefaction Induced Lateral Spread Analysis Using the CIP Method. *Computers and Geotechnics*. Vol. 28, pp. 549–574.

426

Lax, P.D. & Wendroff, B. (1960). Systems of Conservation Laws. *Comm.Pure Appl. Math.* Vol. 13 pp. 217–237. MR 22:11523.

Moriguchi, S. (2005). *CIP-Based Numerical Analysis for Large Deformation of Geomaterials.* PhD Thesis, Gifu University, Japan, 123 p.

Sawada, K., Moriguchi, S., Yashima, A., Zhang, F. & Uzuoka, R. (2004). Large Deformation Analyis in Geomechanics Using CIP Method. *JSME International Journal.* Vol. 47, No. 4, pp. 735–743.

Takewaki, H., Nishiguchi, A. & Yabe, T. (1985). Cubic interpolated pseudo-particle method (CIP) for solving hyperbolic-type equations. *J. Comput. Phys.* Vol. 61, pp. 261–268.

Takewaki, H. & Yabe, T. (1987). The cubic-interpolated pseudo particle (CIP) method: application to nonlinear and multi-dimensional hyperbolic equations, *J. Comput. Phys.* Vol. 70, pp. 355-

Uzuoka, R. 2000. *Analytical Study on the Mechanical Behavior and Prediction of Soil Liquefaction and Flow* (in Japanese). PhD Thesis, Gifu University, Japan.

Xiao, F. (1996) *Numerical Schemes for Advection Equation and Multilayered Fluid Dynamics.* PhD Thesis, Tokyo Institute of Technology, Tokyo.

Yabe, T. & Aoki, T. (1991). A Universal Solver for Hyperbolic Equations by Cubic Polynomial Interpolation. *Comput. Phys. Commun.* Vol. 66, pp. 219–232.

Yabe, T. & Takei, E. (1988) A New Higher-Order Godunov Method for General Hyperbolic Equations. *Journal of the Physical Society of Japan.* Vol. 57, No. 8, pp. 2598–2601.

Applications of Computational Mechanics in Geotechnical Engineering – Sousa,
Fernandes, Vargas Jr & Azevedo (eds)
© 2007 Taylor & Francis Group, London, ISBN 978-0-415-43789-9

Heterogeneous sediment transport model solved by CIP method

A.L.B. Cavalcante, M.M. Farias & A.P. Assis
University of Brasilia

ABSTRACT: This paper presents an alternative method to solve the system of equations that represent the phenomenon of bed load transport of heterogeneous sediments in tailings dams. The field conditions that govern this problem include Navier-Stokes equations for the fluid and the mass balance of sediments, besides a constitutive relation for rate of transported sediments is also required. The Navier-Stokes equations include an advective term. If these equations are solved using traditional approximation techniques, such as the Finite Differences Method (FDM), the numerical solution generates spurious dissipations that are not related to the real physical phenomenon. In order to solve this problem, the equations are split into non-advective and advective parts and the so-called Cubic Interpolated Pseudo-particle method (CIP) is applied to solve the advective phase. The numerical scheme is then used to perform a few parametric analyses in order to gain further insight of the mechanisms that rule the bed load transport in tailings dams.

1 INTRODUCTION

Cavalcante (2004) presented a model to simulate the bed load transport of heterogeneous sediments. This model couples Navier-Stokes equations, representing continuity and equilibrium of the fluid phase, with the equation for the conservation of mass of the solid sediments. Besides, a constitutive relation that describes the rate of transported sediments should also be adopted.

Heterogeneous sediments or tailings, comprising iron and quartz particles, are commonly produced during the extraction and concentration of iron. The bed load transport phenomenon, in which the particles roll when carried by a thin film of water, is the most representative of the actual condition that is observed during the construction of tailings dams using hydraulic deposition techniques. Despite some engineering drawbacks, mainly if the upstream construction method is used, this is the solution for waste disposal generally adopted by most mining industries, due to its relatively low costs.

The coupled model for the conditions of equilibrium and continuity of the fluid phase, plus the continuity of the solid sediments, results in a system of hyperbolic equations. In this kind of problem the material is transported without dissipation. Nevertheless, if this system of equation is solve using classic techniques, such as the explicit Finite Differences Method (FDM), the solution introduces spurious dissipations that are purely numerical without any relation with the real physical problem. In order to overcome this problem, another technique known as the Cubic Interpolated Pseudo-particle method (CIP) is adopted in this paper.

2 COUPLED MODEL

In this section, a mathematical model for the bed load transport of heterogeneous sediments is presented. The model couples the behavior of the fluid phase and the sediments, composed of particles of quartz and iron. The objective when developing this model was to devise a mathematical and numerical tool that could help to forecast the profile of hydraulic deposition, including segregation, in tailings dams. Parametric analyses using this model, are very useful to gain further insight about the mechanisms involved during the deposition process, such as the distribution of density, porosity and grain sizes along the profile, as a function of the production variables (discharge and pulp concentration).

Considering the continuity of the fluid phase, equilibrium in the fluid, continuity of transported sediments and a relation for the rate of sediment transport, respectively, Cavalcante (2004), deduced the following equations:

$$a_t + a \cdot u_x + u \cdot a_x = 0 \tag{1}$$

$$u_t + u \cdot u_x + g \cdot \left(a_x + z_{bx}\right) = -\frac{g \cdot u^2}{C_h^2 \cdot a} \tag{2}$$

$$z_{bt} + s_x = 0 \tag{3}$$

$$s = \sum_k m_k u^{n_k} \tag{4}$$

in which u is the velocity of transport f sediments, a is the film of water above the bed of the deposition profile, z_b is the height of deposited sediments, s is the rate of transported sediments composed of particles of two types k (quartz, Qz and iron, Fe), C_h is the coefficient of Chèzy, m_k and n_k are empirical constants dependent on the properties of the sediments, x is the horizontal distance, as illustrated in Figure 1, and t is time.

Cavalcante et $al.$ (2003) presented a numerical solution for the system of equations (1)–(4), based on the Finite Differences Method using advanced differences in time and central differences in the space domain. The resulting explicit algorithm is summarized as follows:

$$u_k^{n+1} = u_k^n - \frac{g\Delta t}{C_h^2} \frac{\left(u_k^n\right)^2}{a_k^n} - \frac{\Delta t}{2\Delta x}\left\{u_k^n\left[u_{k+1}^n - u_{k-1}^n\right] + g\left(\left[a_{k+1}^n - a_{k-1}^n\right] + \left[z_{bk+1}^n - z_{bk-1}^n\right]\right)\right\} \tag{5}$$

$$a_k^{n+1} = a_k^n - \frac{\Delta t}{2\Delta x}\left\{a_k^n\left[u_{k+1}^n - u_{k-1}^n\right] + u_k^n\left[a_{k+1}^n - a_{k-1}^n\right]\right\} \tag{6}$$

$$z_{bk}^{n+1} = z_{bk}^n - \frac{\Delta t}{2\Delta x}\left[s_{k+1}^n - s_{k-1}^n\right] \tag{7}$$

$$s_k^{n+1} = m_{Qz}\left[u_k^n\right]^{n_{Qz}} + m_{Fe}\left[u_k^n\right]^{n_{Fe}} \tag{8}$$

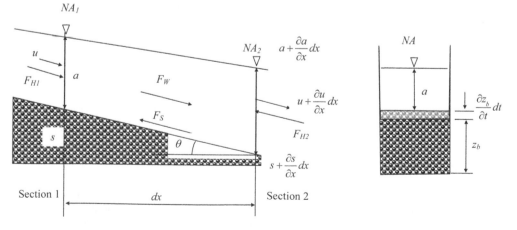

Figure 1.   Details of an infinitesimal element in the upstream slope.

Cavalcante *et al.* (2003) compared the solution obtained with the Finite Differences algorithms in equations (5)–(8) with the experimental data obtained by Ribeiro (2000) from laboratory hydraulic deposition tests. The numerical and empirical results were also compared with analytical solutions obtained by Cavalcante *et al.* (2002) for simplified boundary conditions, expressed as follows:

$$z_b(x,t) = A \cdot \left\{ \sqrt{Bt} \cdot \exp\left(-E\frac{x^2}{t} \right) - D \cdot x \cdot \mathrm{erfc}\left(\sqrt{E}\frac{x}{t} \right) \right\} \tag{9}$$

in which,

$$A = \frac{z_b(0,t)}{\sqrt{Bt}}, \quad B = \frac{60 \cdot s_0 \cdot i_0}{\pi \cdot a_0}, \quad D = \frac{3 \cdot i_0}{a_0} \quad \mathrm{e} \quad E = \frac{9 \cdot i_0}{60 \cdot a_0 \cdot s_0} \tag{10}$$

$$\mathrm{erfc}(x) = 1 - \frac{2}{\sqrt{\pi}} \left(x - \frac{x^3}{3 \cdot 1!} + \frac{x^5}{5 \cdot 2!} - \frac{x^7}{7 \cdot 3!} + \cdots \right) \tag{11}$$

The numerical results were satisfactory for a time-space discretization with a Courant number ($C = u_0 \cdot \Delta t / \Delta x$) equal to one. However, for C values less than unit the numerical solutions showed spurious dissipation and the results did not converge to those obtained analytically and empirically. This numerical error is investigated by Cavalcante & Farias (2007) in another paper in this conference. The solution to such numerical problem may be achieved using the so-called Cubic Interpolated Pseudo-particle (CIP) method.

3 COUPLED METHOD SOLVED BY CIP

The CIP method is efficient to solve advection problems, which results in hyperbolic equations for which dissipation phenomena are not present or may be disregarded. In order to use the CIP method, according to an algorithm known as particle-in-cell (Nishiguchi & Yabe, 1983), the transport model in Eqs. (1)–(4), must be split in two parts: an advective (Lagrangian) phase, and a non-advective (Eulerian) phase.

The non-advective equations governing the model are the following:

$$a_t = -a \cdot u_x \tag{12}$$

$$u_t = -\frac{g \cdot u^2}{C^2 \cdot a} - g\left(a_x + z_{bx} \right) \tag{13}$$

$$z_{bt} + s_x = 0 \tag{14}$$

$$s = \sum_k m_k u^{n_k} \tag{15}$$

The advective equations of the problem are as follows:

$$a_t + u \cdot a_x = 0 \tag{16}$$

$$u_t + u \cdot u_x = 0 \tag{17}$$

Therefore the equation governing continuity of the fluid phase, Eq. (1), was divided into a non-advective part in Eq. (12) and an advective part in Eq. (16). The same process was applied to the equation governing equilibrium of the fluid phase, Eq. (2), which was split into equations (13) and (17).

The CIP method is applied only to the advective equations, thus avoiding the introduction of spurious numerical dissipations in the solution. The non-advective equations may be solved using the Finite Differences Method (FDM). In this case the correspondent non-advective equations (12)–(15), may be re-written as:

$$a_k^{n+1} = a_k^n - \frac{\Delta t}{2 \cdot \Delta x} \left\{ a_k^n \cdot \left(u_{k+1}^n - u_{k-1}^n \right) \right\}$$

(18)

$$u_k^{n+1} = u_k^n - \frac{g\Delta t}{C^2} \frac{\left(u_k^n \right)^2}{a_k^n} - \frac{g\Delta t}{2\Delta x} \left\{ \left(a_{k+1}^n - a_{k-1}^n \right) + \left(z_{b\,k+1}^n - z_{b\,k-1}^n \right) \right\}$$

(19)

$$z_{b\,k}^{n+1} = z_{b\,k}^n - \frac{\Delta t}{2\Delta x} \left[s_{k+1}^n - s_{k-1}^n \right]$$

(20)

$$s_k^{n+1} = m_{Qz} \left[u_k^n \right]^{n_{Qz}} + m_{Fe} \left[u_k^n \right]^{n_{Fe}}$$

(21)

After solving the non-advective equations, these solutions are used as initial conditions for the advective equations. The advective phase, Eqs. (16) and (17), is solved using the CIP method as described by the authors in another paper (Cavalcante & Farias, 2007).

4 EXPERIMENTAL SIMULATION AND MODEL VALIDATION

Considering the importance of predicting the behaviour of hydraulic fill structures in the field and afterwards analyzing the performance of different kinds of laboratory simulation tests, a hydraulic deposition simulation apparatus was developed at the University of Brasilia (Ribeiro, 2000). The apparatus consists of a depositional channel, 6.0 m long, 0.4 m wide and 1.0 m high. The channel was built using steel profiles and panels of tempered glass. This kind of wall permits the observation of the evolution of the deposition process during the entire test. Figure 2 shows a general view of hydraulic deposition simulation tests (HDST), developed at the University of Brasilia.

A series of comparisons were made between the HDST results obtained by Ribeiro (2000) and those forecasted by the coupled model solved by analytical, FDM and CIP techniques, for different Courant numbers. According to Ribeiro (2000), the main characteristics of the material used in the HDST are those presented in Tables 1 and 2. In Tables 1 and 2, Fe is the percentage of iron particles, C_w is the concentration of solid particles (quartz and iron) in the slurry, Q is the slurry flow rate and i_m is the average (global) beach slope.

Figures 3 and 4 present comparisons between the results obtained from tests HDST 1 and 6 (Ribeiro, 2000) and those obtained analytically using the mathematical model proposed by

Figure 2.   General view of the HDST equipment developed at the University of Brasilia.

Cavalcante *et al.* (2002), numerically using the FDM as obtained by Cavalcante *et al.* (2003) and using the CIP technique describe in Cavalcante & Farias (2007) for a Courant number equal to 0.5. Axis x (abscissas) represents the distance from the discharge point and axis y (ordinates) gives the normalized height of the deposited beach.

From Figures 3 and 4, one can notice that the mathematical model is not able to describe the successive erosion and deposition processes, which are clearly observed in the HDST beaches. However, the model describes quite well the basic geometric characteristics of the deposited beaches, such as global slope inclination. The numerical solution using the Finite Differences Method differs drastically from the experimental and analytical solution for the adopted time-space discretization with $C = 0.5$. This is due to spurious dissipation introduced by the numerical solution scheme as explained by Cavalcante & Farias (2007). This problem is solved when the mixed CIP-FDM scheme is adopted and the numerical solution matches satisfactorily the experimental and analytical results for any Courant number, as illustrated in figures 3(b) and 4(b) for $C = 0.5$.

5 PARAMETRIC ANALYSES USING MIXED CIP-FDM SOLUTION

This section shows the results of a series of parametric analyses, using the proposed coupled model of Eqs. (1)–(4), solved with the mixed CIP-FDM scheme, for different initial conditions. A constant

Table 1. Average physical characteristics of the tailings used in the hydraulic deposition simulation test (HDST) carried out by Ribeiro (2000).

	Quartz	Iron
D_{50} (mm)	0.265	0.240
D_{90} (mm)	0.645	0.640
G_s	2.65	5.50

Table 2. HDST controlled variables and final slope inclination of the hydraulically deposited beach (Ribeiro, 2000).

	HDST 1	HDST 6
C_w (%)	8.9	20.4
Fe (%)	23.0	23.0
Q (l/min)	4.8	5.9
i_m (%)	7.7	9.2

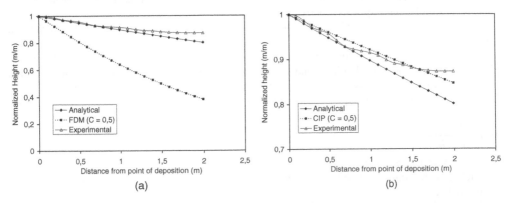

Figure 3. Comparisons between experimental and analytical results for test HDST 1 (Ribeiro, 2000) and the numerical results obtained via (a) Finite Differences and (b) CIP method for Courant number $C = 0.5$.

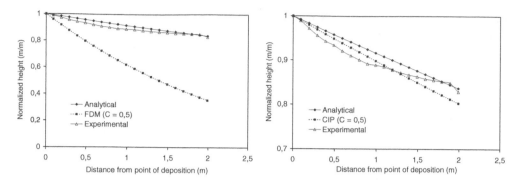

Figure 4. Comparisons between experimental and analytical results for test HDST 6 (Ribeiro, 2000) and the numerical results obtained via (a) Finite Differences and (b) CIP method for Courant number $C = 0.5$.

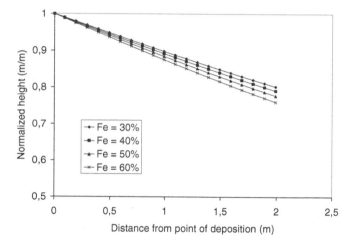

Figure 5. Normalized height versus distance from point of deposition.

slurry concentration ($C_w = 15\%$) was adopted while the iron concentration was varied ($Fe = 30$, 40, 50 e 60%). Parameters presented in Table 1 were adopted. The solutions are used to forecast physical and mechanical properties of the deposited beach, such as inclination, porosity, hydraulic conductivity and friction angle, using a series of analytical and empirical relations described in Cavalcante (2000) and Espósito (2000).

Figure 5 shows the resulting beach inclination for different iron concentration. It may be noticed that the beach slope increases as the iron concentration increases.

According to Cavalcante (2000), the variation of porosity may be obtained from the Theory of Elasticity. In this model, the loading conditions are applied vertically to an element of the deposited beach, assuming one-dimensional condition with null vertical displacements at the bottom of the deposit and null horizontal displacements at the sides of the element. Thus porosity n is given by:

$$n = 1 - \frac{1 - n_0}{1 + d\varepsilon_v} \tag{22}$$

in which n_0 is the initial porosity in each node of a layer before loading and $d\varepsilon_v$ is the increment of volumetric strain.

Figure 6 shows the distribution of porosity along the deposition profile for the different iron contents. Porosity values are lower closer to the discharge points and increase until stabilization for

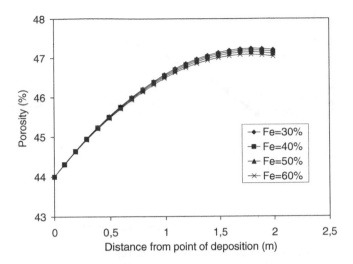

Figure 6. Porosity versus distance from the discharge point.

further points. The iron content decreases the overall porosity, but this effect is not so significant because the adopted sizes of the quartz and iron grains do not differ much in Table 1.

Other geomechanical properties are influenced by the porosity and the iron content in the slurry. These properties in this model are computed from empirical relations. Espósito (2000) proposed the following expression for the hydraulic conductivity:

$$k_x = a \exp(b.n) \tag{23}$$

in which k_x is the hydraulic conductivity, a and b are material coefficients dependent on the properties of the tailings.

Cavalcante (2000) used Eq. (23) and demonstrated that the hydraulic conductivity may be computed from an initial value (k_{x0}) and the variation of porosity as follows:

$$k_x = k_{x0} \exp[b(n - n_0)] \tag{24}$$

Therefore, using the porosity profile shown in Figure 6, the hydraulic conductivity along the profile can be calculated and is illustrated in Figure 7.

Espósito (2000) also proposed an empirical expression for the friction angle ϕ' as a function of the porosity of the deposit:

$$\phi' = p + q \exp(-rn)$$

in which p, q and r are coefficients dependent on the material type.

Cavalcante (2000) used the same empirical observations and deduced an alternative expression for the variation of friction angle as a function of the variation in porosity as follows:

$$\phi' = \phi'_0 + q[\exp(-rn) - \exp(-rn_0)] \tag{26}$$

in which ϕ'_0 is the initial friction angle.

The distribution of different particle types, grain sizes and density, controls the variation of mechanical properties along the deposition beach. This is illustrated for the friction angle ϕ in Figure 8. The variation of friction angle is opposite of that observed for the porosity, with higher

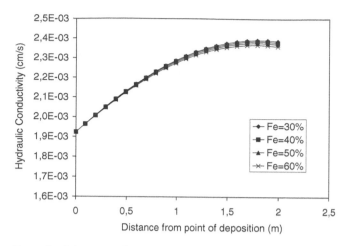

Figure 7. Hydraulic conductivity versus distance from the discharge point.

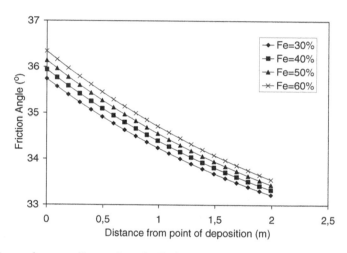

Figure 8. Friction angle versus distance from the discharge point.

ϕ values close to the discharge point. Higher friction angles are also computed as the iron content increases, but then again this variation was less than 1° and is not much relevant from an engineering point of view for the specific parameters adopted for the composition of the tailings.

6 CONCLUSIONS

A coupled system of equations representing the bed load transport of heterogeneous sediments in hydraulically deposited tailings dams was described. This model is governed by the field equations for equilibrium and continuity of the fluids and the condition of continuity of the solid particles. A constitutive model must also be assumed for the rate of transport of the sediments.

The overall system of equations governing the problem may be split in a sub-system of advective equations and a sub-system of non advective equations. Advective equations do not exhibit dissipation during the transport phenomena; however spurious numerical dissipations are introduced in the solution when classical schemes, such as the Finite Differences Method (FDM), are used to solve this kind of hyperbolic equations.

The solution for the unrealistic numerical dissipation problem may be achieved by adopting the so-called CIP (Cubic Interpolated Pseudo-particle) method to solve the advective part of the equations governing the overall problem. The remaining equations may be solved using any other method, such as the FDM, without undesirable numerical problems.

This mixed CIP-FDM solution scheme was used to carry out a series of parametric analyses of the deposition problem varying the residual contents of iron in the slurry to investigate its effect on the properties (inclination, porosity, hydraulic conductivity and friction angle) along the profile of the deposited beach.

The inclination of the deposited beach increases with the increase in the concentration of iron. Porosity and hydraulic conductivity are affected in a similar manner, with lower values close to the discharge point. The distribution of porosity along the profile also affects, in an inverse manner, the variation of friction angle along the beach, resulting in higher ϕ values close to the discharge point.

ACKNOWLEDGEMENTS

The authors acknowledge the support of the following institutions: Brazilian National Research Council (CNPq), University of Brasilia, UPIS University and FINATEC. The authors are also indebted to Dr. Shuji Moriguchi and Professor Takayuki Aoki from Tokyo Institute of Technology, Japan, for their kindness in teaching us the details of the CIP method.

REFERENCES

Cavalcante, A.L.B. (2000). *Effect of the Gradient of Hydraulic Conductivity on Stability of Tailings Dams Constructed using the Upstream Method.* (In Portuguese) Master Dissertation, Department of Civil and Environmental Engineering, University of Brasilia, Brazil, 186 p.

Cavalcante, A.L.B; Assis, A.P. & Farias, M.M. (2002). Numerical Sediment Transport Model of Heterogeneous Tailings. *Proc. 5th European Conference on Numerical Methods in Geotechnical Engineering*, 5th NUMGE, Paris, France, 6 p.

Cavalcante, A.L.B., Assis, A.P. & Farias, M.M. (2003). Bed Load Transport in Tailings Dams – Analytical and Numerical View. *Proc. of the 4th International Workshop on Applications of Computational Mechanics in Geotechnical Engineering*, Brasil, pp. 103–113.

Cavalcante, A.L.B. (2004). *Modeling and Simulation of Bed Load Transport of Heterogeneous Sediments Coupling Stress-Strain-Porepressure Applied to Tailings Dams.* (In Portuguese) PhD Thesis, Department of Civil and Environmental Engineering, University of Brasilia, Brazil, 313 p.

Cavalcante, A.L.B & Farias, M.M (2007). Numerical Schemes for the Solution of Advection Problems. *Proc. of the 5th Int. Workshop on Applications of Computational Mechanics in geotechnical Engineering*, Guimarães, Portugal, 10 p.

Espósito, T.J. (2000). *Probabilistic and Observational Methodology Applied to Tailings Dams Build with Hydraulic Fills.* (In Portuguese) PhD Thesis, Department of Civil and Environmental Engineering, University of Brasilia, Brazil, 363 p.

Nishiguchi, A. & Yabe, T. (1983). Second order fluid particle scheme. *J. Comput. Phys.* Vol. 52, pp. 390-

Ribeiro, L.F.M. 2000. *Physical Simulation of the Process of Formation of Hydraulic Fills Applied to Tailings Dams.* (In Portuguese) PhD Thesis, Department of Civil and Environmental Engineering, University of Brasilia, Brazil, 232 p.

Oil geomechanics

Applications of Computational Mechanics in Geotechnical Engineering – Sousa, Fernandes, Vargas Jr & Azevedo (eds)
© 2007 Taylor & Francis Group, London, ISBN 978-0-415-43789-9

Wellbore response analysis considering spatial variability and fluid-mechanical coupling

A.L. Muller & E.A. Vargas Jr
Pontifical Catholic University of Rio de Janeiro, Rio de Janeiro, Brasil

L.E. Vaz
Federal University of Rio de Janeiro, Rio de Janeiro, Brasil

C. Gonçalves
Cenpes/Petrobras, Rio de Janeiro, Brasil

ABSTRACT: In general, well response analyses are carried out considering that the hydraulic and mechanical parameters of the rock mass are deterministic. However, it is a well known fact, that rock masses and in particular sedimentary rock masses show a large degree of spatial heterogeneity. These spatial heterogeneities are due to the spatial variability in mechanical and hydraulic properties of the rock medium. Therefore, the response in pore pressures, displacements, stresses and plastic regions, also shows variability, and can be expressed in terms of its average values and standard deviation. This paper focuses on the comparison of different methods for performing a wellbore response analysis considering spatial variability of the hydraulic and mechanical parameters of the rock mass, namely, the Monte Carlo method, the Neumann expansion method and the perturbations method. To perform this evaluation a coupled fluid-mechanical finite element analysis is applied which takes into account the spatial variability of the hydraulic and mechanical properties and the variability of the initial pore pressure and stresses. Examples are shown and conclusions are drawn about the effect of the spatial variability of the rock mass parameters on the wellbore response.

1 INTRODUCTION

In general, wellbore response analyses are performed considering that the hydraulic and mechanical parameters of the rock mass are deterministic variables. Nevertheless, rock masses and sedimentary rock masses in particular show a large degree of heterogeneity in micro and macro scales. These heterogeneities produce spatial variability in mechanical and hydraulic properties of the rock medium. This spatial variability can be very pronounced and influence the behavior of the solid phase and the behavior of the fluid phase in the porous media. Therefore, the response of the medium, such as pore pressures, displacements, stresses and plastic region also shows variability and therefore can be expressed in terms of the its average values and dispersion.

The work of Freeze (1975) represent a breakthrough in the probabilistic modeling of hydrological problems although Shvidler (1962) and Matheron (1967), apud Dagan (2002), had presented important theoretical developments about this subject in their work.

According to Daí (2004), the probabilistic modeling of hydrological problems was consolidated by Dagan (1982), which is the basis for the works performed by Gelhar (1993), Neumann (1997), Rubin (1997), Rubin et al (1999) and Zhang (2002).

Dagan (2002) reports in his work that the probabilistic modeling of hydrological problems experienced a considerable development in the two last decades and that a great amount of knowledge has been accumulated lately in this area.

As stated in Glasgow et al (2003), several types of approximations have been proposed to incorporate uncertainties inherent in the hydraulic parameters into the modeling of hydrological problems. Many of these approximations are presented in the work of Gelhar (1993) and in the papers by Rubin (2003) and Zhang (2002). As attested by Jain et al (2002), Lu and Zhang (2003) and Dagan (2002), the more widespread approximations are the simulation methods such as the Monte Carlo Method and those based on perturbation methods.

The simulation methods require a great number of deterministic analyses. Afterwards, probabilistic responses are estimated using these deterministic analyses. The need of a great number of deterministic analyses increases considerably the computational effort, especially for nonlinear and time-dependent problems. The perturbations methods yield reasonable results when the variability of the parameters is small. Besides this limitation, sensitivity analyses must be performed in perturbation methods, which make them computationally very time-consuming for some classes of problems.

A series of alternatives may be derived from the original forms of these approximations aiming at obtaining more efficient and appropriate solutions. Among the existing alternatives for the Monte Carlo simulation method, the simulation method using the Neumann expansions, suggested in Ghanem and Spanos (2003), and the Karhunen-Loeve expansion, as presented by Chen et al (2005) may be quoted. According to Alvarado et al (1998) and Hart (1982), the perturbation methods differ basically on the way the equations are linearized.

It should be mentioned that, when treating the types of analyzed hydrological problems taking into account the spatial variability of the involved parameters, especially when the stochastic finite element method is applied, great emphasis has been given to the modeling of transport or flow problems, being by far, the analyses of coupled problems less explored, mainly these concerning the stability analysis of petrol wellbores.

Zhang and Lu (2003) proposed an approximation technique by applying a high order perturbation for the flow problem of heterogeneous and saturated porous media, comparing the results obtained for different simulation methods. Chen et al (2005) presented a statistical model of the two phase flow problem in porous media considering the variability of the intrinsic permeability and porosity. Ghanem and Dham (1998) presented a two-dimensional model for modeling the NAPL movement in heterogeneous aquiferous considering the variability of the intrinsic permeability. Wu et al (2003), developed a numerical procedure for the three-dimensional analysis of flow and transport of solutes considering a non-stationary behavior for the conductivity of the media. Amir and Neumann (2004) presented an approximation for the solution of the transient flow problem considering the uncertainties of the soil properties. Jain et al (2002) modeled the fluid flow through a porous media using the Monte Carlo simulation. Lu and Zhang (2003) also applied Monte Carlo simulation to analyze flow and transport problems in porous media but used the method of importance which considers that some random proprieties of the media are more important than others.

Foussereau et al (2000) developed analytical solutions to predict the transport of inert solutes in partially saturated heterogeneous porous media submitted to random boundary conditions. Concerning coupled problems, the papers by Frias et al (2001 and 2004) must be quoted. Frias et al (2001) presented a stochastic computational model for compaction and subsidence of reservoirs due to the fluid withdrawing, considering the spatial variability of the intrinsic permeability of the porous media. Frias et al (2004), treated the same problem considering however the assumption of great correlation lengths and fractal characteristics for the intrinsic permeability of the porous media.

The present work proposes the development of a numerical analysis procedure, using finite elements spatial discretization, to analyze coupled fluid-mechanical processes and wellbore response that take into account the spatial variability of hydraulic and mechanical properties and the variability of the initial pore pressure and stresses. In this work, according to Jain et al (2002), Lu e Zhang (2003) and Dagan (2002), the Monte Carlo simulation method, the Neumann expansion method and the perturbations method are used for the determination of the statistical response in pore pressures, displacements and stresses. For describing the nonlinear behavior of the material, the Mohr-Coulomb model, described by Owen e Hinton (1980), is used and for the representation of the random fields a correlation spatial function of exponential type, given for Calvete and Ramirez (1990), is used. The partitioned solution procedure (staggered method), presented by (Lewis and Schrefler (1998), Simoni and Schrefler (1991) and Turska and Schrefler (1992), is adopted for the solution of the mechanical fluid coupling problem. In accordance with Lubliner (1984), Simo and Hughes (1997), the elastoplastic problem at the constitutive level is solved by a mathematical programming algorithm while the the L-BFGS method, see Nocedal (1980) and Zambaldi and Mendonça (2005), is applied to solve the nonlinear equations at the global level.

In following item, the numerical formulation used in the analysis of the coupled hydro-mechanical problem is presented.

2 DEVELOPMENT OF GOVERNING EQUATIONS

In this item the formulation for mechanical fluid coupling in porous media is presented. The equations are described in a domain $\Omega \subset R^3$ with a contour Γ for a time $t \in [0, T]$.

2.1 Equilibrium equations

The equilibrium equation, Equation 1, is obtained through the virtual work principle for quasi static problems. This equation relates the velocities of the static real quantities, as the total stress velocity vector $\dot{\sigma}$ the body forces velocity vector $\dot{\mathbf{b}}$ and the surface forces vector $\dot{\mathbf{t}}$ to the virtual kinematics quantities as the virtual strains $\delta\varepsilon$ and the virtual displacements $\delta\mathbf{u}$.

$$\int_\Omega \delta\varepsilon^T \dot{\sigma}\, d\Omega - \int_\Omega \delta\mathbf{u}^T \dot{\mathbf{b}}\, d\Omega - \int_\Gamma \delta\mathbf{u}^T \dot{\mathbf{t}}\, d\Gamma = 0 \tag{1}$$

The adopted hypotheses are presented in the following equations, Equation 2, Equation 3 e Equation 4.

$$\dot{\sigma} = \dot{\sigma}' + \mathbf{Dm}\frac{\dot{p}}{3K_s} - \mathbf{m}\dot{p} \tag{2}$$

$$\dot{\sigma}' = \mathbf{D}\dot{\varepsilon} + \dot{\sigma}'_0 \tag{3}$$

$$\dot{\varepsilon}_{ij} = \frac{1}{2}\left(\dot{u}_{i,j} + \dot{u}_{j,i}\right) \tag{4}$$

Being $\dot{\sigma}'$ the effective stresses velocities, $\dot{p}$ the pore pressure velocities, $\dot{\varepsilon}$ the skeleton total strain velocity, $\mathbf{D}$ the constitutive tensor, $\dot{\sigma}'_0$ the effective initial stresses velocity, $\mathbf{m}\dot{p}/3K_s$ the volumetric strains velocity caused by uniform compression, K_s the bulk modulus of solid grains and $\mathbf{m} = \{111000\}^T$. Whit this definitions rewrites the Equation 1 as Equation 5.

$$\int_\Omega \delta\varepsilon^T \mathbf{D}\dot{\varepsilon}\, d\Omega + \int_\Omega \delta\varepsilon^T \mathbf{Dm}\dot{p}\frac{1}{3K_s}\, d\Omega + \int_\Omega \delta\varepsilon^T \dot{\sigma}'_0\, d\Omega - \int_\Omega \delta\varepsilon^T \mathbf{m}\dot{p}\, d\Omega - \int_\Omega \delta\mathbf{u}^T \dot{\mathbf{b}}\, d\Omega - \int_\Gamma \delta\mathbf{u}^T \dot{\mathbf{t}}\, d\Gamma = 0 \tag{5}$$

2.2 Flow equations

For the presentation of the equations related to the fluid flow problem the hypothesis of single phase flow is assumed, which is described by the Darcy's law. This formula can be written, according to Lewis e Schrefler (1998) as presented in Equation 6.

$$\left[\mathbf{m}^T - \frac{1}{3K_s}\mathbf{m}^T\mathbf{D}\right]\dot{\varepsilon} + \left[\frac{(1-\phi)}{K_s} - \frac{1}{9K_s^2}\mathbf{m}^T\mathbf{Dm} + \frac{\phi}{K_\pi}\right]\dot{p} - \nabla^T[T_m\nabla(p_\pi + \rho_\pi gh)] = 0 \tag{6}$$

Where π represents a fluid type, ρ_π represents the density of the fluid, ∇ the differentiation operator, $\mathbf{m}^T \dot{\boldsymbol{\varepsilon}}$ represents the rate of change of the total strain, $\phi/K_\pi \dot{p}$ describes the rate of change of the fluid density, K_π the bulk modulus of the fluid π and ϕ the porosity of the medium. $T_m = \mathbf{k}/\mu_\pi$, being, $\mathbf{k}$ the absolute permeability matrix of the porous medium, g the gravity, h the head pressure and μ_π the dynamic viscosity of the fluid π.

3 FINITE ELEMENT FORMULATION

When the spatial distribution of the variables is described by means of finite elements, the following relations may be used: $\mathbf{u} = \mathbf{N}_u \mathbf{u}^*$; $\boldsymbol{\varepsilon} = \mathbf{B}\mathbf{u}^*$; $\mathbf{p} = \mathbf{N}_p \mathbf{p}^*$. Where $\mathbf{N}_u$ and $\mathbf{N}_p$, are respectively the interpolation functions for displacements and pore pressures, $\mathbf{B}$ is the strain-displacement matrix and the symbol $(.)^*$ is a reference to the nodal points. Introducing the above mentioned equations in Equation 5 and Equation 6 and leaving the symbol $()^*$ out for the sake of simplicity, Equation 7 and Equation 8 may be obtained.

$$\int_\Omega \mathbf{B}^T \mathbf{D} \mathbf{B} d\Omega \dot{\mathbf{u}} - \int_\Omega \mathbf{B}^T \mathbf{m} \mathbf{N}_p d\Omega \dot{\mathbf{p}} + \int_\Omega \mathbf{B}^T \dot{\boldsymbol{\sigma}}'_0 d\Omega - \int_\Omega \mathbf{N}_u^T \dot{\mathbf{b}} d\Omega - \int_\Gamma \mathbf{N}_u^T \dot{\mathbf{t}} d\Gamma + \int_\Omega \mathbf{B}^T \mathbf{D} \frac{\mathbf{m}}{3K_s} \mathbf{N}_p d\Omega \dot{\mathbf{p}} = 0 \tag{7}$$

$$\int_\Omega (\nabla \mathbf{N}_p)^T \frac{\mathbf{k}}{\mu_\pi} \nabla \mathbf{N}_p d\Omega \mathbf{p} + \int_\Omega \mathbf{N}_p^T s \mathbf{N}_p d\Omega \dot{\mathbf{p}} + \int_\Omega \mathbf{N}_p^T \left(\mathbf{m}^T - \frac{\mathbf{m} \mathbf{D}}{3K_s} \right) \mathbf{B} d\Omega \dot{\mathbf{u}} +$$

$$\int_\Gamma \mathbf{N}_p^T q d\Gamma + \int_\Omega (\nabla \mathbf{N}_p)^T \frac{\mathbf{k}}{\mu_\pi} \nabla \rho_\pi g h d\Omega = 0 \tag{8}$$

Where $s = (1 - \phi)/K_s + \phi/K_\pi - 1/9K_s^2 \mathbf{m}^T \mathbf{D} \mathbf{m}$ and the integrals over the spatial domain represents the assembly of the integrals over the finite elements which describe the domain.

For the time domain discretization of the variables $\mathbf{u}$ and $\mathbf{p}$ the generalized trapezoidal rule is used. Representing $\mathbf{u}$ and $\mathbf{p}$ generically by $\mathbf{r}$ the time discretization for both variables may be described in Equation 9 and Equation 10.

$$^{t+\theta}\mathbf{r} = (1 - \theta)^t \mathbf{r} + \theta^{t+\Delta t} \mathbf{r} \tag{9}$$

$$^{t+\theta}\dot{\mathbf{r}} = (^{t+\Delta t}\mathbf{r} - {}^t\mathbf{r})\frac{1}{\Delta t} \tag{10}$$

As recommended by Hughes (1977) $\theta = 0,5$ is used in this work. According Hughes this value of θ guarantees good properties of consistency and stability of the solution, as well as a second order precision, both in linear and nonlinear analyses.

From the previously described hypotheses Equation 11 and Equation 12 are obtained from Equation 7 and Equation 8.

$$^{t+\Delta t}\mathbf{F}_u(^{t+\Delta t}\mathbf{u}, {}^{t+\Delta t}\mathbf{p}, t) = \int_\Omega \mathbf{B}^T \mathbf{D} \mathbf{B} d\Omega (\frac{^{t+\Delta t}\mathbf{u} - {}^t\mathbf{u}}{\Delta t}) + \int_\Omega \mathbf{B}^T \mathbf{D} \frac{\mathbf{m}}{3K_s} \mathbf{N}_p d\Omega (\frac{^{t+\Delta t}\mathbf{p} - {}^t\mathbf{p}}{\Delta t})$$

$$- \int_\Omega \mathbf{N}_u^T \dot{\mathbf{b}} d\Omega - \int_\Gamma \mathbf{N}_u^T \dot{\mathbf{t}} d\Gamma - \int_\Omega \mathbf{B}^T \mathbf{m} \mathbf{N}_p d\Omega (\frac{^{t+\Delta t}\mathbf{p} - {}^t\mathbf{p}}{\Delta t}) + \int_\Omega \mathbf{B}^T \dot{\boldsymbol{\sigma}}'_0 d\Omega \tag{11}$$

$$^{t+\Delta t}\mathbf{F}_p(^{t+\Delta t}\mathbf{u}, {}^{t+\Delta t}\mathbf{p}, t) = \int_\Omega (\nabla \mathbf{N}_p)^T \frac{\mathbf{k}}{\mu_\pi} \nabla \mathbf{N}_p d\Omega [(1 - \theta)^t \mathbf{p} + \theta^{t+\Delta t} \mathbf{p}] + \int_\Omega \mathbf{N}_p^T s \mathbf{N}_p d\Omega (\frac{^{t+\Delta t}\mathbf{p} - {}^t\mathbf{p}}{\Delta t}) +$$

$$\int_\Omega \mathbf{N}_p^T \left(\mathbf{m}^T - \frac{\mathbf{m} \mathbf{D}}{3K_s} \right) \mathbf{B} d\Omega (\frac{^{t+\Delta t}\mathbf{u} - {}^t\mathbf{u}}{\Delta t}) + \int_\Gamma \mathbf{N}_p^T q d\Gamma + \int_\Omega (\nabla \mathbf{N}_p)^T \frac{\mathbf{k}}{\mu_\pi} \nabla \rho_\pi g h d\Omega \tag{12}$$

$^{t+\Delta t}\mathbf{F}_u(^{t+\Delta t}\mathbf{u}, {}^{t+\Delta t}\mathbf{p}, t)$ and $^{t+\Delta t}\mathbf{F}_p(^{t+\Delta t}\mathbf{u}, {}^{t+\Delta t}\mathbf{p}, t)$ represent residual vectors $^{t+\Delta t}\mathbf{R}$ at the end of the time interval once the values of $^t\mathbf{u}$ and $^t\mathbf{p}$ at the beginning of the time interval are knows. To solve Equation 11 and Equation 12, the unknown vectors $^{t+\Delta t}\mathbf{u}$ and $^{t+\Delta t}\mathbf{p}$ which makes the residual vectors $^{t+\Delta t}\mathbf{R} \cong 0$, must be found. These equations can be represented in a compact form as described in the following in Equation 13.

$$
\begin{bmatrix} -\mathbf{K} & \mathbf{L} \\ \mathbf{L}^T & \Delta t\theta\mathbf{H} + \mathbf{G} \end{bmatrix}
\begin{Bmatrix} \delta^{i+\Delta t}\mathbf{u}^{i+1} \\ \delta^{i+\Delta t}\mathbf{p}^{i+1} \end{Bmatrix} =
\begin{Bmatrix} -\Delta t^{t+\Delta t}\mathbf{F}_u^i(^{t+\Delta t}\mathbf{u}^i, ^{t+\Delta t}\mathbf{p}^i, t) \\ \Delta t^{t+\Delta t}\mathbf{F}_p^i(^{t+\Delta t}\mathbf{u}^i, ^{t+\Delta t}\mathbf{p}^i, t) \end{Bmatrix}
\tag{13}
$$

$$
\delta^{t+\Delta t}\mathbf{u}^{i+1} = ^{t+\Delta t}\mathbf{u}^{i+1} - ^{t+\Delta t}\mathbf{u}^i
$$
$$
\delta^{t+\Delta t}\mathbf{p}^{i+1} = ^{t+\Delta t}\mathbf{p}^{i+1} - ^{t+\Delta t}\mathbf{p}^i
\tag{14}
$$

$$
\mathbf{K} = \int_\Omega \mathbf{B}^T \mathbf{D}_T \mathbf{B} d\Omega \; ; \quad \mathbf{L} = \int_\Omega \mathbf{B}^T \mathbf{m} \mathbf{N}_p d\Omega - \int_\Omega \mathbf{B}^T \mathbf{D}_T \frac{\mathbf{m}}{3K_s} \mathbf{N}_p d\Omega \; ;
$$

$$
\mathbf{H} = \int_\Omega (\nabla \mathbf{N}_p)^T \frac{k}{\mu_n} \nabla \mathbf{N}_p d\Omega \; ; \quad \mathbf{G} = \int_\Omega \mathbf{N}_p^T \left(\frac{(1-\phi)}{K_s} - \frac{1}{9K_s^2} \mathbf{m}^T \mathbf{D}_T \mathbf{m} + \frac{\phi}{K_\pi} \right) \mathbf{N}_p d\Omega \; ;
\tag{15}
$$

In Equation 13, the variables are the displacements and pore pressures increments as described in Equation 14. The matrices $\mathbf{K}$, $\mathbf{L}$, $\mathbf{H}$ and $\mathbf{G}$ are the so called tangent matrices obtained according to expressions defined in Equation 15.

In Equation 13 and Equation 14 the index i indicates iteration. The matrices $\mathbf{K}$, $\mathbf{L}$, $\mathbf{H}$ and $\mathbf{G}$ are the so called tangent matrices obtained according to expressions defined in Equation 15.

4 SOLUTION PROCEDURES

In accordance with Lewis and Schrefler (1998), the mechanical fluid coupling problems can be solved through coupled or uncoupled strategies. The coupled solutions are divided in fully coupled and partitioned solutions. The uncoupled solutions can be found in Corapcioglu (1984).

In a generalized manner, the results obtained with these methods are similar although the required computational effort for the analysis can be significantly different. The choice of the method of solution depends on the characteristics of the problem, linear or nonlinear problem, number of involved equations, etc.

In this work the partitioned (staggered) solution, presented by Lewis e Schrefler (1998), Simoni and Schrefler (1991) and Turska and Schrefler (1992) is adopted for the solution of the mechanical fluid coupling problem. This procedure was chosen because it requires lesser computational effort for the solution of the kind of problem treated in this work when compared with the fully coupled solution. In the partitioned strategy the coupled matrix $\mathbf{L}$ is present only in the residual vectors and the equations to be solved are represented in Equation 16 and Equation 17.

$$
[\mathbf{K}]\{\delta^{t+\Delta t}\mathbf{u}^{i+1}\} = \{\Delta t^{t+\Delta t}\mathbf{F}_u^i(^{t+\Delta t}\mathbf{u}^i, ^{t+\Delta t}\mathbf{p}^{i+1}, t)\}
\tag{16}
$$

$$
[\Delta t\theta\mathbf{H} + \mathbf{G}]\{\delta^{t+\Delta t}\mathbf{p}^{i+1}\} = \{\Delta t^{t+\Delta t}\mathbf{F}_p^i(^{t+\Delta t}\mathbf{u}^{i+1}, ^{t+\Delta t}\mathbf{p}^i, t)\}
\tag{17}
$$

Notice that when solving Equation 16 and Equation 17 the values of $\mathbf{u}$ and $\mathbf{p}$ at the beginning of the time interval are know from the previous time step and the values of $^{t+\Delta t}\mathbf{p}^{i+1}$ are fixed when solving Equation 16 and $^{t+\Delta t}\mathbf{u}^{i+1}$ are fixed when solving Equation 17. This means that the Equation 16 and Equation 17 must be solved alternatively using the values of $^{t+\Delta t}\mathbf{u}^{i+1}$ determined from the independent iterations of Equation 16 in Equation 17 and the values of $^{t+\Delta t}\mathbf{p}^{i+1}$ determined independent in Equation 17 in Equation 16.

When using the Newton-Raphson method, Equation 16 is solved calculating $\mathbf{K}$ accurately, without using approximate matrices. This approach results in a quadratic convergence for the solution. The accurate calculation of the $\mathbf{K}$ matrix can however have a high computational cost, mainly in problems with many variables. To by-pass this problem the Quasi-Newton methods offer an interesting alternative. In these methods, an approximation of the $\mathbf{K}$ matrix, or of its inverse, is obtained in a given iteration using the gradients of the residual vectors $\mathbf{R}$ and of the increment

Instant $t + \Delta t$, Initial estimate $^{t+\Delta t}\mathbf{p} = {}^{t}\mathbf{p}$	step 1

Iteration: $j + 1$ step 2

$\mathbf{u}^{j+1}$ is evaluated using Eq. (16) and the convergence of the displacements is verified   step 3

$$\left| \frac{|^{t+\Delta t}\mathbf{u}^{j+1}| - |^{t+\Delta t}\mathbf{u}^{j}|}{|^{t+\Delta t}\mathbf{u}^{j+1}|} \right| \leq tol$$

With the vector $\mathbf{u}$ obtained in step 3 $\mathbf{p}^{j+1}$ is evaluated using Eq. (17) and the   step 4
convergence of the pore pressures is verified as in step 3 for $\mathbf{u}$:

$$\left| \frac{|^{t+\Delta t}\mathbf{p}^{j+1}| - |^{t+\Delta t}\mathbf{p}^{j}|}{|^{t+\Delta t}\mathbf{p}^{j+1}|} \right| \leq tol$$

If the inequalities of step 3 and 4 are not satisfied, return to step 2 with the updated   step 5
values of $\mathbf{p}$. If they are satisfied, make $j = 0$ and return to step 1 for a new time step.

---

variable vectors of past iterations. In this work the nonlinear problem in Equation 16 is solved by the Quasi-Newton method L-BFGS, described by Nocedal (1980) and Zambaldi and Mendonça (2005), which uses a inverse update for the stiffness matrix. The equation system of Equation 17 is solved in a similarly as Equation 16 and the following algorithm can be used.

As suggested by Lubliner (1984), Simo and Hughes (1997), the elastoplastic problem is solved as a mathematical programming problem as described in Equation 18.

Minimize : $-\mathrm{D}^{p}(\boldsymbol{\sigma})$

subject to : $F(\boldsymbol{\sigma}) \leq 0$ (18)

D^{p} is the plastic dissipation to be minimized, $F(\boldsymbol{\sigma})$ is the yield function being $\boldsymbol{\sigma}$ the stress vector. The described problem in Equation 18, can be rewritten of the following form.

$$\{\boldsymbol{\sigma}^{i+1}\} = ARG \left[\overset{\text{MINIMIZAR}}{(\boldsymbol{\sigma}) \in \mathcal{E}_{\sigma}} \left\{ \frac{1}{2} \left\| (\boldsymbol{\sigma}^{trial} - \boldsymbol{\sigma}) \right\|^{2} \mathbf{D}^{-1} \right\} \right]$$

subject to : $F(\boldsymbol{\sigma}^{i+1}) \leq 0$ (19)

Where $\mathbf{D}$ is the elasticity tensor, $\mathbf{G}$ is the generalized plastic moduli, both constant, E_{σ} is the admissible stresses space. $\|\boldsymbol{\sigma}\|^{2}\mathbf{D}^{-1} = \sqrt{\boldsymbol{\sigma}\mathbf{D}^{-1}\boldsymbol{\sigma}}$ is the energy norm.

In this work the Mohr-Coulomb criterion, as suggested by Owen and Hinton (1980), is used for the representation of the nonlinear material behaviour and consequently to describe F.

The Sequential Quadratic Programming (SQP) algorithm is used for the solution of the above mentioned mathematical programming problem, see Vanderplaats (1984).

5 RESPONSE ANALYSIS CONSIDERING RANDOM VARIABLES

5.1 *The Monte Carlo simulation method*

The response analysis can be performed for a given response function $f_i(r_j(\mathbf{x}_k), t)$ which can represent the response in pore pressures, displacements and stresses. The functions $f_i(r_j(\mathbf{x}_k), t)$ are explicitly dependent the time point t the random variables $r_j(\mathbf{x}_k)$ (such as the elasticity modulus, the permeability, the cohesion, among others) which in turn are dependent on the position vector $\mathbf{x}_k$ at the point k in the Cartesian space.

In this item the Monte Carlo simulation method is presented for the determination of the statistical response, according to Jain et al (2002), Lu and Zhang (2003) and Dagan (2002). Second order stationary condition is assumed for the random variables as suggested by Calvete and Ramirez (1990) and Gelhar (1993).

The basic step in the Monte Carlo simulation is the generation of m random fields for each one of the $r_j(\mathbf{x}_k)$ variables of the problem. Once these m random fields are generated m responses are obtained in the time domain. Finally, by using the concepts of probability and statistics, the statistical properties of the sampling of the m responses such as mean values, standard deviation, correlation factors etc. can be determined. Once probability density functions and correlation factors are defined for the random variables, the Monte Carlo simulation method can generate a realization m for the response function $f_{i_m}(r(\mathbf{x}_k), t)$. The average response of f_i, $f_i(r(\mathbf{x}_k), t)$, is given by:

$$\bar{f}_i(r(\mathbf{x}_k), t) = \lim_{m \to \infty} \frac{1}{m} \sum_{l=1}^{m} f_{il}(r(\mathbf{x}_k), t) \tag{20}$$

And the its variance by

$$Var(f_i(r(\mathbf{x}_k), t)) = \lim_{m \to \infty} \frac{1}{m} \sum_{l=1}^{m} \left(f_{il}(r(\mathbf{x}_k), t) - \bar{f}_i(r(\mathbf{x}_k), t) \right)^2 \tag{21}$$

Other methods of stochastic analysis, such as the perturbation method, or the Neumann expansion, could be used for estimating the statistical response. However, for the problem treated in this work these methods were proven less efficient than the Monte Carlo method. The perturbation method, besides presenting limitations for the choice of the coefficient of variability of the random variables, demands the calculation of the derivatives of the functions f_i with respect to the random variables of the problem (sensibility analysis). The calculation of these derivatives for pass dependent problems, as is the case of the problem treated in this work, is extremely expensive. Moreover, when using the perturbation method, an expansion of the f_i functions as a function of the random variables in second order Taylor series is needed which requires the knowledge of the fourth order moments of these variables. The Neumann expansion method loses its efficiency mainly due to the strong nonlinear behavior and time dependence of the analyzed problem.

As cited earlier, the Monte Carlo method needs the generation of random fields. In accordance with Zhang (2002) the more popular methods for random field generation are the decomposition method, the turning bands method and the spectral representation method. The decomposition method can generate correlated random fields. The two last methods generate independent random fields demanding a posterior treatment of the generated fields to obtain correlation among them.

Some mechanical properties are assumed to be correlated in this work. Therefore, the decomposition method is used. The decomposition method is so called due to the necessity of a Cholesky decomposition of the covariance matrix of the random variables.

For the determination of the covariance matrix a correlation function of exponential type is used in this work, see Rubin (2003). Two assumptions are assumed for the generation of the random fields, namely, the variables are considered constant in the domain of the element of the mesh of finite elements and the position vectors are taken in the centroids of the finite elements.

5.2 The Neumann expansion method

One of the characteristics of the Monte Carlo method is that it requires a great amount of computational effort, especially for large size problems. The Monte Carlo method with Neumann expansion tries to reduce this computational effort, Ghanem and Spanos (2003) and Araújo and Awruch (1993).

To start with the presentation of the Neumann method applied to the coupled fluid-mechanical problem, the equilibrium equation of the problem for the mth simulation is written in the following:

$$^{t+\Delta t}\mathbf{H}(\mathbf{q}^i)_m \; ^{t+\Delta t}\delta\mathbf{q}_m^{i+1} = {}^{t+\Delta t}\mathbf{R}^i(\mathbf{q}^i)_m \tag{22}$$

447

The Neumann expansion is applied to the inverse of the **H** matrix, which is divided into two parts, one deterministic, obtained with the mean values of the random variables, and the other probabilistic, obtained through simulations. The first part of **H** is denoted by $\mathbf{H}(\bar{\mathbf{r}})$ and the second part by $\Delta\mathbf{H}_m$. Based on this assumption the expression for the displacements and pore pressures increments of the coupled fluid mechanical problem with one phase flow for a *mth* simulation may be written as

$$^{t+\Delta t}\delta\mathbf{q}_m^{i+1} = {}^{t+\Delta t}[\mathbf{H}(\bar{\mathbf{r}}) + \Delta\mathbf{H}_m]^{-1 \, t+\Delta t}\mathbf{R}^i(\mathbf{q}^i)_m \tag{23}$$

where $^{t+\Delta t}\delta q_m^{i+1}$ is the vector of increments for the timepoint $t + \Delta t$, iteration $i + 1$ and simulation m and $^{t+\Delta t}\mathbf{R}^i(\mathbf{q}^i)_m$ is the residuum vector, obtained for the simulation m. Using the Neumann expansion for $^{t+\Delta t}[\mathbf{H}(\bar{\mathbf{r}}) + \Delta\mathbf{H}_m]^{-1}$, one obtains:

$$^{t+\Delta t}\delta\mathbf{q}_m^{i+1} = {}^{t+\Delta t}[\mathbf{I} - \mathbf{P} + \mathbf{P}^2 - \mathbf{P}^3 + ...]^{t+\Delta t}[\mathbf{H}(\mathbf{r})^{-1}]^{t+\Delta t}\mathbf{R}^i(\mathbf{q}^i)_m$$

$$^{t+\Delta t}\delta\mathbf{q}_m^{i+1} = {}^{t+\Delta t}|\mathbf{q}_1 - \mathbf{Pq}_1 + \mathbf{P}^2\mathbf{q}_1 - \mathbf{P}^3\mathbf{q}_1 + ...| \tag{24}$$

$$^{t+\Delta t}\delta\mathbf{q}_m^{i+1} = {}^{t+\Delta t}|\mathbf{q}_1 - \mathbf{q}_2 + \mathbf{q}_3 - \mathbf{q}_4 + ...|$$

where $^{t+\Delta t}\mathbf{P} = {}^{t+\Delta t}[\mathbf{H}(\bar{\mathbf{r}})^{-1}\Delta\mathbf{H}_m]$ and $^{t+\Delta t}q_1 = {}^{t+\Delta t}[\mathbf{H}(\bar{\mathbf{r}})^{-1}]^{t+\Delta t}\mathbf{R}^i(\mathbf{q}^i)_m$. The equation may be rewritten as:

$$^{t+\Delta t}\delta\mathbf{q}_m^{i+1} = {}^{t+\Delta t}\mathbf{q}_1 + {}^{t+\Delta t}[\mathbf{H}(\bar{\mathbf{r}})^{-1}]\sum_{j=2}^{\infty}(-1)^{j-1\,t+\Delta t}[\Delta\mathbf{H}_m]^{t+\Delta t}\mathbf{q}_{j-1} \tag{25}$$

$^{t+\Delta t}q_m^{i+1}$ is obtained by means of $^{t+\Delta t}q_m^{i+1} = {}^{t+\Delta t}q_m^i + {}^{t+\Delta t}\delta q_m^{i+1}$.

After obtaining the response for each simulation, the procedure for determining its statistic proprieties follows the Monte Carlo simulation method. Ghanem & Dham (1998) and Araújo and Awruch (1993), quote that good results are obtained by the Neumann expansion when the random variables present small variability. Otherwise, many terms must be used in the expansion to achieve good results and its application becomes not interesting.

When $\mathbf{H}(\bar{\mathbf{r}})$ is constant, the solution of equation 24 is straightforward, once $\mathbf{H}(\bar{\mathbf{r}})$ may be decomposed by the Cholesky method, the resulting matrices stored and the increment vector obtained using these matrices through forward and backward substitution throughout the whole time analysis. When the matrix $\mathbf{H}(\bar{\mathbf{r}})$ is not constant, which is true in the nonlinear analysis, the use of the Neumann expansion is not effective.

5.3 The perturbation method

The vector of the random variables **r** may be represented by the summation of two terms, namely, a vector of the mean values $\bar{\mathbf{r}}$ and a vector of perturbation values with respect to the mean values $\mathbf{r}'$, or:

$$\mathbf{r} = \bar{\mathbf{r}} + \mathbf{r}' \tag{26}$$

Using (26), a second order Taylor series expansion of the function $f_i(\mathbf{r})$ around $\bar{\mathbf{r}}$ may be written as,

$$f_i(\mathbf{r}) = f_i(\bar{\mathbf{r}}) + \mathbf{Jr}' + \frac{1}{2}\mathbf{r}'^T\mathbf{Hr}' \tag{27}$$

$$J_j = \frac{\partial f_i}{\partial r_j}\bigg|_{r=\bar{r}} \quad \text{and} \quad H_{jk} = \frac{\partial^2 f_i}{\partial r_j \partial r_k}\bigg|_{r=\bar{r}} \tag{28}$$

where **J** is the transpose of the column vector called the gradient of $f_i(\mathbf{r})$ and **H** a matrix named the Hessian of $f_i(\mathbf{r})$ at the point $\bar{\mathbf{r}}$.

Calculating the expected value on both sides of equation 27 the expression for the mean value of $f_i(\mathbf{r})$ is obtained,

$$\bar{f}_i(\mathbf{r}) = E\langle f_i(\mathbf{r})\rangle = f_i(\bar{\mathbf{r}}) + \mathbf{J}E\langle \mathbf{r}'\rangle + \frac{1}{2}E\langle \mathbf{r}'^T \mathbf{H}\mathbf{r}'\rangle$$

$$\bar{f}_i(\mathbf{r}) = f_i(\bar{\mathbf{r}}) + \frac{1}{2}\sum_j \sum_k H_{jk} C_{rr,jk}$$

(29)

where $E\langle \mathbf{r}'\rangle = 0$ and $C_{rr,jk}$ the $(jk)th$ component of the covariant matrix of the random variables $\mathbf{C}_{rr}$.

Obtaining the covariant matrix of is a time consuming task, according to $f_i(\mathbf{r})$ Calvete and Ramirez (1990) and Townley (1984), since it requires the calculation of fourth order moments of the random variables.

When the function $f_i(\mathbf{r})$ is not strongly nonlinear, it is enough to consider a first order Taylor series expansion for representing an approximation of $f_i(\mathbf{r})$ around $\bar{\mathbf{r}}$. Hart (1982) denotes this particular case of the perturbation method as linear statistical method. In this case, the estimate for the mean value of $f_i(\mathbf{r})$ becomes:

$$\bar{f}_i(\mathbf{r}) = E\langle f_i(\mathbf{r})\rangle = f_i(\bar{\mathbf{r}})$$

(30)

Using again a first order Taylor series expansion for $f_i(\mathbf{r})$, the covariant matrix of $f_i(\mathbf{r})$, represented by $\mathbf{C}_{ff}$, may be obtained by:

$$\mathbf{C}_{ff} = \mathbf{J}\mathbf{C}_{rr}\mathbf{J}^T$$

(31)

Applying now the linear statistical method to equation 21 in order to obtain an approximation to the mean value of the vector $^{t+\Delta t}\delta\mathbf{q}^{i+1}$ as a function of the random variables $\mathbf{r}$ yields,

$$E\langle ^{t+\Delta t}\delta\mathbf{q}^{i+1}\rangle = ^{t+\Delta t}[\mathbf{H}(\mathbf{q}^i(\bar{\mathbf{r}}))]^{-1}\,{}^{t+\Delta t}\mathbf{R}^i(\mathbf{q}^i(\bar{\mathbf{r}}))$$

(32)

Using again the symbol ($^-$) for denoting the mean value of a vector, the mean values of displacements and pore pressures at the instant $t + \Delta t$ may be obtained by $^{t+\Delta t}\bar{\mathbf{q}}^{i+1} = {}^{t+\Delta t}\bar{\mathbf{q}}^i + {}^{t+\Delta t}\delta\bar{\mathbf{q}}^{i+1}$. Besides the mean value vectors, the covariant matrix of displacements and pore pressures $\mathbf{C}_{ff}$ at the instant $t + \Delta t$ may also be determined by,

$$^{t+\Delta t}\mathbf{C}_{ff} = \left[\frac{\partial^{t+\Delta t}\mathbf{q}}{\partial\mathbf{r}}\bigg|_{\mathbf{r}=\bar{\mathbf{r}}}\right]\mathbf{C}_{rr}\left[\frac{\partial^{t+\Delta t}\mathbf{q}}{\partial\mathbf{r}}\bigg|_{\mathbf{r}=\bar{\mathbf{r}}}\right]^T$$

(33)

where $\mathbf{C}_{rr}$ is the covariant matrix of the random variables $\mathbf{r}$.

The response in terms of stresses and saturations may also be determined using the same methodology presented above, as shown, for example, in Equation 34 for the mean value of the stress field in instant $t + \Delta t$ and iteration $i + 1$.

$$E\langle ^{t+\Delta t}\boldsymbol{\sigma}^{i+1}\rangle = {}^{t+\Delta t}\boldsymbol{\sigma}^i + {}^{t+\Delta t}\delta\boldsymbol{\sigma}(\bar{\mathbf{r}})^{i+1}$$

(34)

6 EXAMPLES OF DETERMINISTIC ANALYSES

Aiming at validating the suggested numerical procedures, a deterministic analysis of the problem will be presented, comparing the numerical and analytical response. Later the results for the limits of internal pressure will be presented considering a stochastic analysis.

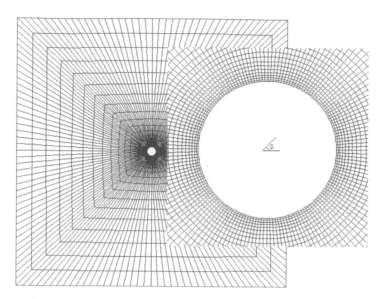

Figure 1. Finite element mesh.

Table 1. Data for the validation of the deterministic analysis.

Parameters	Average	Probability density function type
ν	0.20	–
K_s (MPa)	38000.00	–
K_π (MPa)	2884.00	–
ϕ	0.19	–
G (MPa)	6000.00	lognormal
K (m^2/MPas)	1.9E−6	lognormal
c (MPa)	10.00	lognormal
Φ (degree)	30.00	lognormal
σ_{0xx} (MPa)	−30.00	normal
σ_{0yy} (MPa)	−50.00	normal
p_0 (MPa)	15.00	lognormal

6.1 Deterministic response analysis

For the validation of the considered procedures of deterministic analysis, a vertical well being perforated through a continuous, poreelastic, isotropic and completely saturated media, subject to a state of stresses in situ and initial pore pressures is considered. The analytical solution, proposed by Detournay and Cheng (1988), is used for comparing the numerical the analytical results. The perforation process is modeled considering instantaneous excavation. In the analytical solution the coupled equations are solved using the Laplace transformed space, assuming the plain strain conditions for the plane perpendicular to the axis of the well.

For the numerical modeling of the problem, isoparametric four-node elements are adopted, using the same interpolation function for displacements and pore pressures. The numerical integration is carried out using the Gauss procedure with 2×2 integration points. The finite elements mesh is composed for 4800 elements and 4960 nodes. The Figure 1 shows the finite element mesh used in numerical analyses, the detail corresponds the mesh around the well. The angle β will be used as reference to presented some response.

The diameter of the well is of 0.20 m and the data of the problem are shown in table 1. The internal pressure on the walls of the well is null. The average values of the parameters are used for the deterministic validation example. The results are shown for an angle $\beta = 0$ degrees.

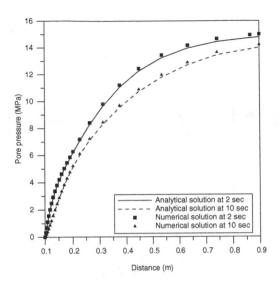

Figure 2. Pore pressure, analytical vs. numerical solutions.

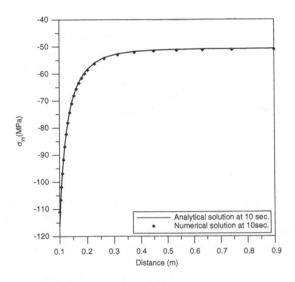

Figure 3. Total stress σ_{yy}, analytical vs. numerical solutions.

In Figure 2 the spatial distributions of the pore pressure corresponding to both the analytical and the numerical solutions at different time instants are depicted. The analytical and numerical results match very well.

Similarly, in Figure 3, the spatial distributions of the analytical and the numerical responses for the total stress σ_{yy} at 10 seconds are compared.

7 EXAMPLES OF STOCHASTIC ANALYSES

A stochastic analysis of the previous example, for one given internal pressure, will be carried out. Perfect plastic behavior is assumed for the material and the Mohr Coulomb strength criterion and plane strains are adopted.

451

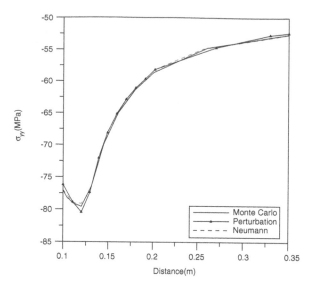

Figure 4.   Mean value of σ_{yy} for $Cv = 0.10$ and $\beta = 0$ degrees.

In this example the transversal young modulus G, the absolute permeability K, the angle of internal friction Φ and the apparent cohesion c in each element are considered as random variables, as well as the initial pore pressure and the initial stresses. Thus, the number of random variables in this example is equal the 19205. The length of correlation adopted is 6.0 m. A coefficient of correlation of 0.7 between G and Φ and between G and c is also assumed. A fluid perforation pressure of 20.0 MPa is applied to the walls of the well.

7.1   Stochastic response analysis for a given internal pressure

In order to estimate the effects of the variability of the random variables on the response calculated by the three methods, two coefficient of variation are adopted, namely, $Cv = 0.10$ and $Cv = 0.20$. For the mesh and random parameters defined previously, 19205 random variables are generated. 1000 simulations are used for the simulations methods and three terms are applied in the Neumann expansion. The results in the graphics presented in the following refer to the time instant of 60 seconds, an angle $\beta = 0$ degrees and distances in a radial direction measured from the center of the well.

In Figure 4, the mean values of the total stress σ_{yy} for $Cv = 0.10$ are plotted. The results are very similar for the three methods.

In Figure 5, the mean values of the total stress σ_{yy} for $Cv = 0.20$ and $\beta = 0$ degrees are presented. The results are similar for the three methods with small differences in the region close to the inner wall of the well, where plastification occurs, when using the Neumann method. In Figure 6, the standard deviations of the total stress field σ_{yy}, for $Cv = 0.10$ and $\beta = 0$ degrees, are depicted along the x-axis. Once again the results for the three methods are very similar in the region where plastification does not occur. On the other hand, in the region near the wall of the well where plastification occurs, the results obtained by the method of the Neumann expansion and the linear statistical method differ from those obtained using the method of Monte Carlo.

Corresponding results to the ones presented in the Figure 6 are depicted in the Figure 7 now for $Cv = 0.20$ instead of $Cv = 0.1$. Once again, in Figure 7 as in Figure 6, the results near to the well inner wall present dispersion, but now, the dispersion is even bigger than before. The linear statistical method does not succeed in representing the same behavior presented by the method of Monte Carlo in the neighborhood of the well wall.

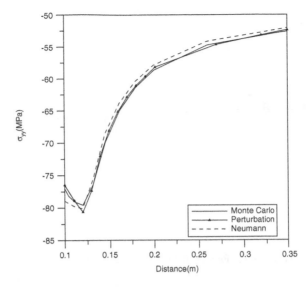

Figure 5.   Mean value of σ_{yy} for Cv $= 0.20$ and $\beta = 0$ degrees.

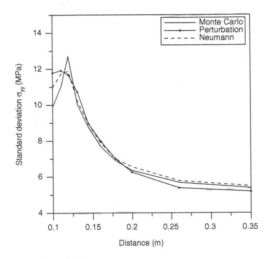

Figure 6.   Standard Deviation of σ_{yy} for Cv $= 0.10$ and $\beta = 0$ degrees.

The mean values of the pore pressures corresponding to Cv $= 0.10$ and Cv $= 0.20$, are presented in the Figures 8 and 9, respectively. Small differences may be observed in the results obtained by the three methods, the most relevant for Cv $= 0.2$ and the linear statistical method.

In the Figure 10 the results for the Standard deviations of the pore pressure for Cv $= 0.10$ are presented. The results are very similar no matter what method is used.

For the condition Cv $= 0.20$ presented in the Figure 11, significant differences in the Standard deviation of the pore pressures in the central region of the domain are observed.

It is worth stressing that the null values for the standard deviation of pore pressures on the wall of the well are due to fact that pressure there is prescribed (20 MPa) and is a deterministic variable.

In general, similar mean values responses were obtained with the different methods of analysis. On the other hand, significant differences were observed among the standard deviations determined by the different methods, especially for the stress field obtained with the linear statistical method.

453

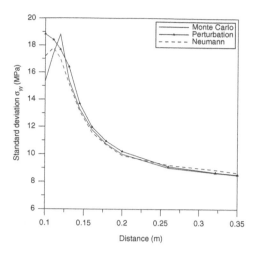

Figure 7. Standard Deviation of σ_{yy} for Cv = 0.20 and $\beta = 0$ degrees.

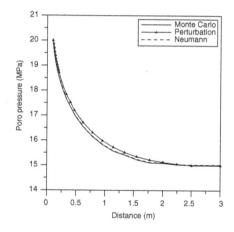

Figure 8. Mean value of pore pressures for Cv = 0.10 and $\beta = 0$ degrees.

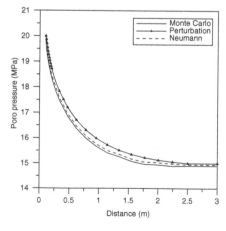

Figure 9. Mean value of pore pressures for Cv = 0.20 and $\beta = 0$ degrees.

454

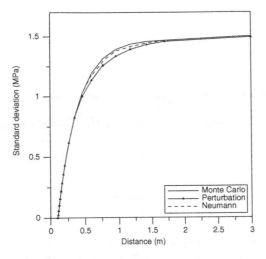

Figure 10. Standard Deviation of pore pressures for $Cv = 0.10$ and $\beta = 0$ degrees.

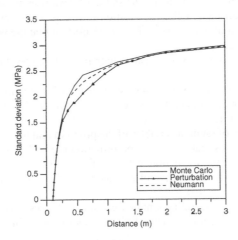

Figure 11. Standard Deviation of pore pressures for $Cv = 0.20$ and $\beta = 0$ degrees.

Table 2. Relative computer time of each analysis.

Method of Analysis	Relative computer time of each analysis
Monte Carlo	1.00
Neumann	1.74
Linear Statistical Analysis	19.89

Another important factor concerns the computer time spent by the analyses performed using the different methods. Due to the great amount of random variables in the problem and the need to perform a sensitivity analysis, the linear statistical method became inefficient. The simulation method based on the Neumann expansion, using only three terms in the expansion, required more computation time than the Monte Carlo simulation method.

The Table 2 presents the relative computer time of each analysis spent in the solution of this problem.

8 CONCLUSIONS

In this work, a comparison among three numerical analysis procedures, namely, the Monte Carlo simulation method, the Neumann expansion method and the perturbation method to analyze coupled hydromechanical processes and wellbore response that takes into account the spatial variability of the hydraulic and mechanical properties and the variability of the initial pore pressure and initial stresses is presented.

The paper examines the computational efficiency and the response accuracy of three methods. These methods are described briefly in the introduction of the work and detailed along the text. Based on the results obtained, the following conclusions may be drawn.

For a deterministic analysis, the conditions of symmetry may be used. As a result, only ¼ of the geometry of the problem can be considered. On the other hand, for a stochastic analysis, due to the heterogeneity of the medium, the total geometry must be taken into account. Nevertheless, the stochastic response, both in terms of pore pressures and stresses, as well as the plastic region, is almost symmetric.

It is also noticed, when performing a stochastic analysis, that the average response in pore pressures and stresses do not change significantly at a given point when changes in the variation coefficient of the stochastic parameters are imposed.

On the other hand, the plastic region and the standard deviation are strongly dependent on the values of the variation coefficient. Other verification is that the consideration of the variability of the initial stress and the initial pore pressure is more important that the variability of the hydraulic and mechanical parameters.

In other words, the responses are very sensitive with respect to the values of the initial stress and of the initial pore pressure.

In general, the response means do not change significantly for the three methods. This is not the case when comparing the responses in standard deviation, mainly when the perturbation method is applied, especially concerning the stress response.

Due to the great number of random variables of the problem and the need of sensitivity analyses with respect to these variables required by the perturbation method, this method is very time consuming when compared with the simulation methods of Monte Carlo and Neumann. The Neumann expansion method, using only three terms of expansion, demanded more computation time than the Monte Carlo simulation method.

Having in hand a tool for performing a numerical stochastic response analysis it is possible to evolve it into a tool for performing a reliability analysis of the stability of wellbores, which is already in progress.

REFERENCES

Alvarado V. Scriven L.E. and Davis H. T., 1998. Stochastic-Perturbation analysis of a one-dimensional dispersion-reaction equation: effects of spatially-varying reaction rates. *Transport in poreus media*, vol. 32, pp. 139–161.

Baecher G. B. and Christian, J. T., 2003. *Reliability and statistic in Geotechnical enginnering* John Wiley & Sons Ltd. England.

Calvete, F.J and Ramirez J., 1990. *Geoestadistica aplicaciones a la hidrologia subterrânea*. Centro Internacional de Métodos numéricos em Ingeniería, Barcelona.

Chen M., Zhang D., Keller A. and Lu, Z., 2005. A stochastic analysis of steady state two-phase flow in heterogeneous media. *Water Resources Research*, vol. 41.

Corapcioglu, M. Y., 1984. Land subsidence – a state of art review. *Fundamentals of transporte phenomena in poreus media*, ed. J. bear and M. Y. Corapcioglu, Nato, A.S.I. Series, E 82, Nijhoff, Dordrecht, pp. 369–444.

Dagan, G., 1982. Stochastic modelling of groundwater flow by unconditional and conditional probabilities, 2. *The solute transport, Water Resource. Res.*, 18(4), pp. 835–848.

Dagan, G., 2002. An overview of stochastic modelling of groundwater flow and transport: from theory to applications, EOS. *Transactions, American Geophysical Union*, 83.

Dai, Z., Ritzi, R. W., Huang C., Rubin Y. and Dominic D., 2004. Transport in heterogeneous sediments with multimodal conductivity and hierarchical organization across scales. *Journal of Hydrology*, 294, pp. 68–86.

Detournay, E.; Cheng, H. D., 1988. Poreelastic Response of a borehole in a hydrostatic stress field. *International Journal of Rock Mechanics and Mining Sciences & Geomechanics abstracts*. Vol. 25(3), p. 171–182.

Foussereau X., Graham W. D., Akpoji G. A., Destouni G., Rao1 P. S. C., 2000. Stochastic analysis of transport in unsaturated heterogeneous soils under transient flow regimes. *Water Resource. Res.* vol. 36, 4, pp. 911–921

Freeze, R.A., 1975. A stochastic-conceptual analysis of one-dimensional groundwater flow in nonuniform homogeneous media. *Water Resource. Res.* 11, pp. 725–741.

Frias, D.G, Murad M, and Pereira F., 2001. Stochastic computational modelling of reservoir compactation due to fluid withdrawal. *Relatórios de pesquisa e desenvolvimento, LNCC*.

Frias, D.G, Murad M, and Pereira F., 2003. A multiscale stochastic poremechanical model of subsidence of a heterogeneous reservoir. *Appl. Comp. Mech. Geoth. Eng. Ouro Preto*, pp. 29–44.

Gelhar, L.W., 1993. *Stochastic subsurface hydrology*. New Jersey, Prentice-Hall.

Ghanem R. and Dham S., 1998. Stochastic Finite Element Analysis for Multiphase Flow in Heterogeneous Poreus Media. *Transport in Poreus Media*, 32: pp. 239–262.

Ghanem R. and Spanos P. D., 2003. *Stochastic Finite Elements – A spectral approach*. New York, Springer-Verlag.

Glasgow, H., Fortney M., Lee J., Graettinger A. and Reeves H., 2003. Modflow 2000 head uncertainty, a first-order second moment method. *Ground water*, vol.41, No.3, pp. 342–350.

Hughes, Thomas J.R., 1977. Unconditionally stable algorithms for nonlinear heat conduction. *Computer methods in applied mechanics and engineering*, North-Holland publishing company, pp. 135–139.

Jain S., Acharya M., Gupta S., Bhaskarwar A. N., 2002. Monte Carlo simulation of flow of fluids through poreus media. *Computers and Chemical Engineering* 27 pp. 385–400.

Lewis, Ronald W. and Schrefler B. A., 1998. *The finite element method in the deformation and consolidation of porous media*. 2nd ed. John Wiley and Sons, Great Britain.

Lu, Z. and Zhang, D., 2003. On importance sampling Monte Carlo approach to uncertainty analysis for flow and transport in poreus media. *Advances in Water Resources*, vol.26, pp. 1177–1188.

Lubliner, J., 1984. On maximum-dissipation principle in generalized plasticity. *Acta Mechanica* 52, pp. 225–237.

Matheron, G., 1967. *Elements pour une theorie des milieux poreux*. Masson et Cie, Paris.

Muller, A. L., Jr. Vargas, E. A., Vaz, L. E., Gonçalves, C. J., 2006. Avaliação através de estudos numéricos de efeitos da variabilidade espacial de propriedades mecânicas e hidráulicas de rochas na estabilidade de poços de petróleo. IV SBMR - *Simpósio Brasileiro de Mecânica das Rochas*, Curitiba, Brazil.

Neumann, S.P., 1997. Stochastic approach to subsurface flow and transport: a view to the future. In: Dagan, G., Neumann, S.P. (Eds.). *Subsurface Flow and Transport: A Stochastic Approach*, Cambridge Press, Cambridge, pp. 231–241.

Nocedal, J., 1980. Updating quasi-newton matrices with limited storage. *Mathematics of Computation*, Vol. 35, pp. 773–782.

Owen, D. R. J and Hinton, E., 1980. *Finite elements in plasticity: theory and practice*, Swansea , Pneridge.

Rubin Y., 2003. *Applied stochastic hidrogeology*, Oxford University Press, University of California, Berkeley.

Rubin, Y., 1997. Transport of inert solutes by groundwater: recent developments and current issues. In: Dagan, G., Neumann, S.P. (Eds.). *Subsurface Flow and Transport: A Stochastic Approach*, Cambridge Press, Cambridge, pp. 115–132.

Rubin, Y., 2003. *Applied stochastic hidrogeology*, Oxford University Press, University of California, Berkeley.

Rubin, Y., Sun, A., Maxwell and R., Bellin, 1999. A. The concept of blockeffective macrodispersivity and a unified approach for grid-scaleand plume-scale-dependent transport. *J. Fluid Mech.*, 395, pp. 161–180.

Shinozuka, M. and Deodatis, G., 1996. Simulation of multidimensional Gaussian stochastic fields by spectral representation. *Appl. Mech. Rev.*, vol 49, No. 1, January, pp. 29–53.

Shvidler, M. I., 1962. *Flow in heterogeneous media* (in Russian), Izv. Akad. Nauk USSR Mekh. Zhidk, Gaza, 3, 185.

Simo, J.C and Hughes T.J.R., 1997. *Computational Inelasticity*, Springer –Verlag, New York.

Simoni, L. and Schrefler B. A., 1991. Staggered finite-element solution for water and gas flow in deforming poreus media. *Communications in Applied Numerical Methods*, v 7, n 3, pp 213–223.

Turska, E. and Schrefler B. A., 1992. On convergence conditions of partioned solution procedures for consolidation problems. *Computer methods in applied mechanics and engineering*, North-Holland publishing company, pp. 51–63.

Vanderplaats, G. N., 1984. *Numerical Optimization Techniques for Enginnering Desing: with Applications*. McGraw Hill.

Wu J., Hu B. X., Zhang D., Shirley C., 2003. A three-dimensional numerical method of moments for groundwater flow and solute transport in a nonstationary conductivity field. *Advances in Water Resources* 26 (2003), pp. 1149–1169.

Zambaldi, M.C. and Mendonça, M., 2005. An efficient approach to restart quasi-newton methods. *Proceedings of the XXVI Iberian Latin-American Congress on Computational Methods in Engineering* (CILAMCE). Guarapari.

Zhang D., Zhiming Lu, 2003. An efficient, high-order perturbation approach for flow in random poreus media via Karhunen–Loeve and polynomial expansions. *Journal of Computational Physics* 194 pp. 773–794.

Zhang, D., 2002. *Stochastic Methods for Flow in Poreus Media: Coping with Uncertainties*, Academic Press.

*Embankments and rail track for
high speed trains*

Applications of Computational Mechanics in Geotechnical Engineering – Sousa,
Fernandes, Vargas Jr & Azevedo (eds)
© 2007 Taylor & Francis Group, London, ISBN 978-0-415-43789-9

Dynamic analysis of rail track for high speed trains. 2D approach

A. Gomes Correia & J. Cunha
University of Minho, Department of Civil Engineering, Civil Engineering Centre, Guimarães, Portugal

J. Marcelino & L. Caldeira
Laboratório Nacional de Engenharia Civil, Lisboa, Portugal

J. Varandas, Z. Dimitrovová, A. Antão & M. Gonçalves da Silva
New University of Lisbon, Lisboa, Portugal

ABSTRACT: In the framework of an ongoing national research project involving the University of Minho, the National Laboratory of Civil Engineering and the New University of Lisbon, different commercial FEM codes (DIANA, PLAXIS and ANSYS) were tested to model the dynamic performance of a high speed train track. Initially a plain strain model is considered in a robin test using experimental data from a well documented instrumented standard ballast rail track under the passage of a HST at 314 km/h.

Numerically predicted results are presented and assessed, and comparison is made between the different codes and also the experimental data.

1   INTRODUCTION

The behaviour of the railway track and infrastructure under the combination of high speed and repetitive axle loadings is affected as a result of a complex soil-structure interaction problem that constitutes a motive of geotechnical and structural R&D. In fact, high speed trains bring some new problems: (1) high train speeds demand tighter tolerances and track alignment for the purpose of safe and passenger comfort; (2) after a critical speed drastic dynamic amplification appears on the deformation of the track, embankment and supporting soft soil. Proper modelling of the dynamic behaviour of the railway track system, the soil and embankment materials and of the loading is essential to obtain realistic results. Additionally, measurements on actual railway sections is necessary for monitoring of the physical behaviour of the rail track and infrastructure for the calibration of the tools. These aspects are being investigated throughout an ongoing national project financed by the Foundation for Science and Technology involving the University of Minho (UM), the National Laboratory of Civil Engineering (LNEC) and the New University of Lisbon (UNL).

This paper presents preliminary results obtained with modelling using different numerical tools available at the different institutions: DIANA (UM), PLAXIS (LNEC) and ANSYS (UNL). A case study of an instrumented section of a high speed line was utilized using monitoring results presented by (Degrande & Schillemans, 2001), in order to evaluate the performance predicted based on the models created with the abovementioned tools.

In these early studies of comparative suitability of these tools to be used in pratical applications a 2D model was assumed for the analysis. As to whether 2D models are suitable for these analysis, there's an almost unanimous agreement between researchers that while they provide some understanding of the problem, 3D models are essential to reach accurate results. Some authors consider a special symmetry, which they call 2.5D. Another interesting simplified approach was proposed by Gardien and Stuit (2003) studying the modelling of soil vibrations from railway tunnels. These authors, instead of creating a three-dimensional model for the dynamic analysis built three complementary models: the first one is three-dimensional, where static loads were applied to obtain equivalent Timoshenko beam parameters, which are used in the second model to calculate the force time history under the sleeper; this force is then introduced in a plain strain model of the tunnel cross section. This type of approach could be interesting to develop for rail-track-foundation modelling.

Anyway, the reliability of all these numerical models depends largely on the accuracy of the input data and the choice of an appropriate underlying theory and can be evaluated through comparison with results from experiments and theoretical analysis. In this respect the results presented in this paper are a first contribution of the projectfor this assessment.

2 MODELLING OF DYNAMIC PERFORMANCE OF RAIL TRACK UNDER HIGH SPEED MOVING LOADS

2.1 *Background*

In the process of modeling and design, material models are an important component. Associated with the material models, it is necessary to take into account the tests needed to obtain their parameters. This chain of operations must be always borne in mind at the design stage in order to ensure a good planning of the tests needed for the models. This process will be completed by choosing the performance models and relevant design criteria (Gomes Correia, 2001, 2005).

For a more practical application, the complexity of all the process is divided in three levels: routine design, advanced design and research based design. An outline of these levels of design were reported in COST 337, action related with pavements and summarized by Gomes Correia (2001). Each process should be object of verification, calibration and validation. Verification is intended to determine whether the operational tools correctly represent the conceptual model that has been formulated. This process is carried out at the model development stage. Calibration refers to the mathematical process by which the differences between observed and predicted results are reduced to a minimum. In this process parameters or coefficients are chosen to ensure that the predicted responses are as close as possible to the observed responses. The final process is validation that ensures the accuracy of the design method. This is generally done using historical input data and by comparing the predicted performance of the model to the observed performance.

Based in this general framework, some particular aspects are developed hereafter in relation with the use of different numerical tools that are available for be used at routine and advanced modeling and design.

Lord (1999) emphasized the empirical rules still used at the construction and design levels of rail track. He also noticed the importance to consider dynamic aspects in design.

Rail track design is probably one of the most complex soil-structure interaction problems to analyse. The various elements in design process comprise (Lord, 1999): (1) multi-axle loading varying in magnitude and frequency; (2) deformable rails attached to deformable sleepers with flexible fixings, with sleeper spacings which can be varied; (3) properties and thickness of ballast, sub-ballast, prepared subgrade (if adopted); (4) properties of underlying soil subgrade layers.

At routine design level, several railway track models are operational and some commercially available. The most popular and simplest model for rail track design represents the rail as a beam, with concentrated wheel loads, supported by an elastic foundation. The stiffness of elastic foundation incorporates the sleepers, ballast, sub-ballast and subgrade, but it is not possible to distinguish between the contribution of the sleeper and underlying layers. This simplified approach has been used to establish dimensionless diagrams in order to quickly assess the maximum track reactions both for a single axle load and a double axle load when the track parameters are changed (Skoglund, 2002).

More sophisticated models have been developed which represent the rails and sleepers as beams resting on a multiple layer system (as in pavements) comprised of the ballast, sub-ballast

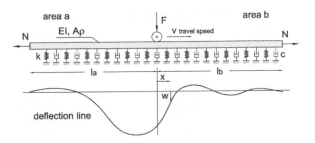

Figure 1.   Sketch of a flexible beam resting on continuous spring-dashpot elements loaded by a moving single load (Koft and Adam, 2005).

and subgrade. These models include the commercial programs ILLITRACK, GEOTRACK and KENTRACK cited by Lord (1999).

In these models incorporating multiple layer systems, the design criterion is identical as for pavements, keeping vertical strain or vertical stress at the top of subgrade soil below a determined limit. This criterion is an indirect verification of limited permanent settlements at the top of the system, having the same drawbacks as mentioned for pavements. Gomes Correia & Lacasse (2005) summarized some values of allowed permanent deformations for rail track adopted in some countries.

To overcome the drawbacks of the previous models, two directions of advanced modelling are identified. The first category of models aim at improving the theory of beam resting on continuous medium by introducing a spring-dashpot to better simulate a multiple layer system. Furthermore, the model was also improved by introducing a moving load at constant speed and also an axial beam force (Koft and Adam, 2005). Figure 1 is a sketch of the model. This model is able to determine dynamic response in different rail track systems due to a load moving with constant speed. A drawback of the model is that it is limited to beams with finite length and consequently only steady state solutions can be provided.

The second group of advanced rail track models use FEM and hybrid methods. The hybrid methods couple FEM and multi-layer systems (Aubry et al., 1999; Madshus, 2001). The track-embankment system is modelled by FEM and the layered ground through discrete Green's functions (Kaynia et al., 2000). The software developed is called VibTrain. The models referred by Aubry et al. (1999) and Madshus (2001) use frequency domain analysis having the drawback to require linear behaviour of materials. However, solutions in the time domain also exist (Hall, 2000 – mentioned by Madshus, 2001).

This last family of models incorporating track-embankment-ground is very powerful as it simulates behaviour at all speeds from low up to the critical speed.

As referred by Madshus (2001), these models need further validation by field monitoring. Dynamic materials characterization should be strongly encouraged, as well as dynamic field observations, mainly for ballasts.

In this paper the second group of advanced models was tested for a case study. These first results only address calculations done in plain strain conditions (2D).

This project also intends to put into operation a numerical model where, by incorporating the global behaviour of the whole of the railway platform and the supporting soil, will serve to quantify advantages and disadvantages, of the methods used at present in the maintenance of ballast platforms. It will be also possible to quantify and predict the consequences that can have, on this type of structures, the increase of the circulation velocity and the axle load in high speed trains.

2.2   Application of different commercial software for a case study

2.2.1   Presentation of the case study

The experimental data used to calibrate de models is obtained from the literature (Degrande & Schillemans, 2001) with material parameters summarized in Table 1.

Table 1. Geometry and material parameters of the HST track (after Degrande & Schillemans, 2001).

Element	Parameter	Value
Sleeper	Poisson	0.2*
	Young Modulus	30 GPa*
	Mass density	2054 Kg/m^3
Rail/sleeper interface	Thickness	0.01 m
	Stiffness	100 MN/m
Rail (UIC60)	Area	76.84 cm^2
	Inertia I$_x$	3055 cm^4
	Inertia I$_z$	512.9 cm^4
	Tortional inertia	100 cm^4*
	Volumic Weight	7800 Kg/m^3
	Poisson	0.3*
	Young Modulus	210 GPa*
Ballast (25/50)	Stiffness	0.3 m
	Mass density	1800 Kg/m^3*
	Poisson	0.1*
	Young Modulus	200 MPa*
	Damping	0.01*
Sub-ballast (0/32)	Stiffness	0.2 m
	Mass density	2200 Kg/m^3*
	Poisson	0.2*
	Young Modulus	300 MPa*
	Damping	0.01*
Capping layer (0/80 a 0/120)	Stiffness	0.5 m
	Mass density	2200 Kg/m^3*
	Poisson	0.2*
	Young Modulus	400 MPa*
	Damping	0.01*
Soil 1	Compression wave velocity	187 m/s
	Mass density	1850 Kg/m^3
	Stiffness	1.4 m
	Shear wave velocity	100 m/s
	Poisson	0.3
	Damping	0.03
Soil 2	Compression wave velocity	249 m/s
	Mass density	1850 Kg/m^3
	Stiffness	1.9 m
	Shear wave velocity	133 m/s
	Poisson	0.3
	Damping	0.03
Soil 3	Compression wave velocity	423 m/s
	Mass density	1850 Kg/m^3
	Stiffness	Infinite
	Shear wave velocity	226 m/s
	Poisson	0.3
	Damping	0.03

*Adopted values.

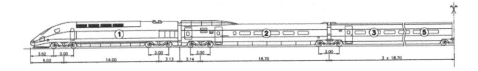

	# carriages	# axles	L_t [m]	L_b [m]	L_a [m]	M_t [kg]
Locomotives	2	4	22.15	14.00	3.00	17000
Side carriages	2	3	21.84	18.70	3.00	14500
Central carriages	6	2	18.70	18.70	3.00	17000

Figure 2. Geometry and load characteristics of the Thalys HST (Degrande & *Lombaert*, 2000).

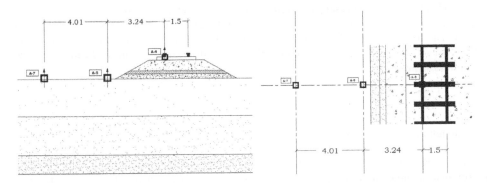

Figure 3. Location of the accelerometers.

These data correspond to vibration measurements made during the passage of a Thalys (high speed train – HST) at 314 km/h on a track between Brussels and Paris, more precisely near Ath, 55 km south of Brussels. The geometry and load characteristics of the HST are presented in Figure 2.

The HST track is a classical ballast track with continuously welded UIC 60 rails fixed with a Pandroll E2039 rail fixing system on precast, prestressed concrete monoblock sleepers of length l = 2.5 m, width b = 0.285 m, height h = 0.205 m (under the rail) and mass 300 kg. Flexible rail pads with thickness t = 0.01 m and a static stiffness of about 100 MN/m, for a load varying between 15 and 90 kN, are under the rail.

The track is supported by ballast and sub-ballast layers, a capping layer and the supporting soil.

Figure 2 shows the configuration of the Thalys HST referred by Degrande & *Lombaert* (2000), consisting of 2 locomotives and 8 carriages; the total length of the train is equal to 200.18 m. The locomotives are supported by 2 bogies and have 4 axles. The carriages next to the locomotives share one bogie with the neighbouring carriage, while the 6 other carriages share both bogies with neighbouring carriages. The total number of bogies equals 13 and, consequently, the number of axles on the train is 26. The carriage length Lt, the distance Lb between bogies, the axle distance La and the total axle mass Mt of all carriages are summarized.

The location of the measurement points (accelerometers) used for this work is presented in Figure 3 and the results are shown hereafter together with numerical predictions.

2.2.2 *Simulation of Thalys HST moving at 314 km/h*
The loads to be considered in the calculations are those that derive from the passing of the train. The loading history depends, naturaly, on the train speed. Because the calculations are to be done in plain strain conditions places some difficulties in the definition of the loading model arise because

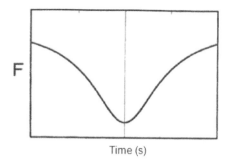

Time (s)

Figure 4.   Load due to a axle at sub-critical speed.

any point load in a plain strain model corresponds to a distributed load in the tree-dimensional model. Therefore, it is necessary to consider some simplifications and assumptions and, in the interpretation of the results, to limit its validity to the distances reported in Gutowski & Dym (1976). The maximum length of the Thalys train is $L_T = 196.7\,m$ (maximum distance between extremity axles) so, the results obtained in 2D model considering a linear load can be considered as valid for distances:

$$d = \frac{L_T}{\pi} = \frac{196.7}{\pi} \approx 62\,m \tag{1}$$

Although the loads transmitted by the axle are discrete, the stiffness of the structural elements of the railroad superstructure provides some distribution of the load. For circulation speeds – V – lower than the critical speed – Vcr – the loading in each point due to the passage of a axle follows approximately the distribution presented in Figure 4. The exact shape of the curve depends on the load speed, on the response of the railroad superstructure and its foundation, being that for higher speeds (but still inferior to the critical one) the curve tends to be thinner. When the speed approaches the critical speed, the curve loses the symmetry and the maximum value occurs after the load.

A possible approach to define the load distribution can be considered admitting a distribution adjusted to the sleeper spacing as is considered in the Japanese regulations. In accordance with this document, a changeable part between 40 and 60% of the load is distributed to the adjacent sleepers.

A simplified way to establish the load distribution consists of using the solution of the Winkler beam for the movement of a load. In accordance with this simplified model, the quasi static response in displacement is given by:

$$w(s) = \frac{Q}{2kL} e^{-|s/L|} \left(\cos\left|\frac{s}{L}\right| + \sin\left|\frac{s}{L}\right| \right) \tag{2}$$

where Q if the applied load, k the reaction module of the foundation, L the characteristic length of the beam and s the coordinate in a moving referential.

It seems reasonable to admit that, for speeds inferior to the critical one, the distribution of load underneath each axle will follow an analogous distribution:

$$F(s) = \frac{F_e}{2L} e^{-|s/L|} \left(\cos\left|\frac{s}{L}\right| + \sin\left|\frac{s}{L}\right| \right) \tag{3}$$

In the previous equation F(s) represents the distribution of the force due to each axle as a function of force F_e correspondent to the axle. The value of characteristic length L can be adjusted to obtain a certain amount of axle load at the point $s = 0$ (underneath the axle).

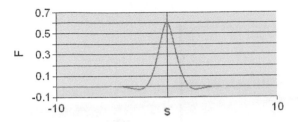

Figure 5.   Load distribution to a unitary axle load.

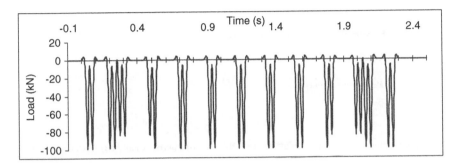

Figure 6.   Loading plan for the Thalys HST at V = 314 km/h.

Transformation between static referential "x" (in global coordinates) and the moving referential "s" is obtained through:

$$s = \frac{1}{L}(x - V_0 t)$$ (4)

where V_0 represents the train speed and t the time.

Thus, admitting that for s = 0 one has 60% of the axle load, L = 0.831 will be obtained. The load distribution corresponding to each axle is presented in Figure 5 for a unitary load.

The effect of the train can now be obtained considering the overlapping of all the axles, in accordance with the load distribution of the Thalys train:

$$F = \sum_{i=1}^{n} F_i$$ (5)

At the speed of 314 km/h (87.22 (2) m/s), the train passes in each section in 2.255 s. The modelling must start a little before the first axle passes in the calculation section because its effect starts before the load passes in that section. The computation must end some time later, in order to stabilize the vibrations in the surrounding media.

Combining the diverse axles in accordance with the geometry and load distribuition of the train and applying the previous expressions, it is possible to establish the loading plan to apply. This plan is represented in Figure 6.

2.2.3   First set of results of accelerations at different observation points

Different commercial softwares were used by different institutions for modelling the behaviour of high-speed rail tracks.

The University of Minho team uses DIANA software. This program is a general finite element code based on the Displacement Method (DIANA = DIsplacement method ANAlyser). It features extensive material, element and procedure libraries based on advanced database techniques.

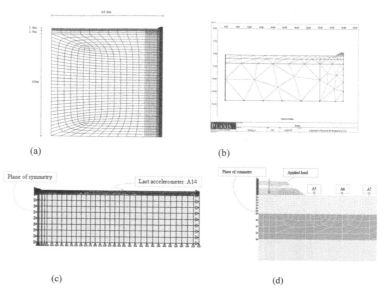

(a)

(b)

(c)

(d)

Figure 7. Finite element mesh for calculations: (a) DIANA, (b) Plaxis, (c) and (d) ANSYS.

The models created with this software are 63.3m width and 65m height, given the symmetry of its geometry and loading, only half of the track was modelled.

The final models comprised 2775 elements, being 12 triangular elements of 6 nodes (named CT12E in DIANA), 2601 quadrilateral elements of 8 nodes (Q8EPS), and 162 bounding elements (L4TB). The bounding elements were used at the limits of the model to take into account the propagation of waves to outer regions. Figure 7a shows the mesh used for numerical simulations.

The material properties are those defined in Table 1. Furthermore, DIANA considers Rayleigh damping according to:

$$[C] = \alpha[M] + \beta[K]$$
(6)

where [C] represents the damping matrix [M] the mass matrix and [K] stiffness matrix. The parameters α and β are the damping coefficients.

In order to establish α and β it is necessary to relate these parameters with the hysteretic damping more adequate to represent soils damping. This can be achieved relating the hysteretic damping coefficient (ξ) to α and β, for two known frequencies. In a separate modal analysis the first two frequencies obtained where (1.71 and 2.55 Hz). Using these frequencies and considering:

$$\xi_i = \frac{1}{2}\left(\frac{\alpha}{\omega_i} + \beta\omega_i\right)$$
(7)

Table 2 is obtained.

The LNEC team decided to use Plaxis Dynamic to perform the numerical simulations. Plaxis Dynamic 8.2, produced by Plaxis BV, Holland, is specially oriented to deal with geotechnical structures.

The model used in Plaxis numerical simulations is represented in Figure 7b.

The model is composed of 135 finite elements of 15 nodes, totaling 1183 nodal points. It represents half of the complete model due to its simmetry. The left boundary is absorbent in order to dissipate the incoming vibrations. The right boundary corresponds to the symmetry axis and

468

Table 2. Damping coefficients for DIANA and Plaxis models.

Material	ξ	α DIANA	α Plaxis	β DIANA	β Plaxis
Ballast, sub-ballast and subgrade	0.01	0.128702	0.290113	0.000747	0.000342
Foundation	0.03	0.386105	0.870339	0.002242	0.001027

therefore has null horizontal displacements. The bottom boundary was considered as fixed in order to avoid an overall movement of the model.

Plaxis, as DIANA, considers Rayleigh damping. The same procedure as for DIANA was done to obtain α and β parameters using the same ξ. Then, in a separate modal analysis the first two frequencies obtained using Plaxis were 4.26 and 5.04 Hz, which lead to the α and β values presented in Table 2. These differences obtained by two different FEM codes using different simplifications in geometry and boundary conditions are under investigation.

The UNL team uses ANSYS code. For the model structural plane, elements of 4 nodes in plane strain condition were selected, with preferential quadrilateral shape used to avoid additional rigidity inherent to triangle shapes. Finite element mesh is shown in Figs. 7c and 7d; in all, 3210 elements with side ranging from 0.05 m to 5.00 m were generated.

Non-reflecting boundary conditions are not available in the code. For that reason, the thickness of the last layer and the soil beyond the last accelerometer is represented to a depth and width of 40 m and 64 m, respectively. On the "infinite" boundaries normal displacements were restrained. Advantage is taken of symmetry. As for the previous models materials are linear elastic and isotropic. In this code, in full transient analysis, only Rayleigh damping, material dependent damping and element damping are accessible. A first exercise was carried out using material dependent damping taking different values for distinct materials, but that cannot change over the analysis. The results obtained were not realistic. At that stage some preliminary work was done by the UNL team, changing some of the features of the ANSYS model by comparing results for different options. Surface displacements, velocities and accelerations along model width for specific time steps were used for comparison at times 0.140, 0.175, 0.301 s, 0.335, 0.373 and 0.407 s that correspond to the first local peaks of the applied load. Calculation was stopped at 0.5 s, when the load stabilized at zero value.

It was verified that the rail, including the interface material, approximated by three springs in the way to preserve the total stiffness 100 MN/m per 0,6 m, does not bring any changes to the results.

It was also confirmed that the size of the elements reported above in the ANSYS model is sufficient, with a total of 66296 elements leading to unaltered values.

The representation of the boundary conditions influence was also examined changing the thickness and width of represented soil layer (extended to 80 m and to 128 m beyond last accelerometer). Slight differences in displacement field were detected while acceleration values were practically unchanged.

The next exercise was to adopt a Rayleigh damping which results are presented hereafter. In this case α and β damping constants are introduced in analysis or load step, but they cannot change for different materials. However, they can vary in each time step. The values adopted for all the materials were $\alpha = 0.870339$ and $\beta = 0.001027$.

Full transient analyses were performed with the different FEM codes and results shown in Figures 8, 9 and 10.

The comparison between the predictions can be summarized as presented in Table 3. The peak accelerations estimated by DIANA, Plaxis and ANSYS are rather close. However, the estimated values have rates very different from unity related with the measured values, particularly for the acceleration on the sleeper. Several reasons can contribute to this discrepancy, the major being:

(a) simplifying assumptions made (dynamic loading, interface between sleeper and ballast rail – rail pad;

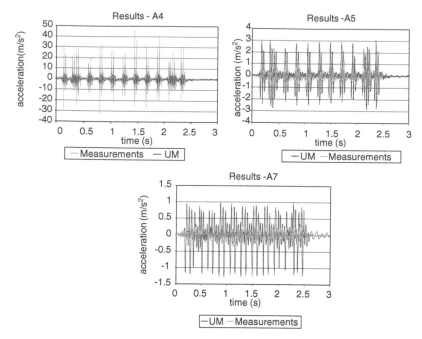

Figure 8.   Results by DIANA (UM).

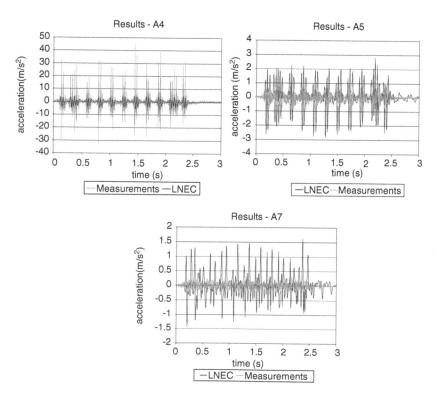

Figure 9.   Results by Plaxis (LNEC).

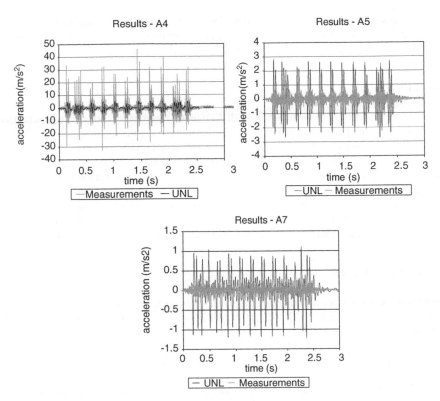

Figure 10. Results by ANSYS (UNL).

Table 3. Rates between estimations by different FEM codes and estimations and measurements for maximum peak accelerations at different points.

	Values m/s^2	DIANA	Plaxis	ANSYS
Rates A4				
Measurements	45.67	0.1	0.1	0.1
DIANA	4.74	–	1.0	1.0
Plaxis	4.82	–	–	1,0
ANSYS	4.96	–	–	–
Rates A5				
Measurements	1.23	2.4	2.3	2.3
DIANA	2.99	–	0.9	0.9
Plaxis	2.80	–	–	1.0
ANSYS	2.84	–	–	–
Rates A7				
Measurements	0.32	3.1	5.0	3.5
DIANA	0.99	–	1.6	1.1
Plaxis	1.61	–	–	0.7
ANSYS	1.11	–	–	–

(b) differences between material values;
(c) differences in the boundary conditions;
(d) simplifying dynamic material behaviour.

These factors need further study to better identify those that need modification. Additional studies on the influence of several parameters are currently under development and the possibility of using different computational codes is being considered.

3 CONCLUSION

The results are presented here for railtrack modelling using three FEM codes. The general trends in prediction of rail track performance and induced vibrations in the nearly area (soil) generally agree between the three models used but there are also some very significant variations in the relative rates for the measured values, particularly for sleeper response.

The ongoing research work in the project involving cooperation between specialists in soil dynamics and structural dynamics is a guarantee of the further improvements in modelling. Some significant aspects considered for further developments include: (a) simulation of dynamic loading, (b) vibrations characteristics and dynamic interaction of concrete sleepers and ballast support system and (c) frequency and strain dependence of the materials stiffness and damping.

ACKNOWLEDGEMENT

The authors are acknowledge the financial support for this research from the Foundation for Science and Technology (FCT) (project POCI/ECM/61114/2004 – Interaction soil-railway track for high speed trains).

REFERENCES

Aubry, A., Baroni, Clouteau, D., Fodil, A. & Modaressi, A. 1999. "Modélisation du comportement du ballast en voie", *Proc. 12th ECSMGE, Geotechnical Engineering for Transportation Infrastructure (Barends et al., eds.)*, Balkema, Rotterdam.
Degrande, G. & Lombaert, G. 2000. High-speed train induced free field vibrations: In situ measurements and numerical modelling, in *Proceedings of the International Workshop Wave 2000, Wave Propagation, Moving Load, Vibration Reduction*, edited by N. Chouw and G. Schmid (Rühr University, Bochum, Germany, pp. 29–41.
Degrande, G. & Schillemans, L. 2001. Free field vibrations during the passage of a Thalys high-speed train at variable speed, *Journal of Sound and Vibration* 247(1), 131–144.
Gardien W. & Stuit H. 2003. Modelling of soil vibrations from railway tunnels, Journal of Sound and Vibration 267, 605–619.
Gomes Correia, A. & Lacasse, S. 2005. Marine & Transportation Geotechnical Engineering. General report of Session 2e. 16th ICSMGE, Osaka, Millpress, Vol. 5, pp. 3045–3069.
Gomes Correia, A. 2001. Soil mechanics in routine and advanced pavement and rail track rational design, *Geotechnics for Roads, Rail tracks and Earth Structures (Gomes Correia and Brandl, eds.)*. Balkema, pp. 165–187.
Gutowski, T. G. & Dym, C. L. 1976. Propagation of ground vibration: A review, *Journal of Sound and Vibration* 49(2), 179–193.
Kopf, F. & Adam, D. 2005. Dynamic effects due to moving loads on tracks for highspeed railways and on tracks for metro lines, 16th ICSMGE, Osaka, Millpress, Vol. 3, Session 2e, pp. 1735–1740.
Lord, J.A. 1999. Railway foundations: Discussion paper, *Proc. 12th ECSMGE, Geotechnical Engineering for Transportation Infrastructure (Barends et al., eds.)*, Balkema, Rotterdam.
Madshus, C. 2001. Modelling, monitoring and controlling the behaviour of embankments under high speed train loads, *Geotechnics for Roads, Rail tracks and Earth Structures (Gomes Correia and Brandl, eds.)*. Balkema, pp. 225–238.
Skoglund, K.A. 2002. Dimensionless sensitivity diagrams in mechanistic railway design, *Bearing Capacity of Roads, Railways and Airfields (Gomes Correia and Branco, eds.)*, Balkema, 2, pp. 1331–1340.

Applications of Computational Mechanics in Geotechnical Engineering – Sousa,
Fernandes, Vargas Jr & Azevedo (eds)
© 2007 Taylor & Francis Group, London, ISBN 978-0-415-43789-9

Dynamic effects induced by abrupt changes in track stiffness in high speed railway lines

A.C. Alves Ribeiro, R. Calçada & R. Delgado
Faculdade de Engenharia da Universidade do Porto, Porto, Portugal

ABSTRACT: The present paper is focused on the study of the dynamic effects induced by abrupt changes in track stiffness. The study has been carried out by means of a dynamic analysis tool which enabled to model the train and the track, taking also into account its interaction. The simulations were carried out for the passage of the ICE2 train, circulating at speeds of 200, 250, 300 and 350 km/h. The results have been assessed in terms of the dynamic amplifications of the wheel-rail interaction force, the stability of the wheel-rail contact and the passengers comfort.

1 INTRODUCTION

One of the complexities associated with the construction of high speed railway lines refers to the transition zones from embankment to bridge, between distinct types of railway, from track in an embankment to track in an excavation, in zones with lower unlevelled crossings, etc.

The changes in the vertical stiffness of the track occurring in those zones induce important variations in the wheel-rail interaction forces, which may contribute to a faster degradation of the track geometry, put at risk the stability of the wheel-rail contact or cause discomfort in passengers (Esveld 2001 & Schooleman 1996).

The present work aims at studying the dynamic effects in zones of abrupt changes in the track stiffness, caused by track foundations on soils with distinct deformability moduli.

Simulation scenarios have been considered corresponding to deformability moduli ratios for the foundation soils equal to 2, 5 and 10, for the circulation of the ICE2 train at the speeds of 200, 250, 300 and 350 km/h.

The analyses were performed based on a numerical analysis program, which enables to appropriately model the track, the train and take into account the dynamic interaction between the two structures and the results have been assessed from the point of view of the safety of the track and the comfort of passengers.

2 DYNAMIC MODEL FOR THE TRACK-TRAIN SYSTEM

The assessment of the dynamic effects induced in zones of track stiffness transition involves the application of analysis tools which enable to model the complexity of the track-train system (Lei & Mao 2004). In this section, the dynamic models adopted within the scope of the present study will be presented, for the train and for the track and how the respective interaction was taken into account.

2.1 *Vehicle model*

The adoption of an analysis methodology with track-train interaction implies the use of vehicle models that enable to appropriately translate its dynamic behaviour. The present study was

Figure 1. ICE2 train circulating in German high speed railway lines.

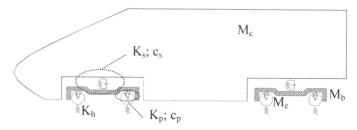

Figure 2. Dynamic model of a locomotive of the ICE2 train.

Table 1. Characteristics of the locomotive of the ICE2 train.

Masses	Box	M_c (kg)	60,768
	Bogie	M_b(kg)	5600
	Axles/Wheels	M_e(kg)	2003
	Load per axle	F (N)	195,000
Suspensions	Primary	K_p(N/m)	48.0×10^5
		c_p(N.s/m)	10.8×10^4
	Secondary	K_s(N/m)	17.6×10^5
		c_s(N.s/m)	15.2×10^4
Wheel-rail connection		K_h(N/m)	19.4×10^8

developed considering the ICE2 train circulating in German high speed railway lines. It is a conventional type of train consisting of two locomotives, one at each end, with loads of 195 kN per axle, and twelve intermediate carriages with axle loads of 112 kN. The total length of the train is 358.6 m.

The dynamic model is presented in Figure 2, which comprises the following elements: (i) rigid body simulating the vehicle box of mass M_c; (ii) springs of stiffness K_s and dashpots with damping ratio c_s simulating the secondary suspensions; (iii) rigid bodies simulating the bogies of mass M_b; (iv) springs of stiffness K_p and dashpots with damping ratio c_p simulating the primary suspensions; (v) masses M_e simulating the axles and wheels, and (vi) springs of stiffness K_h simulating the wheel-rail connection. The stiffness of the wheel-rail contact (K_h) has been determined by the application of Hertz theory (Esveld 2001).

The values for the referred parameters have been obtained in ERRI D214/RP9 (2001) and are listed in Table 1.

By means of the application of an automated calculus software in which the elements available for modelling are 3D beam elements (Euler-Bernoulli formulation), the following has been considered: i) the rigid bodies, referring to the box and the bogies, modelled as beams of high bending

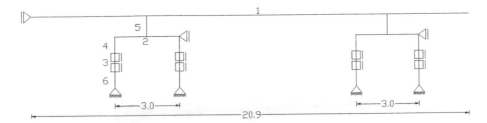

Figure 3. Model of the locomotive of the ICE2 train using beam elements.

Table 2. Characteristics of the beam elements used in the locomotive model.

Locomotive		EI (N.m^2)	EA/l (N/m)	m (Kg/m)	c_1 (s^{-1})	c_2 (s)
Box	(1)	∞	∞	2908	0	0
Bogie	(2)	∞	∞	1867	0	0
Primary suspension	(4)	≈ 0	48.0×10^5	0	0	22.5×10^{-3}
Secondary suspension	(5)	≈ 0	17.6×10^5	0	0	86.4×10^{-3}
Wheel-rail	(6)	≈ 0	1.94×10^9	0	0	0
Axles/Wheels	(3)	≈ 0	1.94×10^{10}	20030	0	0

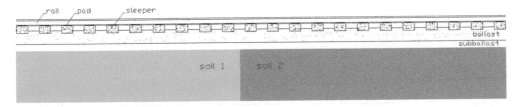

Figure 4. Longitudinal profile of the track in the transition zone.

stiffness and uniformly distributed mass along its length; ii) the suspensions and the wheel-rail connection modelled as nil mass beams, negligible bending stiffness and axial stiffness and damping in accordance with the characteristics of those elements and iii) the masses corresponding to the axles and wheels uniformly distributed as small beams of high axial stiffness (Calçada R., 1995). In Figure 3 is represented the mesh of beam elements relative to the locomotive model as well as the respective support conditions.

The characteristics of the beam elements used in the locomotive model are indicated in Table 2.

The damping of the suspensions has been simulated using a Rayleigh damping matrix $(C = c_1.M + c_2.K)$, where c_1 has taken the value of zero and $c_2 = c_s/K_s$ (or c_p/K_p).

2.2 Track model

In the present work, it is intended to assess the dynamic effects induced by a abrupt change in the railway track stiffness. In Figure 4 is presented a longitudinal profile of the track in such zone and in Figure 5 a cross section illustrating the main elements and features of the track.

In Figure 6, a detail of the dynamic model of the track in the transition zone is presented, which consists of the following elements: (i) continuous beam of bending stiffness $(EI)_r$ and mass per unit length m_r simulating the rails; (ii) springs of stiffness K_p and damping c_p simulating the pads; concentrated masses M_t simulating the sleepers; (iii) continuous beam of bending stiffness $(EI)_b$ and mass m_b simulating the ballast layer; (iv) continuous beam of bending stiffness $(EI)_{sb}$ and

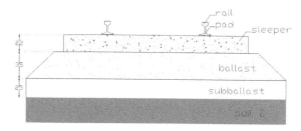

Figure 5.   Cross section of the track.

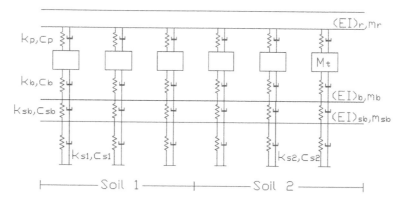

Figure 6.   Dynamic model of the transition zone.

Table 3.   Characteristics of the layer elements of the track.

Track layer	E (N/m^2)	ρ(kg/m^3)	ν
Ballast	70.0×10^6	1529	0.15
Subballast	70.0×10^6	2090	0.30
Soil 1	227.5×10^6	2140.7	0.30

mass m_{sb} simulating the subballast layer; (v) springs of stiffness K_b and damping c_b simulating the ballast; (vi) springs of stiffness K_{sb} and damping c_{sb} simulating the subballast; springs of stiffness K_{s1} (or K_{s2}) and damping c_{s1} (or c_{s2}) simulating the foundation soil.

The characteristics of the ballasted track for high speed railway lines of different European countries can be found in UIC (2001). In the present study have been considered rails of UIC60 type, pads with stiffness equal to 100×10^6 N/m (Pita & Teixeira 2003), mono-block sleepers in pre-stressed reinforced concrete evenly spaced every 0.6 m, with mass equal to 400 kg and dimensions of $0.22 \times 0.30 \times 2.60$ m^3. The characteristics of the ballast, subballast have been taken as identical to those used in a case study of the Supertrack (2005) project, indicated in Table 3. The deformability modulus of the stiffer soil 2 (E_2) has been considered equal to 2, 5 and 10 times the deformability modulus of soil 1 (E_1).

In Figure 7 is presented a detail of the dynamic model of the track in the transition zone using beam elements.

The characteristics of the elements are indicated in Table 4. The mass and bending and axial stiffness characteristics of the elements have been determined in order to comply with the parameters mentioned above.

The stiffness of the foundation soil has been determined by means of expressions of the Theory of Elasticity, considering the track as a flexible foundation set over a half-undefined elastic medium.

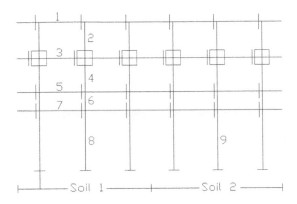

Figure 7. Detail of the track model in the transition zone using beam elements.

Table 4. Characteristics of the beam elements used in the track model.

Track		E (N/m²)	EI (N.m²)	EA/l (N/m)	m (Kg/m)
UIC 60 rail	(1)	200.0×10^9	12.2×10^6	512.5×10^7	120.7
Pads	(2)	60.0×10^6	0	200.0×10^6	0
Sleepers	(3)	0	0	0	666.7
Ballast	(4)	70.0×10^6	0	355.6×10^6	0
Ballast	(5)	70.0×10^6	73.4×10^4	206.5×10^6	1578.7
Subballast	(6)	70.0×10^6	41.6×10^5	554.4×10^6	0
Subballast	(7)	70.0×10^6	30.1×10^4	231.0×10^6	1724.3
Soil 1	(8)	60.0×10^6	0	60.0×10^6	0

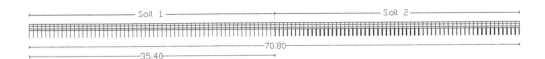

Figure 8. Model of the track in the transition zone using beam elements.

In Figure 8 is presented the complete mesh of beam elements, used for modelling this problem. In total, 70.80 m of track have been simulated, 34.50 m of which are located before the transition zone. The number of elements is equal to 948 and the numbers of nodes is 595.

In Figure 9 are presented the displacements of the 1st axle of the locomotive corresponding to the passage over a transition zone at a very low speed (quasi-static), for the different scenarios of stiffness variation. The observation of the graph enables to conclude that the displacements of the track in the stiffer zone are lower by 23, 39 and 44% to the displacement of the track at the soft zone, for the E_2/E_1 ratios of 2, 5 and 10, respectively.

In Figure 10 can be observed the configurations obtained for the 1st, 3rd, 15th, 16th, 18th, 32nd mode shapes of the track for the case of $E_2 = 2.E_1$. The first modes involve solely movements of the track over soil 1, that is, in the soft zone. Modes after 16th begin to involve movements also of the track over soil 2.

In terms of damping, a global damping ratio has been adopted for the track of 5%. The parameters c_1 and c_2 for the construction of the Rayleigh damping matrix have been obtained by fixing this value of damping for the 1st and 32nd mode shapes, which resulting for the case $E_2 = 2.E_1$, the values of $c_1 = 6.54\,s^{-1}$ and $c_2 = 3.63 \times 10^{-4}\,s$.

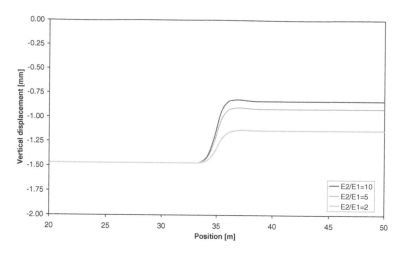

Figure 9.   Quasi-Static displacement of the 1st axle of the locomotive for the different scenarios of stiffness variation.

Mode 1 - f = 16.85 Hz

Mode 3 - f = 17.00 Hz

Mode 15 - f = 22.65 Hz

Mode 16 – f = 23.34 Hz

Mode 18 – f = 23.44 Hz

Mode 32 - f = 26.80 Hz

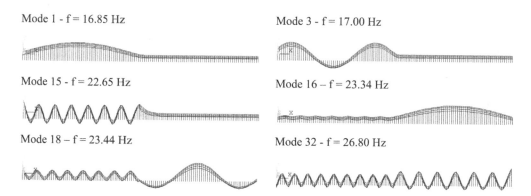

Figure 10.   Schematic representation of some of the mode shapes of the track.

2.3 Track-Train Interaction

The numerical analysis of the track–train behaviour was developed establishing separate equations of motion for the track and the train, and evaluating the dynamic component of the vertical force applied by the train to the track, at each time instant, and solving that set of differential equations using the Newmark method. At each time instant, an iterative procedure was used to make the response of the two structural systems compatible. This procedure is described in detail in Delgado & Cruz (1997) & Calçada (2001).

3 DYNAMIC ANALYSIS

3.1 Simulation scenario

The dynamic analyses were carried out for the passage of the ICE2 train at the speeds of 200, 250, 300 and 350 km/h. As referred in 2.2, three analysis scenarios have been considered, corresponding to the deformability moduli ratios for the foundation soils, E_2/E_1, equal to 2, 5 and 10.

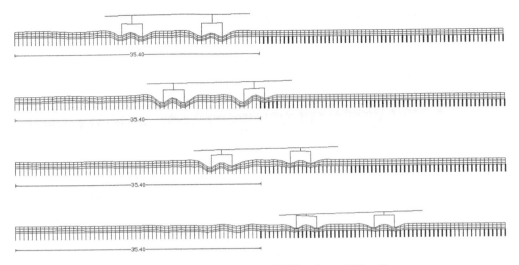

Figure 11.   Deformed meshes of the track-train system ($E_2/E_1 = 5$; v = 300 km/h).

For solving the differential equations system of dynamic equilibrium, the Newmark method has been used, adopting the parameters $\gamma = 1/2$ and $\delta = 1/4$. The time increment (Δt) has been taken equal to 0.00030 s.

3.2 *Results*

In Figure 11 are presented the deformed meshes of the track-train system for diverse time instants of the passage of the ICE2 train at the speed of 300 km/h, for the scenario corresponding to $E_2/E_1 = 5$.

In Figure 12 are presented the values of the dynamic component of the wheel-rail interaction force for the first and second axles of the locomotive, corresponding to the three scenarios of variation of deformability modulus at the speed of 300 km/h. The letters "M" and "m" denote the maximum ($Q_{dyn,M}$) and minimum ($Q_{dyn,m}$) value of such force.

By looking at the graphs it is possible to observe that the largest variations in the dynamic interaction force do not occur at the first, but instead at the second axle of the locomotive. For the second axle, it can also be identified a very significant increase of the values (in absolute terms) of $Q_{dyn,M}$ and $Q_{dyn.m}$ with the increase of the E_2/E_1 ratio.

4   ASSESSMENT OF THE DYNAMIC BEHAVIOUR

In this section, an assessment of the dynamic behaviour of the track-train system is made in terms of track safety and passengers comfort.

4.1 *Track safety*

In Figure 13 are indicated the maximum values of the wheel-rail interaction force, quantified by means of the relation

$$Q_M = Q_{sta} + Q_{dyn,M} \qquad (1)$$

where Q_{sta} is the value of the static load per wheel, equal to 97.5 kN (195/2) for the locomotive of the ICE2 train.

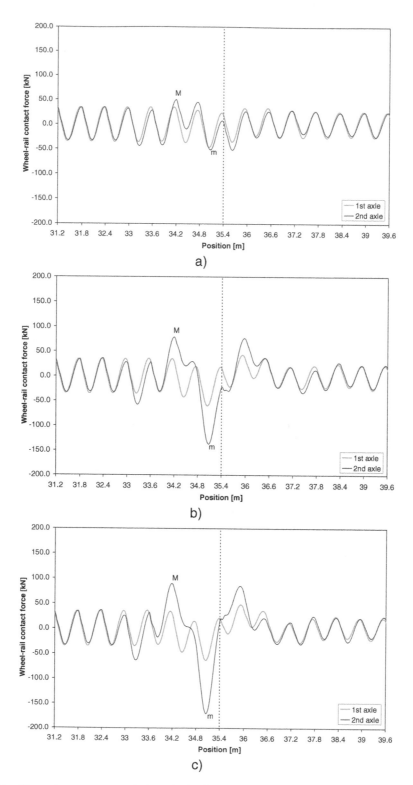

Figure 12. Dynamic component of the wheel-rail interaction force: (a) $E_2/E_1 = 2$; (b) $E_2/E_1 = 5$; (c) $E_2/E_1 = 10$.

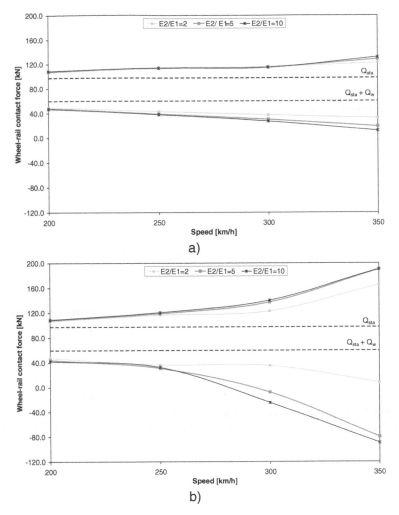

Figure 13. Maximum and minimum values of the wheel-rail interaction force for: (a) 1st axle and (b) 2nd axle of the locomotive, as a function of speed.

In the same figure are also indicated the minimum values of the wheel-rail interaction force, given by the relation

$$Q_M = Q_{sta} + Q_{dyn,M} + Q_w \qquad (2)$$

where Q_w corresponds to the reduction of the wheel-rail contact force due to the wind acting on the train.

Considering, in simplified terms, that the wind action corresponds to a lateral uniformly distributed load of $1.4\,kN/m^2$ in a band with $3.84\,m$ of height, corresponding to a resulting load of $5.4\,kN/m$ at a height of $1.92\,m$, the value of $Q_w = -37.73\,kN$ has been obtained. The stability of the wheel-rail contact is assured for $Q_m \geq 0$.

For the observation of the graph it is possible to draw the following conclusions: (i) the results for $v = 200$ and $250\,km/h$ are nearly coincident; (ii) for $v = 300$ and $350\,km/h$, a very significant increase in the maximum values of the interaction force can be noted, especially for the 2nd axle of the locomotive; (iii) in what concerns the stability of the wheel-rail contact, it can be observed that for $v = 300$ and $350\,km/h$ and for E_2/E_1 ratios equal to 5 and 10, there is loss of contact ($Q_m < 0$)

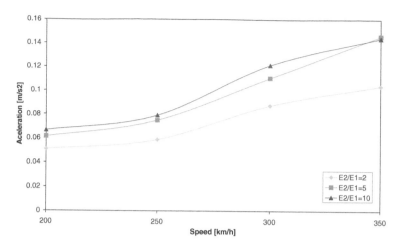

Figure 14. Maximum values of the acceleration in the locomotive box of the ICE2 train, as a function of speed.

for the 2nd axle of the locomotive; (iv) the stability of the wheel-rail contact is guaranteed for all speeds and E_2/E_1 ratios, for the 1st axle, and for $E_2/E_1 = 2$, for the 2nd axle.

4.2 *Passengers comfort*

EN1991-2 (2003) established limits for the peak value of the vertical acceleration inside the carriages equal to $1,0 \, m/s^2$, $1,3 \, m/s^2$ and $2,0 \, m/s^2$ in correspondence with three levels of passengers comfort: Very Good, Good and Fair. In Figure 14 are indicated the peak values of the vertical acceleration in the locomotive box of the ICE2 train for the different speeds and E_2/E_1 ratios. For the observation of the figure it is possible to conclude that the comfort of passengers has remained at the Very Good level for all analysed situations.

5 CONCLUSIONS

The objective of the present work is the study of the dynamic effects in zones of abrupt changes in the stiffness of the track, caused by foundations on soils with distinct deformability moduli. Simulation scenarios have been considered corresponding to deformability moduli ratios for the foundation soils equal to 2, 5 and 10. The dynamic analyses were carried out for the passage of the ICE2 train at the speeds of 200, 250, 300 and 350 km/h.

The analyses were performed based on a numerical analysis program, which enables to appropriately model the track, the train and take into account the dynamic interaction between the two structures and the results have been assessed from the point of view of the safety of the track and the comfort of passengers.

These studies enabled to conclude that: (i) the largest variations of the dynamic interaction force do not occur at the first, but instead at the second axle of the locomotive; (ii) for the second axle, it is noticeable a very significant increase of the maximum and minimum dynamic components of the interaction force with the increase of the deformability moduli ratio.; (iii) in terms of the stability of the wheel-rail contact, it can be seen that for the 2nd axle of the locomotive, $v = 300$ and 350 km/h and for the ratios of soil deformability moduli equal to 5 and 10, there is loss of wheel-rail contact; (iv) the stability of the wheel-rail contact is guaranteed for all speeds and deformability moduli ratios, for the 1st axle, and for ratio 2, for the 2nd axle; (v) the level of passengers comfort remained Very Good for all analysed cases.

ACKNOWLEDGEMENTS

The first author, Ph.D. student, acknowledges the support provided by the European Social Fund, Programa Operacional da Ciência e Inovação 2010.

Ciência.Inovação
2010

Programa Operacional Ciência e Inovação 2010

MINISTÉRIO DA CIÊNCIA, TECNOLOGIA E ENSINO SUPERIOR

REFERENCES

Calçada, R. 1995. Efeitos dinâmicos em pontes resultantes do tráfego ferroviário a alta velocidade. MSc Thesis, Faculty of Engineering of the University of Porto. Porto: FEUP.

Calçada, R. 2001. Avaliação experimental e numérica de efeitos dinâmicos de cargas de tráfego em pontes rodoviárias. Phd Thesis, Faculty of Engineering of the University of Porto. Porto: FEUP.

Delgado, R. & Cruz, S. 1997. Modelling of railway bridge-vehicle interaction on high speed tracks. *Computers and Structures* 63(3): 511–523.

EN1991-2. 2003. *Actions on Structures. Part 2: General actions: Traffic loads on bridges*. European Committee for Standardization (CEN).

ERRI D214/RP9. 2001. *Railway bridges for speed > 200km/h. Final Report*. European Rail Research Institute (ERRI).

ERRI D230.1. 1999. *Bridge ends. Embankment structure transition. RP3, State of the art report*. European Rail Research Institute, Utrecht Netherlands.

Esveld, C. 2001. *Modern railway track*. Zaltbommel: MRT Productions.

Lei, X. & Mao, L. 2004. Dynamic response analyses of vehicle and track coupled system on track transition of conventional high speed railway. *Journal of Sound and Vibration*. 271(3-5):1133–1146.

Pita A. & Teixeira P. 2003. New criteria in embankment-bridge transitions on high-speed lines. *Structures for high speed railway transportation, IABSE Conference, Antwerp, Belgium. CD*.

Schooleman, R. 1996. Overgang kunstwerk-aardebaan voor de hoge-snelheidslijn. MSc Thesis, Technische Universiteit Delft. Deft.

SUPERTRACK – Sustained Performance of Railway Tracks. 2005. *Numerical simulation of train-track dynamics Final Report*. Linköping University.

UIC. 2001. *Design of new lines for speeds of 300-350 km/h State of the art. First Report*. International Union of Railways. High Speed Department.

Applications of Computational Mechanics in Geotechnical Engineering – Sousa,
Fernandes, Vargas Jr & Azevedo (eds)
© *2007 Taylor & Francis Group, London, ISBN 978-0-415-43789-9*

Simplified numerical model for the prediction of train induced vibrations in the vicinity of high speed rail lines

N. Santos, R. Calçada & R. Delgado
Faculdade de Engenharia da Universidade do Porto, Porto, Portugal

ABSTRACT: The present paper is centred on the problem of the propagation of train induced vibrations in the vicinity of high speed railways. General aspects regarding modelling based on the finite element method are introduced, as well as particular aspects regarding modelling of the radiation conditions at the boundaries. An application has been made to the study of vibrations in buildings with reinforced concrete frame structure, located near a railway for the passage of the TGV train at the speeds of 200, 250, 300 and 350 km/h. The analyses were carried out based on a simplified methodology, which makes use of a longitudinal 2D model of the track in order to obtain the vertical displacements at a given section, which will later be imposed in a cross-section 2D model, perpendicular to the track. The results are presented in terms of the vibration speed for points in the ground at different distances to the track and for points in the buildings located at different heights.

1 INTRODUCTION

One of the environmental issues arising from the installation of high speed railways refers to train induced vibrations, since the waves propagating through the soil mass may interact with the adjacent buildings and constructions, hence potentially causing structural damages, malfunctioning of specific sensitive devices or equipments and discomfort to people.

In this context, the development of methodologies for the analysis and prediction of high speed trains induced vibrations is particularly interesting, in order to identify, as early as in the design stage, the zones where problems are most likely to occur, and to enable a well-timed study of the most adequate solutions for its mitigation. This problem has been studied using different methodologies: analytical, experimental, empirical and numerical. According to different authors, best results are obtained through the combination of different approaches.

Analytical methodologies make use of theoretical models to describe the phenomenon (Frýba 1972). Despite its limitations in terms of practical applicability, these models are very useful for understanding the phenomenon, as well as for the calibration and validation of numerical and empirical models.

In situ measurement campaigns are extremely important, as these enable direct recordings of the response of the track, ground, and adjacent structures. The results can be statistically analyzed and treated, serving as basis for the prediction of vibration levels in buildings under similar conditions. Some authors presented cases where, based on a comprehensive number of results, it was possible to study the influence of different parameters in the induced vibrations (Degrande & Schillemans 2001).

Other authors have tried to establish empirical models based on theoretical principles and considering some base hypotheses to simplify the study of the problem. These models intend to reproduce the vibration and propagation stages, considering a small group of parameters which control the phenomenon. Calibration was made through the comparison with results obtained in measurements taken in operating high speed railway lines (Madshus et al. 1996).

The rapid development of information technologies, associated with the appearance and continuous progress of modern computers, has made way for the emergence and evolution of numerical methodologies of analysis. There are various methods for the study of wave propagation problems, namely the boundary elements method, finite differences method and finite elements method. The latter has proven to be the most adequate for the analysis of complex problems. In this method special attention should be given to the modelling of the radiation conditions at the boundaries.

2 GENERAL ASPECTS OF MODELLING BASED ON THE FINITE ELEMENT METHOD

As previously mentioned, there are several numerical methods available for the study of wave propagation problems, in particular, of those induced by railway traffic. Among those methods, the finite element method has proven to be very suitable since, given its great versatility, it enables to analyse complex problems.

The study of the vibration phenomenon can be made essentially by means of two-dimensional models (parallel or perpendicular to the track) or three-dimensional. According to some authors, two-dimensional models can be useful in the analysis of the propagation of vibrations in the embankment (as well as the soil around it) and the dynamic response of the track. However, the results provided by these methods lack quality as the distance in relation to the track increases. On the other hand, three-dimensional enable to obtain better results in zones further away from the track. The observed better behaviour of these models at greater distances in relation to the track is due to a more adequate reproduction of the propagation of the waves, since it is clearly a three-dimensional phenomenon (Hall 2000, Paolucci et al. 2003). The main disadvantage of 3D models is related to the computational resources and the calculation times involved in its application.

An important issue associated with the use of finite element method for the study of this problem is related with the radiation conditions at the boundaries. Given the impossibility of completely modelling the media where these phenomena take place, it becomes necessary to define artificial boundaries. These boundaries should present transparency characteristics and avoid wave reflections which, in reality, would propagate to the exterior. This aspect will be dealt with in greater detail in section 3, where will be demonstrated how the use of viscous dampers at the boundaries enables to overcome this problem.

The waves generated by trains exhibit frequencies within a significantly wide range (0–2000 Hz), where the vibrations and the structural noise occur in the frequency range between 0 and 100 Hz and the audible noise between 30 and 2000 Hz. According to diverse authors, the frequency interval with greater importance in this context is below 200 Hz. In this interval, the higher energetic content is clearly located in the range of the low frequencies, that is, below 20 Hz. The frequencies within 20 Hz and 100 Hz involve yet considerable energetic content and may be associated with vibrations of structural elements (Mateus da Silva 2006).

The vibrations induced by railway traffic depend on the train characteristics (such as suspended and non-suspended loads, stiffness and damping of the suspensions, etc), stiffness and damping characteristics of the track and of the track irregularities and the vehicle wheels. The vibrations in the frequency ranges previously referred have diverse origins. If, in one hand, the bogies and carriages present typically low vibration frequencies (lower than 25 Hz), on the other hand, in order to take into account the irregularities of the rail and vehicle wheels, the involved wavelengths are very low and the frequencies of the generated waves will be considerable higher.

To conclude, a reference should be made to a very important aspect inherent to the development of a finite element model: the dimensions of the elements. These must be small enough to guarantee an acceptable accuracy of the results but, on the other hand, they can not be exaggeratedly small in such a way that the model becomes impracticable in computational terms. Thus, the dimension of the elements must be defined in order to permit modelling the movements associated with the highest frequencies and, consequently, smallest wavelengths of interest. The smallest wavelength to consider will be (Hall 2000):

$$\lambda_{min} = \frac{C_S}{f_{max}} \qquad (1)$$

where C_S = S-waves propagation velocity; f_{max} = highest frequency of interest to the study.

According to the same author, the use of eight elements per wavelength provides an accuracy which is greater than 90% of the amplitude of the wave at high frequencies. As it is easy to understand, it is possible to consider larger elements in zones with higher S-waves velocity or with less interest to the study.

3 ASPECTS REGARDING MODELLING BOUNDARY CONDITIONS

3.1 Radiation conditions at the boundaries

The propagation of waves generated by the passage of trains occurs in media of infinite dimensions, which however are not possible to fully discretize from the computational point of view. It is therefore necessary to define a limit to the medium in study. The artificial boundaries should exhibit transparency properties which will enable its free crossing and avoid the reflexion of waves that would in reality leave the discretized medium (Faria 1994). This objective may be achieved by the application of viscous dampers connected to the faces of the elements that constitute the boundary, as illustrated in the Figure 1 (for a two-dimensional model).

The damping coefficient values that characterize the dampers can be determined according with a weighted damping matrix (for a medium in plane strain), which relates the normal and tangential components of stress and velocity at the boundary (Faria 1994):

$$C'_{pon} = \frac{8\sqrt{\rho G}}{15\pi} \begin{matrix} 5/s-2s+2 & 0 \\ 0 & 2s+3 \end{matrix} \qquad (2)$$

where ρ = density; G = shear modulus; $s = ((1 - 2\nu)/(2 - 2\nu))^{1/2}$; ν = Poisson's ratio.

After performing a shift from the local referential to the general referential and numerically integrating the damping forces along the face of the element, it is possible to obtain the elemental matrix of damping by radiation. The global radiation damping matrix is obtained by grouping the contributions of the different elements of the boundary which, once added to the global material damping matrix, will constitute the global structure damping matrix.

3.2 One-dimensional example of validation

With the aim of evaluating the efficiency of the viscous dampers in the simulation of the transparency conditions at the boundaries, the behaviour of an half-infinite elastic bar has been analysed.

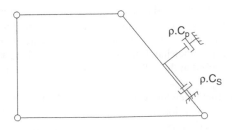

Figure 1.   Transparency condition for an artificial boundary of a finite element.

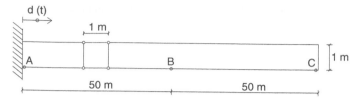

Figure 2.   Studied bar.

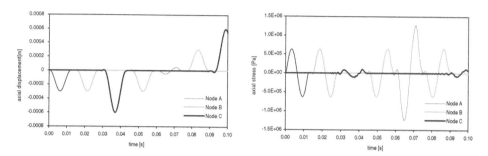

Figure 3.   Time histories of the axial displacements and stresses at three points of the bar (without viscous dampers).

A bar with the following dimensions $100 \times 100 \times 1\,m^3$ was considered, having been discretized with four-node plane-strain finite elements with dimensions of $1 \times 1\,m^2$ (Figure 2). The material that constitutes the bar is characterized by the following parameters: $E = 25\,GPa$, $\nu = 0.2$ and $\rho = 2500\,kg/m^3$ and its behaviour assumed to be linear elastic.

To carry out this study, an axial displacement to the left extremity of the beam was imposed, with a time history defined by the following expression, in order to induce the propagation of a volumetric wave in the bar.

$$d(t)\,[m] = \begin{cases} -1.5 \times 10^{-4}\left[1+\sin\left(\dfrac{2\,\pi\,t}{0.012} - \dfrac{\pi}{2}\right)\right] & ,\ 0 \le t \le 0.012 \\ 0 & ,\ t > 0.012 \end{cases} \tag{3}$$

For the resolution of the dynamic problem, the Newmark method was adopted with the following parameters $\gamma = 0.5$ and $\beta = 0.25$. The results presented had been obtained using a time step $\Delta t = 0.001\,s$.

In the first phase of the study the structure was considered with a free end. Figure 3 shows the time histories of the axial displacements and stresses obtained at three nodes of the bar: A (section near the fixed end), B (section at the middle of the beam) and C (section near the free end), presented in Figure 2.

It is possible to show that the wave produced by the imposed displacement moves with a velocity $C_S = 3333.33\,m/s$. Based on the presented results it was observed that the wave takes $0.015\,s$ to reach the node B and $0.030\,s$ to reach node C, which is in accordance with the value of the velocity theoretically calculated. On the other hand, the evolution of the disturbances at the intermediate node of the bar enables to verify that the reflections do not modify the period nor the amplitude of the wave.

It is also possible to show that when the wave approaches the free section, no stress can be transmitted. In order to satisfy this condition, the displacement of the boundary (transmitted displacement) has twice the amplitude of the incident wave. The reflected wave has the same amplitude as the incident wave but opposite polarity. In other words, the free end reflects the compressional wave as a

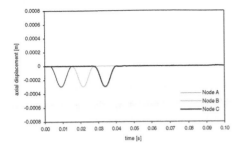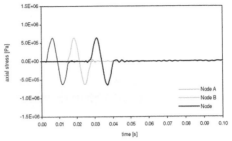

Figure 4. Time histories of the axial displacements and stresses at three points of the bar (with viscous dampers).

tension wave (and vice versa) with equal amplitude and shape. When the wave approaches the fixed end, no displacement can occur. In this case the stress in the boundary is twice the incident stress. On the other hand the reflected wave presents the same amplitude and polarity of the incident wave.

In a second phase of the study, it was tried to attribute adequate characteristics to the boundary (free end) which allow its free crossing, with the intention of simulating an infinitely long bar. To reach this objective, a set of viscous dampers was used linked to the nodes at the boundary and perpendicularly to it. The values of the damping coefficients which characterize the dampers were determined according with the procedure described in the previous section. Figure 4 illustrates the time histories of the axial displacements and stresses at the same nodes previously studied.

Observing these results it is possible to evaluate the efficiency of the viscous dampers. Similarly to the previous case, the wave reaches the node B at instant 0.015 s. However, if any reflections had occurred, the wave would have come back to reach the same section after it had covered a total length of 100 m, being expectable to see not null displacements towards instant $t = 0.045$ s. As it is possible to observe, the referred displacements are different from zero but very small. Therefore it can be concluded that the viscous dampers have exhibited a good behaviour and so the reflection of the incident wave was very small. The same can be concluded from the observation of the time history of the axial stress.

4 APPLICATION

In this section, an application of the simplified numerical analysis methodology will be made to the study of induced vibrations in the vicinity of a railway track by the passage of the TGV train at the speeds of 200, 250, 300 and 350 km/h. The analyses were carried out at a first stage based on a longitudinal 2D model, which enabled to obtain the vertical displacement at a given section of the track, which were at a second stage imposed on a cross-section 2D model, perpendicular to the track at that section, which in turn enables the propagation of the waves in that direction.

4.1 *Longitudinal model*

In Figure 5 is presented the mesh of finite elements used for discretizing the track in the longitudinal direction, which consists of 9432 elements and 27333 nodes.

The profile of the UIC 60 rail has been modelled by beam elements. The rail pads were simulated by spring elements with a constant of $k = 100$ kN/mm. Sleepers with dimensions of $0.3 \times 0.22 \times 2.6$ m^3 were modelled by four-node plane elements ($E = 30$ GPa, $v = 0.2$; $\gamma = 25$ kN/m^3). The characteristics of these elements were obtained from the references (Esveld 2001, Pita & Teixeira 2003, SUPERTRACK 2005, UIC 2001). The ballast, the sub-ballast and the soil were modelled by eight-node plane elements. The ballast between the sleepers was

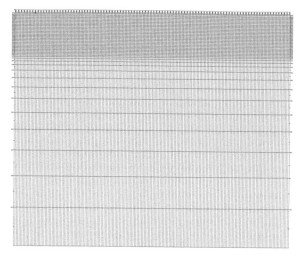

Figure 5.   Finite elements mesh of the longitudinal model.

Table 1.   Assumed properties for ballast, sub-ballast and soil.

Element	E[MPa]	$\nu[-]$	$\rho[\text{kg/m}^3]$	h[m]
Ballast	70	0.15	1530	0.35
Sub-ballast	70	0.30	2090	0.25
Soil	100	0.33	1800	–

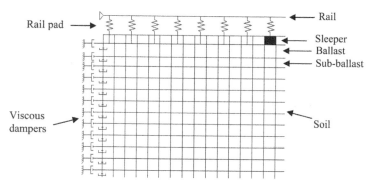

Figure 6.   Detail of the finite elements mesh of the longitudinal model.

modelled by four-node plane elements. The properties assumed for these elements are presented in the Table 1.

A detail of the longitudinal model is illustrated in Figure 6.

The nodes at the lateral boundaries were connected to normal and tangential viscous dampers in order to adequately reproduce the radiation conditions at the boundaries. The nodes at the bottom were fixed in order to reproduce the presence of the bedrock. The elements modelling the soil are 0.30 m high down to a depth of 10 m, and variable between 0.3 and 8 m from 10 to 50 m.

490

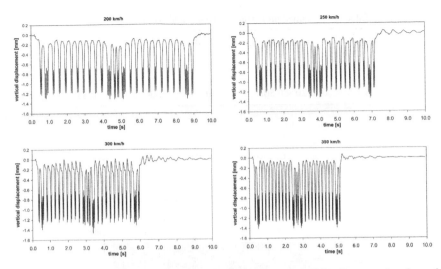

Figure 7.   Time evolution of the vertical displacement of a node at the rail for diverse analysed speeds.

The TGV train was simulated as a set of moving loads of 170 kN. The regular spacing between groups of axles D is equal to 18.7 m. The total length of the train is equal to 468.14 m.

The dynamic analyses were carried by means of the ANSYS software, using the method of modal superposition. The time increment (Δt) was taken equal to 0.002 s. In what concerns the damping, a constant damping coefficient of 3% was adopted for all modes.

In Figure 7 is presented the time evolution of the vertical displacement of a node at the rail for the passage of the TGV train at the speeds of 200, 250, 300 and 350 km/h. The maximum value of the static displacement is equal to 1.29 mm. The maximum values of the dynamic displacements were 1.30 mm, 1.33 mm, 1.46 mm and 1,41 mm, which corresponds to dynamic amplifications of 1, 3, 13 e 9%, respectively.

4.2   Perpendicular model

In Figure 8 is presented the finite elements mesh relative to the cross-section model. Taking into account the symmetry of the loading and of the track in relation to its longitudinal axis, it was decided to model only half of the problem. The model includes, besides the track and the soil, two buildings in the vicinity of the track with four storeys and frame structure in reinforced concrete. The mesh is made of 5172 elements and 15586 nodes.

The nodes on the right boundary were connected to normal and tangential viscous dampers in order to reproduce adequately the radiation conditions. The nodes at the bottom were fixed in all directions to reproduce the bedrock. Finally the nodes at the left boundary were fixed in the horizontal direction with the objective of reproducing the symmetry conditions of the problem.

The frames were modelled by beam elements with the characteristics as in Table 2. The density of the material of section 3 was modified in order to include the mass of the slab corresponding to the influence width of the frame.

By means of a modal analysis, the first 200 mode shapes were determined with frequencies within the range [1.28, 24.71] (Hz). In Figure 9 are presented four of those mode shapes. Mode 1 (f = 1.28 Hz) involves essentially vertical movements of the soil. Mode 4 (f = 2.76 Hz) corresponds to a horizontal mode shape of the frames. The other two modes (f = 5.07 Hz e f = 8.95 Hz) involve combined movements of the soil and the building frames.

The dynamic analyses were performed by the Newmark method, where the parameters $\gamma = 1/2$ and $\beta = 1/4$ have been adopted. The time increment (Δt) was taken equal to 0.002 s. In terms of

Table 2. Assumed properties for beams and columns.

Section	Dimensions (m²)	E[GPa]	ρ[kg/m³]	$v[-]$
1	0.25×0.25	30	2500	0.33
2	0.35×0.35	30	2500	0.33
3	0.20×0.50	30	50000	0.33

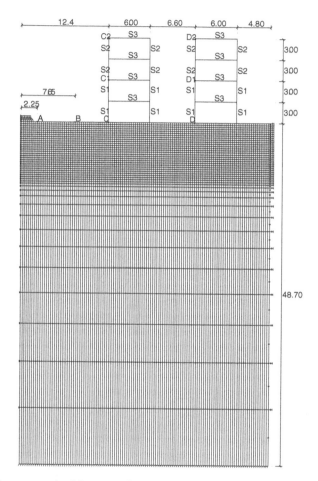

Figure 8. Finite elements mesh of the perpendicular model.

the damping, a global damping coefficient of 3% was adopted for the system. The parameters c_1 e c_2 for the construction of the Rayleigh damping matrix were obtained by fixing this damping value for the 1st and the 200th mode shapes, which resulted in $c_1 = 0.459\,\mathrm{s}^{-1}$ and $c_2 = 3.67 \times 10^{-4}\,\mathrm{s}$. In Figure 10 are presented the time records of the horizontal (v_x) and vertical (v_y) components of the speed in points A and C at the surface of the ground for the passage of the TGV train at the speeds of 200 and 350 km/h.

In Table 3 are presented the rms values of the velocity for the points A, B, C and D at the surface of the ground, as a function of speed.

From the observation of the table it is possible to conclude that the rms values of velocity increase with the increase of the circulation speed. The most significant increases were located at the points

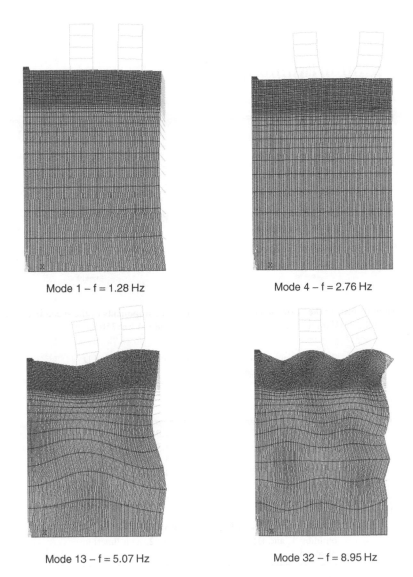

Mode 1 – f = 1.28 Hz

Mode 4 – f = 2.76 Hz

Mode 13 – f = 5.07 Hz

Mode 32 – f = 8.95 Hz

Figure 9. Mode shapes of the perpendicular model.

nearest to the track. For all circulation speeds, it was also identified a decrease in the rms values of velocity with the increase of the distance of the point to the track.

In Figure 11 are presented, in turn, the time records of the horizontal (v_x) and vertical (v_y) components of the speed at points C1 and C2, located at the 2nd floor and at the roof of building 1, for the passage of the TGV train at speeds of 200 and 350 km/h.

The observation of the figure enables to perceive the occurrence of resonance in the structure of the buildings for the passage of the TGV train at the speed of 200 km/h, evidenced by the increase of the horizontal component of velocity with the passage of the train. It should be noted that for v = 200 km/h (200/3.6 = 55.55 m/s) the excitation frequency, given by f = v/D = 55.55/18.7 = 2.97 Hz, almost coincides with the frequency of mode 4 (see Figure 9) which corresponds to a horizontal mode shape of the building frames. It can also be observed that,

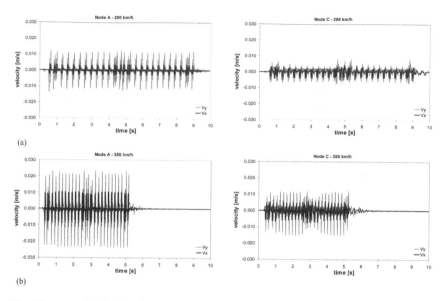

(a)

(b)

Figure 10.   Time records of the horizontal (v_x) and vertical (v_y) components of the speed in points A and C for the passage of the TGV train at the speeds of: (a) 200 km/h and (b) v = 350 km/h.

Table 3.   rms values of velocity in points A, B, C and D, as a function of speed.

Speed [km/h]	A	B	C	D
200	4.8447	3.4085	2.9680	2.4518
250	5.5663	3.9029	3.3770	2.3311
300	6.3738	4.5317	3.8796	2.8790
350	7.3344	4.9178	4.2486	2.9563

Table 4.   rms values of velocity in points C1 and C2, of building 1, and D1 and D2, of building 2, as a function of speed.

Speed [km/h]	C1	C2	D1	D2
200	17.3364	20.3943	16.9264	19.8354
250	7.0381	8.0560	5.5505	6.3990
300	5.2600	5.1594	3.9718	3.9741
350	5.8346	6.0215	4.4451	4.8255

as would be expected, the vertical component of velocity is basically identical for points located at the same vertical alignments, that is, C1 and C2 and D1 and D2.

In Table 4 are presented the rms values of velocity for C1 and C2, of building 1, and D1 and D2, of building 2, as a function of speed.

From the observation of the table it is possible to attest that: (i) the highest values occur for v = 200 km/h; (ii) the values obtained for the roof are higher than those obtained for the 2nd floor;

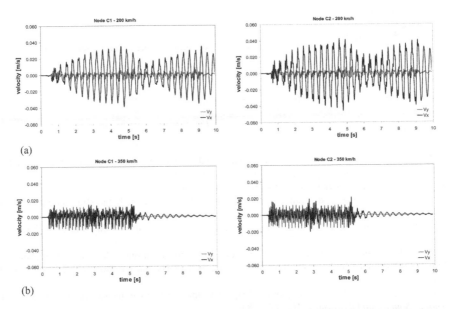

(a)

(b)

Figure 11.   Time records of the horizontal (v_x) and vertical (v_y) components of velocity at points C1and C2, of building 1, for the passage of the TGV train at speeds of: (a) $v = 200$ km/h and (b) $v = 350$ km/h

(iii) the values decrease until $v = 300$ km/h, raising again for $v = 350$ km/h; (iv) the values obtained for building 1 are slightly higher than those obtained for building 2.

5   CONCLUSIONS

The present paper has focused on the issue of train induced vibrations in the vicinity of high speed railways. General aspects regarding modelling based on the finite element method were introduced, along with particular aspects on modelling the radiation conditions at the boundaries. An application has been made to the study of vibrations in buildings located near a railway for the passage of the TGV train at the speeds of 200, 250, 300 and 350 km/h. The analyses were carried out at a first stage based on a 2D longitudinal model, which enabled to obtain the vertical displacements at a given section, which were later imposed in a cross-section 2D model, perpendicular to the track at that section, and which allow for the modelling of the propagation of the waves in that direction. The studies performed enabled to conclude that the rms values of velocity at points on the surface of the ground increase with the increase of the circulation speed. The most significant increases were observed at the points nearest to the track. For all circulation speeds, it was also identified a decrease in the rms values of velocity with the increase of the distance of the point to the track. In what concerns the buildings, the highest rms values of velocity were recorded for $v = 200$ km/h, due to the occurrence of resonance in the buildings by the coincidence of the excitation frequency with the frequency of the 1st horizontal mode shape of the frames. The highest rms values of velocity were obtained in the roof of the buildings. The rms values obtained in building 1, at 12.5 m from the track, were slightly greater than the rms values obtained in building 2, at 25 m from the track.

ACKNOWLEDGEMENTS

The first author, Ph.D. student, acknowledges the support provided by the European Social Fund, Programa Operacional da Ciência e Inovação 2010.

REFERENCES

Degrande, G. & Schillemans, L. 2001. Free field vibrations during the passage of a Thalys high-speed train at variable speed. *Journal of Sound and Vibration*, 247(1): 131–144.

Esveld, C. 2001. *Modern railway track*. Zaltbommel: MRT Productions.

Faria, R. M. M. 1994. Avaliação do comportamento sísmico de barragens de betão através de um modelo de dano contínuo. PhD Thesis, Faculty of Engineering of University of Porto. Porto: FEUP.

Frýba, L. 1972. *Vibrations of solids and structures and moving loads*. Groningen: Noordhoff International Publishing.

Hall, L. 2000. Simulations and analyses of train-induced ground vibrations – A comparative study of two and three dimensional calculations with actual measurements. Dissertation. Stockholm: Royal Institute of Technology.

Madshus, C. et al. 1996. Prediction model for low frequency vibration from high speed railways on soft ground. *Journal of Sound and Vibration*, 193(1):175–184.

Mateus da Silva, J. M. M. & Monteiro, C. 2006. Medição das vibrações em aterros geradas por comboios de alta velocidade. *4as Jornadas Portuguesas de Engenharia de Estruturas*. LNEC, Lisboa, 13–16 December 2006.

Paolucci et al. 2003. Numerical prediction of low-frequency ground vibrations induced by high-speed trains at Ledsgaard, Sweden. *Soil Dynamics and Earthquake Enginnering*, volume 23: 425–433.

Pita A. & Teixeira P. 2003. New criteria in embankment-bridge transitions on high-speed lines. *Structures for high speed railway transportation, IABSE Conference, Antwerp, Belgium. CD*.

SUPERTRACK – Sustained Performance of Railway Tracks. 2005. *Numerical simulation of train-track dynamics - Final Report*. Linköping University.

UIC. 2001. *Design of new lines for speeds of 300–350 km/h State of the art. First Report*. International union of railways high speed department.

Applications of Computational Mechanics in Geotechnical Engineering – Sousa,
Fernandes, Vargas Jr & Azevedo (eds)
© 2007 Taylor & Francis Group, London, ISBN 978-0-415-43789-9

Analytical models for dynamic analysis of track for high speed trains

P. Alves Costa, R. Calçada & A.S. Cardoso
Faculty of Engineering of University of Porto, Porto, Portugal

ABSTRACT: The simulation of the dynamic behaviour of railway tracks under high-speed traffic is a complex problem and it has constituted one challenge for many researchers in the last years. Although in recent years complex numerical models in this domain have appeared, the use of analytical models continues to have its importance, mainly in the preliminary stages of design or in the validation of numerical models.

In this paper, different analytical models to simulate the track response under high-speed traffic are presented and discussed and a new methodology for determination of input parameters is proposed. Some aspects of the dynamic behaviour of the railway track are analyzed by means of the application of the model to a case study and by the development of a parametric study.

1   INTRODUCTION

Development of high-speed train lines is growing rapidly throughout the Europe and East Asia. This rapid increase has stimulated researchers and engineers to develop and calibrate predictive tools for analysis of track behaviour, especially on the demand of dynamic behaviour and ground vibrations induced by the traffic. In the case of Portugal, it is expected that the high-speed line may render to cross soft soil areas. These areas are particularly susceptible to excessive vibrations from high-speed trains because the ground constitutes a very flexible foundation and the embankment height over these areas is, in general, reduced. The international experience on this field has shown that the deflections of the track can present very high amplifications as the traffic speed increases.

Nowadays, several formulations have been proposed for the prediction of train-induced ground vibrations, with different levels of accuracy and, obviously, with different stages of complexity and time cost. However, for preliminary stages of design, the analytical solutions still have some importance and some advantages in relation to more complex numerical models consequence of: (i) the existence of a closed-form solution; (ii) the reduced number of input parameters; (iii) the simplicity of application. However, just some factors of the generation of ground vibrations can be simulated by resource of these models and the application is limited to some particular conditions that not always correspond to the complexity of the reality.

The present study is limited to the analytical models that include the embankment and the railway track. So, only two dimensional models are analyzed where all the components of track-embankment system are simulated by means of an infinite Euler beam resting on an elastic foundation and trains are replaced by a set of moving loads. Three models are presented: the Winkler model, the Kelvin model and the Pasternak model. The difference between them resides essentially on the way of modeling the foundation.

In this paper the analytical formulation of the models is presented and the main differences between those formulations are emphasized. Also a methodology to predict the input parameters is proposed and discussed. To evaluate the possibility of practical use of these models, it is presented the simulation of the track vibrations on a real case study. Moreover a parametric study is developed with the intuit of evaluate the main factors that affect the system.

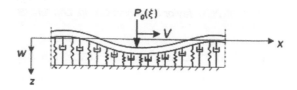

Figure 1.   Schematical representation of the model of Kelvin, solicited by the passage of a point load with constant speed (Wenander, 2004).

2   ANALYTICAL FORMULATION

2.1   *Winkler/Kelvin foundation – moving point load*

The most common way of modeling the ground foundation of a railway track is using the Winkler model. The original form of this model was developed for an elastic foundation which is simulated by means of continuous springs. An evolution of this model is provided by the Kelvin model where the damping forces are also included, so the foundation is simulated by continuous system of spring-dampers as it is shown in Figure 1.

The Winkler and Kelvin models are very similar so they are presented together in the present paper.

Consider that an infinite Euler beam is resting on an elastic/visco-elastic foundation and that it is solicited by a moving point load. The differential equation that describes the equilibrium of the system is given by:

$$EI\frac{\partial^4 u(x,t)}{\partial x^4} + k\,u(x,t) + c\,\dot{u}(x,t) + \rho\,\ddot{u}(x,t) = P\delta(x - V_0 t) \tag{1}$$

where EI is the flexural stiffness of the beam, u is the vertical displacement, x is the distance along the beam, t is time, k is the subgrade reaction of the Winkler foundation, c is the damping coefficient (this parameter just exist in the Kelvin foundation), ρ is the mass per unit length of the beam, P is load and $\delta(x - V_0 t)$ is the Dirac function of a point unit load moving with the velocity V_0.

The solution of equation 1 can be obtained in closed form (Kenney, 1954; Fryba, 1972). However, in the present study, it was opted to solve the equation by the method proposed by Esveld (2001) that considers a moving referential with load. By this way, introducing a new variable s:

$$s = \lambda(x - V_0 t) \tag{2}$$

where λ is the characteristic length of the beam given by:

$$\lambda = \sqrt[4]{\frac{k}{4EI}} \tag{3}$$

The traffic velocity can be defined by its relation with the critical velocity of the beam (embankment and track for this case), that corresponds to the velocity of propagation of bending waves along the beam. So, the variable α is defined by:

$$\alpha = \frac{V_0}{V_{cr}} \quad and \quad V_{cr} = \sqrt[4]{\frac{4EIk}{\rho^2}} \tag{4}$$

Considering the change of referential and the news constants above presented, equation 1 can be written in this form:

$$\frac{d^4 u(s)}{ds^4} + 4\alpha^2 \frac{d^2 u(s)}{ds^2} - 8\alpha\beta \frac{du(s)}{ds} + 4u(s) = P\delta s \tag{5}$$

498

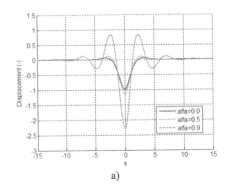 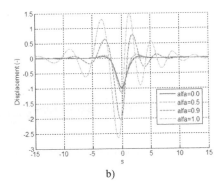

a)   b)

Figure 2.   Adimensional displacements of a generic beam on Winkler/Kelvin foundation for different values of α: a)Winkler model, $\beta = 0$; b) Kelvin model, $\beta = 0.1$.

where β is defined by:

$$\beta = \frac{c}{2\rho}\sqrt{\frac{\rho}{\kappa}} \tag{6}$$

The solution of the equation is:

$$s \geq 0 \qquad u(s) = A_1 e^{\gamma_1 s} + A_2 e^{\gamma_2 s} \tag{7}$$

$$s < 0 \qquad u(s) = A_3 e^{\gamma_3 s} + A_4 e^{\gamma_4 s} \tag{8}$$

The coefficients A_i depends of the boundary conditions of the problem. The roots γ_i and the coefficients A_i are, in general rule, complex numbers but the final solution just present real part.

In Figure 2 is presented a generic adimensional deformation of a beam on elastic foundation for different values of α.

The analysis of the figure presented above permit to identify some particular aspects of the behaviour of the system as: (i) in the Winkler model ($\beta = 0$), when the traffic velocity approaches to the critical velocity the amplifications tend to infinite so it occurs a resonance like phenomena where the stability of the beam can not be guaranteed; (ii) if $\beta \neq 0$, it exists a solution for $\alpha = 1$, where the amplification of displacement reaches the maximum value; (iii) for $\beta = 0$, the maximum displacement occurs exactly below the point load application and for $\beta \neq 0$ it is verified that the maximum displacement occurs behind those point.

2.2   Pasternak foundation – moving point load

The Pasternak model is an evolution of the models presented above. One limitation that is appointed to the Winkler and Kelvin models is the fact of the springs are uncoupled, that means that the displacement of one spring is independent of the displacements of the adjacent springs, and it is known that the soil presents shear resistance what implies displacements interaction (Hall, 2000; Kerr, 1961; Wang et al., 2005; Kargarnovin and Yousenian, 2004). The Pasternak model exceeds this limitation of the models referred above. In this model the foundation is simulated by means of springs, dampers (if the damping is considered) and shear blocks to include the interaction between the springs as it is shown in Figure 3 (Wenander, 2004).

In this case, when the system is solicited by a moving load, the equation that establishes the equilibrium is given by:

$$EI\frac{\partial^4 u(x,t)}{\partial x^4} - G'\frac{\partial^2 u(x,t)}{\partial x^2} + k u(x,t) + c\dot{u}(x,t) + \rho\ddot{u}(x,t) = P\delta(x - V_0 t) \tag{9}$$

499

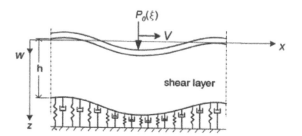

Figure 3.   Pasternak model submitted to one moving point load.

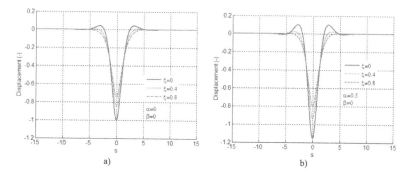

a)   b)

Figure 4.   Adimensional displacements of a generic beam on Pasternak foundation for different values of ξ: (a) $\alpha = 0$ (static conditions); (b) $\alpha = 0.5$.

In this equation the shear stiffness of the soil appears trough the constant G'. Considering the same referential change used on the last deduction, introducing a new constant to include the shear stiffness, ξ, and knowing that the solution is of the type presented by equations 7 and 8, the equation can be wrote in this form:

$$\gamma^4 + 4(\alpha^2 - \xi)\gamma^2 - 8\alpha\beta\gamma + 4 = P\delta(s) \tag{10}$$

where:

$$\xi = \frac{G'}{\sqrt{4kEI}} \tag{11}$$

Figure 4 shows a generic deformation of a beam on Pasternak foundation for several values of ξ and α.

Analyzing Figure 4a, that refers to static conditions, it is possible to identify that the increase of ξ is accompanied by the decrease of the maximum settlement and also by the decrease of the uplift of the beam. For dynamic conditions, the influence of this parameter is more evident as it is showed in Figure 4b. Taking into account the results showed in the Figure 4 and analyzing equation (9) it is possible to conclude that the critical velocity in the Pasternak model is higher than in Winkler model. The relation between the critical velocities of these two models is expressed by:

$$V_{cr\,Pasternak} = V_{cr\,Winkler}\sqrt{1 + \xi} \tag{12}$$

2.3   Determination of input parameters

Concerning to the determination of the input parameters of the models, it should be remembered that there are several methodologies proposed on the bibliography, nor always consensual. In this paper is presented a new methodology proposed by the authors to determine those parameters.

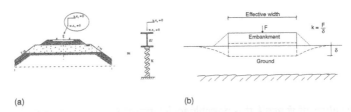

(a)   (b)

Figure 5. Modelling the track trough a beam resting on the ground: (a) schematic representation; (b) methodology for determine k.

The first issue is related to the definition of the beam. There are several proposals about this topic. The most common is to assume that the beam is composed by all the components of the track (rails, sleepers, ballast and sub-ballast) and of the embankment (if it exists.). This "beam" is resting on the foundation composed by the natural ground as it is shown in Figure 5a).

In relation to the determination of the subgrade reaction modulus, k, several proposals exist in the bibliography. The more commons are the methods proposed by Biot (1937) or by Vesic (1963), expressed by equations (13) and (14), respectively.

$$k = \frac{0.95 E_s}{(1 - v^2)} \left[\frac{E_s b^4}{(1 - v^2) EI} \right]^{0.108} \tag{13}$$

$$k = 1.23 \times \left[\frac{1}{(1 - v_s^2)} \frac{E_s \times b^4}{EI} \right]^{0.11} \times \frac{E_s}{(1 - v_s^2)} \tag{14}$$

where E_s is the Young modulus of the soil, v is the coefficient of Poisson of the soil and b is the width of the beam. These formulations are based on the assumption that the soil is homogeneous and semi-indefinite.

A more accurate estimation of the value of k can be made through a FEM analysis in static conditions and considering a strain plane state. The "beam" should be assumed as a rigid block and the interface between the beam and the ground should be admitted as rough. An illustrative example of this method is shown in Figure 5b).

Another input parameter that has to be estimated is the damping of the system. This parameter is usually determined trough experimental testing.

For the Pasternak model it is also necessary an estimative of the value of G'. Several authors point that this is the major disadvantage of this model. Hovarth (1989) proposes the relation expressed below:

$$G' = \frac{h_s^2}{4} k \tag{15}$$

where h_s is the thickness of the soil layer. The same author observed that this equation could give origin to unrealistic values of G'.

The authors suggest that the parameter G' can be determined by the following methodology:

- model the ground by finite elements with all relevant layers;
- impose a vertical point displacement on the surface of the ground;
- analyze the vertical displacements along the surface of the ground;
- adjust the shape of displacements through the equation:

$$u(x) = e^{-\psi x} \tag{16}$$

501

being :

$$\psi = \sqrt{\frac{k}{G'}}$$ (17)

- ψ will take the value that conduce to a better correlation between analytical and numerical results;
- knowing the values of ψ and k it is possible to determine G'.

2.4 Train moving – multiple point loads

The models presented can be used to study the response of the track when solicited by a train by superposition of the effects of the passage of all axles. This is possible if it is assumed elastic behaviour for the system.

Supposing that the system beam-ground is perfectly elastic and linear, the response to the passage of a train with n axles can be expressed as:

$$u\,(x,t)_{Train} = \sum_{i=1}^{n} u_i\,(x,t)$$ (18)

In almost cases, when the velocity of traffic is considerably inferior to the critical velocity, that assumption is verified. However, there are some cases illustrated in the bibliography where the deformation induced by the high-speed train was so significant that the involved material showed non linear behaviour.

3 APPLICATION TO A CASE STUDY

3.1 Description of the site and train

To evaluate the performance of the analytical models presented, a real case has been simulated and some results are confronted with the measures made in the local.

The case of Ledsgard, in Sweden, had been studied by many researchers and had permit to take several lessons about the dynamic behaviour of ballasted railway tracks (Hall, 2000; 2003; Kaynia et al., 2000; Madshus & Kaynia, 2000; Takemiya, 2003; Holm et al., 2002). For these reasons this was the case selected for this study.

In 1997, when the new line was inaugurated, it was observed excessive vibrations during the passage of the train X2000 at 200 km/h. To mitigate this effects the Swedish National Rail Administration started investigations to diagnose the problem and to find a solution. More details about this can be found in Kaynia et al. (2000) or Hall (2000).

Concerning to the geotechnical characteristics of the local, it was done a vast investigation program, with field and laboratory tests, and it was observed the existence of one layer of organic clay with a shear velocity of 40 m/s. The main properties of the ground are resumed in Figure 6. In Figure 7a geometry of the local is shown.

In order to perform the measurements of train-induced ground movements a high speed train of the type X2000 (see Figure 8) was used. The measurements of the vibrations were made by different methods as is shown in Figure 7b. The analysis of the measurements and numerical simulations performed by other authors revealed that the ground and embankment dynamic strains for high speed train passages were so high that these materials behave non-linearly.

In this paper the shear modulus of the granular materials were reduced to attend to those effects.

3.2 Comparison between experimental and analytical results

For the simulation of this case history, it was selected the Pasternak model by the fact that it is an evolution of Winkler and Kelvin model's. The comparison of results was done in terms of displacements of the track for the passage of the train X2000 at 70 km/h.

502

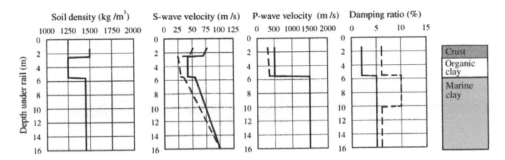

Figure 6. Dynamic soil properties for test site (continuous line – small strain values; dashed line – values accounting for the non linearity) (Madhus and Kaynia, 2000).

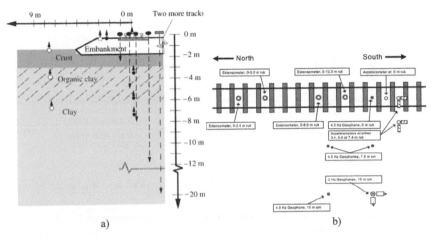

Figure 7. Test site and instrumentation: a) Transversal section (Madshus and Kaynia, 2000); b) Localization of several instruments (Hall, 2003).

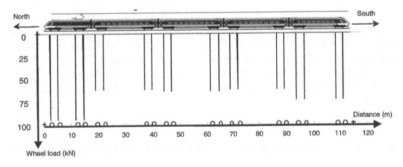

Figure 8. Wheel loads of the train X2000 used in Ledsgaard tests (Hall, 2003).

The elastic properties of the ground were reduced to attend to the shear modulus degradation with the dynamic strain. Table 1 summarizes the soil parameters adopted (Kaynia et al., 2000).

Concerning to the properties of the track-embankment it was assumed that the effective width take the value of 3,0 m and the bending stiffness, EI, the value of 80 MN.m^2. The definition of the mass of the beam was complex. The analytical models presented consider a beam resting on a "massless" foundation. This is a rough simplification, because there are mass of the ground that is excited during the passage of the train. To consider the effect of this mass, knowing that the critical

503

Table 1. Soil parameters used in simulations.

Soil layer	Thickness (m)	Mass density (kg/m^3)	S – waves velocity (m/s)	P- waves velocity (m/s)
Crust	1.1	1500	65	500
Organic Clay	3.0	1260	33	500
Clay 1	4.5	1475	60	1500
Clay 2	6.0	1475	85	1500

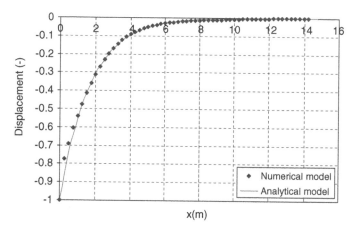

Figure 9. Adjust between the analytical equation and the results provided from the numerical model.

velocity for this case study is around 230 km/h (from the measurements and analyses provided by other authors), and taking into account equation (12), the mass of the beam was adjusted. This procedure resulted in the value of 21000 kg per meter along the track.

The input parameters were determined by methodology presented in 2.3.

Considering the properties of the ground above mentioned, the value of the subgrade reaction modulus was obtained by finite elements analysis and it takes the value of 10.7 MPa.

The shear stiffness constant, G', was obtained trough the adjustment of the results provided by equation 16 to the numerical results, permitting to conclude that $\psi = 0.59$ conducts to a good adjustment of the numerical results, as it can be observed in Figure 9.

Figure 10a shows the time history of the recorded vertical track displacements for a train passage at 70 km/h in the south direction. In the Figure 10b) the simulated results for the passage at same speed are presented.

Comparing the results expressed in the Figure 10 it is possible to observe that the computed displacements are lower in about 20% than the measured ones. However, the configuration of the time histories is very similar, witch permit to conclude that the computed results are satisfactory.

3.3 *Parametric study*

In order to evaluate the influence of some parameters in the system behaviour, a parametric study was developed. Analytical models can be very useful in the development of parametric studies because: the calculation time is low and they premit to obtain an idea of the influence of several parameters in the system behaviour. These characteristics made these models very useful for the preliminary stages of design or for feasibility studies of vibration mitigation countermeasures.

Figure 11 shows the evolution of the dynamic amplification factor with the velocity for the passage of the train X2000 for different values of the damping coefficient.

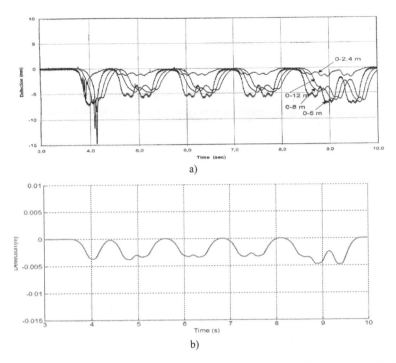

Figure 10. Time histories of the track deflection: a) measured results (Adolfsson et al., 1999); b) computed results.

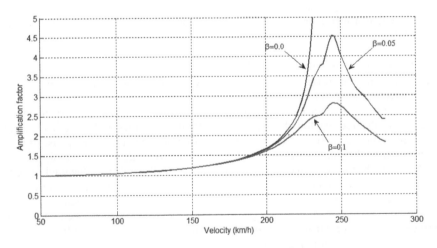

Figure 11. Dynamic amplification factor versus speed of traffic – Train X2000.

Analyzing the Figure 11 it is possible to conclude that the amplifications suffer a great increase for velocities higher to 150 km/h. For velocities higher than 200 km/h the damping of the system manifests a great influence in the response of the system. It is also possible to observe the existence of a critical velocity for which the amplification of the displacements reaches the maximum value (for undamped system this value tends to infinite). For velocity smaller than the critical velocity the response is designated as sub-critical and super-critical for velocities higher than the critical value.

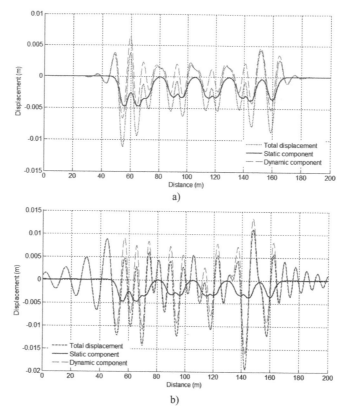

Figure 12. Deflection of the track for different speeds of passage and $\beta = 0.05$: (a) $V_0 = 220\,km/h$; (b) $V_0 = 250\,km/h$.

It should be remembered that the results presented in Figure 11 are just valid for this case, because the dynamic amplifications depends of several factor, as the axle loads and the load sequence (Kaynia et al., 2000).

To better discern the influence of the velocity on the dynamic response of the track, in Figure 13 the deflection of the track is plotted versus distance for the velocities of 220 km/h and 250 km/h.

Analyze of Figure 12a permit to verify that for a speed of 220 km/h, the response of the system occurs in sub-critical regime. However, for this speed the dynamic part of the displacement has great influence in the total displacements than the static part. The uplifts of the track are obviously a consequence of the dynamic response. In Figure 12b, the Doppler Effect's is observed, so the response of the system occurs in supercritical regime. The wave that propagates in front of the train has minor magnitude and wave length than the wave that propagates behind the train. For supercritical response the dynamic part of the deformation is out of phase with static part of the deformation, this well patent in Figure 12b.

The stiffness of the foundation is another parameter of major influence in the response of the system. Obviously, as the stiffness of the foundation increases the static displacements decreases. Concerning to the dynamic response, the increase of the stiffness of foundation generate an increase of the critical velocity so, for the same speed of passage of the train, the amplifications decreases with the increase of the stiffness of the foundation. Figure 13 shows the time histories of the displacement of the track for different values of k and to the traffic velocities of 140 km/h and 210 km/h.

So, the increase of the foundation stiffness is one way of mitigate the vibration induce by the train. The efficiency of this type of countermeasure increases when the traffic occurs on very high speed, inducing a great decrease on the dynamic amplifications.

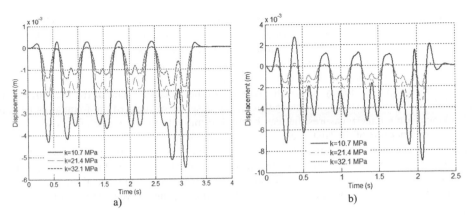

Figure 13. Influence of the subgrade reaction modulus on the temporal track response: (a) $V_0 = 140$ km/h; (b) $V_0 = 210$ km/h.

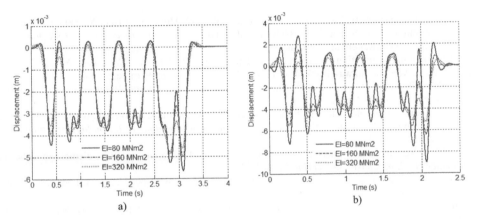

Figure 14. Influence of the flexibility of the embankment on the track response: (a) $v = 140$ km/h; (b) $v = 210$ km/h.

The bending stiffness of the "beam" or, by other way, the flexibility of the embankment is usually appointed as a parameter that has influence on the behaviour of the system. This parameter was analyzed, and in Figure 14 is presented the time histories for different values of EI and for the traffic velocities of 140 km/h and 210 km/h.

Comparison of Figures 14a and 14b shows that the influence of the flexibility of the embankment on the system response increases with the increase of the traffic velocity. This effects permit to conclude that this parameter has more influence on the dynamic response of the system than on the static response. For this case and for the speed of 210 km/h, the maximum displacement of the track can be reduced in more than 40% doubling the stiffness of the embankment.

4 CONCLUSIONS

This paper presented analytical methods that can be used as a predictive tool for analysis of track behaviour for high speed trains. The analytical formulation of those methods was presented and a new methodology for determination of input parameters was proposed.

The methodology proposed was applied to a case study and some computed results compared with the results provided by instrumentation. This comparison showed a satisfactory agreement between those results. However, it was necessary to do an adjustment of the mass of the system to take accounting the mass of the ground. This procedure just was possible because the critical velocity of the system was known. In almost cases, the critical velocity is an unknown parameter and, for these conditions, the use of this type of models can be limited.

A parametric study was developed for the purpose of to analyze the influence of the train speed, of the damping of the system, of the stiffness of the foundation and of the flexibility of the embankment. It was possible to conclude that the increases of stiffness of the foundation or of the embankment are countermeasures that can be used to mitigate the dynamic effects of the track, especially the increase of the first one.

This paper constituted a preliminary study in the domain of ground vibrations induced by high-speed trains developed in the context of the PhD thesis of the first author of the present paper. The implementation, calibration and experimental validation of advanced numerical models that allow including the train-track-ground interaction and the complex constitutive behaviour of the soils are purposes of this work.

REFERENCES

Adofsson, K; Andréasson, B; Bengtsson, P.; Zackrisson, P. 1999. High speed train X2000 on soft organic clay – measurements in Sweden". In Barends et al. (eds), *12th European Conf. on Soil Mechanics and Foundations (1999)*. Rotterdam: Balkema.

Biot. 1937. Bending of an infinite beam on an elastic foundation. *J. of Applied Mechanics:* A1–A7.

Esveld C. 2001. Modern Railway Track. *MRT-Productions*. Delft University of Technology.

Frýba 1972. Vibrations of solids and structures and moving loads. Noordhoff Internanional Publishing, Groeningen, The Nedherlands.

Hall L. 2000. Simulation and analyses of train-induced ground vibrations. A comparative study of two- and three dimensional calculations with actual measurements. *Royal Institute of Technology*. Doctoral Thesis 1034. Stockholm 2000.

Hall L. 2003. Simulation and analyses of train-induced ground vibrations in finite element models. *Soil Dynamics and Earthquake Engineering 23*: 403–413.

Holm, G. et al. 2002. Mitigation of track and ground vibrations by high speed trains at Ledsgard, Sweden. Report 10, Svensk Djupstabilisering, Linkoping, Sweden.

Hovarth, J. 1989. Subgrade models for soil-structure interaction analysis. ASCE (ed.), *Foundation engineering: current principles and practice:*. 599–612.

Kargarnovin, M. & Younesian, D. 2004. Dynamics of Timoshenko beams on Pasternak foundation under moving load. *Mechanics Research Communications 31:* 713–723.

Kaynia, A.; Madshus, C.; Zackricsson, P. 2000. Ground vibrations from high-speed trains: prediction and countermeasure. *J. Geotechnical and Geoenvironmental Engineering, Vol. 126, No. 6:* 531–537.

Kenney 1954. Steady-State vibrations of a beam on elastic foundation for moving load. *Journal of Applied Mechanics, Vol. 76:* 359–364.

Kerr A. 1961. Viscoelastic Winkler foundation with shear interactions. *J. Engng. Mech 87, EM3:* 13–20.

Madshus C., & Kyania, A. 2000. High-speed railway lines on soft ground: dynamic behaviour at critical train speed. *Journal of Sound and Vibration 231(3):* 689–701.

Takemiya, H. 2003. Simulation of track-ground vibrations due a high-speed train: the case of X2000 at Ledsgaard. *Journal of Sound and Vibration 261:* 503–526.

Wang, Y.; Tham, L.; Geung, Y. 2005. Beams and plates on elastic foundations: a review. *Prog. Struct. Engng. Mater.* (7): 174–182.

Wenander, K. 2004. Models of train induced vibrations in railway embankment – Analytical solutions and practical applications. *Royal Institute of Technology*. Master of Science Thesis. Stockholm 2004.

Author index

Printed and bound by CPI Group (UK) Ltd, Croydon, CR0 4YY

21/10/2024

01777098-0007